G. Wayne Hott
&
E. Dale Pinkhusin

Natural Toxicants in Feeds and Poisonous Plants

Frontispiece:

1. A tansy ragwort (*Senecio jacobaea*) seedling.
2. The rosette stage of tansy ragwort, formed in the first growing season.
3. The second-year plant, from which a flowering stalk emerges.
4. Flowering tansy ragwort has very distinctive flowers. Each plant may produce up to 150,000 viable seeds, which are dispersed long distances by wind.
5. Tansy ragwort leaves are deeply lobed and are usually a very dark green. Under conditions of low soil nitrogen, a purplish color of the leaf stalk and stem is evident, as is slightly visible in (**1**).
6. Larvae of the cinnabar moth feed on tansy ragwort and other plants containing pyrrolizidine alkaloids. Coevolution of poisonous plants and herbivores has resulted in evolution of detoxification mechanisms and in some cases a dietary requirement for the plant toxins by herbivores.
7. Flowers of common groundsel (*Senecio vulgaris*) lack the petal rays of tansy ragwort flowers.
8. Common tansy (*Tanacetum vulgare*), an herb, is sometimes confused with tansy ragwort. It is desirable to use botanical names to prevent confusion of this type.

Natural Toxicants in Feeds and Poisonous Plants

Peter R. Cheeke
Department of Animal Science
Oregon State University
Corvallis, Oregon

Lee R. Shull
Department of Environmental Toxicology
University of California
Davis, California

AVI PUBLISHING COMPANY, INC.
Westport, Connecticut

*Frontispiece illustration courtesy of the
Oregon State University Extension Service.*

*Appreciation is expressed to Nutrition Chemicals Division of
Monsanto Co., St. Louis, Missouri, for financial assistance in
making possible the use of this color illustration.*

Copyright 1985 by
THE AVI PUBLISHING COMPANY, INC.
P.O. Box 831
250 Post Road East
Westport, Connecticut 06881

All rights reserved. No part of this work covered by the copyright hereon may be reproduced or used in any form or by
any means—graphic, electronic, or mechanical, including
photocopying, recording, taping, or information storage and
retrieval systems—without written permission of the publisher.

Library of Congress Cataloging in Publication Data

Cheeke, Peter R.
   Natural toxicants in feeds and poisonous plants.
   Bibliography: p.
   Includes index.
   1. Veterinary toxicology. 2. Livestock poisoning
plants—Toxicology. 3. Feeds—Contamination. 4. Toxins.
I. Shull, Lee R. II. Title.
SF757.5.C44   1985   636.089'59   85-3866
ISBN 0-87055-482-4

Printed in the United States of America
A B C D 4321098765

Contents

Preface   ix

1 Natural Toxicants and Their General Biological Effects   1
    Historical Aspects of Natural Toxicants in Livestock Production   2
    Toxicant Problems Around the World   4
    Classification of Natural Toxicants by Chemical Structure   5
    Classification of Natural Toxicants by Their Occurrence in Feeds   19
    Classification of Toxicants By Their Site of Action and/or
        Metabolic Effect   20
    References   28

2 Techniques and Calculations in Toxicology   30
    Isolation and Identification of Plant Toxins   30
    Bioassay of Toxicants   31
    Assessment of Tissue Damage   33
    Techniques for Studying Toxicant Metabolism   36
    Expression of Biological Responses to Toxicants   42
    Management of Toxicoses   43
    Plant Identification and Use of Botanical Names   44
    References   45

3 Metabolism of Toxicants   47
    Metabolic Fate of Toxicants   48
    References   62

4 Effects of Toxicants on Livestock Production and Human Health   64
    Impacts of Toxicants in Feedstuffs   64
    Impacts of Poisonous Plants   70
    Range and Pasture Management   71
    Interrelationships Between Poisonous Plants and Grazing Animals   76
    Effects of Toxicants on Wild Animals   80
    Species Differences in Response to Toxins   81
    Reducing Deleterious Factors Through Plant Breeding   82
    Poisonous Plants Hazardous to Humans   85
    Ethnobotany   87
    References   87

5 Alkaloids — 92

Pyrrolizidine Alkaloids	93
Piperidine Alkaloids	115
Pyridine Alkaloids	119
Indole Alkaloids	120
Quinolizidine Alkaloids	127
Steroid Alkaloids	131
Polycyclic Diterpene Alkaloids	140
Indolizidine Alkaloids	142
Tryptamine Alkaloids	148
Tropane Alkaloids	152
Fescue Alkaloids	154
Miscellaneous Alkaloids	161
References	162

6 Glycosides — 173

Cyanogens	173
Glucosinolates	180
Coumarin	186
Steroids and Triterpenoids	190
Nitro-Containing Glycosides	199
Vicine (Favism)	204
Calcinogenic Glycosides	209
Azoxyglycosides	211
Carboxyatractyloside	212
Isoflavones and Coumestans	214
Jojoba Glycosides	223
Ranunculin	224
References	226

7 Proteins and Amino Acids — 235

Trypsin (Protease) Inhibitors	235
Amylase Inhibitors	239
Carboxypeptidase Inhibitors	239
Allergenic Proteins	239
Hemagglutinins (Lectins)	240
Enzymes	245
Bloat-Producing Proteins	252
Mimosine	257
Tryptophan	261
Selenoamino Acids	265
Lathyrogens	268
Linatine	274
Indospecine	274
Canavanine	275
Brassica Anemia Factor	276
Hypoglycin	280
Biogenic (Pressor) Amines	282
References	284

8 Carbohydrates, Lipids, and Conjugates — 292
Carbohydrates — 292
Fatty Acids — 300
Glycolipids — 304
Glycoproteins — 308
References — 309

9 Metal-Binding Substances and Inorganic Toxicants — 314
Oxalates — 314
Phytates — 321
Trimethylamine Oxide and Formaldehyde — 322
Silica Urolithiasis (Urinary Calculi) — 325
Nitrate–Nitrite Toxicity — 327
Nitrosamines and Nitrosamides — 328
References — 329

10 Tannins and Polyphenolic Compounds — 332
Sorghum Tannins — 337
Phenolics in Protein Supplements — 340
Oak Poisoning — 342
Tannins in Forage Legumes — 344
Gossypol — 345
Hypericin — 348
"Protected" or "Bypass" Protein — 350
Phenolic Compounds as Reproduction Inhibitors — 351
Black Walnut (Juglone) Toxicity — 352
Resorcinols — 352
References — 353

11 Other Plant Toxins and Poisonous Plants — 358
Sleepy Grass (*Stipa robusta*) — 358
Sesquiterpene Lactones — 359
Cicuta (Water Hemlock) — 363
Tetradymia–Artemisia Poisoning — 365
Amaranthus Poisoning — 368
Tremetone Toxicity — 370
Pine Needle Abortion — 373
Fluoroacetate (1080) — 375
Bracken Poisoning — 376
Buckwheat Toxicity — 381
Alsike Clover Poisoning — 382
Yellow Star Thistle (*Centaurea solstitialis*) — 383
Stypandra imbricata (Blind Grass) — 384
Weed Seeds in Grains and Screenings — 385
Blue-Green Algae — 385
References — 386

12 Mycotoxins	**393**
General Elements of Mycotoxicology	396
Mycotoxins Associated with Concentrate Feeds	402
Aflatoxin	402
Zearalenone	422
Trichothecenes	425
Ochratoxins	436
Citrinin	442
Rubratoxins	445
Patulin	448
Mycotoxins Associated with Roughages	450
Fungal Tremorgens	450
Lupinosis	453
Sporidesmin	456
Stachybotryotoxins	460
Moldy Straw-Induced Photosensitization	461
Kikuyu Poisoning	461
Decreasing Exposure to Mycotoxins	462
References	468
General References	476
Index	**479**

Preface

This book has evolved from a course on Toxicants in Feeds and Poisonous Plants taught for several years at Oregon State University. In the development of this course, a void was encountered which this book will attempt to fill. Existing textbooks on poisonous plants and natural toxicants are oriented either toward the botanical aspects or to a veterinary perspective with emphasis on clinical and pathologic signs of toxicity. This book is written from the perspective of animal scientists who are concerned with an appreciation of all aspects of toxicants that can influence animal production. These include consideration of botanical characteristics of crop plants, range and pasture weeds; the chemical nature and metabolism of toxicants; the pathology induced by particular toxicants; and an over-all appreciation for the significance of particular toxicants in animal production. We have endeavored to integrate these aspects in a manner that will make the book useful and interesting to a diverse readership. The intended readership includes students in animal science, range management and veterinary medicine, livestock and range extension specialists, toxicologists, veterinarians, and livestock producers. A major effort has been made to write in a style and technical language that will be in keeping with the diversity of the target audience.

It is planned that this book will be revised at intervals and will continue to be available for many years. I would appreciate comments and criticisms, including suggestions for improvement of the next edition. I would also appreciate slides and photographs that would enhance future editions.

I plan to make available slide sets covering each of the categories of toxicants covered in the book. These should be very useful to instructors using this book as a text. You may contact me directly for information on slides.

Many people have made important contributions to the preparation of this manuscript. I would like to thank Grace Hayes and Halcyon Hambleton for typing the manuscript, and Donald G. Kirsch for preparing the artwork. Many scientists responded to my request for photographs for use in the book; their helpfulness is gratefully acknowledged. I regret that because of space limitations many of their illustrations could not be included in the volume.

I am most appreciative of the time spent with a number of poisonous plant researchers in Australia during my trip there in 1981; discussions with them were most valuable in enlarging my horizons. In particular, I thank Dr. C.C.J. Culvenor for his efforts in showing me a number of toxicological problems in Australia. It is appropriate to acknowledge the assistance of the staff of the USDA Poisonous Plants Research Laboratory in Logan, Utah in providing information for use in this book.

It is a pleasure to have had the collaboration of Lee R. Shull in the preparation of this text. Dr. Shull wrote the chapter on mycotoxins and assisted with several others. I also want to recognize his contributions to the development of my research program in the toxicology area, during the conduct of his graduate work at Oregon State University. I take pleasure in acknowledging the contributions of other former graduate students, including G.W. Buckmaster, M.L. Pierson-Goeger, D.E. Goeger, R.A. Swick, B.J. Garrett, R.D. White, and M.A. Grobner, whose work on pyrrolizidine alkaloids has in large part been responsible for my continued involvement in toxicology. The efficient and thorough work of Harriet D. Shields in facilitating the editing process is much appreciated. Finally, I want to express my gratitude to my family for their help and understanding while I was home on sabbatical writing the manuscript.

PETER R. CHEEKE

Related AVI Books

ADVANCES IN MEAT RESEARCH
   Pearson and Dutson
BREEDING FIELD CROPS, 2nd Edition
   Poehlman
FIELD CROP DISEASES HANDBOOK
   Nyvall
FUNDAMENTALS OF ENTOMOLOGY AND PLANT PATHOLOGY, 2nd Edition
   Pyenson
HORTICULTURAL REVIEWS
   Janick
INTRODUCTION TO FRESHWATER VEGETATION
   Riemer
INTRODUCTION TO PLANT DISEASES
   Lucas, Campbell, Lucas
LEAF PROTEIN CONCENTRATES
   Telek and Graham
PLANT BREEDING REVIEWS
   Janick
SUSTAINABLE FOOD SYSTEMS
   Knorr

1

Natural Toxicants and Their General Biological Effects

A toxicant is a substance which under practical circumstances can impair some aspect of animal metabolism and produce adverse biological or economic effects in animal production. This is a broad definition, but encompasses those aspects that are relevant in livestock production. Virtually everything is toxic, including oxygen, water, and all nutrients, if given in a large enough dose. Thus, the term "toxicant" refers only to those substances which might normally be encountered at toxic levels. Other terms used synonymously with toxicant are "poison" and "toxin."

Toxicants can be classified in various ways, including by chemical structure. An outline of the various chemical categories is presented in this chapter, with a brief description of their general biological effects. The remainder of the book is concerned with a more detailed treatment of each toxicant, including historical and contemporary perspectives on livestock problems, the chemical nature of the toxins, their biochemistry and mode of action, the pathological signs of toxicity, and the treatment and prevention of toxicity.

Toxicants discussed have been selected on the basis of their importance in livestock production in North America and to some extent in the rest of the world. Also, consideration has been given to including representative examples of toxicants of particular chemical types or unique biochemical metabolism in order to illustrate the diversity of structure and biological effects of toxicants. The compounds discussed are overwhelmingly of plant origin, with only a few toxins of animal origin considered. This is not a book on poisonous plants per se, as many of the important toxins affecting livestock production occur in common feedstuffs. In fact, very few feedstuffs do not contain one or more substances that have deleterious effects.

Toxicants can influence animal agriculture in several ways. They can directly intoxicate animals, resulting in mortality or decreased production of animal products. Toxicants may be implicated in reducing the wholesomeness of meat, poultry products, and dairy products due to the presence of hazardous residues. Natural toxins may reduce the availability or usability of nutritious feedstuffs, or may necessitate the use of costly feed processing techniques to eliminate their effects.

In the past few years, the focus of toxicology has widened dramatically. The importance of chronic toxicoses has been realized, with greater emphasis placed on delayed responses such as mutations, cancer, birth defects, and neurological and immunological effects. Modern toxicology seeks to define the disruption of the homeostatic condition by determining the specific site (receptor) of the disturbance, the dose–effect relationship, and the specific chemical nature of the toxicant–receptor interaction. Thus large animal toxicology is an extremely broad discipline, incorporating pharmacology, botany, physiology, pathology, chemistry, biochemistry, immunology, range and animal management, veterinary medicine, and economics.

The advances in the study of toxicants have in large part been due to developments in analytical techniques and instrumentation. Sophisticated methodology allows detection of toxicants at extremely low levels and the identification of short-lived and reactive metabolites. In spite of these advances, there are still important natural toxicoses, such as fescue foot, summer fescue toxicosis, bracken poisoning, and vetch poisoning of cattle, for which the toxic agents have not been conclusively identified. There are also several natural toxicants for which no method of analysis exists. Many opportunities exist for further advances in the understanding of the roles of toxicants in animal production.

HISTORICAL ASPECTS OF NATURAL TOXICANTS IN LIVESTOCK PRODUCTION

Natural toxicants have had significant effects on livestock production in North America. In the frontier days, extensive stock losses occurred due to consumption of poisonous plants. In the settlement of new areas, both the livestock and the ranchers were inexperienced as to the toxicity of native plants, and some spectacular losses occurred as this experience was gained. In the late 1800s and early 1900s, some of the major toxicity problems in the U.S. included milk sickness in humans caused by milk transfer of a toxin from white snakeroot, and livestock poisonings from consumption of such plants as locoweed, larkspur, water hemlock, lupine, death camas, bitterweed, and sleepy grass. An early function of the United States Department of Agriculture (USDA), which was organized in 1862, was the investigation of toxic plants. Numerous field stations in the western U.S. were established by the USDA in the early 1900s for the study of specific local problems (Fig. 1.1). Some of these included stations in Hugo, Colorado to study locoweed, in Gunnison National Forest, Colorado to study larkspur, lupine, and water hemlock poisoning, in Greycliff, Montana to investigate lupine and death camas, and in Salina, Utah to study oak, sneezewood, locoweed, and milkweed problems. In 1955, the USDA Poisonous Plants Research Laboratory was established in Logan, Utah. In cooperation with Utah State University in Logan, scientists at this laboratory have conducted in-depth investigations of many plant toxins and poisonous plants, including locoweed, larkspur, lupine, halogeton, pyrrolizidine alkaloid-containing plants, and numerous others. Considerable effort has been placed on the study of teratogens in plants, such as those in lupine and poison hemlock responsible for crooked calf disease, and *Veratrum* alkaloids which cause birth defects in sheep. Numerous other institutions have embarked on significant investigations of natural toxicants affecting livestock. With the inevitable risk of omission, some of these are: Texas A&M Univer-

FIG. 1.1. USDA poisonous plant investigators performing an autopsy—1906.
Courtesy of R. F. Keeler.

sity, southwestern toxic plants with emphasis on bitterweed; University of California, Davis, pyrrolizidine alkaloids, milkweeds, and cyanogens; Washington State University, acute bovine pulmonary emphysema; Oregon State University, pyrrolizidine alkaloids; Cornell University, calcinogenic glycosides; University of British Columbia, phytoestrogens, pine needle abortion; Agriculture Canada, Kamloops, British Columbia, timber milk vetch, cyanogens, and pine needle abortion; and the University of Alberta and University of Saskatchewan, rapeseed glucosinolates. In addition, many universities in the U.S. and Canada have programs on mycotoxins, which since 1960 have become increasingly recognized as major problems.

TOXICANT PROBLEMS AROUND THE WORLD

Natural toxicants are and have been responsible for livestock problems in many parts of the world. Australia might be regarded as the land of poisonous plants. Its arid, hostile environment and extensive pastoral industries contribute to the scope of the problems. In times of drought, stockmen might regard the availability of poisonous plants as being better than nothing at all. This could account for the common names of Paterson's curse and Salvation Jane for *Echium plantagineum,* a pyrrolizidine alkaloid-containing plant. Among the significant Australian toxicity situations are disorders in all livestock species caused by pyrrolizidine alkaloids, and fluoroacetate toxicity, lupinosis, ryegrass staggers, annual ryegrass toxicity, clover disease caused by phytoestrogens, *Swainsona* poisoning, oxalate poisoning, and numerous others. In New Zealand, some of the major problems have involved photosensitization in which severe skin lesions occur with exposure to sunlight. Probably the most important of these is facial eczema in dairy cattle and sheep, caused by a mycotoxin, sporidesmin. Another mycotoxin problem of significance in New Zealand is perennial ryegrass staggers. Brassica poisoning occurs in those areas where *Brassica* spp. such as rape and kale are used as forage. In the past, *Senecio jacobaea* poisoning has been a significant problem in parts of New Zealand. In Southeast Asian countries such as Indonesia, the Philippines, Papua New Guinea, and tropical areas of Australia, toxins in tropical forages such as *Leucaena leucocephala* are of concern. In South Africa, numerous poisonous plant problems have affected the extensive livestock industries of that country; many of these problems are similar to those seen in Australia. Some of the conditions reported from South Africa include annual ryegrass toxicity, poisoning from various *Senecio* spp., and mycotoxin contamination of feedstuffs. Photosensitization caused by consumption of a variety of plants is a particular problem in South Africa. Cardiac glycoside-containing plants are widespread and cause considerable losses. Vomiting disease caused by sesquiterpene lactones in *Geigeria* spp. is significant. Fluoroacetate toxicity is one of the major problems. In equatorial Africa, mycotoxins in grains and protein supplements are major problems because of high temperature and humidity, and poor crop storage facilities. Aflatoxin contamination of foodstuffs is believed to be a major contributing factor to liver cancer in humans in tropical African countries. In northern Africa, tannins in sorghum and millets are important. In Great Britain and western Europe, some of the toxicants of interest include glucosinolates in rapeseed, the brassica anemia factor, pyrrolizidine alkaloids, and a variety of mycotoxins in feeds. Acute bovine pulmonary emphysema and bracken poisoning are other important European problems. In Latin American countries, a number of toxicants and poisonous plant problems have been reported, including bracken poisoning, mycotoxicoses, and toxins such as fluoroacetate, oxalates, and thiaminases.

While this is a very brief overview of the situation, it is apparent that many toxicants affect livestock production in all parts of the world.

Natural toxicants can be classified on the basis of their chemical structure. With some exceptions, most are distributed among the following broad categories: alkaloids; glycosides; proteins; amino acids and derivatives; carbohydrates; lipids; glycoproteins; glycolipids; metal-binding substances; resins; phenolics; and sesquiterpene lactones.

In addition to these, it is convenient to consider mycotoxins as a specific category.

Each of these categories will be defined, and a few examples given. In some cases, as with alkaloids, they can be subclassified. Listing of these toxicant groups at this point is useful to provide a broad perspective on the nature of toxicants, as well as providing an outline of what is covered in this book.

CLASSIFICATION OF NATURAL TOXICANTS BY CHEMICAL STRUCTURE

1. Alkaloids

Alkaloids ("alkali-like") are compounds that contain nitrogen, usually in a heterocyclic ring, and are generally basic substances. They are usually bitter, and most are toxic. They are subclassified on the basis of the chemical type of the nitrogen-containing heterocyclic ring.

A. Pyrrolizidine Alkaloids

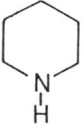

The nucleus consists of two five-membered rings. Pyrrolizidine alkaloids are responsible for the toxicity of various *Senecio* spp. (tansy ragwort, common groundsel), crotalaria, *E. plantagineum* (Paterson's curse), and *Heliotropium europaeum*. The latter two are particularly important in Australia. Pyrrolizidine alkaloids cause irreversible liver damage. Examples are monocrotaline, heliotrine, lasiocarpine, and senecionine.

B. Piperidine Alkaloids

The most important piperidine alkaloids in animal production are coniine and related alkaloids found in *Conium maculatum* (poison hemlock). These alkaloids affect the central nervous system and are also teratogens.

C. Pyridine Alkaloids

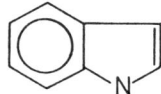

An example of a pyridine alkaloid is nicotine, in *Nicotiana* spp. (cultivated and wild tobacco).

D. Indole Alkaloids

Indoles are derivatives of the amino acid tryptophan. Examples are the ergot alkaloids, the alkaloids such as perloline in tall fescue, and 3-

methylindole, an alkaloid implicated in bovine pulmonary emphysema.

E. Quinolizidine Alkaloids

The quinolizidine nucleus consists of two six-membered rings. Lupines contain these alkaloids, which cause acute poisoning in sheep and teratogenic effects in calves (crooked calf disease).

F. Steroid Alkaloids
i. Solanum type (e.g., solanidine)

These are found in green potatoes, tomatoes, and nightshade. They are central nervous system poisons and cholinesterase inhibitors. Solanidine in potatoes has human health implications; an upper limit for alkaloid in new potato cultivars in the U.S. has been established by regulatory agencies.

ii. Veratrum type (e.g., veratramine)

False hellebore (*Veratrum californicum*) contains veratrum alkaloids that produce teratogenic effects in lambs (cyclops lamb) and prolonged gestation in ewes. Death camas, a source of extensive sheep losses in the American West in the past, contains alkaloids of this type.

G. Polycyclic Diterpene Alkaloids

These complex alkaloids are found in *Delphinium* spp., commonly known as larkspurs. Larkspurs are responsible for more cattle losses in the U.S. than all other toxic plants combined. They cause acute central nervous system effects.

H. Indolizidine Alkaloids

Indolizidine alkaloids, such as swainsonine, have been identified as the toxic components of *Swainsona* spp. in Australia and *Astragalus* spp. (locoweed) in the U.S. They are inhibitors of α-mannosidase, resulting in the accumulation of mannose deposits in lysosomes of nerve cells. Slaframine is an indolizidine alkaloid produced on red clover by a fungus; it causes profuse salivation.

I. Tryptamine Alkaloids

Tryptamine alkaloids are found in *Phalaris tuberosa*, a forage grass grown in Australia. Phalaris poisoning results in acute neurological signs and chronic muscular incoordination.

J. Tropane Alkaloids

Atropine, found in *Datura* spp. (Jimsonweed), is an example of a tropane alkaloid. It has pronounced effects on the central nervous system.

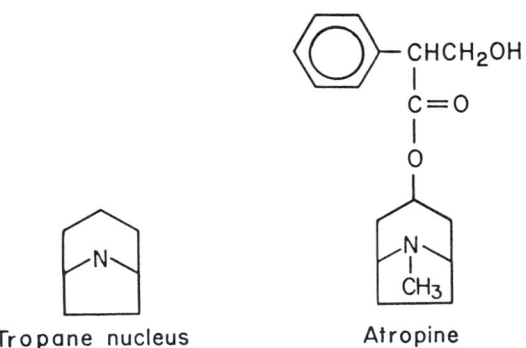

Tropane nucleus   Atropine

2. Glycosides

Glycosides are ethers containing a carbohydrate moiety and a non-carbohydrate moiety (aglycone) joined with an ether bond. They are usually bitter substances. Often the aglycone is released by enzymatic action when the plant tissue is damaged, as by wilting, freezing, mastication, or trampling. They are classified on the basis of the structure and/or properties of the aglycone.

A. Cyanogenic Glycosides

These yield hydrocyanic (prussic) acid when hydrolyzed. An example is laetrile (amygdalin) found in the kernels of almonds, apricots, peaches, apples, and the leaves of chokecherries. It is hydrolyzed by β-glycosidase to release hydrogen cyanide, glucose, and benzaldehyde as follows:

$$\text{C}_6\text{H}_5\text{-CH(O-C}_6\text{H}_{10}\text{O}_4\text{-C}_6\text{H}_{11}\text{O}_5\text{)-CN} \longrightarrow \text{C}_6\text{H}_5\text{-CHO} + \text{HCN} + 2\,\text{C}_6\text{H}_{12}\text{O}_6$$

The hydrogen cyanide is a potent inhibitor of cytochrome oxidase, a respiratory enzyme.

B. Goitrogenic Glycosides

Goitrogens decrease production of the thyroid hormones by inhibiting their synthesis by the thyroid gland. As a result, the thyroid enlarges to compensate for reduced thyroxine output, producing a goiter.

Goitrogenic glycosides are thioethers, containing an organic aglycone:

$$R-C(\text{S-glucose})=N-O-SO_2-O^- \; K^+$$

where R (the aglycone) is an alkyl group. These compounds, called glucosinolates (formerly called thioglucosides) are commonly found in *Brassica* spp. such as cabbage, kale, and rape. They are hydrolyzed by glucosinolases to β-D-glucose, HSO_4^-, and derivatives of the aglycone. These include isothiocyanates, nitriles, and thiocyanates. The glucosinolates are anions, usually found as potassium salts.

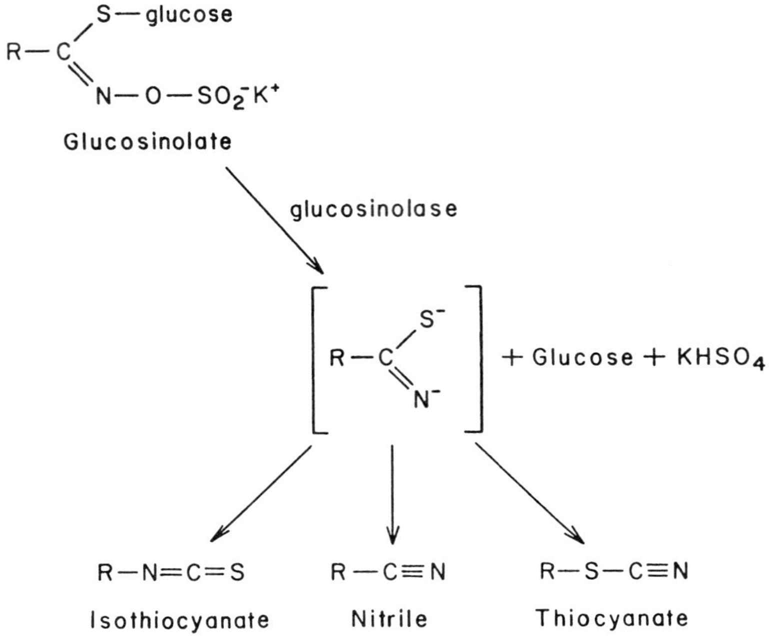

$$R-C\begin{cases}S-\text{glucose}\\ N-O-SO_2^-K^+\end{cases}$$

Glucosinolate

↓ glucosinolase

$$\left[R-C\begin{cases}S^-\\ N^-\end{cases}\right] + \text{Glucose} + KHSO_4$$

$R-N=C=S$ $R-C\equiv N$ $R-S-C\equiv N$

Isothiocyanate   Nitrile   Thiocyanate

The glucosinolase is released from plant tissue by crushing (mastication) and is also produced by rumen microorganisms. Isothiocyanates may form ring structures such as goitrin (5-vinyl-2-oxazolidinethione):

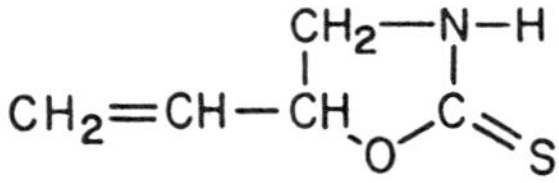

$$CH_2=CH-CH\begin{smallmatrix}CH_2-N-H\\ |\quad\quad|\\ O\quad\; C=S\end{smallmatrix}$$

C. Coumarin Glycosides

Coumarin is found in sweet clover (*Melilotus* spp.) as melilotoside.

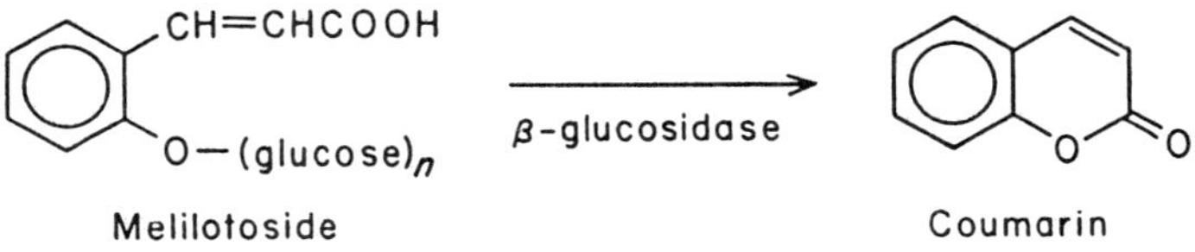

Melilotoside (with CH=CHCOOH and O–(glucose)$_n$ substituents on benzene ring) →β-glucosidase→ Coumarin

Coumarin is converted by mold growth to dicoumarol, an antagonist of vitamin K. Sweet clover poisoning, caused by feeding moldy sweet clover hay, is therefore an induced vitamin K deficiency.

D. Steroid and Triterpenoid Glycosides

i. Cardiac glycosides

The best known cardiac glycoside is digitonin, contained in foxgloves (*Digitalis* spp.). As the name suggests, these glycosides contain a sterol group in their structure. Physiologically, they are potent stimulators of heart rate and are used medicinally. Milkweeds (*Asclepias* spp.) contain cardiac glycosides (cardenolides) and are extremely toxic to livestock. As little as 100 g of milkweed can kill a sheep.

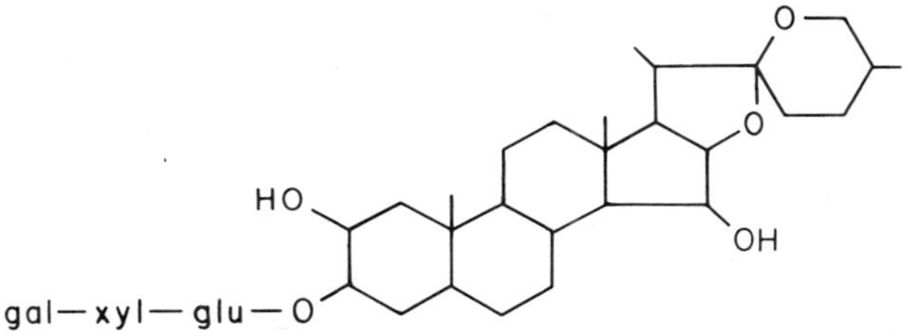

Digitonin

ii. Saponins

Saponins are glycosides which have profuse foaming properties, producing a distinctive honeycombed stable foam when an aqueous solution is shaken. They are widely distributed in plants and, in animal nutrition, are particularly important in temperate legume forages. They are bitter compounds, affecting palatability and feed intake. They have growth-depressing properties in poultry and swine, and have been implicated in bloat in ruminants. Saponins contain a polycyclic aglycone (steroid or triterperoid) and a side chain of sugars attached by an ether bond to C_3 as follows:

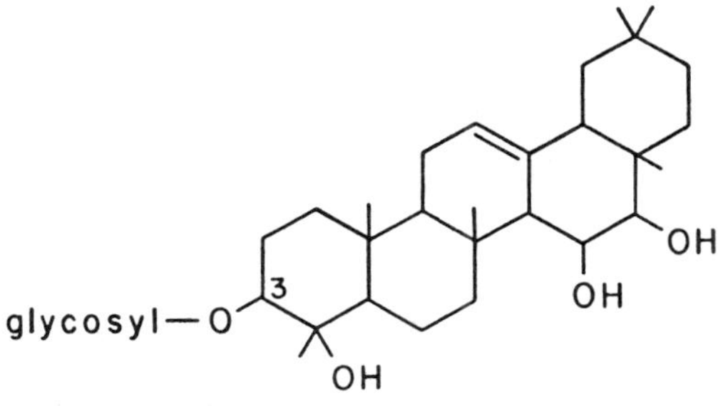

Soyasapogenol A

E. Nitropropanol Glycosides

Many *Astragalus* species, such as timber milk vetch (*Astragalus miser*), owe their toxicity to this class of glycoside. They are metabolized to 3-nitro-1-propanol (3NPOH) in ruminants, and 3-nitropropionic acid in nonruminants. These compounds, especially 3NPOH, are acutely toxic, producing methemoglobinemia. They also produce chronic toxicity symptoms, involving permanent nerve damage.

$$\text{glucose}-O-CH_2-CH_2-CH_2-NO_2$$
Miserotoxin

nonruminants ↙ ↘ ruminants

$HOOC-CH_2-CH_2-NO_2$ $HO-CH_2-CH_2-CH_2-NO_2$
3-nitropropionic acid   3-nitro-1-propanol

This class of glycosides is also found in crown vetch (*Coronilla* spp.).

F. Vicine

Vicine is a glycoside in fava beans (*Vicia faba*). It causes hemolytic anemia (favism) in people who have a genetic deficiency of glucose-6-phosphate dehydrogenase activity in their red blood cells. Fava beans are being utilized, especially in Canada, as a protein supplement for livestock.

G. Calcinogenic Glycosides

Some plants, such as *Cestrum diurnum, Solanum malacoxylon,* and *Trisetum flavescens,* have glycosides that contain the active metabolite of vitamin D (1,25-dihydroxycholecalciferol). The consumption of these plants by cattle and horses results in excessive levels of active vitamin D in their tissues, overriding the feedback control mechanisms involved in calcium homeostasis. This results in excessive calcium absorption and the calcification of soft tissues such as arteries and kidney.

1,25-dihydroxycholecalciferol glycoside

H. Carboxyatractyloside

A glycoside called carboxyatractyloside has been identified as the toxicant in cocklebur (*Xanthium strumarium*). It produces hepatic lesions, convulsions, and severe hypoglycemia.

I. Isoflavones

These compounds, which are called phytoestrogens, contain a flavone nucleus. Examples of isoflavones are genistein, formononetin, and coumestrol.

Flavone nucleus   Genistein

Isoflavones are particularly important in subterranean clover and have caused extensive reproductive problems in sheep in Australia.

3. Proteins

Several important inhibitors in plants are proteins. Of interest is that in some cases the effect of these is to inhibit the utilization of other proteins by animals.

A. Protease (Trypsin) and Amylase Inhibitors

Soybeans, most other legume seeds, and some grains (e.g., rye and triticale) contain trypsin inhibitors. These are small protein molecules which combine with and inactivate the digestive enzyme trypsin in the small intestine. They cause reduced growth and pancreatic hypertrophy. Amylase inhibitors occur in beans and have been commercialized as "starch blockers" to reduce obesity in humans.

B. Lectins (Hemagglutinins)

These are glycoproteins of 60,000–100,000 MW that cause agglutination (clumping) of red blood cells in vitro. They are found in most types of beans, including soybeans. They cause reduced growth, diarrhea, and interfere with nutrient absorption.

C. Enzymes

An example of an enzyme toxin is thiaminase, found in bracken fern (*Pteridium aquilinum*) and certain fish such as carp. The enzyme cleaves the B vitamin thiamine, thereby inactivating it. Consumption of bracken fern causes thiamine deficiency in some animals. Use of carp and other types of thiaminase-containing fish in mink diets has produced thiamine deficiency (Chastek's paralysis).

Other enzymes in feeds which produce deleterious effects in livestock include lipoxidases in soybeans and alfalfa, which degrade fat-soluble vitamins.

D. Plant Cytoplasmic Proteins

Many leguminous forages such as alfalfa and numerous clovers may cause bloat in ruminants. This is primarily due to formation of a stable foam in the rumen, involving cytoplasmic proteins in the plant cell contents.

4. Amino Acids and Amino Acid Derivatives

A. Amino Acids

There are over 300 amino acids in plants. Not surprisingly, some of these are toxic. The best known is mimosine, which is structurally similar to tyrosine.

$$O= \text{(pyridinone ring with HO)} -N-CH_2-CH(NH_2)-COOH$$

Mimosine occurs in the tropical forage *L. leucocephala,* which produces protein-rich leaves that have considerable potential as livestock feed. Mimosine causes reduced growth and alopecia (loss of hair) in nonruminants and is metabolized in the rumen to a goitrogenic compound, producing goiter in ruminants.

Tryptophan is a dietary essential amino acid. It can also be regarded as a toxicant because under some conditions it is metabolized in cattle to 3-methylindole, a compound responsible for acute bovine pulmonary emphysema.

Dihydroxyphenylalanine (dopa) occurs in fava beans and has pharmacological effects.

Selenoamino acids, such as selenocystine, methylselenocysteine, selenocystathionine, and selenomethionine, which contain selenium in place of sulfur, are implicated in selenium toxicity due to consumption of selenium accumulator plants such as *Astragalus* spp.

There are several lathyrogenic amino acids, including β-cyano-L-alanine in *Vicia* spp. and β-amino propionitrile in *Lathyrus* spp. Lathyrus seeds are consumed as food in India, causing a major public health problem, as permanent paralysis and skeletal deformity may occur. In livestock, consumption of *Lathyrus odoratus* seeds by cattle and horses causes paralysis, aortic aneurysm, and skeletal deformity. Lathyrism is the term used to describe the symptoms; both neurolathyrism (paralysis) and osteolathyrism (skeletal deformity) are observed.

Another toxic amino acid is 1-amino-D-proline. It is produced from linatine, found in linseed meal. 1-Amino-D-proline is an antagonist of pyridoxine (vitamin B_6).

Indospecine is a toxic amino acid found in *Indigofera spicata,* a potentially useful tropical pasture legume. It is an antagonist of arginine, resulting in depressed incorporation of arginine into liver-synthesized proteins.

The brassica anemia factor, S-methylcysteine sulfoxide, is an amino acid derivative. It leads to red blood cell hemolysis and anemia. It is found in forage brassicas such as kale and turnips.

B. Polypeptides

Amanita mushrooms are extremely poisonous and owe their toxicity to cyclic polypeptides. An unusual livestock poisoning problem in Australia is caused by the larvae of the sawfly, which contain an octapeptide called lophyrotomin. The sawfly larvae congregate in mounds beneath the silver-leaf ironbark tree (*Eucalyptus melanophloia*) and are avidly consumed by cattle. Affected animals show incoordination, trembling, and liver damage.

5. Carbohydrates

There are few toxicity problems due to carbohydrates. Xylose, a hexose sugar, causes reduced growth and eye cataracts in pigs and poultry. Certain oligosaccharides, such as raffinose, are not digested in the small intestine, and so promote bacterial growth in the hindgut. These are the flatulence factors in beans. The β-glucans in certain barley varieties sometimes cause nutritional problems in poultry.

6. Lipids

Several fatty acids are toxic. These include erucic acid in rapeseed, which may result in myocardial lesions in rats. Cyclopropenoid fatty acids, such as sterculic and malvalic acids in cottonseed, have toxic properties and cause pink albumins to develop in stored eggs. They are also cocarcinogens, increasing the carcinogenicity of aflatoxins.

7. Glycoproteins

Lectins (discussed under Proteins) are glycoproteins.

Avidin, a glycoprotein in egg albumin, is an antagonist of the B vitamin biotin. Raw eggs can be used to induce biotin deficiency in experimental animals. Biotin deficiency has occurred in fur animals (foxes and mink) fed raw eggs or poultry-processing plant wastes containing unlaid eggs.

8. Glycolipids

The cause of annual ryegrass toxicity (ARGT) has been identified as a glycolipid(s) called corynetoxin. This toxin is synthesized by a corynebacterium which colonizes galls, produced by a nematode, in the ryegrass seed head. The toxin affects the brain, leading to incoordination and staggering.

9. Metal-Binding Substances

A. Oxalates

Oxalic acid is a chelating agent which chelates calcium very effectively. Plants with a high oxalate content, such as the U.S. range weed *Halogeton glomeratus,* may produce acute metabolic calcium deficiency (hypocalcemia) when consumed by livestock.

$$\begin{array}{c} COOH \\ | \\ COOH \end{array} \qquad \text{Calcium oxalate structure with Ca}$$

Oxalic acid   Calcium oxalate

B. Phytates

Phytic acid in cereal grains and soybean meal causes reduced mineral availability, particularly of zinc, through the formation of unabsorbable phytates. Organic phosphorus (phytin phosphorus) is of low availability to nonruminant animals.

1. NATURAL TOXICANTS AND THEIR EFFECTS   17

$$\begin{array}{c}\text{inositol hexaphosphate structure with six } OPO_3H_2 \text{ groups}\end{array}$$

C. Mimosine

Mimosine, the toxic amino acid in *L. leucocephala*, is reputed to have metal-binding properties.

10. Resins

These compounds do not share any particular structural features, but have certain common physical characteristics. They are soluble in a number of organic solvents, are insoluble in water, and do not contain nitrogen. An example of a resin is cicutoxin, the poisonous principle of *Cicuta* spp. (water hemlock). It is one of the most spectacular poisons known, acting directly on the central nervous system to produce violent convulsions.

11. Phenolic Compounds

Phenolics contain an aromatic ring(s) with one or more hydroxyl groups.

A. Hypericin

Hypericin is a phenolic compound in *Hypericum perforatum* (St.-John's-wort). It is a primary photosensitizing agent, producing dermatitis and skin lesions in light-skinned animals by reacting with ultraviolet light at the skin surface to produce a photodynamic reaction.

B. Gossypol

Cottonseed meal contains a toxic polyphenol called gossypol. Gossypol causes reduced growth and feed intake, cardiac lesions, and male infertility.

C. Tannins

Tannins are phenolic compounds that react with proteins. The term was originally used for plant extracts that were used in tanning leather. They are astringent and adversely affect feed intake. Tannins are important in oak poisoning and in nutritional problems with sorghum grain (milo).

12. Sesquiterpene Lactones

Sesquiterpene lactones are derivatives of the germacranolide nucleus:

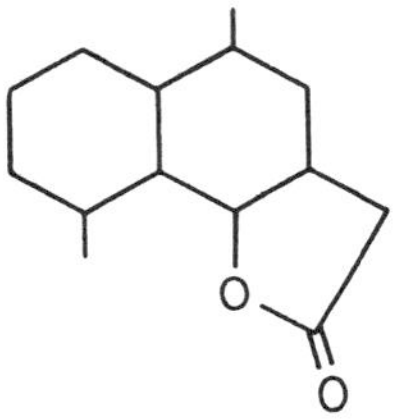

These compounds are the toxic constituents of sneezeweeds (*Helenium* spp. and bitterweeds (*Hymenoxys* spp.). They cause severe irritation of nasal and intestinal membranes.

13. Mycotoxins

Mycotoxins are metabolites of fungi (molds) that are toxic to animals. There are at least 25 specific disease entities in livestock that can be attributed to mycotoxins. Some of the mycotoxins involved are aflatoxins, phomopsin, tremorgens, T-2 toxin, citrinin, ochratoxin, sporidesmin, and zearalenone. They are implicated in diverse conditions such as acute death of poultry (turkey X disease), liver cancer in trout, lupinosis, fescue foot of cattle, sweet clover poisoning, facial eczema of sheep, ryegrass staggers, and ergotism.

14. Other Toxins

A. Plant Carcinogens

The best known example of carcinogenic effects in livestock due to consumption of a poisonous plant is bladder and intestinal cancer in cattle consuming bracken fern. The carcinogen(s) has not been conclusively identified. Another carcinogen is safrole, which in the past has been used as a flavoring agent in soft drinks. Some of the pyrrolizidine alkaloids have carcinogenic properties.

B. White Snakeroot Toxin

In the pioneer days in the U.S., a disease known as milk sickness affected many people and caused many fatalities. It was due to the consumption of milk from cows that had grazed white snakeroot (*Eupatorium rugosum*). The plant contains a toxic ketone called tremetone.

C. Fluoroacetate (Organofluorine Compounds)

A variety of plants in Australia and South Africa contain organofluorine compounds, principally fluoroacetate. Fluoroacetate is a potent metabolic toxin inhibiting the conversion of citrate to isocitrate in the tricarboxylic acid cycle.

D. *N*-Propyl Disulfide

Onions contain *N*-propyl disulfide, an inhibitor of glucose-6-phosphate dehydrogenase in red blood cells. The result is anemia, similar to that induced by the brassica anemia factor. In some parts of the U.S., cull onions are an important livestock feed.

E. Trimethylamine Oxide and Formaldehyde

These compounds occur in the flesh of certain types of marine fish and impair iron absorption when the fish are used in the diets of fur animals. Iron deficiency signs such as reduced growth, anemia, and loss of hair pigmentation (achromatrichia) are observed.

CLASSIFICATION OF NATURAL TOXICANTS BY THEIR OCCURRENCE IN FEEDS

Most feedstuffs and forages used in the feeding of livestock contain potentially deleterious factors. Indeed, it is rare to find a feedstuff in which a deleterious factor cannot be identified. Even the ubiquitous chlorophyll can have adverse effects on animals, causing photosensitization reactions under certain conditions. Some of these factors are of academic interest only, while others are of considerable concern in animal production.

A listing of some of the more common and important toxicants in feedstuffs is provided to illustrate their widespread distribution and to provide ready access to potential deleterious factors in common feeds. (Table 1.1).

TABLE 1.1. Natural Toxicants in Common Feedstuffs

Feedstuff	Toxicant
Grains	
All	Phytates, mycotoxins
Rye, triticale	Trypsin inhibitors, ergot
Milo	Tannins
Grain amaranth	Oxalates, saponins
Buckwheat	Fagopyrin
Tubers	
Potatoes	Solanum alkaloids
Cassava	Cyanogenic glycosides
Protein supplements	
Soybeans	Trypsin inhibitors, lectins, goitrogens, saponins, phytates, mycotoxins
Cottonseed	Gossypol, tannins, cyclopropenoid fatty acids, mycotoxins
Rapeseed	Glucosinolates, tannins, erucic acid, sinapine
Linseed meal	Linatine, linamarin
Fava beans	Trypsin inhibitors, vicine, lectins
Field beans	Trypsin inhibitors, lectins
Forages	
Legumes	
Alfalfa	Saponins, phytoestrogens, bloating agents
White clover	Cyanogens, phytoestrogens, bloating agents
Red clover	Slaframine, phytoestrogens, bloating agents
Alsike clover	Photosensitizing agents
Sweet clover	Coumarin
Subterranean clover	Phytoestrogens
Crown vetch	β-Nitropropanol glycosides
Leucaena spp.	Mimosine
Indigofera spp.	Indospecine
Grasses	
Forage sorghums	Cyanogens
Tall fescue	Fescue alkaloids
Tropical grasses	Oxalates
Others	
Forage brassicas	Brassica anemia factor

CLASSIFICATION OF TOXICANTS BY THEIR SITE OF ACTION AND/OR METABOLIC EFFECT

It is remarkable that for practically all major metabolic functions in animals, there exists in the plant world inhibitors of these functions. For virtually every organ, endocrine gland, and metabolic pathway in animals, there is a corresponding inhibitor in plants. Considerable speculation has been made as to the reasons why plants contain such a wide variety of deleterious (to animals) compounds. The following is by no means a complete listing, but is intended to give an appreciation for the diversity of effects that plant substances can have on animals.

1. Mouth
   A. Proteolytic Enzymes
      Bromelain (from pineapple) and papain (from papaya) are proteolytic enzymes that digest cellular proteins in the buccal cavity.
   B. Oxalate Crystals
      The common houseplant *Dieffenbachia sequine* (dumb cane) contains calcium oxalate crystals which penetrate the mucous membranes of the mouth and throat and cause intense irritation.
   C. Phenolics (Tannins)
      Tannins have an astringent effect due to their binding with proteins on the tongue and in the oral cavity.
2. Digestive Tract
   A. Rumen
      i. Nitrate and nitrite
         Nitrates are metabolized to nitrites by rumen microorganisms.
      ii. Rumen stasis
         Mesquite poisoning, lupinosis, and oxalate toxicity cause rumen stasis.
      iii. Bloat
         Cytoplasmic proteins and saponins have been implicated in causing bloat. Tannins may have protective effects.
      iv. Rumenitis
         Oxalate poisoning results in damage to the lining of the rumen.
   B. Intestine
      i. Irritation
         Saponins, tannins, and seleno amino acids cause irritation and damage to the intestinal mucosa.
      ii. Enzyme inhibitors
         Trypsin and amylase inhibitors in feeds exert their inhibitory effects on digestive enzymes in the small intestine.
      iii. Nutrient absorption
         Lectins alter intestinal permeability and reduce the absorption of nutrients.
   C. Diarrhea
      Diarrhea is commonly observed in pyrrolizidine alkaloid toxicity. Nitrates also cause diarrhea.
   D. Prolapsed Rectum
      This is a symptom of pyrrolizidine alkaloid poisoning.

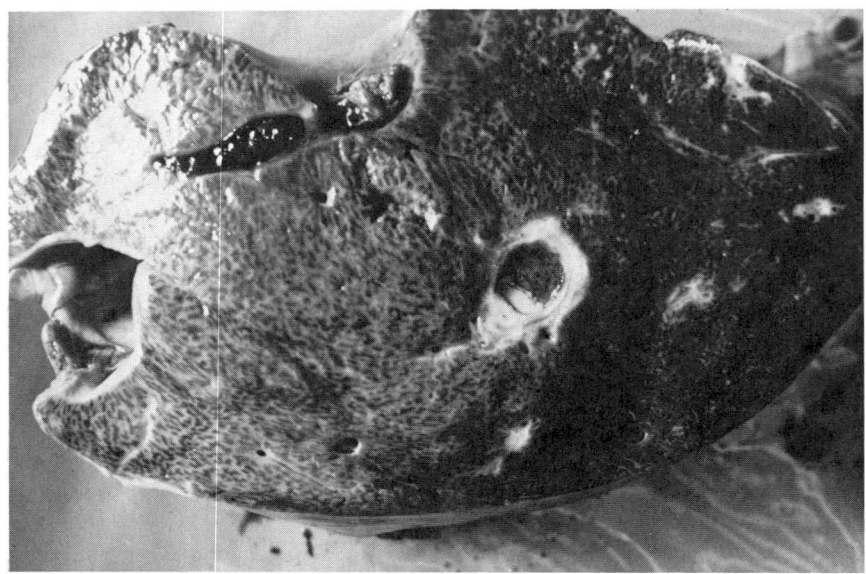

FIG. 1.2. The liver of a horse poisoned by *Echium plantagineum*, showing fibrosis and loss of normal structure.
Courtesy of John Seaman.

3. Liver
   A. Hepatotoxins
      Pyrrolizidine alkaloids cause irreversible liver damage (Fig. 1.2). Lupinosis results in cirrhotic, fatty livers. Facial eczema in sheep and cattle, caused by the mycotoxin sporidesmin, is accompanied by liver damage.
   B. Mineral Metabolism
      Pyrrolizidine alkaloid poisoning and lupinosis result in elevated liver copper concentrations and alterations in hepatic zinc and iron metabolism.
   C. Cholestasis
      Lantana poisoning causes cholestasis, or cessation of bile flow, producing an enlarged green liver. *Lantana camara* is the poisonous plant involved.
4. Lung
   A. Pyrrolizidine alkaloids cause pulmonary damage.
   B. Acute bovine pulmonary emphysema is caused by indoles (e.g., 3-methylindole) which are products of rumen metabolism of tryptophan.

been used as a flavoring agent in soft drinks. Some of the pyrrolizidine alkaloids have carcinogenic properties.

B. White Snakeroot Toxin

In the pioneer days in the U.S., a disease known as milk sickness affected many people and caused many fatalities. It was due to the consumption of milk from cows that had grazed white snakeroot (*Eupatorium rugosum*). The plant contains a toxic ketone called tremetone.

C. Fluoroacetate (Organofluorine Compounds)

A variety of plants in Australia and South Africa contain organofluorine compounds, principally fluoroacetate. Fluoroacetate is a potent metabolic toxin inhibiting the conversion of citrate to isocitrate in the tricarboxylic acid cycle.

D. *N*-Propyl Disulfide

Onions contain *N*-propyl disulfide, an inhibitor of glucose-6-phosphate dehydrogenase in red blood cells. The result is anemia, similar to that induced by the brassica anemia factor. In some parts of the U.S., cull onions are an important livestock feed.

E. Trimethylamine Oxide and Formaldehyde

These compounds occur in the flesh of certain types of marine fish and impair iron absorption when the fish are used in the diets of fur animals. Iron deficiency signs such as reduced growth, anemia, and loss of hair pigmentation (achromatrichia) are observed.

CLASSIFICATION OF NATURAL TOXICANTS BY THEIR OCCURRENCE IN FEEDS

Most feedstuffs and forages used in the feeding of livestock contain potentially deleterious factors. Indeed, it is rare to find a feedstuff in which a deleterious factor cannot be identified. Even the ubiquitous chlorophyll can have adverse effects on animals, causing photosensitization reactions under certain conditions. Some of these factors are of academic interest only, while others are of considerable concern in animal production.

A listing of some of the more common and important toxicants in feedstuffs is provided to illustrate their widespread distribution and to provide ready access to potential deleterious factors in common feeds. (Table 1.1).

TABLE 1.1. Natural Toxicants in Common Feedstuffs

Feedstuff	Toxicant
Grains	
All	Phytates, mycotoxins
Rye, triticale	Trypsin inhibitors, ergot
Milo	Tannins
Grain amaranth	Oxalates, saponins
Buckwheat	Fagopyrin
Tubers	
Potatoes	Solanum alkaloids
Cassava	Cyanogenic glycosides
Protein supplements	
Soybeans	Trypsin inhibitors, lectins, goitrogens, saponins, phytates, mycotoxins
Cottonseed	Gossypol, tannins, cyclopropenoid fatty acids, mycotoxins
Rapeseed	Glucosinolates, tannins, erucic acid, sinapine
Linseed meal	Linatine, linamarin
Fava beans	Trypsin inhibitors, vicine, lectins
Field beans	Trypsin inhibitors, lectins
Forages	
Legumes	
Alfalfa	Saponins, phytoestrogens, bloating agents
White clover	Cyanogens, phytoestrogens, bloating agents
Red clover	Slaframine, phytoestrogens, bloating agents
Alsike clover	Photosensitizing agents
Sweet clover	Coumarin
Subterranean clover	Phytoestrogens
Crown vetch	β-Nitropropanol glycosides
Leucaena spp.	Mimosine
Indigofera spp.	Indospecine
Grasses	
Forage sorghums	Cyanogens
Tall fescue	Fescue alkaloids
Tropical grasses	Oxalates
Others	
Forage brassicas	Brassica anemia factor

CLASSIFICATION OF TOXICANTS BY THEIR SITE OF ACTION AND/OR METABOLIC EFFECT

It is remarkable that for practically all major metabolic functions in animals, there exists in the plant world inhibitors of these functions. For virtually every organ, endocrine gland, and metabolic pathway in animals, there is a corresponding inhibitor in plants. Considerable speculation has been made as to the reasons why plants contain such a wide variety of deleterious (to animals) compounds. The following is by no means a complete listing, but is intended to give an appreciation for the diversity of effects that plant substances can have on animals.

5. Kidney

   Kidney damage occurs in poisonings from pyrrolizidine alkaloids, oxalates, and sesquiterpene lactones.

6. Circulatory System
   A. Aortic aneurysm occurs in lathyrism.
   B. Erythrocyte hemolysis is caused by saponins, brassica anemia factor, favism, copper toxicity, and pyrrolizidine alkaloids.
   C. Hematopoesis is impaired in pyrrolizidine alkaloid poisoning.
   D. Anemia occurs in favism and in poisonings due to pyrrolizidine alkaloids, brassica anemia factor, and copper toxicity.
   E. Delayed blood clotting occurs in sweet clover poisoning because of the properties of dicoumarol as a vitamin K antagonist.
   F. Agglutination of red blood cells in vitro is caused by hemagglutinins or lectins found in beans.
   G. Vasoconstriction is caused by ergot alkaloids and the toxins involved in "fescue foot," resulting in loss of blood supply to the extremities, and gangrene.
   H. Vasodilation is caused by veratrum alkaloids.
   I. Hypocalcemia is observed with oxalate poisoning.
   J. Hypercalcemia is caused by calcinogenic glycosides.
   K. Hypomagnesemia may be induced by *trans*-aconitic acid in early spring grass, leading to the condition of grass tetany.
   L. Hypoglycemia occurs in cocklebur poisoning due to the action of carboxyatractyloside. Severe hypoglycemia is also noted in "vomiting sickness" induced by the toxic amino acid hypoglycin in ackee (*Blighia sapida*).
   M. Hyperglycemia occurs in fluoroacetate (1080) poisoning. The 1080 compound occurs in several poisonous plants in Australia and South Africa.
   N. Hypercholesterolemia is induced by cyclopropenoid fatty acids in cottonseed oil.

7. Heart
   A. Heart Lesions
      Erucic acid in rapeseed oil and gossypol in cottonseed meal are known to produce cardiac lesions.
   B. Increased heart rate is caused by digitonin from foxglove and coniine, a piperidine alkaloid in poison hemlock.
   C. Decreased heart rate is characteristic of the toxic effects of veratrum alkaloids.
   D. Cardiac irregularity is caused by gossypol.

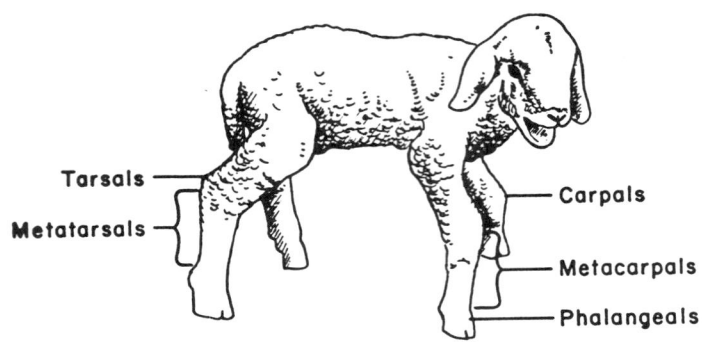

FIG. 1.3. Bones and joints affected by toxicants. For example, lambs from ewes consuming *Veratrum califòrnicum* may have shortened metatarsal and metacarpal bones. Lambs and calves from dams consuming locoweed may have a permanent flexure of the carpal joints. *Cestrum diurnum* intoxication in horses results in hyperextension of the metacarpophalangeal and metatarsophalangeal joints.

8. Bone
   A. Skeletal Deformity (Fig. 1.3)
      A wide variety of teratogenic agents, such as lupine and *Astragalus* toxins, cause skeletal deformities in fetuses. Skeletal deformity is also observed in lathyrism.
   B. Destruction of bone marrow occurs in bracken fern poisoning of cattle.
   C. Fibrosis of bone occurs in nutritional secondary hyperparathyroidism. The condition, called "bighead," is seen in horses consuming tropical grasses that contain oxalates.
9. Eye
   A. Disturbed vision occurs in animals consuming Jimsonweed, which contains atropine.
   B. Blindness and impaired vision occur in *Astragalus* poisoning due to degeneration of the optic nerve. Blind grass (*Stypandra imbricata*), an important problem in Western Australia, causes degeneration of the optic nerve. Selenium toxicity, from consumption of *Astragalus* spp. and other selenium accumulators, causes "blind staggers." Sheep grazing bracken fern may develop a degeneration of the neuroepithelium of the retina causing blindness.

10. Nervous System
    A. Nerve Cell Dysfunction
       The indolizidine alkaloids in *Swainsona* and *Astragalus* spp. cause accumulation of mannose in nerve cell lysosomes, causing axonal dystrophy.
    B. Degeneration of the spinal cord occurs in lathyrism.
    C. Cholinesterase inhibitors, such as solanine in green potatoes, inhibit breakdown of acetylcholine at neuromuscular junctions.
    D. Somnolence Effects
       Sleepy grass (*Stipa robusta*) induces a profound somnolence in horses; they may remain in deep slumber for several days.
    E. Incoordination (ataxia) occurs with locoweed poisoning, in pyrrolizidine alkaloid poisoning in horses, and in ryegrass staggers and annual ryegrass toxicity in sheep.
    F. Polyneuritis occurs in thiamine deficiency, induced by thiaminase in bracken fern. It also is observed in mink fed fish that contain thiaminase.
11. Muscle
    Nutritional myopathy, similar to white muscle disease in sheep due to selenium deficiency, has been observed associated with lupinosis in Western Australia.
12. Thyroid Gland
    A variety of goitrogenic agents in plants inhibit the synthesis of thyroxine. These include glucosinolates in *Brassica* spp., thiocyanate produced from detoxification of cyanogens, and dihydropyridone (DHP), a rumen metabolite of mimosine produced in animals grazing *L. leucocephala*.
13. Reproductive System
    A. Various mycotoxins and phytoestrogens in clovers and alfalfa have estrogenic effects.
    B. Female infertility is caused by isoflavones (phytoestrogens) in subterranean clover.
    C. Male infertility is caused by gossypol in cottonseed meal.
    D. Teratogenic effects have been observed with a number of toxic plants, including lupines, poison hemlock, and veratrum.
    E. Abortifacients or agents that induce abortion are found in ponderosa pine needles.
14. Milk-Transferred Toxins
    The best known example is white snakeroot toxin, which produced milk sickness in pioneer communities in the U.S. Pyrrolizi-

dine alkaloids and mycotoxins such as aflatoxin may be transferred in milk. Bitterweed (*Helenium amarum*) toxins give a bitter taste to milk, reducing its consumer acceptance.
15. Immune System
    Some toxicants affect the immune system and resistance to disease. Examples are lectins in beans, and impairment of the immune system in chronic toxicoses of aflatoxin and trichothecenes.
16. Hair, Skin, and Extremities
    A. Photosensitization
       Primary photosensitization is caused by photodynamic agents, such as hypericin in St.-John's-wort (*H. perforatum*), which are absorbed and react with light at the skin's surface, producing lesions. Secondary photosensitization is caused by phylloerythrin, a metabolite of chlorophyll. It is normally excreted in the bile, but in cases of liver damage, such as with pyrrolizidine alkaloid toxicity, or facial exzema caused by sporidesmin, phylloerythrin enters the general circulation and causes skin lesions when it reacts with ultraviolet light.
    B. Achromatrichia
       Two examples of this in mink are cotton fur in which a white band develops in the fur due to iron-deficiency anemia induced by trimethylamine oxide and formaldehyde in fish, and a graying of the fur caused by biotin deficiency, induced by avidin in raw eggs (turkey waste graying).
    C. Alopecia
       This occurs with mimosine and selenium toxicities.
    D. Hoof deformities occur with mimosine and selenium toxicities.
    E. Sloughing off of extremities, due to vasoconstriction and gangrene, occurs with ergot poisoning and tall fesuce toxicity.
17. Energy and Protein Metabolism
    A. Protein
       i. Digestion
          Protein digestion is impaired by protease inhibitors, such as the trypsin inhibitor in soybeans.
       ii. Amino acid metabolism
           Indospecine, an analog of arginine, induces arginine deficiency in animals grazing *I. spicata* (creeping indigo). Pyrrolizidine alkaloid toxicity causes impairment of liver functions, reducing the liver's ability to deaminate amino acids and synthesize urea from ammonia.
       iii. Protein synthesis by cells is inhibited by pyrrolizidine

alkaloids, which interfere with DNA and RNA metabolism.
B. Carbohydrates
  i. Starch digestion is inhibited by α-amylase inhibitors in wheat, oats, rye, navy beans, and kidney beans.
  ii. Mannosidosis
  Accumulation of α-mannose in nerve tissue occurs as a result of an α-mannosidase inhibitor in *Swainsona* spp. and *Astragalus* spp. (locoweed).
C. Lipids
  i. Fatty livers occur in lupinosis in sheep.
  ii. Fat absorption is inhibited by aflatoxins, which cause steatorrhea (lipid in feces).
D. Energy Metabolism
  i. Fluoroacetate, found in many Australian toxic plants, inhibits the tricarboxylic acid cycle by inhibiting aconitase.
  ii. Nitropropionic acid in crown vetch has been suggested to be an inhibitor of succinate dehydrogenase.
  iii. Hydrogen cyanide (HCN) from cyanogens is a potent inhibitor of cytochrome oxidase.

18. Cell Division
Pyrrolizidine alkaloids inhibit the prophase stage of mitosis. Metaphase arrest is observed in lupinosis.

19. Mineral Metabolism
A. Absorption
Chelating agents such as phytic acid and oxalates impair mineral absorption.
B. Liver storage of copper and zinc is altered by hepatotoxins. Liver copper is dramatically increased in animals poisoned with pyrrolizidine alkaloids, whereas the concentration of zinc is decreased.
C. Blood levels of minerals are influenced by toxicants. Oxalates cause hypocalcemia, while calcinogenic glycosides cause hypercalcemia.

20. Vitamin Metabolism
Various vitamin antagonists occur in feeds. These include avidin in raw eggs, which is an antagonist of biotin, and linatine, a pyridoxine antagonist, in linseed meal. Thiaminases in bracken and certain fish cause thiamine deficiency. Calcinogenic glycosides provide the active metabolite of vitamin D, and so interfere with the homeostatic effects of vitamin D on calcium metabolism. Dicoumarol causes a vitamin K deficiency. Lipoxidases destroy carotene, the precursor of vitamin A.

This list of animal tissues and systems influenced by plant toxicants is by no means complete, but gives an indication of the remarkably diverse effects that feed toxicants have on animal metabolism.

In this text, extensive reference is made to the classic book *Poisonous Plants of the United States and Canada*, by J. M. Kingsbury (1964). Kingsbury's book provides an excellent historical background to poisonous plant problems in North America. Liener's *Toxic Constituents of Plant Foodstuffs* (1980) is an excellent reference for a detailed treatment of some of the toxicants found in feeds, such as protease inhibitors, lectins, and glucosinolates. *Effect of Poisonous Plants on Livestock*, edited by Keeler et al. (1978), is the proceedings of a joint U.S.–Australia symposium on poisonous plants and is a source of more detailed information on specific poisonous plant problems. *Casarett and Doull's Toxicology: The Basic Science of Poisons*, by Doull, Klaassen, and Andur (1980), provides a comprehensive coverage of the principles of toxicology. *Clinical and Diagnostic Veterinary Toxicology*, by Buck et al. (1976), deals with practical aspects of toxicology from a veterinary perspective.

The study of natural toxicants has to some extent taken a back seat to research on agricultural chemicals, drugs, feed additives, and other synthetic toxicants which are the focus of attention of many toxicology programs. While the investigation of natural toxicants, particularly those in poisonous range and pasture plants, is not a "glamor" area of research, at least in North America, many pressing research needs remain. The widespread distribution of natural toxicants in food plants and the search for new crops to maximize efficient utilization of resources for the ever-increasing human population suggest that the study of natural toxins will be of increasing importance. Ames (1983) estimated that the human dietary intake of natural toxicants ("nature's pesticides") is several grams per day, or at least 10,000 times the dietary intake of man-made pesticides. It is hoped that this text will help to focus the attention of those involved in animal agriculture on the remarkable diversity of natural toxicants and their metabolic effects, and the exciting challenges remaining in research in this field.

REFERENCES

AMES, B. N. 1983. Dietary carcinogens and anticarcinogens. Science *221*, 1256–1264.

ARENA, J. M. 1974. Poisoning. Toxicology, Symptoms, Treatments, 3rd Edition. Charles C. Thomas Publishers, Springfield, IL.

BUCK, W. B., OSWEITER, G. D., and VAN GELDER, G. A. 1976. Clinical and

Diagnostic Veterinary Toxicology, 2nd Edition. Kendall-Hunt Publishing Co., Dubuque, IA.

DOULL, J., KLAASSEN, C. D., and AMDUR, M. O. (Editors) 1980. Casarett and Doull's Toxicology: The Basic Science of Poisons. Macmillan Publishing Co., NY.

EVERIST, S.L. 1981. Poisonous Plants of Australia. Angus & Robertson, Sydney, Australia.

HALLIBURTON, J. C., and BUCK, W. B. 1983. Animal poison control center: Summary of telephone inquiries during first three years of service. J. Am. Vet. Med. Assoc. *182*, 514–515.

KEELER, R. F., and TU, A. T. (Editors) 1983. Handbook of Natural Toxins, Vol. 1. Plant and Fungal Toxins. Marcel Dekker, NY.

KEELER, R. F., VAN KAMPEN, K. R., and JAMES, L. F. (Editors) 1978. Effects of Poisonous Plants on Livestock. Academic Press, NY.

KINGHORN, A. D. (Editor) 1979. Toxic Plants. Columbia Univ. Press, Irvington, NY.

KINGSBURY, J. M. 1964. Poisonous Plants of the United States and Canada. Prentice-Hall, Englewood Cliffs, NJ.

KIRK, R. W. 1983. Current Veterinary Therapy, 8th Edition. W. B. Saunders Co., Philadelphia, PA.

LIENER, I. E. (Editor) 1980. Toxic Constituents of Plant Foodstuffs, 2nd Edition. Academic Press, NY.

ROBINSON, T. 1980. The Organic Constituents of Higher Plants, 4th Edition. Cordus Press, North Amherst, MA.

SEAWRIGHT, A. A. 1982. Animal Health in Australia, Vol. 2. Chemical and Plant Poisons. Australian Government Publishing Service, Canberra.

SEAWRIGHT, A. A., HRLICKA, J., LEE, J. A., and OGUNSAN, E. A. 1982. Toxic substances in the food of animals: Some recent findings of Australian poisonous plant investigators. J. Appl. Toxicol. *2*, 75–82.

SHULL, L. R., and CHEEKE, P. R. 1983. The effects of synthetic and natural toxicants on livestock. J. Anim. Sci. *57* (Suppl. 2), 330–354.

VAHRMEIJER, J. 1981. Poisonous Plants of Southern Africa. Tafelberg Publ. Ltd., Cape Town, RSA.

2

Techniques and Calculations in Toxicology

ISOLATION AND IDENTIFICATION OF PLANT TOXINS

In studying the chemical nature of plant toxins and their metabolism in animals, it is necessary to extract and purify them. While specific techniques and methods are published for many toxicants, a few general guidelines are of interest. In cases where the chemical nature of the toxicant is not yet known, it may be necessary to employ a bioassay technique to identify the location of the toxic component(s) during the fractionation and purification procedures. For example, in fractionation of tall fescue extracts to attempt to identify the causative agent of fescue foot, fractions produced by extraction and ion-exchange chromatography were tested for biological activity by injection into calves and the subsequent measurement of skin temperature in the area of the coronary band of the hoof (Garner et al. 1982).

The stability of toxicants in plant tissue varies with the chemical nature of the specific compounds involved. In many cases, no special handling is required, and samples of hay or sun-cured or oven-dried plant material may be used. In some cases, rapid freezing in liquid nitrogen followed by freeze-drying may be desirable to prevent changes from occurring postharvest. Enzymatic action, resulting in hydrolysis and oxidation, may occur as breakdown in cell structure occurs. Special techniques, such as homogenization with polyvinyl pyrrolidone to bind tannins, may be used to prevent such action (Loomis and Battaile 1966).

Extraction of plant tissue can be performed in several ways. Green plant tissue can be macerated in a Waring blender. Dried material may be extracted with an appropriate solvent in a Soxhlet extraction unit. For extraction of large quantities, the dried plant tissue can be soaked in drums or large vats, with the solvent exchanged several times, followed by evaporation of the solvent. Selection of the appropriate solvent is based upon the chemical characteristics of the specific toxicant. For example, alkaloids normally occur in plant tissue as salts of organic acids. They can be extracted with an acidic, aqueous solvent; the aqueous extract may be made basic and the alkaloids extracted with an organic solvent, leaving neutral and acidic water-soluble compounds behind. Alternatively, the plant material can be made alkaline, and all the free alkaloid bases extracted with organic solvents.

Following extraction of the plant tissue, various techniques are employed to further isolate and identify specific plant toxins. Extracts may be passed through cation or anion exchange columns, followed by elution with various solvents. Various chromatographic techniques can be used to effect separations, including paper, thin-layer, column, or gas–liquid chromatography (GLC), and high-performance liquid chromatography (HPLC). The HPLC technique is now widely employed for isolation and identification of plant toxins. Detection of individual compounds is often based on colorimetric procedures. For example, pyrrolizidine alkaloids are commonly detected by the Ehrlich reagent test. Ehrlich reagent (4-dimethylaminobenzaldehyde) reacts with pyrroles to produce a purple chromophore, which may be detected visually with paper or thin-layer chromatography, or measured spectrophotometrically. Other methods of identification include infrared, ultraviolet, and nuclear magnetic resonance (NMF) spectra, and mass spectra. Various derivatives may be made to improve the gas chromatography separation, mass spectra, and detector sensitivity. Trimethylsilyl derivatives are commonly made for gas chromatography–mass spectrometry (GC–MS) analysis. (Fig. 2.1). The GC procedure allows separation of extracts into specific peaks based on retention time; the peaks can be subjected to MS analysis for identification of functional groups.

Specific techniques for analysis and identification of most plant toxicants have been published. A detailed consideration of analytical techniques is beyond the scope of this book. Pertinent literature should be consulted by those who need to pursue this aspect.

BIOASSAY OF TOXICANTS

Coincident with chemical analysis, or sometimes in place of it, bioassays are often employed in the study of toxicants. A few examples will

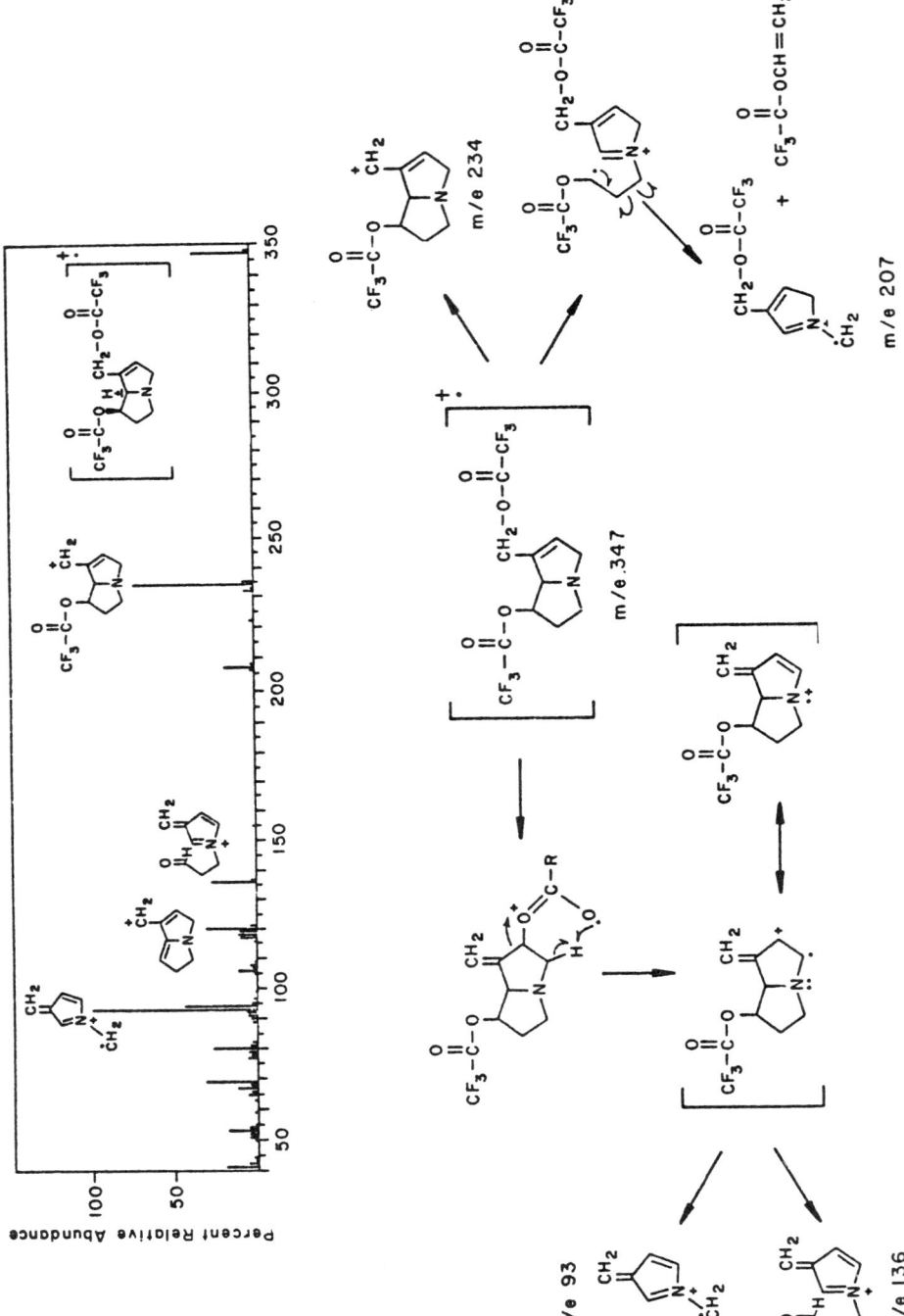

illustrate their application. Saponins in alfalfa have been measured using a soil fungus, *Trichoderma viride*. The fungus is grown on a culture medium containing various levels of alfalfa extract; the fungal growth is proportional to the saponin content (Livingston et al. 1977). Another saponin bioassay has involved the use of guppies or minnows. Fish are extremely sensitive to the detergent properties of saponins, so that fish paralysis can be correlated with saponin content of the solution. Saponins can also be measured by erythrocyte hemolysis. Suspensions of red blood cells are exposed to serial dilutions of alfalfa extracts, and the minimum dilution causing complete hemolysis is determined.

Another bioassay technique, alluded to earlier, is that for the study of the toxin in tall fescue responsible for fescue foot. During the fractionation of tall fescue extracts with ion-exchange chromatography (Garner et al. 1982), the activity of the various fractions has to be tested to determine if it contains the fescue foot toxin. The fractions are injected intraperitoneally into calves and the temperature of the coronary band of the rear hoof measured by an infrared technique. Those fractions containing the fescue foot toxin produce an elevation in the coronary band temperature.

A final example of a bioassay technique is the mouse or rat uterine weight bioassay for plant estrogens (Fig. 2.2). Plant extracts are injected into immature female mice or rats; 24 hr later the uterine weight is measured. With plant extracts containing phytoestrogen activity, the uterine weight is elevated. A standard curve is prepared using an estrogenic hormone such as diethylstilbestrol or estradiol.

ASSESSMENT OF TISSUE DAMAGE

Various procedures are used to assess pathology when animals are fed toxicants or poisonous plants. The specific techniques used to assess tissue damage depend, of course, on what tissues are affected by the particular toxicant being studied. In the case of hepatotoxins, liver biopsies can be used to assess sequential changes in liver histology. A number of tests can be performed to assess liver function. The sulfobromophthalein (BSP) clearance rate is widely used; this test measures the ability of the liver to remove a dye (BSP) from the blood. The BSP is injected intravenously, and sequential blood samples are taken over

FIG. 2.1. Mass spectrometry of di(trifluoroacetyl) ester of the pyrrolizidine alkaloid retronecine, showing association of MS peaks with functional groups.
From Deinzer et al. (1979).

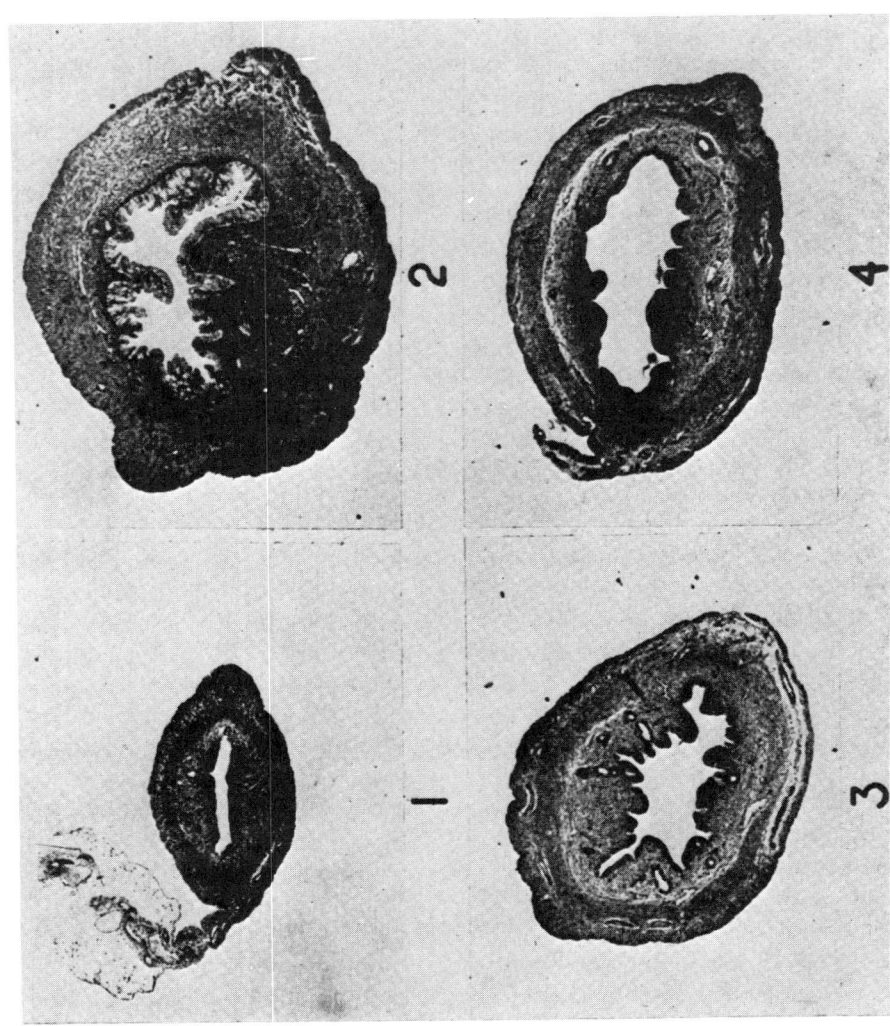

a period of 2.5–15 min. In hepatic insufficiency, the removal of BSP from the blood is impaired. The BSP is conjugated with glutathione and excreted in the urine as a mercaptide (see Chapter 3). The serum bilirubin level may be measured to assess liver function. In the colorimetric procedures used, "direct-reacting" and "indirect-reacting" bilirubin are measured. Direct-reacting bilirubin is measured without the addition of alcohol to the serum, whereas the indirect-reacting form is measured following addition of alcohol. The direct-reacting bilirubin is a bilirubin glucuronide which can react directly with the reagent without being solubilized in alcohol. In obstruction jaundice, conjugated bilirubin may be returned to the blood rather than excreted in the bile, so the serum direct-reacting bilirubin increases. The indirect-reacting bilirubin is unconjugated bilirubin enroute to the liver from the reticuloendothelial tissues where it was produced from the breakdown of heme porphyrins. It is not water soluble and has to be solubilized in alcohol to react with the azo dye used in the bilirubin assay. The indirect type increases in hemolytic jaundice.

Another serum test used to assess liver function is the measurement of serum albumin. Albumin is synthesized in the liver, so hypoalbuminemia reflects impaired liver function. Because serum albumin has a major function in maintaining osmotic balance between the blood and tissues, hypoalbuminemia is often accompanied by edema and ascites (fluid accumulation in body cavities). A variety of serum enzymes is measured in assessment of liver function. These include alkaline phosphatase, glutamic-oxaloacetic transaminase (GOT), lactic dehydrogenase (LDH), γ-glutamyl transpeptidase (GGT), and several others. These enzymes are released from liver tissue during hepatic necrosis. The enzymes are compartmentalized into functional units within liver cells, so when the cell membranes are disrupted by liver damage, the enzymes are released. The primary site of liver damage can sometimes be located by observing differential changes in serum enzymes. Alkaline phosphatase and γ-glutamyl transpeptidase are increased in biliary hyperplasia and blockage of bile excretion. Sorbitol dehydrogenase (SDH) is found mainly in liver and is a good indicator of hepatic damage. Lactic dehydrogenase (LDH) and glutamate dehydrogenase (GLDH) are also useful indicators of

---

FIG. 2.2. Photomicrographs of histological cross sections of uteri obtained from prepubertal (1) and pubertal normal intact rats (2) and from prepubertal intact rats receiving a red clover estrogenic compound (3) and coumesterol acetate (4).
From Ostrovsky and Kitts (1963). Courtesy of W. D. Kitts.

liver damage, as well as damage to other tissues. Liver parasites such as liver flukes can cause elevations in serum levels of these enzymes.

The pyrrolizidine alkaloids are typical examples of hepatotoxins. They cause liver necrosis, with accompanying elevation in serum bilirubin, decreased serum albumin (resulting in ascites), impaired BSP clearance, and elevations in serum enzymes such as GOT and GGT.

γ-Glutamyl transpeptidase is found mainly in the kidney, but its activity in the liver is relatively high in cattle, horses, sheep, and goats (Braun et al. 1983), and very low in dogs, cats, rabbits, and birds. Measurement of this enzyme in serum is useful in hepatobiliary diseases of cattle, sheep, goats, and cholestatic disorders of dogs. The urinary activity is a good test for kidney damage.

TECHNIQUES FOR STUDYING TOXICANT METABOLISM

Isolated Cellular Preparations and Whole Animal Studies

The metabolism of toxicants is often studied in isolated liver tissue preparations. This reflects the importance of the liver in the metabolism of toxicants. Preparations of other tissues are used when appropriate. For example, the metabolism of 3-methylindole in lung tissue has been investigated by Bray and Carlson (1979) in studies of acute bovine pulmonary emphysema. The tissue is obtained immediately upon sacrifice of the animal and placed in cold isotonic saline in ice. Homogenization should be conducted as soon as possible. Homogenates are prepared by mincing the tissue in a tissue homogenizer equipped with a Teflon pestle. The tissue is homogenized in a buffer solution containing nutrients and cofactors to allow continued metabolism. Further fractionation of the homogenate to produce fractions of nuclei and debris, mitochondria, lysosomes, microsomes, and the cytosol is accomplished by selective ultracentrifugation. Typical fractionation of homogenates involves the following sequential centrifugation of a homogenate: nuclei and debris, 600 g; mitochondria, 4100 g; lysosomal fraction 24,500 g; and microsomes, 105,500 g. The cytosolic fraction is the supernatant remaining. The microsome fraction consists primarily of endoplasmic reticulum in which drug-metabolizing enzymes are located. While fractionation of tissue is very useful in isolating enzyme systems and studying the metabolism of toxins without extraneous complications, it is necessary to maintain the perspective that the intact animal is what is ultimately important. Sometimes there is a tendency to regard an animal as a giant liver, or even a microsomal preparation. The use of isolated hepatocytes is an example of a technique that more closely correlates with intact liver metabolism than the study of individual preparations. Studies involving the whole animal employ techniques such as the distribution of isotope-labeled compounds and their metabolism and excretion of metabolites.

For ruminant animals, study of metabolism of toxins in the rumen is often desirable. In vitro rumen fermentation techniques are useful. For example, Swick *et al.* (1983) studied the metabolism of pyrrolizidine alkaloids in an artificial rumen to determine if the alkaloids are detoxified by rumen microorganisms. Similarly, Allison *et al.* (1981) used a continuous-flow rumen fermenter to investigate the degradation of oxalate by rumen microorganisms. These techniques allow duplication of the events occurring in the rumen without the expense and complications of utilizing an intact animal.

Mutagenesis

Other techniques sometimes employed in large animal toxicology include the study of mutagenesis, carcinogenicity, and teratogenesis. Mutagenesis is the alteration of DNA, involving chromosomal changes such as breaks or rearrangements. Types of mutagenesis tests commonly used include the cytogenetic test, the dominant-lethal test, the host-mediated microbial assay, and the Ames test. In the cytogenetic test, mammalian cells are treated with the test substance. Either cell cultures or intact animals may be used. Bone marrow, regenerating liver cells, or reproductive cells are often used. After an appropriate period, mitosis is suspended by treatment with colchicine, and then the chromosomes are karyotyped and examined for abnormalities. In the dominant-lethal test, male rats or mice are given the test substance and are subsequently mated to groups of untreated virgin females. Time of mating is established by examination of vaginal smears for sperm. At a defined period postmating (e.g., 14 days), the females are sacrificed and evaluated for number of corpora lutea, implantation sites, and dead and live fetuses. Dominant lethality is assessed by preimplantation or postimplantation loss compared to untreated controls. In the host-mediated test, microorganisms are injected into a host animal (e.g., mouse) and then the test substance is administered to the animal. The organisms, such as a strain of *Salmonella,* are injected intraperitoneally, and then samples of fluid are withdrawn periodically. Organisms in the fluid are cultured and examined for mutagenic changes. An in vitro modification of this procedure is the Ames test. Histidine-dependent *Salmonella* organisms are cultured on a histidine-deficient medium containing the test substance and animal tissue microsomes (Fig. 2.3). The microsomal fraction (usually rat liver microsomes) is used to mimic mammalian metabolism. The test measures the extent of mutagenesis producing a wild-type nonhistidine-dependent *Salmonella* which can grow on the histidine-deficient medium. This test (Maron and Ames, 1983), developed by Ames et al. (1975), is widely used as a short-term predictive screening test to detect possible carcinogens. There is a high correlation between mutagenic activity in the Ames test and carcinogenicity (McCann et al. 1975). An example of its application in large animal toxicology is its use to assess the public health significance of the transfer of pyrrolizidine alkaloids in milk (White et al. 1984).

Carcinogenesis

A tumor (neoplasm) is an abnormal mass of tissue whose growth exceeds and is uncoordinated with that of normal tissue. Benign tumors are those with cells structurally identical to those of normal tissues, and they are confined to the area of origination, without invasion of neighboring tissue. A malignant tumor (cancer) contains cells that are not typical of the structures from which they arise, and they have a tendency to invade neighboring tissue. They also have a tendency to metastasize; that is, they form secondary tumors at a site distant to the primary tumor. Carcinogens are substances that induce cancer. Primary or direct-acting carcinogens are those that do not require metabolic activation. They generally act at the point of application. An example is mustard gas. Secondary carcinogens require metabolic activation to a reactive (electrophilic) form. This is usually accomplished by the mixed function oxidases (see Chapter 3). They usually affect a specific target organ and may not act at the point of application. Examples of secondary carcinogens are pyrrolizidine alkaloids, aflatoxin B_1, and carbon tetrachloride. Pyrrolizidine alkaloids are bioactivated to reactive pyrrole derivatives (see Chapter 5) while aflatoxin B_1 is converted to the active carcinogen, aflatoxin B_1 2,3-epoxide, by liver enzymes. Cocarcinogens (promoters) are substances that potentiate or promote the effects of carcinogens. For example, cyclopropenoid fatty acids (see Chapter 8) potentiate the effects of aflatoxin so that with exposure to both cyclopropenoids and aflatoxin, the tumor incidence is greater than what would be caused by aflatoxin alone.

There is a difference between other toxins and carcinogens in the dose–response relationship. With other toxins, the dose needs to exceed the capacity of detoxification mechanisms for toxicity signs to

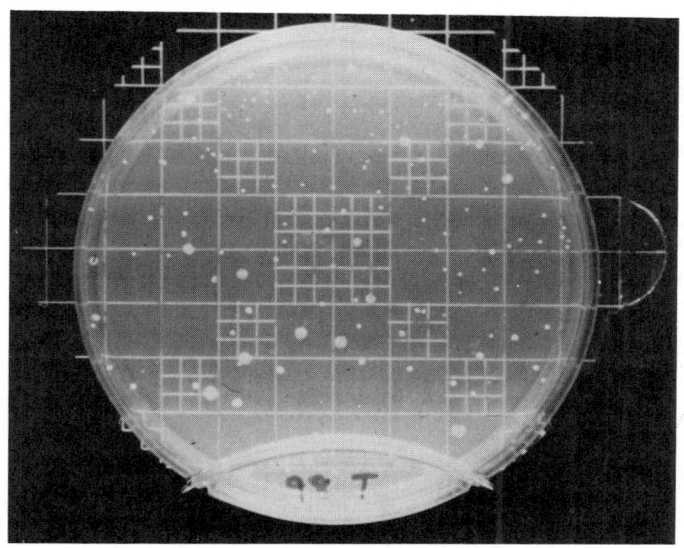

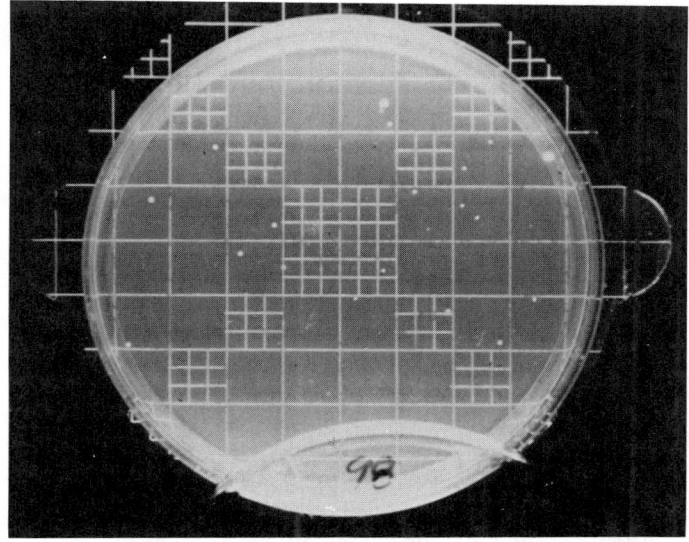

FIG. 2.3. The Ames *Salmonella* mammalian microsome mutagenicity test. An example of a negative response with the Ames test (top), showing a few bacterial colonies resulting from spontaneous mutation. On the bottom is an example of a positive Ames test response, showing a large number of colonies on the media. The mutagenic agent added to the media caused an increased rate of mutation, resulting in an increased quantity of bacteria capable of synthesizing histidine and thus able to grow on the histidine-free medium.
Courtesy of R. D. White.

occur. Carcinogens react with specific tissue receptors and may produce a permanent change. Subsequent doses add to such a change. After a sufficient number of cells have been altered, a visible neoplasm may result. Thus it is difficult to establish if there is a "no-effect" level for carcinogens. There is generally a latent period between exposure and occurrence of neoplasms. Increasing doses of carcinogens increase the number of tumors and reduce the latent period.

Various factors may modify carcinogenesis, including dietary factors. In some cases, different carcinogens that act on the same target organ have either additive or synergistic effects. Factors that modify microsomal mixed function oxidase activity may modify carcinogenesis. For example, feeding phenothiazine, an enzyme inducer, to rats consuming bracken fern reduced tumor incidence by 60% in the studies of Pamukcu et al. (1971). The carcinogenicity of aflatoxin is increased by stimulation of microsomal enzymes.

Natural toxicants that have carcinogenic activity include aflatoxins, pyrrolizidine alkaloids, and bracken carcinogens. Except for bracken, livestock problems are not normally encountered with these carcinogens because most livestock are marketed or culled before cancer would normally be seen. Probably of greater significance to animal production than direct carcinogenesis is the possibility of transfer of carcinogens to humans through animal products. Aflatoxin and pyrrolizidine alkaloids in milk are examples.

Testing for carcinogenic activity is usually performed with laboratory animals such as rats. Several dosage levels of the test substance are used and fed in diets for a prolonged period, such as 24 months for rats and 21–24 months for mice. At the termination, tissues are examined by a pathologist for tumors. At least 20 animals per dosage level is recommended.

Teratogenesis

The term "teratology" is derived from the Greek word "teras," meaning monsters. In more common terms, teratology is the study of congenital malformations. Teratogens are agents that induce fetal abnormalities.

The study of the effect of plant teratogens on livestock has been quite limited, mainly because of the expense of conducting this type of research with large animal species. There are a number of significant teratogenic problems in livestock, which in the U.S. include crooked calf disease, induced by lupines and poison hemlock, abnormalities associated with the consumption of locoweeds, and the bizarre "cyclops lambs" produced when pregnant ewes consume *Veratrum californi-*

cum. Keeler (1984) has reviewed some general principles of teratology. These will be enumerated briefly.

Principle 1. Genotype determines susceptibility.

There is species variation in response to teratogens. For example, consumption of lupines by pregnant cows may cause crooked calf disease, but no teratogenic effects are seen when lupines are consumed by pregnant ewes.

Principle 2. A teratogen must reach the fetus or produce a metabolite which does.

The teratogen exerts its effect within the fetal tissue.

Principle 3. Deformities are dose dependent.

The incidence of malformations varies according to the amount of teratogen reaching the fetus. This ranges from a no-effect level where no observable incidence of teratogenesis occurs to toxic levels where fetal death occurs.

Principle 4. A teratogen can produce death rather than deformities.

In livestock, it is possible that a problem with plant teratogens could be manifested mainly in a high incidence of abortions or fetal resorptions.

Principle 5. The fetus must be exposed at the susceptible period in gestation.

The fetus is susceptible to the effects of a teratogen during the sharply defined period when the particular tissue affected is undergoing development. For example, the cyclops lamb condition occurs only if a ewe consumes *V. californicum* on the fourteenth day of pregnancy. The time in days of gestation when an insult induces a specific defect varies greatly among species because of differences in the length of the gestation period. The approximate periods when teratogens produce specific defects in a number of animal species are shown in Table 2.1.

Principle 6. Teratogens exert their effects by specific mechanisms.

Dissimilar teratogens may induce the same metabolic defect to give rise to similar deformities. For example, crooked calf disease may be induced by the quinolizidine alkaloid anagyrine in lupines and the piperidine alkaloid coniine in poison hemlock.

Plants implicated in teratogenesis can be grouped into three categories (Keeler 1984): those with known teratogens, teratogenic plants with as yet unidentified teratogens, and suspected teratogenic plants.

TABLE 2.1. Approximate Gestation Period When Teratogens Produce Defects in Five Animal Species[a]

Type of fetal defect	Days of gestation when teratogenic effects are produced				
	Humans	Hamsters	Sheep	Cattle	Pigs
Microphthalmia	16–17	6.5	12–13	16–18	10–11
Cyclopia	21–23	7.0	14	21	12
Exencephaly	26	7.5	16.5	26–27	15.5
Spina bifida	28–29	8–8.5	18	29–31	17
Shortened limbs	36	9.5	29–31	42	20
Total gestation period	267	16	147	283	115

[a] Adapted from Keeler (1978).

Included in the first group with known teratogens are plants in the *Lupinus, Veratrum, Conium,* and *Leucaena* genera. Examples of teratogenic plants for which the specific teratogenic agent has not been identified are *Astragalus* and *Nicotiana* spp. Suspected teratogenic plants, for which definitive feeding trials have not yet been conducted or for which field cases suggest an involvement, include *Datura, Prunus,* and *Sorghum* spp.

In general, the critical period for most birth defects is the first trimester of pregnancy. Chemical attack by a teratogen during early organogenesis is most likely to produce a malformation; during the fetal growth period, after the organs are fully formed, changes induced by exogenous substances are almost always toxic rather than teratologic effects.

EXPRESSION OF BIOLOGICAL RESPONSES TO TOXICANTS

The expression of toxicity is generally made with respect to body size, for example, in milligrams of compound per kilogram body weight. The most commonly used measure of acute toxicity is the lethal dose (LD_{50}) which is the dosage of the toxin that on the average will kill 50% of the test animals. Several methods for calculation of the LD_{50} are available. The most widely used are those of Miller and Tainter (1944), Litchfield and Wilcoxon (1949), and Weil (1952). The method of Weil is probably the simplest to use. Its application involves the use of four or more dosage levels, and 2, 3, 4, 5, 6, or 10 animals per dose level. The LD_{50} and a 95% confidence interval are derived from tables published by Weil (1952), using the number of mortalities at each dosage level.

Acute toxicity is usually used to describe the effects seen following a single dose or doses within a 24-hr period. Chronic toxicity refers to effects produced by prolonged exposure to the test substance over a period of 3 months or longer. Subchronic toxicity refers to effects of exposure of more than 24 hr but less than 3 months.

In expressing the chronic toxicity of poisonous plants, the lethal dosage as a percentage of body weight is often given and is useful for rough comparisons. For example, the lethal dose of tansy ragwort (*Senecio jacobaea*) is about 5–10% of body weight for cattle and horses, and 100–200% for sheep and goats. A problem with this mode of expression for chronic toxicity is deciding what body weight to use. With growing animals, the body weight may change considerably during a chronic toxicity study. Also, there may be a marked decline in body weight near the terminal stages of toxicity. Methods to compensate for these difficulties include using either initial body weight or average body weight for the period of the study.

MANAGEMENT OF TOXICOSES

Toxicoses in livestock may involve spectacular acute effects or insidious conditions of a chronic nature. Detailed consideration of antidotes and therapeutic agents is beyond the scope of this text. Veterinary practitioners should consult appropriate sources, such as Kirk (1983), Arena (1974), and Buck *et al.* (1976). In the U.S. and Canada, assistance may be obtained from the Animal Poison Control Center, University of Illinois. The center was established in 1978 to provide telephone consultation to veterinarians seeking assistance with suspected animal poisonings and to provide information on the potential hazard and toxicity of pharmaceutical products, agrichemicals, and environmental pollutants. A description of the activities of the center is provided by Haliburton and Buck (1983). A rapid-response investigation service is available; a team of toxicology specialists will travel to assist with the identification and containment of poisoning outbreaks. The address and phone number of the center are

> Animal Poison Control Center
> University of Illinois
> College of Veterinary Medicine
> 2001 S. Lincoln Avenue
> Urbana, IL 61801
> (217) 333-3611

Buck et al. (1976) described a management plan for toxicologic emergencies, as follows:

1. Institute the necessary emergency and supportive therapy to keep the animal alive.
2. Establish a tentative clinical diagnosis on which to base therapy.
3. Institute the appropriate remedial and antidotal procedures.
4. Identify the toxic agent as rapidly as possible.
5. Determine the source of the toxin.
6. Counsel the livestock owner on the hazards of the implicated toxicant, and provide instruction for the avoidance of the problem in the future.

The services of a veterinarian should be sought immediately when a toxicosis is suspected. The livestock owner should provide a description of the clinical signs, and any information on the suspected source of the toxin that might be available. Therapeutic measures should be attempted by the owner only following veterinary advice.

PLANT IDENTIFICATION AND USE OF BOTANICAL NAMES

All plant material used in toxicology studies should be identified by genus and species. Publications arising from experimental work should indicate the authority by which the plant was identified. Voucher specimens of the material should be deposited in an herbarium for future verification if necessary. Kingsbury (1980) notes that herbaria can be regarded as libraries, being depositories of materials in good condition which can be recalled at a later date if necessary. The importance of the use of botanical names is obvious when one considers the multiplicity of common names used for the same plant and the multiplicity of plants identified by the same common name. Tansy (*Tanacetum vulgare*) and tansy ragwort (*Senecio jacobaea*) are both commonly referred to as tansy. *Tanacetum vulgare* is an herb used safely in cooking, whereas *Senecio jacobaea* is a dangerous poisonous plant causing irreversible liver damage. "Hemlock" is a common name for water hemlock (*Cicuta maculata*), poison hemlock (*Conium maculatum*), the hemlock tree (*Tsuga canadensis*), and ground hemlock (*Taxus canadensis*). One of these "hemlocks" is deadly poisonous (*Cicuta maculata*), while another is harmless (*Tsuga canadensis*). Thus, confusion as to the identity of a particular plant can be avoided if the botanical names are used.

REFERENCES

ALLISON, M. J., COOK, H. M., and DAWSON, K. A. 1981. Selection of oxalate-degrading rumen bacteria in continuous cultures. J. Anim. Sci. *53*, 810–816.
AMES, B. N., McCANN, J., and YAMASAKI, E. 1975. Methods for detecting carcinogens and mutagens with the *Salmonella* mammalian microsome mutagenicity test. Mutat. Res. *31*, 347–364.
ARENA, J. M. 1974. Poisoning, Toxicology, Symptoms, Treatments, 3rd Edition. Charles C. Thomas Publishers, Springfield, IL.
BECKMAN, D. A., and BRENT, R. L. 1984. Mechanisms of teratogenesis. Annu. Rev. Pharmacol. Toxicol. *24*, 483–500.
BRAUN, J. P., BENARD, P., BURGAT, V., and RICO, A. G. 1983. Gamma glutamyl transferase in domestic animals. Vet. Res. Commun. *6*, 77–90.
BRAY, T. M., and CARLSON, J. R. 1979. Covalent binding of 3-methylinodole metabolites in goat lung microsomal preparations. Fed. Proc., Fed. Am. Soc. Exp. Biol. *38*, 1329.
BUCK, W. B., OSWEITER, G. D., and VAN GELDER, G. A. 1976. Clinical and Diagnostic Veterinary Toxicology, 2nd Edition. Kendall-Hunt Publishing Co., Dubuque, IA.
DEINZER, M. L., THOMSON, P. A., GRIFFIN, D. A., and BURGETT, D. M. 1979. The analysis of pyrrolizidine alkaloids in agricultural food products. *In* Symposium on Pyrrolizidine (*Senecio*) Alkaloids: Toxicity, Metabolism and Poisonous Plant Control Measures. P. R. Cheeke (Ed.), pp. 95–106. Nutrition Research Institute, Oregon State Univ., Corvallis.
GARNER, G. B., CORNELL, C. N., YATES, S. G., PLATTNER, R. D., ROTHFUS, J. A., and KWOLEK, W. F. 1982. Fescue foot: Assay of extracts of toxic tall fescue herbage. J. Anim. Sci. *55*, 185–193.
HALIBURTON, J. C., and BUCK, W. B. 1983. Animal poison control center: Summary of telephone inquiries during first three years of service. J. Am. Vet. Med. Assoc. *182*, 514–515.
JOHNSON, E. M., and KOCHHAR, D. M. (Editors) 1983. Teratogenesis and reproductive toxicology. Handbook of Experimental Pharmacology, Vol. 65. Springer-Verlag, Berlin, 1983.
KEELER, R. F. 1978. Reducing incidence of plant-caused congenital deformities in livestock by grazing management. J. Range Manage. *31*, 355–360.
KEELER, R. F. 1984. Teratogens in plants. J. Anim. Sci. *58*, 1029–1039.
KINGSBURY, J. M. 1980. Phytotoxicology. *In* Doull, J., C. D. Klaassen, and M. O. Amdur (Editors). Casarett and Doull's Toxicology: The Basic Science of Poisons. Macmillan Publishing Co., NY.
KIRK, R. W. 1983. Current Veterinary Therapy, 8th Edition. W. B. Saunders Co., Philadephia, PA.
LITCHFIELD, J. T., and WILCOXON, F. 1949. A simplified method of evaluating dose-effect experiments. J. Pharmacol. Exp. Ther. *96*, 99–113.
LIVINGSTON, A. L., WHITEHAND, L. C., and KOHLER, G. O. 1977. Microbiological assay for saponin in alfalfa products. J. Assoc. Off. Anal. Chem. *60*, 957–960.

LOOMIS, W. D., and BATTAILE, J. 1966. Plant phenolic compounds and isolation of plant enzymes. Phytochemistry 5, 423–438.

MARON, D. M., and AMES, B. N. 1983. Revised methods for the *Salmonella* mutagenicity test. Mutat. Res. *113,* 173–215.

McCANN, J., CHOI, E., YAMASAKI, E., and AMES, B. N. 1975. Detection of carcinogens as mutagens in the *Salmonella* microsome test: Assay of 300 chemicals. Proc. Natl. Acad. Sci. U.S.A. *72,* 5135–5139.

MILLER, L. C., and TAINTER, M. L. 1944. Estimation of the ED_{50} and its error by means of logarithmic-Probic graph paper. Proc. Soc. Exp. Biol. Med. *57,* 261–264.

OSTROVSKY, D., and KITTS, W. D. 1963. The effect of estrogenic plant extracts on the uterus of the laboratory rat. Can. J. Anim. Sci. *43,* 106–112.

PAMUKCU, A. M., WATTENBERG, L. W., PRICE, J. M., and BRYAN, G. T. 1971. Phenothiazine inhibition of intestinal and urinary bladder tumors induced in rats by bracken fern. J. Natl. Cancer Inst. (U.S.) *47,* 155–159.

SWICK, R. A., CHEEKE, P. R., RAMSDELL, H. S., and BUHLER, D. R. 1983. Effect of sheep rumen fermentation and methane inhibition on the toxicity of *Senecio jacobaea*. J. Anim. Sci. *56,* 645–651.

WEIL, C. 1952. Tables for convenient calculation of median-effective dose (LD_{50} or ED_{50}) and instructions in their use. Biometrics *8,* 249–263.

WHITE, R. D., KRUMPERMAN, P. H., CHEEKE, P. R., DEINZER, M. L., and BUHLER, D. R. 1984. Mutagenic responses of tansy ragwort (*Senecio jacobaea*) plant, pyrrolizidine alkaloids and metabolites in goat milk with the *Salmonella* mammalian-microsome mutagenicity test. J. Anim. Sci. *58,* 1245–1254.

3

Metabolism of Toxicants

The principal route of exposure of most natural toxicants to livestock is through the diet. Ingested toxins may be subjected to a number of metabolic processes in the digestive tract and in various tissues prior to excretion. Animals have been exposed to toxic constituents of plants during the long period of their coevolution and have developed numerous biochemical strategies for detoxification of poisonous compounds. This capacity is shared by all plant predators, including vertebrate herbivores, insects, and microorganisms. Once a toxicant has been consumed, there are several barriers it must surmount before reaching its critical target. These include chemical and microbiological detoxification mechanisms in the gastrointestinal tract, a host of detoxifying enzymes in the liver, and similar enzymes in all other tissues. Probably the ability to overcome the effects of toxicants is most developed in the insect world, where there are numerous examples of specialist feeders which feed with impunity on poisonous plants. The cinnibar moth (*Tyria jacobaea*) larvae feed only on plants such as *Senecio* spp. that contain pyrrolizidine alkaloids; the larvae have evolved a metabolic requirement for these alkaloids. In Australia, numerous species of two butterfly families and one moth family require pyrrolizidine alkaloids for synthesis of a phermone. The *Chrysolina* beetle was introduced into California and the Pacific Northwest to control St.-John's-wort (*Hypericum perforatum*), a toxic weed. The larvae of the monarch butterfly feed on milkweeds and accumulate the toxic milkweed cardenolides, making the larvae repellent to birds. Many more examples are known of insects that have evolved mechanisms to detoxify specific toxicants to the point where in some cases they now have a metabolic requirement for them. Among mammals and specifically livestock, there are few examples of specialist feeders. One example of a wild species is the koala, which specializes in eating eucalyptus leaves that are rich in phenolics and terpenoids. Vultures have developed immunity to the deadly botulinus toxin, produced in rotting carcasses which they consume. Many mammals, including some livestock species, do have specialized resistance to some toxicants. Sheep and goats can tolerate more than ten times as much pyrrolizidine alkaloid as can cattle and horses. Sheep will avidly consume foliage of plants containing these alkaloids, whereas cattle and horses will avoid them. Sheep are more resistant to larkspur and poison hemlock toxicity than are cattle. Horses are not affected by the toxin which causes fescue foot in cattle. Sheep are not affected by the teratogenic effects of quinolizidine alkaloids in lupines which cause birth defects in cattle. Many more examples of these differences in livestock response to poisonous plants are discussed in this book.

METABOLIC FATE OF TOXICANTS

After an animal has ingested a toxic plant, numerous metabolic processes occur before the toxicant involved exerts its toxic effect or is excreted. These processes may occur in the gastrointestinal tract, in the liver, or in other tissues. Liver tissue has a very high level of toxin-metabolizing enzymes. This is of obvious significance, as absorbed substances are taken to the liver by the portal circulation before entering the general circulatory system of the body. Thus, the liver is a first line of defense and can detoxify many poisons before other tissues are exposed to them. A consequence of this activity is that the liver is particularly likely to be the target organ of many toxicants.

Some generalizations can be made concerning the overall fate of toxicants. They are absorbed in the lipid-soluble form and are excreted as water-soluble metabolites. Thus metabolism of toxicants involves enzymatic reactions to convert fat-soluble substances to water-soluble compounds. These metabolic processes, known as biotransformation, may either increase or decrease the toxicity of the ingested toxicant.

Reactions in the Gastrointestinal Tract

Most toxicants do not produce their toxic effects in the gut, with the exception of irritants such as saponins, selenium-containing amino acids, and oxalates, and toxicants which affect digestive processes, such as trypsin inhibitors and lectins.

The rate of absorption of toxicants is largely determined by their lipid solubility, which is commonly expressed as the lipid/water partition coefficient. Nonionic compounds are more readily absorbed than ionized substances, as they are more lipophilic. Thus, organic acids are more likely to be absorbed in an acid environment (the stomach) while organic bases are absorbed in a basic environment (small intestine). The extent of dissociation of a compound will determine the proportion that is ionized at a given pH. Most toxicants are absorbed by simple diffusion.

An example of the effect of water solubility and base strength on toxicity is shown in Table 3.1, in which the toxicity of numerous pyrrolizidine alkaloids is compared to these characteristics. The lower the pK_a, the lower the base strength. The most toxic alkaloids, such as lasiocarpine, have a low water solubility and low base strength (low degree of ionization) while alkaloids with a low toxicity, such as heliotridine, have a high water solubility and a high base strength (high degree of ionization). Of course, other factors influence the toxicity of these alkaloids, but there is a clear association between toxicity and lipid solubility.

In ruminant animals, considerable metabolism of toxicants occurs in the rumen. This aspect is discussed separately. In nonruminants, there may be metabolism of toxicants by microorganisms in the hindgut. Because of the anatomical location of these processes in the posterior region of the gut, metabolism by cecal organisms generally has limited effects in livestock. In humans, this aspect is of interest as there may be microbial metabolism producing carcinogens. In some breeds of chickens, cecal metabolism of sinapine in rapeseed meal can result in formation of trimethylamine, causing a fishy flavor of the eggs.

The degree of absorption of toxicants can be influenced by several other factors. Anion exchange resins, bentonite and zeolites (clays with ion-exchange capacity), and alfalfa meal in the diet reduce the absorption of mycotoxins such as zearalenone. Factors affecting gastrointesti-

TABLE 3.1. Comparison of Acute Toxicity of Pyrrolizidine Alkaloids with Water Solubility and Base Strength[a]

Alkaloid	LD_{50} in in male rats (mg/kg)	Solubility in water	pK_a
Lasiocarpine	72	Low	7.6
Seneciphylline	77	Low	7.6
Senecionine	85	Low	7.7
Monocrotaline	175	Medium	7.9
Heliotrine	300	Medium	8.5
Echinatine	350	High	8.4
Intermedine + lycopsamine	>1000	High	8.5
Heliotridine	1200	High	9.0
Heliotrine N-oxide	5000	High	>9.0

[a] Adapted from Bull et al. (1968).

(a)
$$\text{glucose}-\underset{\underset{CH_3}{|}}{\overset{\overset{CH_3}{|}}{C}}-CN \xrightarrow[H_2O]{\beta\text{-glucosidase}} \text{glucose} + \underset{\underset{CH_3}{|}}{\overset{\overset{CH_3}{|}}{C}}=O + HCN$$

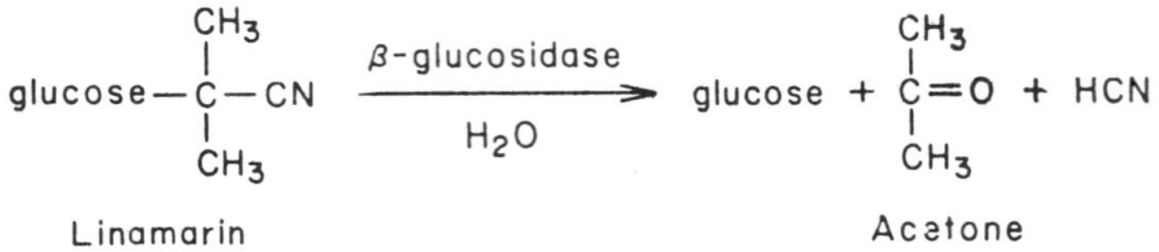

Linamarin — Acetone

(b) Mimosine

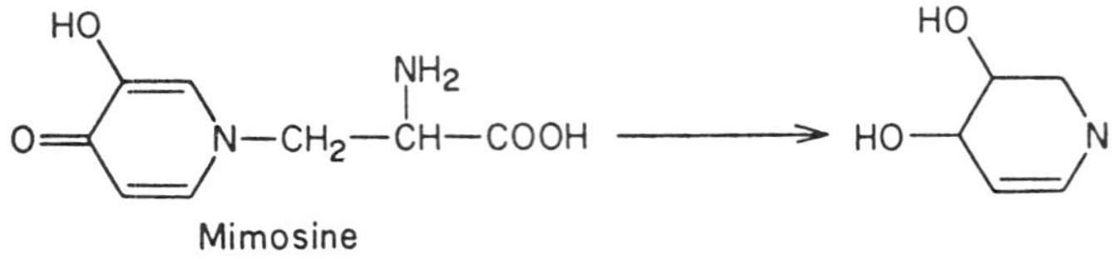

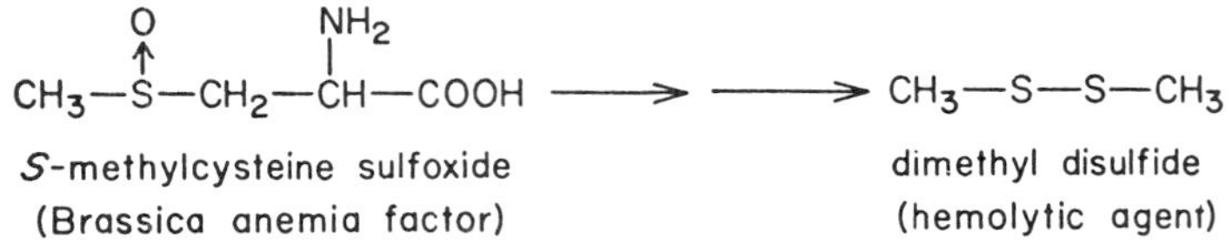

(c)
S-methylcysteine sulfoxide (Brassica anemia factor) ⟶ ⟶ $CH_3-S-S-CH_3$ dimethyl disulfide (hemolytic agent)

(d) Formononetin ⟶ Daidzein ↓ Equol

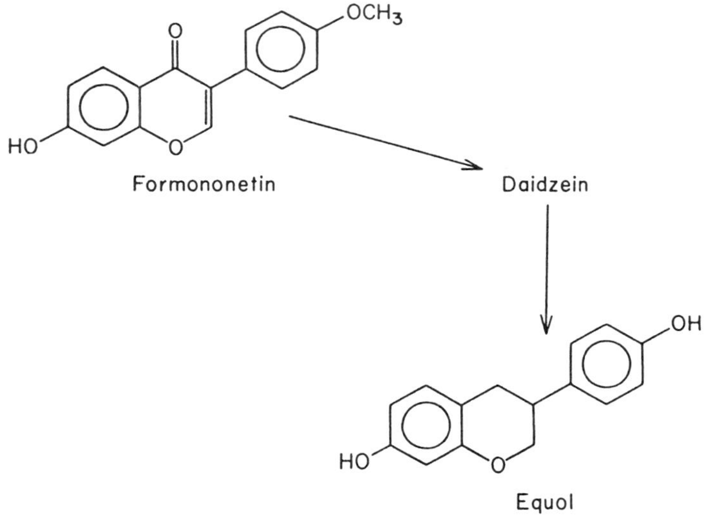

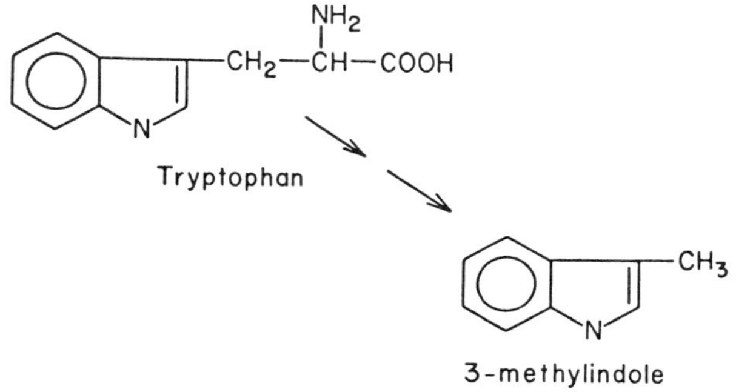

(e) Tryptophan ⟶ 3-methylindole

nal motility affect absorption; in general, reduction in motility results in a longer period during which absorption can occur, and greater absorption.

Ruminal Metabolism of Toxicants. The rumen is a specialized area that may have pronounced effects on ingested toxicants in ruminant animals. It generally has a slightly acid pH and has an immense population of diverse types of microorganisms. Ingested material remains in the rumen for a much longer period than is the case for ingesta in the gut of nonruminants. It is an area where small water-soluble molecules such as the volatile fatty acids are absorbed.

Ruminal metabolism by microorganisms has diverse effects on toxicants. Some of the potential interactions of toxicants and ruminal metabolism are: (1) toxicity may be increased as a result of ruminal metabolism; (2) toxicity may be decreased; (3) rumen microorganisms may produce toxins; and (4) dietary toxicants may inhibit rumen fermentation. Examples of each of these will be discussed briefly here and in more detail later. Examples of toxicants for which an increase in toxicity due to rumen metabolism occurs include nitrates, cyanogens, mimosine, the brassica anemia factor (S-methylcysteine sulfoxide), and phytoestrogens (Fig. 3.1). Nitrate is converted to the more toxic nitrite in the rumen. Cyanogenic glycosides are hydrolyzed more rapidly in ruminants than in nonruminants, and thus are more toxic because the enzymatic hydrolysis of the glycosides is favored by the higher pH of the rumen than the highly acid nonruminant stomach. The toxic amino acid mimosine in *Leucaena leucocephala* is converted by rumen metabolism to dihyropyridone, a more toxic metabolite. The brassica anemia factor is metabolized in the rumen to dimethyl disulfide, which when absorbed causes hemolytic anemia. The phytoestrogen formononetin in subterranean clover is metabolized in the sheep rumen to a more potent estrogen. These are a few examples of how rumen metabolism can increase toxicity. Alternatively, detoxification can occur in the rumen (Fig. 3.2). Oxalates can be degraded by rumen microorganisms. There is evidence that pyrrolizidine alkaloids in *Heliotropium europaeum* are detoxified in the sheep rumen with the formation of methylated derivatives. Gossypol, a phenolic substance in cottonseed meal, is detoxified in the rumen, as are trypsin inhibitors in soybeans and glucosinolates in rapeseed meal. Some phytoestrogens,

---

FIG. 3.1. Ruminal metabolism which increases toxicity of toxicants. (a) Cyanide formation. (b) Mimosine metabolism. (c) Metabolism of brassica anemia factor. (d) Metabolism of formononetin. (e) Formation of 3-methylindole.

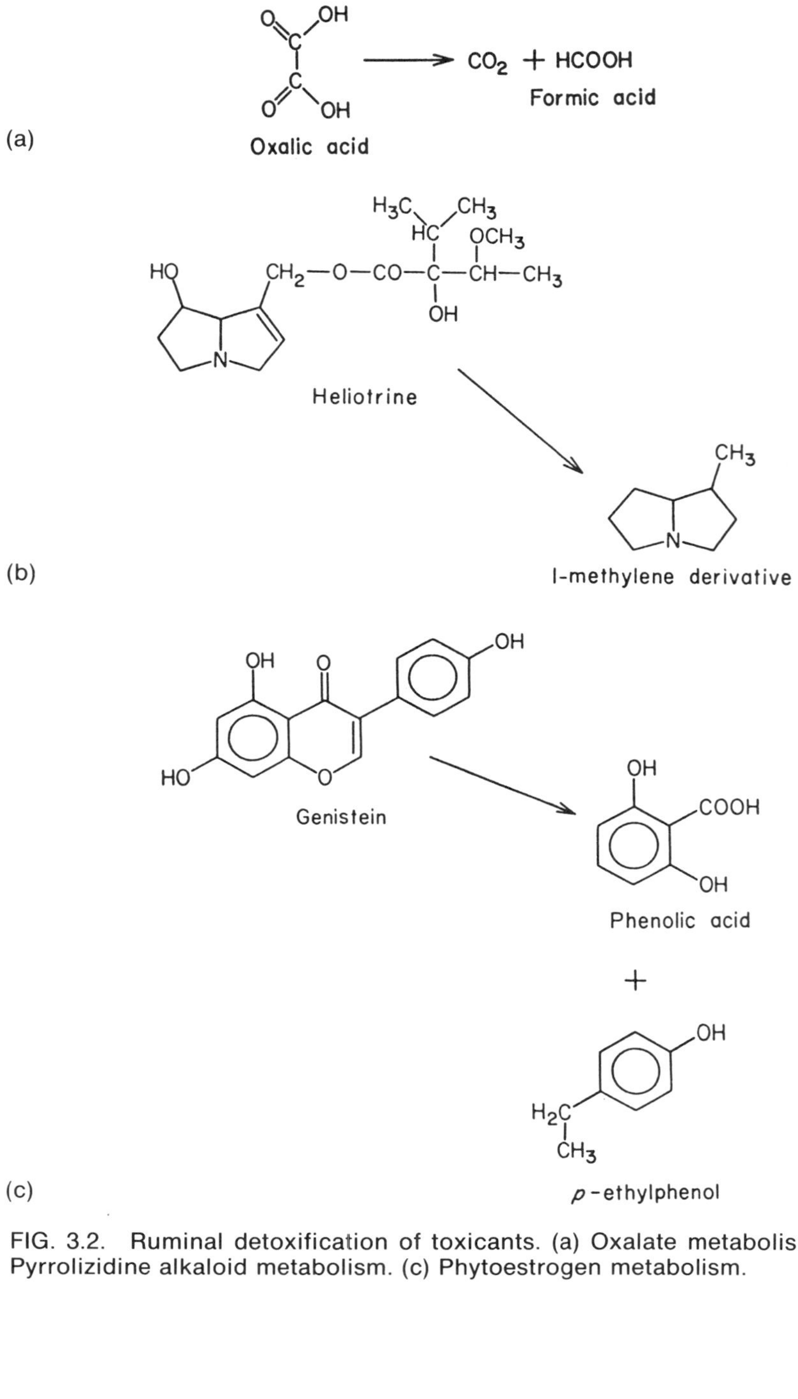

FIG. 3.2. Ruminal detoxification of toxicants. (a) Oxalate metabolism. (b) Pyrrolizidine alkaloid metabolism. (c) Phytoestrogen metabolism.

such as biochanin A and genistein in subterranean clover, are converted to *p*-ethylphenol, a nonestrogenic compound, in the sheep rumen. A mycotoxin, ochratoxin A, is detoxified in the rumen. Thus, it is apparent that in many cases the rumen has a favorable effect in detoxifying many of the toxic substances that ruminant animals are exposed to in their typically diverse types of diets. In some instances, usually associated with a dietary change, rumen microorganisms may produce toxins. An example is the production of lactic acid by *Streptococcus bovis,* causing acute acid indigestion when a sudden shift to a high starch diet is made. The condition of acute bovine pulmonary emphysema is caused by the abnormal rumen metabolism of tryptophan to 3-methylindole, following an abrupt dietary change. Polioencephalomalacia is a thiamine deficiency in ruminants, provoked by a thiamine-degrading enzyme, thiaminase I, produced by rumen microorganisms when high concentrate diets are fed. Conversely, ruminants are normally protected against thiamine deficiency due to thiaminase in bracken fern because of the abundant synthesis of thiamine by the rumen bacteria.

In some cases, dietary toxicants may inhibit rumen function. Oxalates and mesquite bean toxicity result in rumen stasis, perloline in tall fescue reduces cellulose digestibility, tannins reduce protein digestibility, aflatoxins may inhibit rumen fermentation, and fungi on soil or herbage may have antimicrobial effects.

It is apparent that there are many interactions between toxicants and rumen microorganisms. In many cases, the result is detoxification, with the ability of ruminants to utilize feedstuffs that are toxic to monogastric animals.

Postabsorptive Metabolism of Toxicants

Toxicants are absorbed as lipid-soluble substances and are metabolized in the tissues to water-soluble metabolites that can be excreted in the urine. This metabolism is largely carried out in the liver by a complement of enzymes referred to collectively as the mixed function oxidases (MFO). This name is derived from the double role of the oxygen molecule in MFO-catalyzed reactions, both as an oxidizing (to form water) and as an oxygenating agent. The MFO system has a remarkable degree of nonspecificity, in contrast to most enzymes, so it metabolizes many diverse compounds. The MFO enzymes catalyze numerous oxidative reactions, producing more polar, water-soluble metabolites. Another attribute of MFOs in detoxification is that they respond very quickly to the presence of dietary toxicants with a marked increase in level or activity. This process is called enzyme induction, whereby the presence of a toxicant induces increased activity of the enzymes that detoxify it. Along with the MFO system, a variety of other enzymes such as esterases, reductases, and transferases are involved in detoxification.

Mixed-Function Oxidases. The MFO system can metabolize a great variety of foreign lipophilic substances that are commonly encountered by animals in the course of their normal life processes. Many of these are natural toxicants in plants, since presumably the MFO system in animals and insects has developed or evolved to allow the organisms to cope with these compounds in their food supply. Much of the knowledge of specific MFO reactions is concerned with man-made chemicals such as pesticides, drugs, and industrial chemicals, rather than with natural toxicants.

All MFO reactions are basically oxidations. The MFO system is attached to the endoplasmic reticulum of cells, particularly the smooth endoplasmic reticulum. When tissue is homogenized and subjected to ultracentrifugation, the endoplasmic reticulum fragments are fractionated out as a pellet called the microsomal pellet or microsomes. The MFO system is often referred to as microsomal enzymes. The MFO system has several components: cytochrome P_{450}, NADPH, a flavoprotein enzyme called NADPH-cytochrome P_{450} reductase, and phosphatidylcholine.

Cytochrome P_{450} is the terminal oxidase of the MFO system. It is a b-type cytochrome which binds carbon monoxide. The reduced cytochrome–carbon monoxide complex has an absorption peak at 450 nm, hence its name. The mechanism of action of the MFO system is that the toxicant reacts with the oxidized cytochrome P_{450}, producing a complex which then is reduced by picking up hydrogen from the reduced flavoprotein, and subsequently reacts with molecular oxygen to produce water, reoxidized cytochrome P_{450}, and the toxicant with a hydroxyl group attached. These reactions are shown in Fig. 3.3. The net result is that an aromatic substrate has been hydroxylated, so it can then be conjugated with polar compounds to produce a water-soluble metabolite.

Specific MFO Reactions. The MFO reactions are referred to as biotransformation because the substrate is transformed into metabolites. Examples of oxidation reactions include the metabolism of safrole to hydroxysafrole (Fig. 3.4); this is a hydroxylation reaction. Safrole is a component of essential oils such as sassafras oil; it was widely used as a flavoring agent in soft drinks until the discovery that it has carcinogenic activity in rodents. Another type of oxidation is epoxide forma-

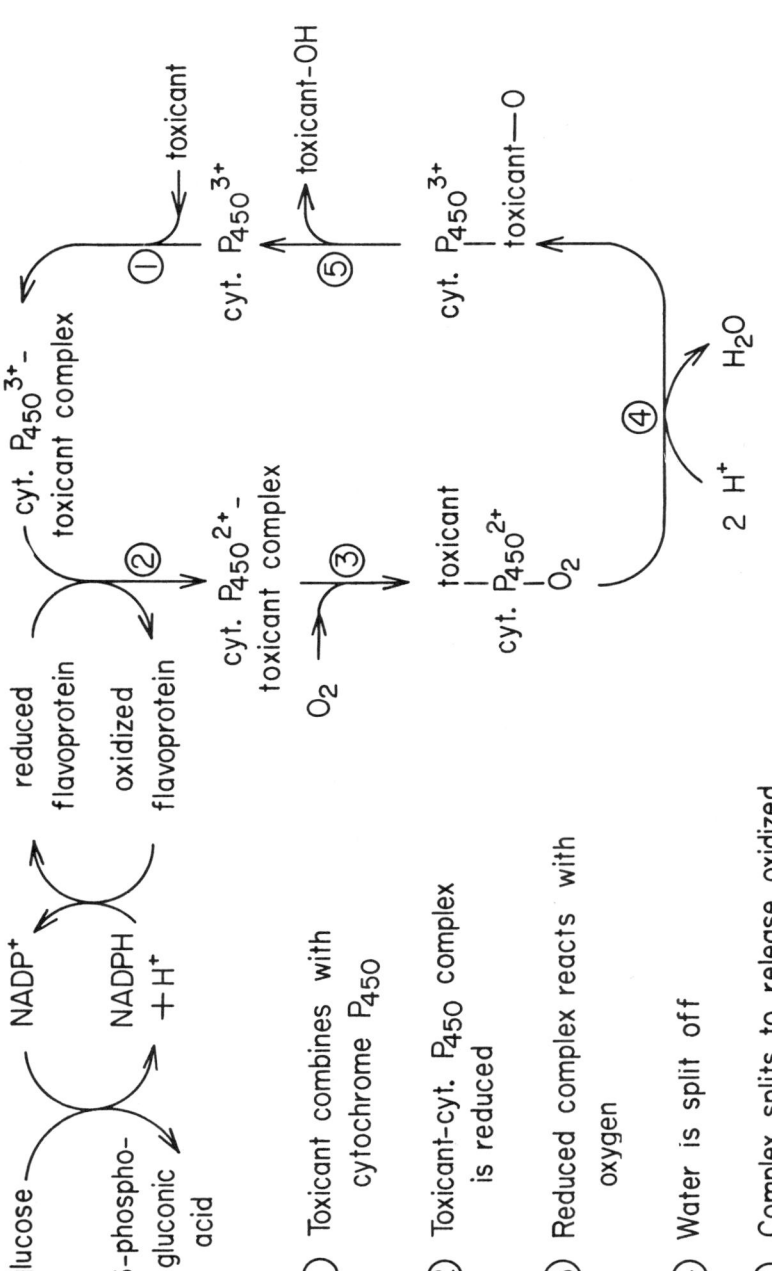

FIG. 3.3. The mixed function oxidase (MFO) enzyme system.

(a) Safrole —MFO enzymes→ 1'-hydroxysafrole

(b) Aflatoxin B₁ —MFO enzymes→ Aflatoxin B₁-2,3-oxide

(c) Nicotine —MFO enzymes→ Nicotine N-oxide

(f) Benzoic acid + UDP-sugar → Benzoic acid glucuronide + UDP

(g) 1'-hydroxysafrole + SO_4^{2-} → Safrole-1-sulfate

(h) Benzoic acid + Glycine → Hippuric acid

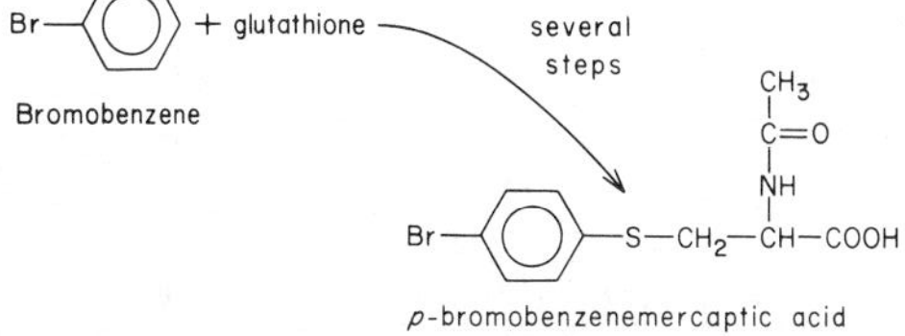

FIG. 3.4. Examples of typical metabolic reactions in the conversion of toxicants to metabolites. (a) Hydroxylation. (b) Epoxidation reaction. (c) N-Oxide formation. (d) N-Demethylation. (e) Reduction. (f) Glucuronide formation. (g) Sulfation. (h) Glycine conjugation. (i) Esterase activity. (j) Glutathione conjugation.

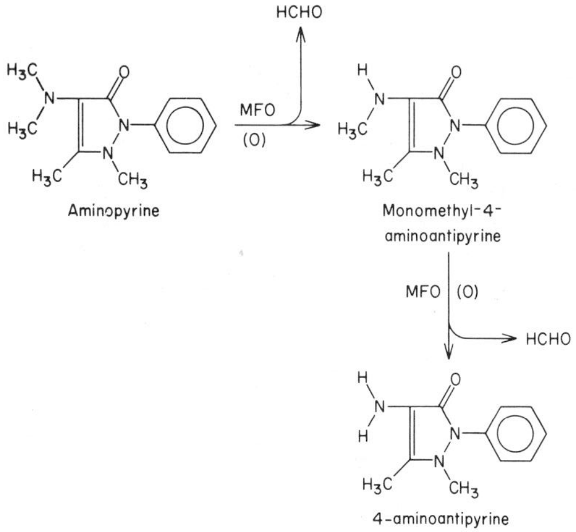

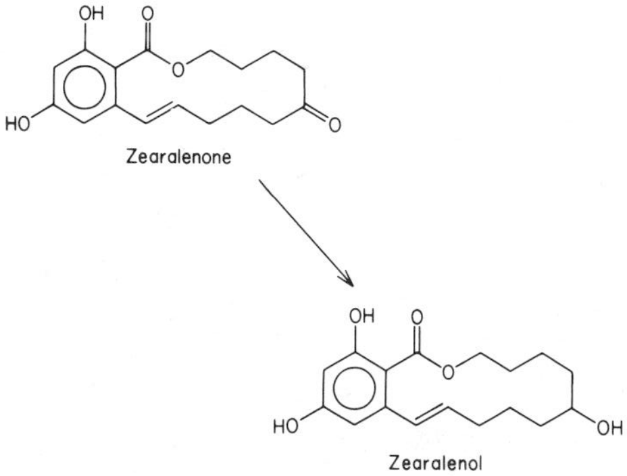

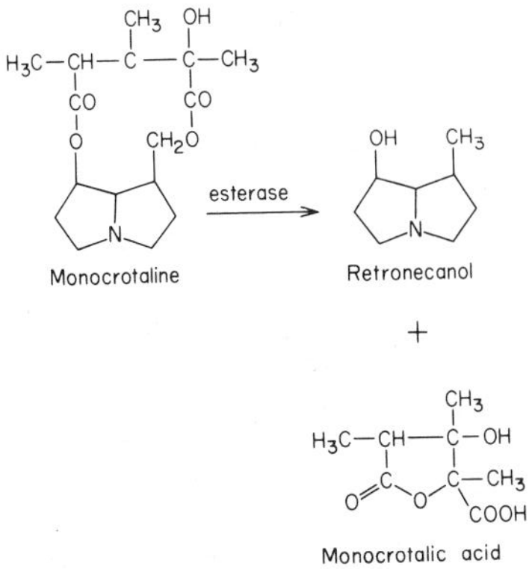

tion, such as the metabolism of aflatoxin B_1 (Fig. 3.4) to produce aflatoxin 2,3-epoxide. Another group of reactions involves formation of N-oxides, such as from nicotine and pyrrolizidine alkaloids (Fig. 3.4) from tertiary amines. Oxidation can also result in oxidation of thioethers to sulfoxides and sulfones, as in the formation of the brassica anemia factor (S-methylcysteine sulfoxide). N-Demethylation is another example of an MFO-catalyzed reaction. The activity of aminopyrine-N-demethylase is often measured as an indicator of drug-metabolizing ability. This is easily accomplished by measuring the rate of formation of formaldehyde, an end product of the reaction (Fig. 3.4). Reductions are another detoxification reaction, as in the formation of toxic pyrroles from pyrrolizidine alkaloids and conversion of the mycotoxin zearalenone to zearalenol. These types of reactions are often referred to as phase I reactions. Phase II involves conjugation of the metabolites produced in the MFO-mediated phase I reactions.

Biotransformation in general is a detoxification process, converting toxic lipophilic compounds to nontoxic water-soluble metabolites. Sometimes this scheme backfires and the metabolites are more toxic than the original compounds. Examples include the conversion of aflatoxin to the carcinogenic epoxide and the conversion of the relatively nontoxic pyrrolizidine alkaloids to very toxic pyrroles.

Factors Affecting MFO Activity. The MFO enzyme activities are generally much higher in the liver than in other organs, so that the liver is the major site of detoxification. All other tissues do possess MFO activity. Thus in acute bovine pulmonary emphysema, for instance, 3-methylindole is metabolized by the lung to an active metabolite that attacks lung tissue. Both the liver and lung are important sites for detoxification enzymes, since the portal vein and the lungs are the major routes for entrance of toxicants into the internal environment.

Fetal and newborn animals have very low MFO activity, and thus lack the ability to metabolize many toxicants and drugs. There are numerous sex-related differences in MFO activity, presumably related to steroid hormone balance. Steroid hormones are metabolized by the MFO enzymes.

There are pronounced species differences in microsomal enzyme activities. Ducklings and trout are extremely sensitive to aflatoxins because of a rapid rate of metabolism to the epoxide, whereas sheep and rats metabolize aflatoxin more slowly and are much more resistant to it. Animals that are resistant to the toxic effects of pyrrolizidine alkaloids, such as sheep, guinea pigs, and Japanese quail, have a much lower rate of pyrrole production than susceptible species such as cattle, horses, and rats (Shull et al. 1976).

Nutritional factors influence MFO activity. Low dietary protein levels generally result in reduced MFO function. Thus the toxicity of aflatoxin is reduced on low-protein diets because less of the active metabolite is produced. Vitamin and mineral deficiencies may affect MFO status.

Factors that stimulate MFO activity are called inducers or inducing agents. Most toxicants are inducers and therefore exposure to toxicants results in an increased ability of animals to detoxify them. This, of course, is a sound strategy for coping with environmental hazards. Certain drugs are widely used for research purposes to induce MFO activity. Ths most common is phenobarbital. Pretreatment of animals with barbiturates markedly increases their microsomal enzyme activities. This increases the toxicity if the product of biotransformation is more toxic than the original toxicant. For example, guinea pigs are resistant to the pyrrolizidine alkaloid monocrotaline (Swick et al. 1982), but if pretreated with phenobarbital, they become susceptible (White et al. 1973). Other drugs are used to inhibit MFO activity. These include SKF 525-A (from Smith, Kline, & French Co.), and piperonyl butoxide (Fig. 3.5). Pretreatment of animals with these drugs blocks the formation of active metabolites. Thus by studying the effect of inducers and inhibitors on the toxicity of a compound it is possible to determine if it is bioactivated by the MFO enzymes.

One procedure used to assess MFO activity is phenobarbital sleeping time. Animals are given a dose of barbiturate, and the time taken for them to awaken is measured (sleeping time). The more rapidly the phenobarbital is metabolized, the more rapidly they awaken. Therefore a high MFO activity will result in a short sleeping time.

Other Detoxifying Enzyme Systems. Various enzymes besides the MFO system are involved in detoxification. Epoxide hydrolases are enzymes that convert epoxides to diols. For example, aflatoxin B_1 2,3-epoxide is detoxified by an epoxide hydrolase. The pyrrolizidine alkaloid jacobine in *Senecio jacobaea* (tansy ragwort) is an epoxide. Swick et al. (1983) have shown that sheep liver has high epoxide hydrolase activity, which may account in part for the resistance of sheep to tansy ragwort. Phenobarbital induces epoxide hydrolase activity.

Mullin and Croft (1984) suggest a role of epoxide hydrolase in plant–herbivore interactions. Plants produce a variety of fatty acids, phenolics, alkaloids, and terpenoids which may be metabolized in mammalian and insect herbivores to epoxides. Mullin and Croft (1984)

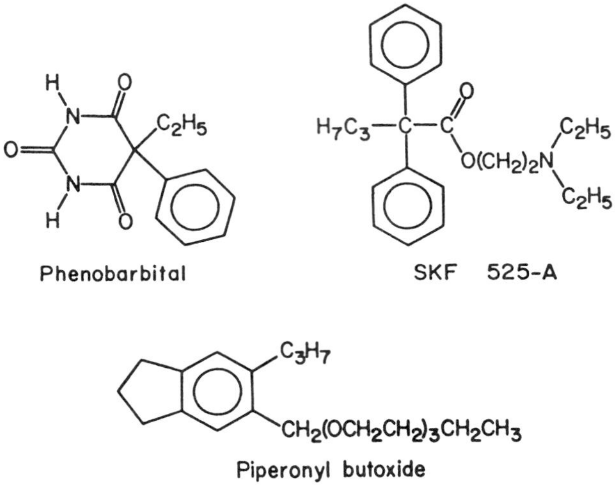

FIG. 3.5. Structures of drugs commonly used to induce or inhibit MFO activity for experimental purposes.

hypothesized that transepoxide hydrolase may function in a detoxification role in herbivores. Some plants counter insect herbivory by synthesizing mimics of insect juvenile hormones or antagonists of the hormone's synthesis or degradation within insects. Some of these compounds are epoxides. The insects evolve countermeasures such as epoxide hydrolase (the coevolutionary arms race!).

Esterases are important in some detoxifications. Pyrrolizidine alkaloids have esterified side chains on the pyrrolizidine ring; the presence of the ester groups is necessary for toxicity of these alkaloids. Esterases may play a role in pyrrolizidine alkaloid detoxification. For example, White and Cheeke (1984) found that esterase inhibition with the organophosphate coumaphos slightly increased the toxicity of *Senecio* alkaloids in rats.

Group transferase enzymes or transferases are involved in the secondary phase (phase II) of toxicant metabolism in conjugation reaction. Glutathione S-transferases are examples. In essence, these enzymes result in the reaction of an hydroxylated metabolite with glutathione:

$$\underset{\text{Hydroxylated compound}}{\text{R—OH}} + \underset{\text{Reduced glutatione}}{\text{GSH}} \rightarrow \underset{\text{Conjugated metabolite}}{\text{R—S—G}} + H_2O$$

UDPglucuronosyltransferase catalyzes the formation of a β-glucuronide:

$$\text{ROH} + \text{UDP-glucuronic acid} \rightarrow \text{RO-}\beta\text{-glucuronide} + \text{UDP} + \text{H}_2\text{O}$$

Sulfotransferases result in the formation of water-soluble sulfates. Rhodanese is a transferase involved in the detoxification of cyanide. Its function will be further discussed in the section on cyanogenic glycosides.

Conjugation Reactions. Metabolites produced by MFO activity are generally excreted in the urine as conjugated compounds; that is, they are conjugated with other substances such as glutathione, glycine, glucuronic acid, and sulfate to increase their water solubility. Transferases, discussed in the previous section, are involved. There are species differences in activities of these enzymes. Cats lack glucuronosyltransferase and so cannot form glucuronides. Therefore they are more susceptible than other species to intoxication by benzoic acid and phenol. Guinea pigs lack the ability to synthesize mercaptic acid metabolites from glutathione conjugation, perhaps explaining the lack of protection against pyrrolizidine alkaloid toxicity by dietary cysteine (Swick et al. 1982) as compared to the effectiveness of cysteine in rats (Buckmaster et al. 1976).

The activity of conjugation mechanisms can be altered by various additives. This is useful for experimental purposes and in some cases may present possibilities for diet supplementation programs to reduce toxicities. Dietary cysteine stimulates increased glutathione production, whereas administration of diethyl maleate blocks glutathione conjugation. Salicylamide blocks glucuronide conjugation. These compounds can be given to test animals prior to administration of a toxicant, and the effects on toxicity can be observed. An example of the use of the technique is that (Table 3.2) of Jennings et al. (1978), who showed the effect of various pretreatments on mice given a dose of tetradymol. Tetradymol is the toxic agent of *Tetradymia* spp. (horsebrush, rabbit brush) which causes photosensitization in sheep. These data indicate that tetradymol is activated by MFO activity to a more

TABLE 3.2. Effect of Various Pretreatments on Toxicity of Tetradymol in Mice[a]

Pretreatment	Survival time (hr)	Interpretation
None	7.5	—
Phenobarbital	5.5	Increased MFO, increased toxicity
SKF 525-A	13.6	Decreased MFO, decreased toxicity
Piperonyl butoxide	17.9	Decreased MFO, decreased toxicity
Cysteine	9.2	Increased glutathione, decreased toxicity
Diethyl maleate	6.3	Decreased glutathione, increased toxicity

[a] Adapted from Jennings et al. (1978).

toxic metabolite, and that glutathione conjugated tetradymol or its metabolites, resulting in reduced toxicity. Blocking glutathione conjugation with diethyl maleate increased toxicity. Piperonyl butoxide and SKF 525-A inhibited MFO activity, and so reduced the tetradymol toxicity, showing that it was bioactivated.

The activity of various drug-metabolizing enzymes can be influenced by other additives and by exposure to toxicants. Miranda et al. (1980A) found that administration of jacobine, a pyrrolizidine alkaloid with an epoxide group, to rats resulted in a significant increase in epoxide hydrolase and glutathione S-transferase activity, and reductions in cytochrome P_{450} and aminopyrine demethylase activity. Feeding a plant, S. jacobaea, which contains jacobine and other pyrrolizidine alkaloids, resulted in similar changes, with elevated epoxide hydrolase and glutathione S-transferase activities (Miranda et al. 1980B). Thus these alkaloids induce detoxification mechanisms. Synthetic antioxidants such as ethoxyquin and butylated hydroxyanisole (BHA) also induce conjugating enzymes. Miranda et al. (1981A) found that dietary ethoxyquin caused increases in liver glutathione, glutathione S-transferase, and cytochrome P_{450} levels, and reduce the toxicity of pyrrolizidine alkaloids in mice. Other studies (Miranda et al. 1981B, 1982) have shown protective effects of synthetic antioxidants against pyrrolizidine alkaloid toxicity, while Kim et al. (1981, 1982) have observed similar protection with synthetic antioxidants against bitterweed (Hymenoxys odorata) toxins.

Much of the information on the metabolism of toxicants has been obtained using pesticides and other man-made chemicals. This is primarily because of the economic incentives and the potential toxic effects of these substances in humans. For many of the natural plant toxicants affecting livestock, knowledge of their metabolism in large animals is nonexistent or skimpy.

REFERENCES

BUCKMASTER, G. W., CHEEKE, P. R., and SHULL, L. R.   1976.   Pyrrolizidine alkaloid poisoning in rats: Protective effects of dietary cysteine. J. Anim. Sci. 43, 464–473.

BULL, L. B., CULVENOR, C. C. J., and DICK, A. T.   1968.   The Pyrrolizidine Alkaloids. American Elsevier, NY.

CARLSON, J. R., and BREEZE, R. G.   1984.   Ruminal metabolism of plant toxins with emphasis on indolic compounds. J. Anim. Sci. 58: 1040–1049.

FRANKLIN, R. B., ELCOMBE, C. R., VODICNIK, M. J., and LECH, J. J.   1980. Comparative aspects of the disposition and metabolism of xenobiotics in fish and mammals. Fed. Proc., Fed. Am. Soc. Exp. Biol. 39, 3144–3149.

JENNINGS, P. W., REEDER, S. K., HURLEY, J. C., ROBBINS, J. E., HOLIAN, S. K., HOLIAN, A., LEE, P., PRIBANIC, J. A., and HULL, M. 1978. Toxic constituents and hepatotoxicity of the plant *Tetradymia glabrata*. In Effects of Poisonous Plants on Livestock. R. F. Keeler, K. R. Van Kampen, and L. R. James (Editors), pp. 217–228. Academic Press, NY.

KETTERER, B., COLES, B., and MEYER, D. J. 1983. The role of glutathione in detoxification. Envion. Health Perspect. *49*, 56–70.

KIM, H. L., ANDERSON, A. C., TERRY, M. K., and BAILEY, E. M. 1981. Protective effect of butylated hydroxyanisole on acute hymenoxon and bitterweed poisoning. Res. Commun. Chem. Pathol. Pharmacol. *33*, 365–368.

KIM, H. L., ANDERSON, A. C., HERRIG, B. W., JONES, L. P., and CALHOUN, M. C. 1982. Protective effects of antioxidants on bitterweed (*Hymenoxys odorata* DC) toxicity in sheep. Am. J. Vet. Res. *43*, 1945–1950.

MIRANDA, C. L., CHEEKE, P. R., and BUHLER, D. R. 1980A. Effect of pyrrolizidine alkaloids from tansy ragwort (*Senecio jacobaea*) on hepatic drug-metabolizing enzymes in male rats. Biochem. Pharmacol. *29*, 2645–2649.

MIRANDA, C. L., CHEEKE, P. R., and BUHLER, D. R. 1980B. Comparative effects of the pyrrolizidine alkaloids jacobine and monocrotaline on hepatic drug-metabolizing enzymes in the rat. Res. Commun. Chem. Pathol. Pharmacol. *29*, 573–587.

MIRANDA, C. L., CARPENTER, H. M., CHEEKE, P. R., and BUHLER, D. R. 1981A. Effect of ethoxyquin on the toxicity of the pyrrolizidine alkaloid monocrotaline and on hepatic drug metabolism in mice. Chem.-Biol. Interact. *37*, 95–107.

MIRANDA, C. L., REED, R. L., CHEEKE, P. R., and BUHLER, D. R. 1981B. Protective effects of butylated hydroxyanisole against the acute toxicity of monocrotaline in mice. Toxicol. Appl. Pharmacol. *59*, 424–430.

MIRANDA, C. L., BUHLER, D. R., RAMSDELL, H. S., CHEEKE, P. R., and SCHMITZ, J. A. 1982. Modifications of chronic hepatotoxicity of pyrrolizidine (*Senecio*) alkaloids by butylated hydroxyanisole and cysteine. Toxicol. Lett. *10*, 177–182.

MULLIN, C. A., and CROFT, B. A. 1984. Trans-epoxide hydrolase: A key indicator enzyme for herbivory in arthropods. Experientia *40*, 176–178.

SHULL, L. R., BUCKMASTER, G. W., and CHEEKE, P. R. 1976. Factors influencing pyrrolizidine (*Senecio*) alkaloid metabolism: Species, liver sulfhydryls and rumen fermentation. J. Anim. Sci. *43*, 1247–1253.

SMITH, G. S., WATKINS, J. B., KLAASSEN, C. D., ROZMAN, K., and THOMPSON, T.N. 1984. Oxidative and conjugative metabolism of xenobiotics by livers of cattle, sheep, swine and rats, J. Anim. Sci. *58*, 386–395.

SWICK, R. A. 1984. Hepatic metabolism and bioactivation of mycotoxins and plant toxins. J. Anim. Sci. *58*, 1017–1028.

SWICK, R. A., CHEEKE, P. R., GOEGER, D. E., and BUHLER, D. R. 1982. Effect of dietary *Senecio jacobaea* and injected *Senecio* alkaloids and monocrotaline on guinea pigs. J. Anim. Sci. *55*, 1411–1416.

SWICK, R. A., MIRANDA, C. L., CHEEKE, P. R., and BUHLER, D. R. 1983. Effect of phenobarbitol on toxicity of pyrrolizidine (*Senecio*) alkaloids in sheep. J. Anim. Sci. *56*, 887–894.

WHITE, I. N. H., MATTOCKS, A. R., and BUTLER, W. H. 1973. The conversion of the pyrrolizidine alkaloid retrorsine to pyrrolic derivatives *in vivo* and *in vitro* and its acute toxicity to various animal species. Chem.-Biol. Interact. *6*, 207–218.

WHITE, R. D. and CHEEKE, P. R. 1984. Organophosphate effects on pyrrolizidine alkaloid toxicosis of rats. J. Anim. Sci. *59* (Suppl. 1), 311–312.

4

Effects of Toxicants on Livestock Production and Human Health

IMPACTS OF TOXICANTS IN FEEDSTUFFS

The impact and significance of toxicants on livestock and humans depend on a variety of factors. Intentional or coincidental changes in these factors can result in a previously minor problem becoming a major one or conversely can eliminate a toxicity problem. A few examples will illustrate these points. In the Middle Ages, ergotism was a major public health problem in Europe. Two factors were primarily responsible for it becoming a minor problem. A change in cultivation techniques, namely, the introduction of deep plowing, resulted in the fungal spores being buried too deep for them to germinate. The new cultivation methods, accompanied by improved soil drainage, allowed wheat to grow where previously only rye could be grown as a grain crop. Wheat is much less susceptible than rye to ergot infestation. Thus these agronomic developments had far-reaching impacts on human health in markedly reducing ergotism. In the pioneer days of the U.S., a disease called milk sickness decimated villages and was a major public health problem. The disease was caused by a toxin secreted in the milk of cows consuming the weed white snakeroot. Most families had their own cow, which frequently grazed in waste areas, roadsides, or stream banks, where the white snakeroot grew. With greater intensification of the dairy industry, the problem was virtually eliminated by changes in grazing management and the use of improved pastures. Interestingly, the problem of white snakeroot toxicity has reappeared because of the current interest in small-scale, part-time farming with

4. TOXICANTS AND LIVESTOCK PRODUCTION

its use of less intensive production systems and weed control programs than are common in commercial agriculture (Stotts 1984).

Geographic differences in toxicity problems may be associated with cropping or pasture management systems. Perennial ryegrass staggers and annual ryegrass toxicity are significant livestock poisoning problems in Australia. In Oregon, ryegrass is extensively grown and grazed by sheep, but poisoning problems are very uncommon. Ryegrass staggers is caused by fungal tremorgens, while annual ryegrass toxicity is due to a bacterial toxin produced in nematode-infected seed heads. In Oregon, ryegrass is grown as a seed crop, with pasturing by sheep of the vegetative stage during the winter. Therefore, sheep are not exposed to the seed heads except when the grass seed screenings are used. An almost universal practice with Oregon ryegrass fields is burning of the stubble to destroy fungi, nematodes, and other pests (Fig. 4.1). The causative agents of the two ryegrass-associated toxicities are kept in check by field burning, while in Australia the fungi, nematodes, and bacteria are not controlled and build up large infestations.

Another example of an agricultural change that has virtually elimi-

FIG. 4.1. Field burning in western Oregon. Field burning destroys the nematodes which in Australia cause annual ryegrass toxicity.

nated many toxicity problems in the U.S. is the introduction and almost universal use of herbicides to control broad-leaved weeds in grain fields. In the past, before herbicides were used, grain fields were frequently heavily contaminated with many weeds, and both human and livestock toxicities occurred because of contamination of the grain with toxic weed seeds (Fig. 4.2). For example, poisoning of livestock due to tarweed, which contains pyrrolizidine alkaloids, and poison hemlock, containing the alkaloid conium, occurred because of grain contamination. In the Pacific Northwest, contamination of grain with vetch seed caused losses of poultry. This type of problem still exists in areas where herbicides are not routinely used. Several outbreaks of pyrrolizidine alkaloid poisoning in Australia occurred in the 1970s due to contamination of grain with crotalaria seeds, which contain pyrrolizidine alkaloids. Hooper (1978) described a situation in which pigs which were fed sorghum grain died from pyrrolizidine alkaloid poisoning. This was shown to be due to *Crotalaria retusa* seeds in the sorghum at a level of 0.004%, which, according to P. T. Hooper (personal communication, Alice Springs, Australia, 1977), was equivalent to 1 crotalaria seed per

FIG. 4.2. A field of wheat in Oregon heavily contaminated with tarweed (*Amsinckia intermedia*), a weed containing pyrrolizidine alkaloids. Prior to the almost universal use of herbicides in wheat production in the U.S., significant livestock poisoning occurred due to the contamination of grain with seeds of poisonous plants.
Courtesy of A. P. Appleby.

65,000 sorghum seeds. This batch of sorghum was destined for export for human consumption until the poisoning outbreak in pigs occurred. In modern grain-handling processes, most weed seeds are removed in the cleaning process and constitute the screenings to be fed to livestock. Hence, any toxic weed seeds in the grain become concentrated in the fraction fed to livestock. There is a certain amount of irony associated with efforts in the U.S. to ban the use of herbicides. There is no known case of injury or death from residues on food, while prior to their use, public health and livestock problems existed because of natural toxins in grain. This situation still exists in many parts of the world. In Afghanistan and India in 1976, epidemics of pyrrolizidine alkaloid poisoning in humans occurred because of *Heliotropium* and *Crotalaria* contamination of wheat used to make bread. Wheat from the affected areas contained about 300 mg of *Heliotropium* seeds per kilogram of wheat. Chronic contamination of foodstuffs has been a problem in South Africa, with *Senecio* and *Crotalaria* the major offending species. Huxtable (1979) (see Fig. 4.3) aptly states,

> This type of exposure no longer occurs in the advanced industrialized nations, due to the widespread use of chemicals for controlling unwanted plants in grain fields. Twenty-five years ago, it was not an uncommon sight in the English countryside to see fields of grain speckled with bright red poppies and blue cornflowers. However this bucolic picture can be seen no more.

FIG. 4.3. The widespread use of herbicides in crop production has markedly reduced livestock poisoning problems due to contamination of grain with seeds of toxic plants.

The intentional consumption of toxic plants sometimes occurs without the intent of self-poisoning. "Bush tea" is used in a number of tropical areas, such as in the Carribbean region. A variety of wild leaves is picked and brewed into a tea for medicinal purposes. Frequently the leaves contain toxins, such as pyrrolizidine alkaloids that cause irreversible liver damage. Two current trends in the U.S. that increase human exposure to toxicants are the use of herbal teas and "foraging," or collecting edible wild plants. Herbal teas have been implicated in several human health problems. Huxtable (1979) cited several examples. In Arizona and other parts of the Southwest, an herbal tea, called gordolobo yerba, is widely used for medicinal purposes by Mexican-Americans. Several mortalities in children administered this tea have been reported and are due to the presence in the tea of *Senecio longilobus,* a plant containing a very high concentration of highly toxic pyrrolizidine alkaloids. Comfrey (*Symphytum officinale*) is widely used to make tea; it is of interest that comfrey contains at least eight pyrrolizidine alkaloids (Culvenor *et al.* 1980) and both the leaves and roots have been shown to be carcinogenic (Hirono *et al.* 1978). Huxtable (1979) reports a case of two people who thought they were collecting comfrey when in fact they collected *Digitalis* (foxglove) and died from drinking a tea prepared from it. Another case cited by Huxtable was the death of a person drinking *Datura* (Jimsonweed) tea on a desert survival exercise. The practice of foraging for edible wild plants may increase exposure to toxicants. The fiddleheads of bracken fern are an example of a plant frequently collected. Bracken fern, especially in the young frond stage, is carcinogenic, has caused many tumors in livestock, and is suspected as a cause of human cancer in Japan (Hirono 1981). Finally, there is the deliberate ingestion of poisonous plants to experience a disturbance of the central nervous system. Many deaths have occurred from the ingestion of poisonous wild mushrooms that were picked for their hoped-for hallucinogenic properties.

The recognition of toxicants in some feeds has allowed an easy solution of an existing problem by changes in feed processing procedures. Raw soybeans or soybean meal are unsatisfactory as poultry or swine feeds unless they are heat treated because of the presence of heat-labile toxins such as trypsin inhibitors and lectins. Routine heat treatment of these products prevents any problems due to these factors. The widespread use of soybean meal in the 1950s resulted in a high incidence of parakeratosis in swine. This was found to be due to dietary zinc being rendered unavailable by the high phytic acid content of soybeans. The solution was to increase the level of supplementary zinc used with soybean meal-containing diets. The effects of gossypol in cottonseed meal can be overcome by the addition of iron salts during processing.

Another approach to the problems caused by toxicants in feeds is the modification of crops and forages by plant breeding. Perhaps the best example of this is the work by Canadian scientists to improve rapeseed meal. They have produced a marked drop in the glucosinolate and erucic acid contents of rapeseed to virtually eliminate toxicity problems associated with the feeding of rapeseed meal to livestock. Other plant-breeding developments include low gossypol cotton, low saponin alfalfa, and low alkaloid reed canarygrass. A case where plant breeding has had adverse effects was the development of bird-resistant sorghum which reduced wild bird damage to sorghum. However, this was due to an increase in the tannin content of the grain, which markedly lowered its nutritional value. Plant breeders have had to rescind these high-tannin varieties and develop strains with a lower tannin content but still with some bird resistance.

While problems with many plant toxicants are less now than in the past because of recognition of the cause, changes in agronomic practices, plant breeding, feed processing, and so on, problems caused by mycotoxins are in many situations becoming of increasing importance. Factors responsible for this include continuous cropping with one crop which allows fungal populations to build up, increased corn production in the U.S. South and southeast where the climate is favorable to fungal growth, and greater recognition of and concern for the effects of mycotoxins. Higher energy costs may prompt farmers to alter crop-drying procedures and store corn at a higher moisture content, allowing fungal growth during storage. Mycotoxin problems also show a marked seasonal variation. Epidemic years, such as 1977 when the Food and Drug Administration (FDA)-approved levels of aflatoxin contamination had to be raised because much of the U.S. corn crop exceeded the tolerated aflatoxin level, are due to climatic factors such as drought, predisposing the stressed plant to fungal infection.

In tropical areas of the world, mycotoxins in grains, protein concentrates, and other feedstuffs are a major problem. Warm humid conditions favor fungal growth, and farming practices in many tropical areas are not sophisticated. Crop storage conditions are frequently inadequate. Epidemiological studies in Africa and Asia have shown a strong positive association between liver cancer rates in humans and dietary aflatoxin intake (Wogan and Busby 1980), providing circumstantial evidence for a causal relationship between aflatoxin intake and liver cancer incidence.

IMPACTS OF POISONOUS PLANTS

The economic impact of livestock losses from poisonous plants is very difficult to assess. In range areas, with extensive grazing, losses may be unnoticed, or the cause of death of a dead animal may be unknown. If the cause is known by the rancher, the loss will be unreported unless it is of such major proportion that professional help is solicited. Losses are not merely those of dead animals. Losses from poisonous plants include decreased animal productivity, decreased resistance to other stresses, abortions, birth defects and other reproductive problems, and the cost of poisonous plant control, which may include spraying, reseeding, fencing, increased herding, and abandoning of certain ranges. Nielsen (1978) has estimated that in the 17 western states of the U.S. the economic loss from poisonous plants is about $17 million annually.

Correct grazing management of livestock can control many poisonous plant problems by grazing particular species when their toxicity is lowest, by using certain ranges when the toxic plants they contain are unpalatable and will not likely be grazed, by avoiding the herding of sheep or trailing of cattle through areas infested with poisonous plants, and by avoiding certain plants when livestock are physiologically most susceptible. This latter consideration is of particular relevance with plants that contain teratogens or abortifacients. Keeler (1978) has provided a good discussion of ways to reduce the incidence of congenital deformities by grazing management. The fetus is susceptible to effects of a teratogen during a fairly well-defined period when a particular tissue is undergoing rapid development. Thus a particular type of deformity would be produced by the maternal ingestion of a plant at a specific period, and that period is not necessarily the same for each class of livestock. The approximate period when teratogens produce certain defects is shown in Table 2.1.

A few examples will illustrate the relationships between teratogens, fetal susceptibility, and animal management. Various lupine species on western ranges contain quinolizidine alkaloids (e.g., anagyrine) that have teratogenic effects, producing a condition called crooked calf disease. The calves are born with twisted or bowed limbs, spinal curvatures, and cleft palates. The susceptible period of gestation is between 40 and 70 days. The alkaloid content of lupines is very high early in the season and decreases markedly with maturity, except for a brief period while mature seeds are in the pods. Therefore, the likelihood of teratogenicity is highest when young plants or plants in the mature seed stage are grazed by cows whose gestation stage is between 40 and 70 days. By adjusting grazing managment accordingly, the teratogenic effects of lupines can be avoided, while the high-quality lupine forage can still be utilized. Poison hemlock (*Conium maculatum*) also causes crooked calf disease when consumed by cows between days 40 and 70 of gestation. However, the piperidine alkaloid content of the plant is highly variable, so there is little likelihood of lowering the dose by selective grazing during a low hazard period. In this case, animal management involves keeping pregnant cows from *Conium* exposure by control of the weed or by fencing off weed patches. Another example of how management and teratogenicity interrelate is the case of *Veratrum californicum* and the cyclops eye condition in lambs. When ewes consume the plant on the fourteenth day of pregnancy, the lambs may be born with gross facial deformities. Veratrum has a restrictive habitat, growing in dense, sharply defined stands in high moist meadows or other moist lightly wooded areas. Sheep ranchers can prevent teratogenicity from veratrum by keeping ewes in areas where the plant does not grow until at least 15 days after the rams have been removed, thus avoiding exposure to the plant during the 1-day period of susceptibility. A final example is the teratogenic effect of locoweed (*Astragalus* spp.). Locoweed produces a variety of deformities in lambs, and the susceptibility periods for each deformity are different. Consumption by ewes of significant amounts of locoweed during almost any stage of gestation may produce birth defects, so there is no safe grazing period for pregnant ewes. The only successful grazing management procedure is to remove sheep from loco-infested ranges, or if this not feasible, to be prepared to stand a certain amount of loss. Since locoweed is addictive, provision of adequate amounts of other pasture or of supplementary feed does not prevent losses.

RANGE AND PASTURE MANAGEMENT

Management to reduce livestock losses from poisonous plants can be either plant management of the range or pasture, or management of the grazing animals. If a pasture contains a high degree of infestation of one or more poisonous plants, thereby posing a significant risk to grazing animals, it is likely that pasture renovation will be economical. On overgrazed ranges where infestation with poisonous plants has occurred, chemical spraying and reseeding with desirable vegetation may be warranted. Most poisonous plants are broad-leaved forbs or shrubs and can be killed by phenoxy herbicides without damage to grasses. Caution should be used when pastures and ranges are sprayed for the control of poisonous plants. There is often an increase in palatability of plants following spraying associated with the wilted condition or perhaps an increase in sugars in the sprayed material. Livestock may consume poisonous plants after spraying, whereas previously they did not. It is advisable to remove animals from pastures for 1–2 weeks after spraying for weed control. Williams and James (1983) have reviewed the effects of herbicides on the concentration of toxins in poisonous plants. In some cases, as with *Astragalus* spp. and spring parsley (*Cymopterus watsonii*), spraying with phenoxy herbicides results in a rapid decline in toxicity of the plants, associated with bleaching of the leaves. Within 4–5 weeks of spraying, these plants have lost their toxicity. Conversely, the alkaloid concentration of Barbey larkspur (*Delphinium barbeyi*) is markedly increased for 3 weeks following spraying with Silvex or 2,4,5-T; grazing of treated larkspur should be delayed for at least 4 weeks following spraying. Tansy ragwort (*Senecio jacobaea*) is normally unpalatable to cattle and horses, but, following spraying, may be consumed. Hence, in range and pasture management, consideration should be given to the types of poisonous plants present and the effects of herbicides on their toxicity and on grazing habits of livestock.

In a few cases, biological control of the poisonous plants may be feasible. Tansy ragwort (*S. jacobaea*) has in certain situations been controlled with larvae of the cinnibar moth (*Tyria jacobeae*) (Fig. 4.4).

FIG. 4.4. Larvae of the cinnibar moth, showing defoliation of a tansy ragwort plant. The cinnibar moth is used in biological control of tansy ragwort. The larvae are yellow and black.

St.-John's-wort (*Hypericum perforatum*) has been very successfully controlled in California and Oregon by two beetles, *Chrysolina gemellata* and *C. hyperici*. In a few cases, it may be feasible to use alternate types of livestock to avoid losses. Sheep can be used to control tansy ragwort in cattle pastures because of their marked resistance to the toxic effects of pyrrolizidine alkaloids. Sheep are less susceptible than cattle to larkspur toxicity, so if other conditions (e.g., type of range and predators) are satisfactory, sheep may be used instead of cattle in areas with larkspur.

Poisonous plants are often localized in their distribution, and sometimes a relatively small stand of poisonous plants can have a major effect on livestock because of its location. Chokecherry poisoning often occurs when the plant is located near water sources. Animals may consume chokecherry leaves as they are congregating to drink, and when they drink, the water in the rumen speeds up the hydrolysis of glycosides to yield free cyanide. Larkspur often occurs in patches, and chemical control is often economically advantageous. Control of poisonous plants has numerous advantages besides the obvious one of eliminating losses. There may be greater flexibility in seasonal grazing patterns, and there may be a longer total grazing period. When there are no toxic plants present, it is often feasible to increase the stocking rate, giving a greater economic return from the pasture.

Animal management to avoid losses, as reviewed by Krueger and Sharp (1978), makes use of general characteristics of poisonous plants: they are usually unpalatable, frequently become less toxic as they mature, usually make up a greater proportion of the available forage in the early spring, and become proportionately less abundant as forage species begin growth. Hence, animal management on early spring ranges is especially critical. Schuster (1978) listed the following suggestions for management of livestock to minimize losses from poisonous plants:

1. Learn identification and toxic principles of poisonous plants.
2. Use good grazing management to maintain range in a condition that is not conducive to the development of high densities of poisonous plants.
3. Adjust stocking rates so that animals have ample availability of forage relative to the amount of poisonous plants present.
4. Supplement with salt, minerals, and other nutrients as needed.
5. Avoid grazing livestock in areas where toxic plants are abundant either by herding or fencing off the infested areas.
6. Use a class of livestock not generally poisoned by the plants present.

7. Avoid turning hungry animals onto ranges containing poisonous plants. This is especially pertinent when sheep or cattle are released after shipping.
8. Provide adequate watering facilities to prevent nonselective grazing following water deprivation and subsequent watering.
9. Reduce poisonous plant populations by use of mechanical, chemical, biological, or other control methods.

Hungry animals may eat toxic plants that they would normally avoid. Therefore, when livestock are trailed or shipped during shearing or branding or during other times when they may be hungry or stressed, management is particularly important to avoid exposure to poisonous plants. Sheep are especially likely to consume a toxic dose of halogeton when they are unloaded hungry and then herded along a road lined with the plant. Sheep can adapt to halogeton; if allowed a few days of exposure to it, a population of oxalate-degrading microorganisms proliferates in the rumen. A management system has been developed in Texas for grazing sheep on ranges heavily infested with sneezeweed (*Helenium hoopesii*). If the range is fenced, a portion of the range can be treated to control sneezeweed and the sheep removed from sneezeweed areas for a few days every 3 weeks. Similarly, if the sheep are under the control of a herder, they can be taken off sneezeweed-infested areas for a few days. Apparently this prevents accumulation of the toxins. In Australia, the darling pea (*Swainsona* spp.) has toxic effects similar to those of locoweed. A high stocking rate of short duration may be used to remove the palatable darling pea before any animals can be permanently affected. If a lower stocking rate is used, certain sheep may tend to concentrate on grazing the *Swainsona* spp. and develop permanent brain injury.

In some cases, provision of dietary supplements to livestock may help to control losses from toxicants. It is often believed, although definitive proof seems to be lacking, that lack of salt and mineral supplements may increase the likelihood of animals consuming poisonous plants. Livestock deficient in phosphorus may develop a depraved appetite (pica) which could result in them consuming toxic plants. Palfrey *et al.* (1967) observed a positive correlation between the incidence of ragwort (*S. jacobaea*) poisoning of cattle and the lack of use of mineral supplements, and suggested that cattle consumed ragwort to obtain the phosphorus lacking in their diet. Another explanation of this finding is that farmers who provide mineral supplements probably also provide superior pasture management. Duby (1975) reported field trial results of an "alleviator" of ragwort poisoning, containing a mixture of 19 components, mostly mineral elements. Johnson (1982) examined the effects of a number of these components on ragwort toxicosis in cattle and found no protective activity. Keeler et al. (1977) studied the effect of provision of a mineral supplement to cattle on the incidence of lupine-induced fetal malformations and observed no protective effect. Ranchers using the supplement had believed it to be effective, but the apparent protective activity was due to variation in alkaloid content of the lupines and the nature of the management practices used. Protein and mineral supplements in the diet of sheep fed astragalus had no preventative value against the onset and severity of locoweed poisoning (James and Van Kampen 1974). The above examples are indicative of the general failure of dietary supplements to protect against poisonous plant losses. Synthetic antioxidants, which protect against pyrrolizidine alkaloid poisoning in laboratory animals (Garrett and Cheeke 1984), showed evidence of only slight protective activity when given to horses (Garrett et al. 1984) and cattle (Cheeke et al. 1985) fed tansy ragwort. Positive examples of effectiveness of supplements include the use of supplementary zinc to protect against sporidesmin toxicity (Smith et al. 1978), lupinosis (Allen and Masters, 1980), and pyrrolizidine alkaloid poisoning (Miranda et al. 1982). Dicalcium phosphate as a supplement may in some cases help to prevent halogeton (oxalate) poisoning in sheep. A dietary supplement, containing synthetic antioxidants and methionine, has been developed to help control sneezeweed poisoning (Kim et al. 1981).

Adaptation of livestock to poisonous plants prior to exposure is a technique that has sometimes been employed with varying degrees of success. Farmers in South Africa have traditionally predosed cattle orally with blue tulp (*Morea polystachya*) prior to allowing them to graze on tulp-infested pastures. Blue tulp contains cardiac glycosides. Strydom and Joubert (1983) compared the toxicity of tulp in cattle that had been predosed with the plant to those that had not and found no influence of the predosing on toxicity. Another example of adaptation is the increased resistance of sheep to halogeton poisoning following prior exposure to oxalate. In those instances where plants cause irreversible, cumulative damage, as with those containing pyrrolizidine alkaloids, preexposure would, of course, be undesirable.

Toxicity of some plants varies with the site on which they grow. Majak et al. (1977) in British Columbia found that timber milk vetch (*Astragalus miser*) in shaded, moist timbered sites was lower in toxin (miserotoxin) than milk vetch growing on open, dry grasslands. Management programs can be devised to account for this type of variation. Soil fertility can influence the concentration of toxicants in plants. In New Zealand, the condensed tannin content of *Lotus pedunculatus* may reach very high levels (8–11% of dry matter), markedly reducing grazing animal productivity. The tannin content is several times higher in lotus grown on acid soils without fertilizer application than when phosphorus and sulfur fertilizers are applied (Barry and Forss 1983). In contrast, the content of S-methylcysteine sulfoxide (brassica anemia factor) in kale and other forage brassicas is increased following nitrogen and sulfur fertilization (McDonald et al. 1981).

Some plants have properties that make animal management difficult for control of the problem. Water hemlock (*Cicuta* spp.) is of such high toxicity that livestock must be kept from contact with it, either through eradication of the plant or fencing off of infested areas. Larkspurs are quite palatable, while locoweeds are addictive, making it difficult to control losses when these plants are present.

Control of poisonous plants can sometimes be controversial, depending upon competing interests. In Australia, *Echium plantagineum*, or Paterson's curse, causes pyrrolizidine alkaloid poisoning of cattle and horses. Attempts to introduce biological agents to control echium infestations have been vigorously opposed by beekeepers who fear the loss of a major food source for their bees. Therefore, the desirability of controlling echium is largely a matter of the relative importance of the plant to livestock producers and beekeepers. This particular controversy also illustrates the importance of accurate data on toxicologic properties of poisonous plants. In spite of a long-held belief and clear field evidence that echium is toxic, at the time the controversy on its control erupted, there was little data in the literature showing echium to be toxic to livestock. The situation was further confused because a subsequent extensive feeding trial with sheep showed little evidence of toxicity of echium (Culvenor et al. 1984). Hence, simple feeding trials with poisonous plants, while seemingly mundane to some, can be of critical importance in policymaking, and the livestock industry can be negatively affected by the absence of this type of data.

INTERRELATIONSHIPS BETWEEN POISONOUS PLANTS AND GRAZING ANIMALS

There is considerable interest in the interactions of plant factors with herbivores (Laycock 1978; Rosenthal and Janzen 1979), particularly regarding evolutionary aspects. An intriguing question concerns the reason(s) for the existence of the myriad toxic secondary plant metabolites. While it is impossible to prove why plants have these constituents, a reasonable hypothesis is that poisonous compounds have evolved as defense mechanisms against herbivores. Evidence in support of this hypothesis includes the following: (1) many plant toxins do confer resistance to mammalian, fungal, and bacterial pests (e.g., saponins, tannins, sesquiterpene lactones, and alkaloids); (2) in some plants, the concentration of the toxicants is so high or they are so structurally complex that the energy cost of producing and storing them seems too high unless they had a function that was selectively advantageous compared to a nontoxicant-containing plant of the same species; (3) the great number, wide structural diversity, and widespread distribution of these compounds in plants appear to be too high for their evolution to have been accidental; (4) few of the known toxicants can be classified as either essential compounds for plant metabolism or as waste products of plant metabolism; and (5) insects and some large herbivores have developed resistance to plant toxicants by developing ways to detoxify specific toxins, suggesting coevolution of the poisonous plants and the herbivores. In some cases, it may be that a plant toxicant may have no particular physiological function and merely represents "flotsam and jetsam on the biochemical beach!" However, increasing evidence points to a major role of toxicants in plant defense. Interestingly, symbiotic relationships between plants and fungi may fill this role also. Endophyte infestation of perennial ryegrass, responsible for ryegrass staggers (Chapter 12), confers protection to the ryegrass from harmful insects such as the Argentine stem weevil (Prestridge *et al.* 1982). Insect resistance of endophyte-infected plants probably explains why old ryegrass pastures in New Zealand contain a very high incidence of infected plants. The endophyte associated with tall fescue, implicated in fescue-induced disorders such as summer fescue toxicosis (Chapter 5), may also have a role in protection of the fescue from insect damage. These observations add a further dimension to the interrelationships of plant toxicants, mycotoxins, pest resistance, and livestock poisoning.

Coevolution of plants and herbivores is a subject of much discussion (Rosenthal and Janzen 1979). Plants contain thousands of chemical compounds, which can have diverse effects on organisms consuming them. These compounds, with the exception of nutrients, are referred to as allelochemicals. Allelochemicals are nonnutritive compounds produced by one organism that affect the growth, health, behavior, or population biology of other species of organisms. They can be subclassified into kairomones and allomones. Kairomones are allelochemicals presumed to be useful to the receiving organism, such as attractants and feeding stimulants. There are numerous examples of insects that utilize secondary plant compounds in their own defense mechanisms, such as the monarch butterfly larvae which sequester milkweed cardenolides as protection against avian predators. Allomones are allelochemicals that are deleterious to the receiving organism, such as repellants, feeding inhibitors, and inhibitors of digestive enzymes. A detailed treatment of allelochemicals and the coevolution of plants and herbivores is provided by Rosenthal and Janzen (1979).

The continuum of evolutionary changes in synthesis of secondary compounds by plants, followed by the evolution of comparable detoxification mechanisms in animals, has been aptly termed the coevolutionary arms race (Berenbaum and Feeny 1981).

Besides the apparent role of secondary substances in protecting plants against herbivory, some plants produce compounds which protect them from the competition of other plants. These compounds, which inhibit the growth of other plants, are termed allelopathic substances (Rice 1974; 1979). Juglone secreted from the roots of black walnuts inhibits the growth of many other plants, including common garden vegetables and fruit trees (Coder 1983). Diffuse knapweed (*Centaurea diffusa*) is an important range weed in western North America. It contains allelopathic substances (Miur and Majak 1983) which may aid in its successful invasion of grassland. Black knapweed (*Centaurea nigra*) also has allelopathic effects (Vezina and Doyon 1983). Nicollier *et al.* (1983) suggest that Johnson grass secretes phytotoxins which inhibit the growth of other plants, allowing it to establish pure stands.

Towers (1980) reviewed the photodynamic action of photosensitizers in plants. Some plants produce phototoxic compounds which, when eaten by insects, result in the insects being killed by subsequent exposure to sunlight. Many of these compounds are polyacetylene compounds. The roots of marigolds (*Tagetes* spp.) produce thiophene derivatives which have nematocidal activity; their toxic effects on nematodes are tremendously enhanced by light.

Another aspect of coevolution of plants and herbivores is that in some instances, herbivore predation has beneficial effects on plants. McNaughton (1979) has discussed interrelationships between plants and wild ruminants on the Serengeti Plain of southern Africa. Some plant species in this community require animal grazing for their survival. Sod-forming species with a prostrate growth form depend on heavy grazing pressure on more upright species to prevent their being crowded out. In pasture management, it is well known that close grazing is required to maintain white clover in swards; with low grazing intensity, grasses become dominant. Herbivores have been suggested to have a direct stimulatory effect on grass productivity arising from plant growth-promoting agents in ruminant saliva. Reardon *et al.* (1972) and Reardon (1974) reported favorable effects of bovine saliva on plant growth, while Johnston and Bailey (1972) found no influence of saliva on grass growth. It is probable that any effect of saliva on plants is slight.

Some of the ways that herbivores could minimize the effects of plant toxins include consuming a generalized diet rather than feeding exclusively on just a few species, having the ability to detect and avoid poisonous plants, and having the ability to detoxify ingested plant toxins. Livestock do consume a generalized diet, in contrast to species such as the koala which feeds exclusively on a few species of eucalyptus. There is evidence that livestock can detect and avoid most plant toxins. Most poisonous plants are bitter and not generally consumed when other forage is available in adequate quantities. This is not invariably true; some toxic plants, such as *Leucaena leucocephala, Delphinium* spp., and fluoroacetate-containing plants are highly palatable. For those forage species that do contain toxicants, livestock show a preference for nontoxic or less toxic strains (Table 4.1).

There are possible interactions between naturally occurring toxicants. Sheep in Australia may graze on heliotrope for one season and on lupine stubble containing phomopsins the next. Since hepatotoxins are involved in each case, it is likely that the two different toxicants have an additive effect and that sheep with heliotrope-induced liver damage may be more susceptible than others to lupinosis. Bracken, St.-John's-wort, and tansy ragwort grow in similar habitats, so that animals could be simultaneously exposed to several types of toxicants. However, Garrett *et al.* (1982) found no evidence of an interaction between these plants. It is likely that numerous interactions between toxic plants in their effects on livestock may occur that have not yet been identified.

Seawright *et al.* (1972) reported an incident in which sheep were noted to be unusually sensitive to toxicity of carbon tetrachloride, which is used as an anthelmintic. The sheep were on a sparse pasture

TABLE 4.1. Examples of Studies in Which Livestock Exhibited Distinct Preferences for Nontoxic or Less Toxic Plant Species or Strains[a]

Plant species	Toxicant	Animal	Reference
Blue lupine	Alkaloid	Sheep	Laycock (1978)
Reed canarygrass	Alkaloid	Sheep	Laycock (1978)
Crotalaria	Alkaloid	Cattle	Laycock (1978)
Bracken fern	Cyanogenic glycoside	Sheep and deer	Laycock (1978)
Sorghum and Sudan grass	Cyanogenic glycoside	Cattle and sheep	Laycock (1978)
Sericea lespedeza	Tannin	Cattle	Laycock (1978)
Alfalfa	Saponin	Swine	Cheeke *et al.* (1978)
Rapeseed meal	Glucosinolate	Ruminants swine, poultry	Van Etten and Tookey (1979)

[a] Adapted from Laycock (1978).

and had browsed on foliage of *Eucalyptus caleyi. Eucalyptus* spp. have essential oils containing compounds such as cineol, which are MFO inducers. Therefore, the increase in toxicity of carbon tetrachloride was probably due to its more rapid metabolism to the toxic metabolites following MFO induction. White *et al.* (1983) found that pretreatment of rats with 10% *Eucalyptus globulus* leaves in the diet prior to oral dosing with *Senecio longilobus* pyrrolizidine alkaloid increased the toxicity of the alkaloid. Hence, exposure of livestock to plants that affect MFO activity but are not themselves overtly toxic may influence the toxicity of poisonous plants.

EFFECTS OF TOXICANTS ON WILD ANIMALS

Evolutionary adaptation to toxicants appears to have occurred in wild species that have been exposed to particular plants for many generations. Domestic animals have been removed from their native habitat and therefore have had less opportunity to adapt to specific plants. Several examples of adaptation of wild species can be cited. In Western Australia, sheep readily consume plants containing fluoroacetate (*Gastrolobium* and *Oxylobium* spp.); just a few leaves are enough to kill a sheep. Some of the native kangaroos avoid these plants and are not poisoned. Other types of kangaroos, and some other marsupials have developed mechanisms to detoxify fluoroacetate (see Chapter 11) and so can tolerate fluoracetate (1080) levels hundreds of times higher than can be tolerated by the same species native to eastern Australia, where 1080-containing plants do not occur. Native big game animals (such as mule deer, pronghorn antelope, elk, and giraffes) also appear to be able to safely consume large quantities of poisonous plants native to their habitats, although there are documented cases of poisonings (Fowler 1983). Locoism has been observed in elk and pronghorn antelope feeding on *Astragalus* spp. and pyrrolizidine alkaloid poisoning has been noted in deer (Fowler 1983), although Dean and Winward (1974) found that black-tailed deer were resistant to tansy ragwort toxicosis. Wolfe and Lance (1984) described locoweed poisoning in elk in New Mexico. Under normal range conditions, locoweed poisoning was not a significant mortality factor, but could affect population dynamics of elk herds on ranges severely infested by locoweed.

There are a number of behavioral and other adaptations which may account for the general paucity of toxicological problems with poisonous plants in wild animals. Many wild animals have fastidious grazing habits and tend to nibble small quantities of feed from a variety of different plants, minimizing the likelihood of consuming an acutely toxic dose of toxins. Large wild herbivores tend to range over extended areas and are not forced to consume poisonous plants because of lack of other feed, as sometimes occurs with confined domestic animals. Fencing can affect wild species; Fowler (1983) cited an incident in Texas in which the use of woven wire rather than barbed wire fencing resulted in pronghorn antelope being confined to specific ranches rather than being allowed to follow a traditional migratory pattern. During a drought in 1964–1965, a large number of pronghorns on these ranches died from consumption of tarbush (*Fluorensia cernua*) because of a lack of other feed. In South Africa, wild animals confined to game ranches by fencing have suffered mortality from consumption of tannin-containing browse plants which would not be consumed by free-ranging animals (Van Hoven 1984). Thus, overgrazing, range deterioration, and fencing may enhance the risk of poisoning of both livestock and wild species.

There seems to be little evidence that large wild herbivores have evolved specific anatomical or metabolic defenses against poisonous plants. Their grazing and browsing behavior probably is the major factor limiting poisoning. The extent of problems associated with natural toxicants in free-ranging wild animals is extremely difficult to assess, but in general it appears that plant poisoning is a minor problem.

SPECIES DIFFERENCES IN RESPONSE TO TOXINS

With domestic livestock, there are a number of examples of species differences in susceptibility to plant toxins. While these will be considered in more detail in the coverage of specific toxicants, it is useful to list some to illustrate the variety of factors that may influence whether livestock are affected by certain plants. These factors may include simple differences in management or feeding programs, digestive tract differences, palatability variations, or differences in metabolism.

There are many situations where a toxicant is important in nonruminants (swine, poultry) and not in ruminants, and vice versa. This may simply be a reflection of the type of diet; nonruminants are fed high-concentrate diets and so are not likely to be exposed to large quantities of some of the poisonous plants that ruminants consume. Many toxicants are degraded or otherwise rendered inactive in the rumen. These include trypsin inhibitors, lectins, gossypol, glucosinolates, or certain pyrrolizidine alkaloids. Some toxins are made more potent or their toxic nature altered as a result of rumen metabolism. For example, *L. leucocephala* is goitrogenic in ruminants but not in nonruminants because the toxicant it contains, mimosine, is converted in the rumen to dihydropyridone, a goitrogen. There are a number of species differences in susceptibility to pyrrolizidine alkaloids. Cattle, horses, chickens, rats, and swine are quite susceptible, whereas sheep, goats, Japanese quail, guinea pigs, and rabbits are very resistant. These differences appear to be due to variations in the way that the alkaloids are bioactivated and metabolized in the liver.

Animal management can affect losses. Sheep on western U.S. ranges are herded, whereas cattle are free-ranging. Hence, sheep have less opportunity to avoid poisonous plants, and the occurrence or lack of losses may depend as much on the skill of the herder as on other factors. Grazing behavior and palatability differences are important. Sheep tend to consume short grasses, forbs, and shrubs, whereas cattle consume more coarse vegetation. Cattle may develop fescue foot on tall fescue pastures; sheep won't usually eat tall fescue, whereas it is readily consumed by cattle. Horses can graze without ill effect tall fescue pastures that are toxic to cattle. Sheep are much more likely than cattle to be killed by acute lupine toxicity; this is largely because they eat the toxic lupine pods in large quantities while cattle do not. On the other hand, consumption of lupines by cattle can produce teratogenic effects (crooked calf disease); sheep, however, are not affected. Sheep are much more resistant than cattle to the polycyclic diterpene alkaloids in larkspur. Cattle are more susceptible to *Conium* alkaloids than are horses, while sheep are highly resistant. The teratogenic effects of *Conium* alkaloids (crooked calf disease) are seen in calves, but not in lambs or foals. *Tetradymia* consumption causes photosensitization in sheep (bighead), but cattle are not affected. Many other examples could be cited; the point is that for a variety of reasons there are many species differences among livestock in response to dietary toxicants.

There may be breed differences in susceptibility to plant poisoning. In Australia, echium or Paterson's curse seems to be more toxic to British breeds of sheep than to merinos (Culvenor *et al.* 1984). Seawright (1982) suggests that this is because British breeds and their merino crosses graze plants with pyrrolizidine alkaloids, such as echium and heliotrope, much more readily than do merinos, and so consume a larger dose of the toxins.

REDUCING DELETERIOUS FACTORS THROUGH PLANT BREEDING

A variety of crops contain toxic factors, and much progress has been made in reducing their significance through plant breeding and selection. A few examples will be given as illustrations. Rapeseed contains glucosinolates and erucic acid which have detrimental effects on livestock. Canadian plant breeders have made great progress in reducing the content of these substances to the extent that "double zero" cultivars are available which are very low in both substances. The meal from these has been called "canola meal" to help overcome resistance in the feed trade to the use of rapeseed meal, which formerly had the reputation of containing toxins. Active plant-breeding programs exist to reduce the level of trypsin inhibitors in triticale, a feed grain with much potential. The trypsin inhibitor content of soybeans has been lowered through plant breeding. Low-gossypol types of cotton have been developed; unfortunately, these are of increased susceptibility to pest damage and so are not grown commercially to a major extent. Sorghum grain is an interesting example; plant breeders developed bird-resistant cultivars of sorghum which are of lower feeding value to livestock because of the deleterious effects of high tannin content. Another example is the potato cultivar "Lenape," which had to be withdrawn from production because of its excessive alkaloid content. The nutritional value of cassava, a tropical feed and forage plant, has been improved by the selection of low cyanogenic glycoside strains.

In the case of forages, potential exists for improvement in their feeding value through selection for reduced toxicant content. Low-saponin alfalfa has been developed which is of improved value for nonruminants. Progress is being made in the selection of bloat-free alfalfa by reduction in the bloat-producing cytoplasmic protein fractions. An alternative approach showing evidence of success is the incorporation of protein precipitants such as phenolics (tannins) into bloat-producing legumes. Since some of the genera that cause bloat do not have tannin-containing species, it is necessary to use mutagenic techniques to attempt to introduce the desired factor. Changes in the alkaloid content of tall fescue have been accomplished; unfortunately, these have intensified rather than decreased summer fescue toxicosis. Low-phytoestrogen cultivars of subterranean clover in Australia have reduced the problems of "clover disease" in that country.

Because toxicants in plants may play a role in resistance to pests and diseases, it is not surprising that selection for low toxicant levels may increase susceptibility of plants to insects and diseases. Low-saponin alfalfa is more susceptible to insect damage (Fig. 4.5) than unselected alfalfa. Low-gossypol cotton is of increased susceptibility to pests and diseases. Low-tannin milo is susceptible to bird damage. Thus, in some cases, a certain level of toxicants in plants may be necessary so that they can be grown effectively. Plant breeders must balance these requirements with the desirability of reducing toxicant levels.

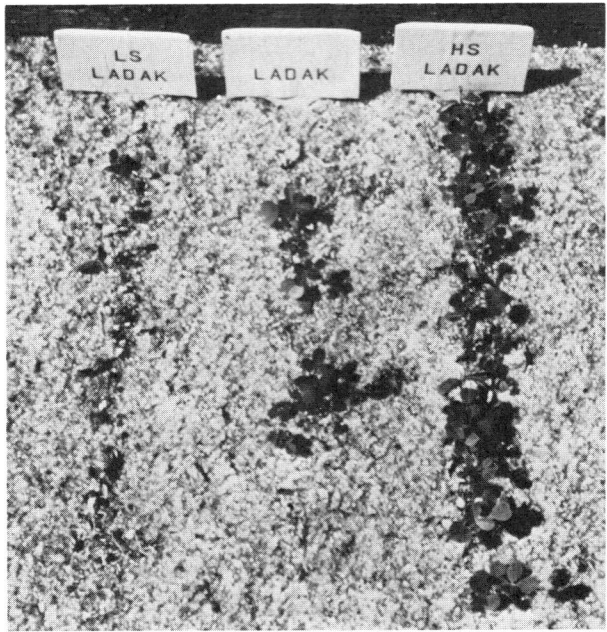

FIG. 4.5. Alfalfa seedlings of low-saponin (LS ladak), commercial (ladak), and high-saponin (HS ladak) selections of ladak alfalfa. The low-saponin seedlings have been decimated by pea aphids, whereas the high-saponin selection is untouched, suggesting that saponins have a role in protection of plants against insect pests.
Courtesy of E. L. Sorensen.

There is much interest in increased exploitation of tropical forage legumes to improve livestock production in tropical areas. Unfortunately, many of these contain deleterious factors. In some cases, plant breeders have worked on improving agronomic characteristics of new forage legumes, only to find that they are toxic. Australian researchers are now routinely screening new accessions through the use of a rat toxicity test to identify those that are frankly toxic. For example, several *Indigofera* species that show promise as tropical legumes are free of indospecine, while others are toxic. With the use of the rat toxicity test, plant breeders can begin their selection of improved agronomic features with nontoxic types rather than using the traditional approach of not discovering if a new strain is toxic until after a great deal of work has been invested in it. *Leucaena* is a tropical forage with a tremendous potential for protein production, but it is toxic because of its mimosine content. Efforts to produce low-mimosine cultivars are in progress.

In the case of range plants, it might appear that there would be little opportunity for genetic modification, but in some cases there may be. Williams and Davis (1982) discussed the use of screening for aliphatic nitro compounds of *Astragalus* spp. before they are introduced into the U.S. from other parts of the world. For example, sicklepod milk vetch (*A. falcatus*) has been introduced to western ranges; it is highly toxic to cattle and sheep. It would be wise in the future to test such plants for toxicants before they are introduced. The U.S. Soil Conservation Service has developed a nontoxic strain of sickle-keeled lupine for seeding on roadsides and rangelands as a replacement for toxic lupine species. In Australia, it has been suggested that it might be feasible to select a low alkaloid type of Paterson's curse that could be overseeded on areas where the wild plant is widespread.

Thus, there are many opportunities and challenges to reduce the effects of toxicants on livestock through plant modification by plant breeding.

POISONOUS PLANTS HAZARDOUS TO HUMANS

Kingsbury (1964, 1980) has summarized much of the scientific knowledge on plants poisonous to humans. Although many common house and ornamental plants are regarded as poisonous, the evidence to support that contention is lacking in many cases. Kingsbury (1980) cites several examples. Holly (*Ilex* spp.) berries are viewed as being toxic to children, although the only recorded reference to *Ilex* toxicity is a second-hand French report from 1889 with no authority stated. This report has been carried forward in books to the present time. Rodrigues *et al.* (1984) presented a case report on two children who consumed a "handful" of holly berries. Moderate to severe gastrointestinal disturbance resulted. Poinsetta (*Euphorbia pulcherrima*) has been reported (Kingsbury 1964) to have caused the death of a child, whereas feeding experiments (Stone and Collins 1971) have shown a lack of toxicity. Oehme (1978) dosed rats with several common ornamental plants including chrysanthemum (*Chrysanthemum morifolium*), Jerusalem cherry (*Solanum pseudocapsicum*), dieffenbachia (*Dieffenbachia picta*), geranium (*Pelargonium clonesticum*), and philodendron (*Philodendron cordatum*). The dosage was 5 g of fresh material per kilogram of body weight. Little evidence of toxicity was observed with any of the plants or plant parts. According to Oehme (1978), it is unlikely that adults or children would voluntarily consume these plants at levels equivalent to the dosage used. For example, a 40-lb child would be required to chew and swallow over 100 g of the fresh plant material to take in an equivalent amount.

Some of the plants that have been implicated in human poisoning will be briefly discussed. The majority of the plant intoxications reported at poison control centers in the U.S. result from the ingestion of plants containing gastrointestinal tract irritants. Symptoms range from burning of the mouth and throat because of chewing (usually by toddlers) houseplants such as *Philodendron* spp. and *Dieffenbachia* spp. to severe vomiting, intestinal cramping, and diarrhea from consumption of such plant parts as fresh roots and stems of pokeweed (*Phytolacca americana*), *Wisteria* spp. seeds, the berries of *Daphne* spp., or leaves of buttercups (*Ranunculus* spp.).

A number of human fatalities have occurred from the consumption of plants containing cardiac glycosides. These include foxglove (*Digitalis purpurea*), lily of the valley (*Convallaria majalis*), and oleander (*Nerium oleander*). Poisonings have resulted from consumption of the berries, chewing the leaves or flowers, or drinking water from vases containing the flowers of these plants. There is initial irritation to the mouth and vomiting, followed by abdominal pain and diarrhea. The consumption of tea brewed from foxglove leaves, mistaken for comfrey (*Symphytum officinale*), has caused fatalities. The first-year growth of foxglove has a superficial resemblance to a clump of comfrey, a commonly grown herb.

A number of plants contain pyridine and piperidine alkaloids, including wild tobacco (*Nicotiana* spp.) and poison hemlock (*Conium maculatum*). Poisonings result from nibbling on the leaves, using them in "wild salads," or eating the seeds of conium.

Atropine and its related alkaloids have been implicated in a number of poisonings. Jimsonweed (*Datura stramonium*) is the most important source. Contamination of a nationally distributed brand of comfrey tea with an atropine-containing plant(s) caused atropine intoxication in a heavy user of the tea in 1983, resulting in a recall of the tea nationwide. Comfrey contains carcinogenic pyrrolizidine alkaloids; their hazard to users of comfrey products has not been fully assessed.

The seeds of the popular ornamental tree, the horse chestnut (*Aesculus hippocastanum*), are poisonous. They contain a glycoside, aesculin (Williams and Olsen 1984). The toxic nuts should be collected and removed from areas where children and livestock might gain access to them. Signs of toxicity include depression, incoordination, paralysis, coma, and death. The seeds of Ohio buckeye (*A. glabra*) and yellow buckeye (*A. octandra*) are of much lower toxicity than those of the horse chestnut.

The major source of toxicosis involving convulsions is consumption of the roots of water hemlock (*Cicuta* spp.).

Mushrooms are responsible for a large number of intoxications, particularly following the growth of popularity of hallucinogenic mushroom consumption in the U.S. A discussion of toxic mushrooms is beyond the scope of this book, but because of the great similarity in appearance between many edible mushrooms and very deadly ones, the importance of consulting an authoritative source before collecting wild mushrooms cannot be overemphasized.

While this is a brief treatment of common poisonous plants that have been implicated in toxicoses in humans, it is noteworthy that in many instances, particularly for some common ornamental plants widely assumed to be toxic, definitive studies on their toxicity are lacking.

ETHNOBOTANY

Ethnobotany involves the study of the use of plants by different ethnic groups. In many cases, these interactions involve toxicants in foodstuffs and knowledge of poisonous and medicinal plants. Many native peoples have developed knowledge of harmful plants and often have evolved methods of food processing which minimize their toxic effects. Cassava is widely used as a human food in tropical countries. Indigenous peoples developed a variety of processing methods, including pulping or smashing the tubers, soaking the pulped material, and then washing it prior to consumption. This process results in hydrolysis of cyanogenic glycosides and washing out of the cyanide. The seeds of *Cycas media,* which contain a carcinogen, have been commonly used as a food by tropical peoples. A variety of processing methods, including fermentation, heating, water extraction, and sun drying, have been used to detoxify the seeds to make a nutritious flour. The aboriginal people of Australia had an awareness of poisonous plants and developed strategies for avoiding their toxic effects. For example, the seeds of nardoo (*Marsilea drummondii*) were widely used. Nardoo contains potent thiaminase activity, which is destroyed by heat treatment.

Ethnobotany is an active area of scientific investigation, and a journal, *The Journal of Ethnopharmacology* (Elsevier), is devoted to the subject.

REFERENCES

ALLEN, J. G., and MASTERS, H. G. 1980. Prevention of ovine lupinosis by the oral administration of zinc sulphate and the effect of such therapy on liver and pancreas zinc and liver copper. Aust. Vet. J. 57, 212–215.

ARNASON, J. T., FORTIER, G., CHAMPAGNE, D., and PHILOGENE, B. J. R. 1983. The phytotoxic action of plant secondary products on insects. Rev. Can. Biol. Exp. 42, 205-208.

BARRY, T. N., and FORSS, D. A. 1983. The condensed tannin content of vegetative Lotus pedunculatus, its regulation by fertilizer application, and effect upon protein solubility. J. Sci. Food Agric. 34, 1047-1056.

BERENBAUM, M., and FEENY, P. 1981. Toxicity of angular furanocoumarins to swallow tail butterflies: Escalation in a coevolutionary arms race? Science 212, 927-929.

CHEEKE, P. R., PEDERSON, M. W., and ENGLAND, D. C. 1978. Responses of rats and swine to alfalfa saponins. Can. J. Anim. Sci. 58, 783-789.

CHEEKE, P. R., SCHMITZ, J. A., LASSEN, E. D., and PEARSON, E. G. 1985. Effects of dietary supplementation with ethoxyquin, magnesium oxide, methionine hydroxy analog, and B vitamins on tansy ragwort (Senecio jacobaea) toxicosis in beef cattle. Am. J. Vet. Res. 46 (in press).

CODER, K. D. 1983. Seasonal changes of juglone potential in leaves of black walnut (Juglous nigra L.) J. Chem. Ecol. 9, 1203-1212.

CULVENOR, C. C. J., CLARKE, M., EDGAR, J. A., FRAHN, J. L., JAGO, M. V., PETERSON, J. E., and SMITH, L. W. 1980. Structures and toxicity of the alkaloids of Russian comfrey (Symphytum × uplandicum Nyman), a medicinal herb and item of human diet. Experientia 36, 377-379.

CULVENOR, C. C. J., JAGO, M. V., PETERSON, J. E., SMITH, L. W., PAYNE, A. L., CAMPBELL, D. G., EDGAR, J. A., and FRAHN, J. L. 1984. Toxicity of Echium plantagineum (Paterson's Curse). I. Marginal toxic effects in Merino wethers from long-term feeding. Aust. J. Agric. Res. 35, 293-304.

DEAN, R. E., and WINWARD, A. H. 1974. An investigation into the possibility of tansy ragwort poisoning of blacktailed deer. J. Wildl. Dis. 10, 166-169.

DUBY, G. D. 1975. Tansy ragwort. Mod. Vet. Pract. 56, 183-188.

FOWLER, M.E. 1983. Plant poisoning in free-living wild animals—a review. J. Wildl. Dis. 19, 34-43.

FRAENKEL, G. S. 1959. The raison d'être of secondary plant substances. Science 129, 1466-1470.

GARRETT, B. J., and CHEEKE, P. R. 1984. Evaluation of amino acids, B vitamins and butylated hydroxyanisole as protective agents against pyrrolizidine alkaloid toxicity in rats. J. Anim. Sci. 58, 138-144.

GARRETT, B. J., CHEEKE, P. R., MIRANDA, C. L., GOEGER, D. E., and BUHLER, D. R. 1982. Consumption of poisonous plants (Senecio jacobaea, Symphytum officinale, Pteridium aquilinum, Hypericum perforatum) by rats: Chronic toxicity, mineral metabolism and hepatic drug-metabolizing enzymes. Toxicol. Lett. 10, 183-188.

GARRETT, B. J., HOLTAN, D. W., CHEEKE, P. R., SCHMITZ, J. S., and ROGERS, Q. R. 1984. Effects of dietary supplementation with butylated hydroxyanisole, cysteine and B vitamins on tansy ragwort (Senecio jacobaea) toxicosis in ponies. Am. J. Vet. Res. 45, 459-464.

HIRONO, I. 1981. Natural carcinogenic products of plant origin. CRC Crit. Rev. Toxicol. 8, 235-276.

HIRONO, I., MORI, H., and HAGA, M. 1978. Carcinogenic activity of Symphytum officinale. JNCI, J. Natl. Cancer Inst. 61, 865-869.

HOOPER, P. T. 1978. Pyrrolizidine alkaloid poisoning—pathology with particular reference to differences in animal and plant species. In Effects of Poisonous Plants

on Livestock. R. F. Keeler, K. R. Van Kampen, and L. R. James (Editors), pp. 161–176. Academic Press, NY.
HUXTABLE, R. J. 1979. Herbal teas and pyrrolizidine alkaloids. *In* Symposium on Pyrrolizidine (*Senecio*) Alkaloids: Toxicity, Metabolism and Poisonous Plant Control Measures. P. R. Cheeke (Editor), pp. 87–93. Nutrition Research Institute, Oregon State Univ., Corvallis.
JAMES, L. F., and VAN KAMPEN, K. R. 1974. Effect of protein and mineral supplementation on potential locoweed (*Astragalus* spp.) poisoning in sheep. J. Am. Vet. Med. Assoc. *164*, 1042–1043.
JOHNSON, A. E. 1982. Failure of mineral-vitamin supplements to prevent tansy ragwort (*Senecio jacobaea*) toxicosis in cattle. Am. J. Vet. Res. *43*, 718–723.
JOHNSTON, A., and BAILEY, C. B. 1972. Influence of bovine saliva on grass regrowth in the greenhouse. Can. J. Anim. Sci. *52*, 573–574.
KEELER, R. F. 1978. Reducing incidence of plant-caused congenital deformities in livestock by grazing management. J. Range Manage. *31*, 355–360.
KEELER, R. F., JAMES, L. F., SHUPE, J. L., and VAN KAMPEN, K. R. 1977. Lupine-induced crooked calf disease and a management method to reduce incidence. J. Range Manage. *30*, 97–102.
KIM, H. L., ANDERSON, A. C., TERRY, M. K., and BAILEY, E. M. 1981. Protective effect of butylated hydroxyanisole in acute hymenoxon and bitterweed poisoning. Res. Commun. Chem. Pathol. Pharmacol. *33*, 365–368.
KINGHORN, A. D. (Editor) 1979. Toxic Plants. Columbia Univ. Press, Irvington, NY.
KINGSBURY, J. M. 1964. Poisonous Plants of the United States and Canada. Prentice-Hall, Englewood Cliffs, NJ.
KINGSBURY, J. M. 1980. Phytotoxicology. *In* Casarret and Doull's Toxicology: The Basic Science of Poisons. J. Doull, C. D. Klaassen, and M. O. Amdur (Editors), pp. 578–590. Macmillan Publishing Co., NY.
KRUEGER, W. C., and SHARP, L. A. 1978. Management approaches to reduce livestock losses from poisonous plants on rangeland. J. Range Manage. *31*, 347–350.
LAYCOCK, W. A. 1978. Coevolution of poisonous plants and large herbivores on rangelands. J. Range Manage. *31*, 335–342.
MAJAK, W., PARKINSON, P. D., WILLIAMS, R. J., LOONEY, N. E., and VANRYSWYK, A. L. 1977. The effect of light and moisture on Columbia milkvetch toxicity in lodgepole pine forests. J. Range Manage. *30*, 423–427.
MARTEN, G. C. 1978. The animal-plant complex in forage-palatability phenomena. J. Anim. Sci. *46*, 1470–1477.
McDONALD, R. C., MANLEY, T. R., BARRY, T. N., FORSS, D. A., and SINCLAIR, A. G. 1981. Nutritional evaluation of kale (*Brassica oleracea*) diets. 3. Changes in plant composition induced by soil fertility practices, with special reference to SMCO and glucosinolate concentration. J. Agric. Sci. *97*, 13–23.
McNAUGHTON, S. J. 1979. Grazing as an optimization process: Grass-ungulate relationships in the Serengeti. Am. Nat. *113*, 691–703.
McNAUGHTON, S. J., and TARRANTS, J. L. 1983. Grass leaf silicification: Natural selection for an inducible defense against herbivores. Proc. Natl. Acad. Sci. U.S.A. *80*, 790–791.
MIRANDA, C. L., HENDERSON, M. C., REED, R. L., SCHMITZ, J. A., and BUHLER, D. R. 1982. Protective action of zinc against pyrrolizidine alkaloid-induced hepatotoxicity in rats. J. Toxicol. Environ. Health *9*, 359–366.

MUIR, A. D., and MAJAK, W. 1983. Allelopathic potential of diffuse knapweed (*Centaurea diffusa*) extracts. Can. J. Plant Sci. *63,* 989–996.
NICOLLIER, G. F., POPE, D. F., and THOMPSON, A. C. 1983. Biological activity of dhurrin and other compounds from Johnson grass (*Sorghum halepense*). J. Agric. Food Chem. *31,* 744–748.
NIELSEN, D. B. 1978. The economic impact of poisonous plants on the range livestock industry in the 17 western states. J. Range Manage. *31,* 325–328.
OEHME, F. W. 1978. The hazard of plant toxicities to the human population. *In* Effects of Poisonous Plants on Livestock. R. F. Keeler, K. R. Van Kampen, and L. F. James (Editors), pp. 67–80. Academic Press, NY.
PALFREY, G. D., MACLEAN, K. S., and LANGILLE, W. M. 1967. Correlation between incidence of ragwort (*Senecio jacobaea* L.) poisoning and lack of mineral in cattle. Weed Res. *7,* 171–175.
PRESTRIDGE, R. A., POTTINGER, R. P., and BARKER, G. M. 1982. An association of *Lolium* endophyte with ryegrass resistance to Argentine stem weevil. Proc. 35th, N. Z. Weed, Pest Control Conf., pp. 199–222.
PUTMAN, A. R., and DUKE, W. B. 1978. Allelopathy in agroecosystems. Annu. Rev. Phytopathol. *16,* 431–451.
REARDON, P. O. 1974. Response of sideoats grama to animal saliva and thiamine. J. Range Mange. *27,* 400–401.
REARDON, P. O., LEINWEBER, C. L., and MERRILL, L. B. 1972. The effect of bovine saliva on grasses. J. Am. Sci. *34,* 897–898.
RHOADES, D. F. 1979. Evolution of plant chemical defense against herbivores. *In* Herbivores: Their Interaction with Secondary Plant Metabolites. G. A. Rosenthal and D. H. Janzen (Editors), pp. 4–54. Academic Press, NY.
RICE, E. L. 1974. Allelopathy. Academic Press, NY.
RICE, E. L. 1979. Allelopathy—an update. Bot. Rev. *45,* 15–109.
RODRIGUES, T. D., JOHNSON, P. N., and JEFFREY, L. P. 1984. Hollyberry ingestion. Case report. Vet. Hum. Toxicol. *26,* 157–158.
ROSENTHAL, G. A., and JANZEN, D. H. (Editors) 1979. Herbivores: Their Interaction with Secondary Plant Metabolites. Academic Press, NY.
SCHUSTER, J. L. 1978. Poisonous plant management problems and control measures on U.S. rangelands. *In* Effects of Poisonous Plants on Livestock. R. F. Keeler, K. R. Van Kampen, and L. F. James (Editors), pp. 23–34. Academic Press, NY.
SEAWRIGHT, A. A. 1982. Animal Health in Australia, Vol. 2. Chemical and Plant Poisons. Australian Government Publishing Service, Canberrra.
SEAWRIGHT, A. A., STEELE, D. P., and MENRATH, R. E. 1972. Seasonal variation in hepatic microsomal oxidative metabolism *in vitro* and susceptibility to carbon tetrachloride in a flock of sheep. Aust. Vet. J. *48,* 488.
SMITH, B. L., COE, B. D., and EMBLING, P. P. 1978. Protective effect of zinc sulphate in a natural facial eczema outbreak in dairy cows. N. Z. Vet. J. *26,* 314–315.
STONE, R. P., and COLLINS, W. J. 1971. *Euphorbia pulcherrima*: Toxicity to rats. Toxicon *9,* 301–302.
STOTTS, R. 1984. White snakeroot toxicity in dairy cattle. Vet. Med. Small Anim. Clin. *79,* 118–120.
STRYDOM, J. A., and JOUBERT, J. P. J. 1983. The effect of predosing *Homeria pallida* Bak. to cattle to prevent tulp poisoning. J. S. Afr. Vet. Assoc. *54,* 201–203.
TOWERS, G. N. N. 1980. Photosensitizers from plants and their photodynamic action. Prog. Phytochem. *6,* 183–202.
VAN ETTEN, C. H., and TOOKEY, H. L. 1979. Chemistry and biological effects of glucosinolates. *In* Herbivores: Their Interaction with Secondary Plant Metabolites.

G. A. Rosenthal and D. H. Janzen (Editors), pp. 471–500. Academic Press, NY.

VAN HOVEN, W. 1984. Tannins and digestibility in greater kudu. Can. J. Anim. Sci. *64,* (Suppl. 1), 177–178.

VEZINA, L., and DOYON, D. 1983. A note on the inhibitor effect of black knapweed (*Centaurea nigra*) residues. Phytoprotection *64,* 77–81.

WHITE, R. D., SWICK, R. A., and CHEEKE, P. R. 1983. Effect of microsomal enzyme induction on the toxicity of pyrrolizidine (*Senecio*) alkaloids. J. Toxicol. Environ. Health *12,* 633–640.

WILLIAMS, M. C., and DAVIS, A. M. 1982. Nitro compounds in introduced *Astragalus* species. J. Range Manage. *35,* 113–115.

WILLIAMS, M. C., and JAMES, L. F. 1983. Effects of herbicides on the concentration of poisonous compounds in plants: A review. Am. J. Vet. Res. *44,* 2420–2422.

WILLIAMS, M. C., and OLSEN, J. D. 1984. Toxicity of seeds of three *Aescubus* spp. to chicks and hamsters. Am. J. Vet. Res. *45,* 539–542.

WOGAN, G. N., and BUSBY, W. F. 1980. Naturally occurring carcinogens. *In* Toxic Constituents of Plant Foodstuffs. I. E. Liener (Editor), 2nd Edition, pp. 329–369. Academic Press, NY.

WOLFE, G. J., and LANCE, W. R. 1984. Locoweed poisoning in a northern New Mexico elk herd. J. Range Manage. *37,* 59–63.

5

Alkaloids

Alkaloids are basic substances that contain nitrogen in a heterocyclic ring. They are widely distributed in the plant kingdom; it has been estimated that 15–20% of all vascular plants contain alkaloids. Most alkaloids are derived from amino acids. In their synthesis by plants, amino acids are decarboxylated to amines, and the amines are converted to aldehydes by amine oxides. Condensation of the aldehyde and amine groups then yields the heterocylic ring (Fig. 5.1). Their role(s) in plant tissue is uncertain; they have been regarded either as metabolic by-products or substances that serve to repel insects and herbivorous predators. There is no doubt that in many cases their bitter taste does repel grazing animals. In some cases, complex ecological relationships are noted between alkaloids and insects. For example, the cinnabar

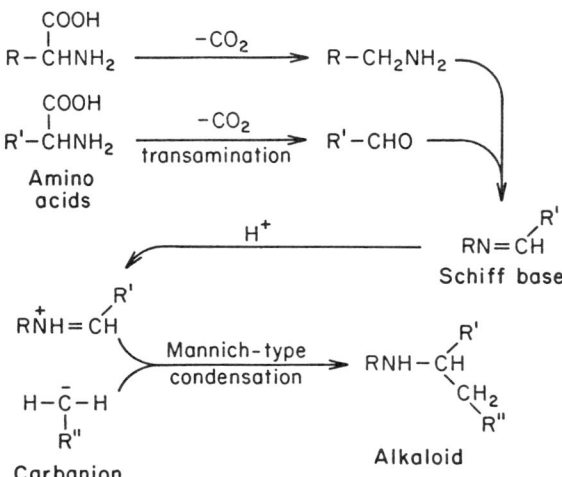

FIG. 5.1. General reactions of alkaloid biosynthesis.

moth (*Tyria jacobaeae*) larval stage is dependent on the presence of pyrrolizidine alkaloids (PAs), and the moth only lays eggs on plants containing these alkaloids. Certain danoid butterflies require PAs for their synthesis of pheromones.

PYRROLIZIDINE ALKALOIDS

Pyrrolizidine alkaloids contain the pyrrolizidine nucleus (Fig. 5.2). The structures of senecionine and heliotrine (Fig. 5.2) are representative of the toxic PAs important in livestock nutrition. All hepatoxic PAs share the characteristics that these exemplify in having a 1,2 double bond in the ring and esterification of the CH_2OH group in the side chain(s). Senecionine is representative of a closed ester, while heliotrine is an example of an open-ester PA. The PAs are biosynthesized from amino acids such as ornithine (Fig. 5.3). Most of the PA-

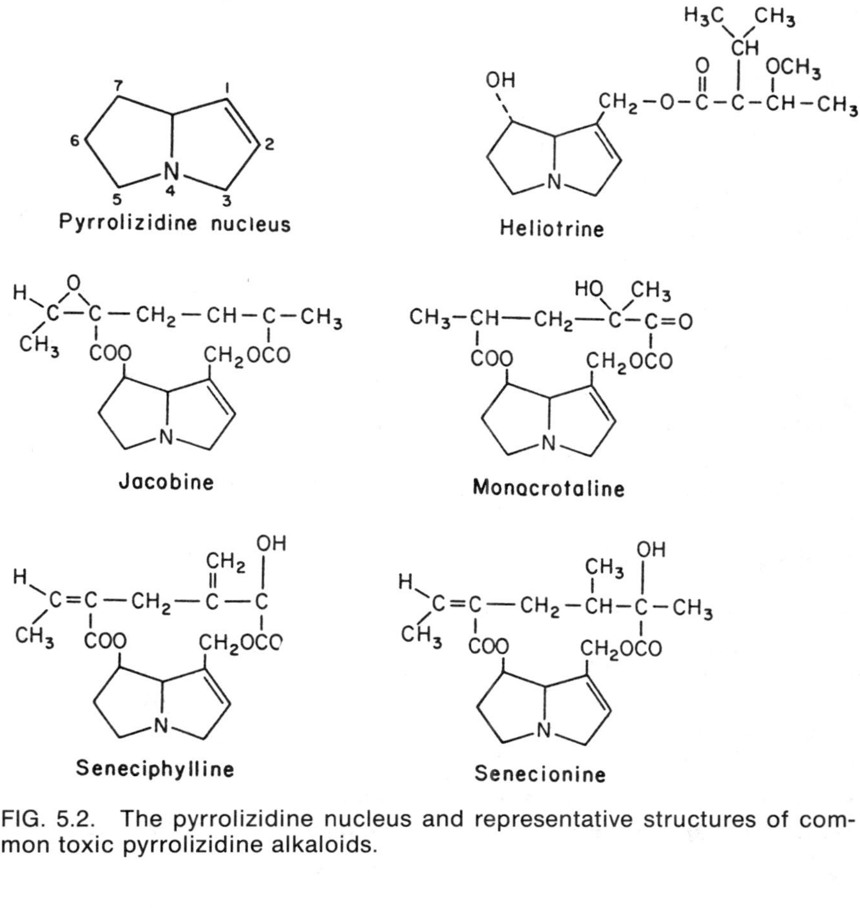

FIG. 5.2. The pyrrolizidine nucleus and representative structures of common toxic pyrrolizidine alkaloids.

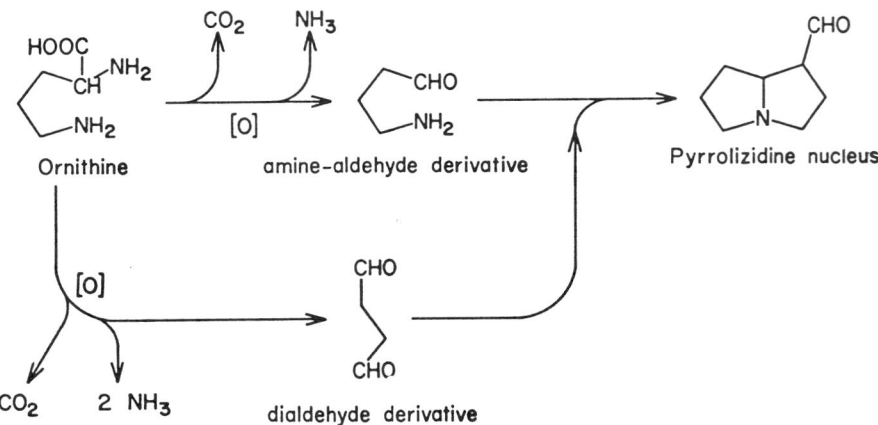

FIG. 5.3. Simplified example of pyrrolizidine alkaloid biosynthesis from amino acids.

containing plants that have been implicated in livestock poisonings are in the genera *Senecio, Crotalaria, Heliotropium,* and *Echium.* While PAs produce a diversity of biological effects, the principal pathology is irreversible liver cirrhosis with pronounced fibrosis and biliary hyperplasia. Most of the other tissue lesions and signs of toxicity, including mortality, can be related to impaired liver function.

Senecio Poisoning

The major livestock losses due to *Senecio* poisoning have been with *Senecio jacobaea,* commonly called tansy ragwort (Fig. 5.4). This plant is a native of the British Isles. It flourishes in humid temperate regions and has been inadvertently introduced to Western Europe, South Africa, Australia, New Zealand, the Pacific Northwest of the U.S., and eastern Canada (Ontario, Quebec, and Nova Scotia). Other *Senecio* species that have caused significant livestock losses are *S. vulgaris, S. longilobus, S. riddelli,* and *S. latifolius.*

Demonstration of the role of *Senecio* spp. and PAs in livestock poisoning occurred in the early 1900s, but livestock producers had suspected the plants of being poisonous much earlier. A problem in Wales called stomach staggers in the late 1700s was thought by farmers to be due to consumption of *S. jacobaea* (Bull et al. 1968). Two major outbreaks of livestock poisonings occurred in the period from 1860 to 1900 in Nova Scotia and New Zealand. They were called Pictou disease and Winton disease, respectively. Cattle and horses developed liver cirrho-

FIG. 5.4.   The rosette or first-year stage of tansy ragwort (*Senecio jacobaea*).

sis, jaundice, and, in the case of horses, staggering, pressing the head against objects, and walking in a straight line regardless of obstructions. Livestock owners in Nova Scotia suspected ragwort consumption as the cause, but agricultural officials believed it to be a bacterial infection. For nearly 20 years, the Canadian government ordered slaughter of affected herds and disinfection of the premises. Finally, in 1906, feeding trials definitely established that Pictou disease was caused by *S. jacobaea*. This was largely due to the reports from New Zealand on the etiology of Winton disease. There, based on association of losses with the presence of ragwort in pastures and feeding trials with the plant, it was demonstrated that *S. jacobaea* was the causative agent. Similar conditions were noted in other parts of the world. In South Africa, Molteno horse disease was traced to the consumption of *S. latifolius* and other *Senecio* species. In the U.S., walking disease of horses in Nebraska, Colorado, and Wyoming was a considerable economic problem in the early 1900s; in 1929–1933, feeding trials established *S. riddelli* as the cause. Other *Senecio* poisonings have occurred, including losses of cattle to *S. longilobus* (Fig. 5.5) in Arizona and

FIG. 5.5. *Senecio longilobus,* a toxic range plant containing pyrrolizidine alkaloids.
Courtesy of Earl Johnson.

S. vulgaris in California and Oregon. In the 1960s, *S. jacobaea* became a significant problem in the Pacific Northwest. Several million dollars of livestock losses have occurred since that time. As in the case of the weed in Nova Scotia, the introduction of ragwort to Oregon was traced to contaminated straw ballast used in shipping.

Tansy ragwort is normally a biennial. In the Pacific Northwest, it germinates in the autumn following the start of the fall rains, and the seedlings develop into rosettes the following year. The next year, a flowering stalk appears in late spring (Fig. 5.6), and the plant sets seeds and dies. One plant can produce many thousands of seeds which are readily dispersed by wind. If the flowering stalk is cut or mown, the plant may become a perennial, living until it has had the opportunity to produce seeds. If ragwort is cut in the early flowering stage, the cut plant will still set viable seeds, so control mowing should be done before flowering. Tansy ragwort is unpalatable and is generally not voluntarily consumed by cattle and horses even if other pasture is scarce. Poisonings generally occur as a result of contamination of hay and silage, or in the spring when the leaves of the plant in a dense pasture are not readily differentiated by animals from grass and clo-

FIG. 5.6. Flowers of tansy ragwort (Senecio jacobaea).

vers, thus causing inadvertent consumption. All parts of the plant are toxic, with the flowers having about four times the PA concentration as the leaves. The stems are quite low in toxicity.

Senecio poisoning is a significant human health problem in several parts of the world. In the 1920s "bread poisoning" of low-income Europeans in South Africa was caused by contamination of wheat with the flowers and seeds of *Senecio* spp, with many fatalities. In Africa and the West Indies, "bush tea" is prepared to combat various illnesses. Leaves from a variety of plants collected in the bush are steeped in water and the resulting tea used for medicinal purposes. *Crotalaria* and *Senecio* spp. are commonly used as leaf sources, with the result that the tea contains PAs. Consumption of bush tea has been linked to veno-occlusive disease in which occlusion of the hepatic veins occurs. Huxtable (1979) reported an outbreak of PA poisoning in Arizona in which children died as a result of consuming herbal tea made from *S. longilobus*. Culvenor (1983) made a comprehensive evaluation of outbreaks of PA poisoning in humans and compared the estimated PA intakes with dose rates causing tumors in rats. In several instances, the PA intake in chronic poisonings in humans was comparable to a carcinogenic dose in rats. He suggested that long-term observation of survivors of PA poisonings be made to evaluate the carcinogenic potential in humans.

Tansy ragwort is commonly referred to as tansy in many areas. This is unfortunate because it is sometimes confused with *Tanacetum vulgare,* which is also commonly called tansy. Tanacetum is an herb that has been used in herbal medicines. An example of the possible consequences of confusing these two tansys is a letter to a farm paper (Oregon Farmer-Stockman, May 17, 1973) in which the following suggestion was made for helping to control tansy ragwort:

> Tansy is an herb that has been used for years. Dried you can mix it with your tea and drink it hot or cold. A sprinkle of the dried tansy can be added to bread, rolls or even cookies. Tansy control is possible if all of us recognize this golden peril and attack it from every angle instead of waiting for someone to legislate it out of existence.

It is to be hoped that this well-intentioned advice to consume tansy ragwort was not followed by anyone! If the full name is not used, it is preferable to refer to the plant as ragwort rather than as tansy to avoid this confusion. Another observation of this type was made by Burns (1972), who determined the effect of tansy ragwort consumption on the heart and pulmonary arteries of rats. He stated,

> A disturbing feature that came to light during the course of the present study was the fact that I was able to purchase a large quantity of dried chopped *Senecio jacobaea* from a "Health Stores" in the Liverpool area. The material purchased was meant to be taken in the form of an infusion to cure a variety of ailments.

Crotalaria Poisoning

The loss of large numbers of horses grazing on Missouri River bottomland in the late 1800s (Missouri bottom disease) was due to consumption of *Crotalaria sagittalis* (see Fig. 5.7). In the Pacific Northwest, walking disease occurred in horses grazing wheat stubble infested with *C. sagittalis.* A similar problem in South Africa was caused by consumption of *Crotalaria dura.* In the U.S., *Crotalaria spectabilis* was widely grown in the South as a green manure crop. Contamination of grain with seeds of *C. spectabilis* has resulted in numerous poisonings of swine and poultry. The problem is still a significant one in the southern states. In Western Australia, Kimberley horse disease, or "walkabout," was responsible for the death of large numbers of horses. *Crotalaria retusa* was identified as the causative agent. In the early 1970s, hundreds of people in India and Afghanistan died in

FIG. 5.7. An example of a *Crotalaria* sp. showing leaves and seed pods.

an outbreak of veno-occlusive disease caused by contamination of grain with *Crotalaria* seeds (Mohabbat *et al.* 1976; Tandon *et al.* 1976). *Crotalaria* contamination of grains continues to be a significant animal and human health problem, particularly in developing countries. Routine spraying of grain fields for braod-leaved weeds in North America and Europe has virtually eliminated problems of grain contamination with toxic seeds in these areas.

Heliotropium and *Echium* Poisoning

Extensive poisoning of sheep and cattle in Australia due to consumption of *Heliotropium europaeum* (Fig. 5.8) has occurred. The plant continues to be a major problem. Sheep are highly resistant to the hepatoxic effects of PAs, but exposure to PAs causes the liver to accumulate excessive copper levels. Death is frequently due to the hemolytic crisis of copper toxicity. Usually more than one grazing season on heliotrope is necessary for sheep losses to occur. *Echium plantagineum* (Fig. 5.9) grows extensively in Victoria, southern New South Wales, and South Australia. It is called Paterson's curse or Salvation Jane, depending on one's point of view. While it provides feed in the spring when other feed

FIG. 5.8. The leaves and flowers of *Heliotropium europaeum*, a pyrrolizidine alkaloid-containing plant which causes extensive losses of sheep.
Courtesy of C. C. J. Culvenor.

may be short, it contains PAs. Feeding trials with sheep (Culvenor *et al.* 1984) have shown that they experience little detrimental effect from eating large quantities of dried echium. However, since echium and heliotrope frequently grow in the same areas, they may both contribute to poisoning problems. Prolonged grazing of echium may result in chronic copper poisoning of sheep (St. George-Grambauer and Rac 1962) with signs of severe jaundice and hemoglobinuria. The fat of affected animals is bright yellow, and the liver is enlarged, fibrotic, and bright yellow.

Other Plants with Pyrrolizidine Alkaloids

Amsinckia intermedia (tarweed, fiddleneck) is an annual weed growing in waste areas and grain fields of California and the Pacific Northwest. It contains a number of PAs, including intermedine, lycopsamine, echiumine, and sincamidine (Bull *et al.* 1968). At one time, up until the 1940s, *Amsinckia* toxicity of livestock was quite common in the Pacific Northwest, affecting horses, swine, and cattle. The conditions were called "walking disease" in horses and "hard liver disease" in cattle and swine. Poisoning was due mainly to the presence of the weed seeds in grain and grain screenings fed to livestock. Following the widespread use of herbicides to control broad-leaved weeds in grain, the problems have disappeared.

FIG. 5.9. *Echium plantagineum,* a pyrrolizidine alkaloid-containing plant that is widespread in Australia. It is commonly referred to as Paterson's curse or Salvation Jane, depending on whether or not other forage is available.

Comfrey (*Symphytum officinale*) is a vigorous, deep-rooted perennial herb with large, coarse prickly leaves (Fig. 5.10). It has been promoted at various times as a forage crop for livestock and is also widely used as an herb with reputed medicinal properties. Comfrey contains at least eight PAs (Culvenor *et al.* 1980), including echimidine, symphytine, lycopsamine, and intermedine. Administration of comfrey PAs to rats resulted in mortality and liver pathology characteristic of PA toxicosis. Hirono *et al.* (1978) fed comfrey roots and leaves to rats and found that liver cancer was produced. Dietary levels of 0.5% root and 8% leaves produced hepatomas. Mutagenicity in the Ames test has been noted for comfrey (White *et al.* 1983) and comfrey extracts (Furmanowa *et al.* 1983). The public health significance of comfrey PAs has not been fully established. It is a commonly used herb and is readily available in "health food" stores. The comments of Culvenor *et al.* (1980) are pertinent:

> The pyrrolizidine alkaloid content of Russian comfrey provides grounds for concern at the human consumption of this plant, especially by children because of the

FIG. 5.10. Comfrey, a succulent plant grown as an herb and forage. It contains hepatotoxic pyrrolizidine alkaloids.

greater sensitivity to young animals of the effects of this type of alkaloid. The lack of reports of toxicity of this plant despite claims of dietary use over many years is not necessarily an indicator of safety. The effects of such alkaloids are cumulative and overt damage may be long delayed, thus preventing association with the plant cause.

Another herb containing PA is borage (*Borago officinalis*); the PA is in low concentration and probably is not a significant health hazard (Larson *et al.* 1984).

Cynoglossum officinale (hound's tongue) is an annual or biennial herbaceous plant containing PA, including heliosupine and echinatine (Bull *et al.* 1968). A native of Europe, it has become naturalized in various parts of the U.S. and Canada. The green plant has a distinctive unpleasant odor which discourages consumption, but is is more palatable in hay, and poisoning may occur when contaminated hay is fed to livestock. Knight *et al.* (1984) reported classic signs of PA toxicity in horses fed contaminated hay in Colorado. The level of PA in the *C. officinale* was about 0.6–2.1% of the dry matter.

Tall fescue contains PAs. They are discussed in the section on fescue alkaloids.

These brief summaries indicate the scope of PA poisoning problems that have been experienced in several parts of the world. For a more extensive discussion of the history of PA toxicities, Bull et al. (1968) and Kingsbury (1964) are excellent references. Allen et al. (1979) and Huxtable (1979) have presented reviews of more recent problems. In several parts of the world, including the western U.S., New Zealand, and Australia, PA-containing plants still cause extensive livestock losses. Culvenor (1978) has estimated that up to 100 million sheep in southeastern Australia may be exposed to *H. europaeum* and that the productive life of animals so exposed may be reduced by as much as 30%.

Structure of Pyrrolizidine Alkaloids

About 150 PAs have been identified and their structures determined. Of the hepatoxic PAs, most are esters of the bases retronecine and heliotridine:

Retronecine   Heliotridine

These bases are called amino alcohols. Retronecine and heliotridine are diastereomers, with opposite configurations at C7. The PAs in toxic plants are of three general types; examples of each are retrorsine, lasiocarpine, and heliotrine (Fig. 5.11). For toxicity, several structural features are necessary (Culvenor et al. 1976). There must be a 1,2 double bond, and there must be a branch in the ester group. The amino alcohols are not toxic. The acute toxicity is generally such that the cyclic diesters such as retrorsine are most toxic, the noncyclic diesters are of intermediate toxicity, and the monoesters are the least toxic. Esters of heliotridine are more toxic than retronecine esters; esters of these two bases are the main groups of hepatoxic PAs. The PAs in several genera and species of toxic plants and their relative toxicity to rats (acute LD_{50}) are shown in Table 5.1. Representative structures are in Fig. 5.2. The PAs are also found in plants as N-oxides; these are of the same order of toxicity as the PAs.

TABLE 5.1. Distribution of Pyrrolizidine Alkaloids (PA) and Their Comparative Toxicities

Toxic plant	Amino alcohol	PA	LD_{50}[a]
Compositae			
Senecio jacobaea			
(tansy ragwort, stinking Willie)			
(British Isles,	Retronecine	Seneciphylline	77
Western Europe,	Retronecine	Senecionine	85
South Africa, Australia,	Retronecine	Jacobine	77 (mouse)
New Zealand, North	Retronecine	Jaconine	168
America)			(female rat)
	Retronecine	Jacoline	NA[b]
	Retronecine	Jacozine	NA
Senecio vulgaris			
(common groundsel)			
(North America)	Retronecine	Senecionine	85
	Retronecine	Seneciphylline	77
	Retronecine	Retrorsine	38
Senecio longilobus			
(woolly or threadleaf groundsel)			
(North America)	Retronecine	Seneciphylline	77
	Retronecine	Retrorsine	38
	Retronecine	Riddelliine	105 (mouse)
Leguminosae			
Crotalaria spectabilis;			
C. retusa (rattlebox)			
(North America,	Retronecine	Monocrotaline	175
Asia, Australia)	Retronecine	Spectabiline	220
	Turneforcidine	Retusine	NA
Boraginaceae			
Heliotropium europaeum			
(heliotrope)			
(Australia)	Heliotridine	Heliotrine	300
	Heliotridine	Lasiocarpine	72
	Heliotridine	Europine	>1000
	Supinidine	Supinine	450
	Supinidine	Heleurine	140
Echium plantagineum			
(Paterson's curse, Salvation Jane)			
(Australia)	Retronecine	Echiumine	NA
	Retronecine	Echimidine	200
Amsinckia intermedia			
(tarweed)			
(U.S. Pacific Coast)	Retronecine	Intermedine	1500
	Retronecine	Echiumine	NA
	Retronecine	Lycopsamine	1500
Symphytum officinale			
(comfrey)			
(temperate regions)	Retronecine	Symphytine	300
	Retronecine	Echimidine	200
Cynoglossum officinale	Heliotridine	Heliosupine	60
(hound's tongue)	Heliotridine	Echinatine	350

TABLE 5.1. (Continued)

Toxic plant	Amino alcohol	PA	LD_{50}[a]
Gramineae			
Festuca arundinacea			
(tall fescue)			
(temperate regions)	(no named amino alcohol for these alkaloids)	Loline	NA
		N-Acetyl loline	NA
		N-Formyl loline	NA

[a] All LD_{50} are for male rats unless otherwise indicated.
[b] NA = not available.

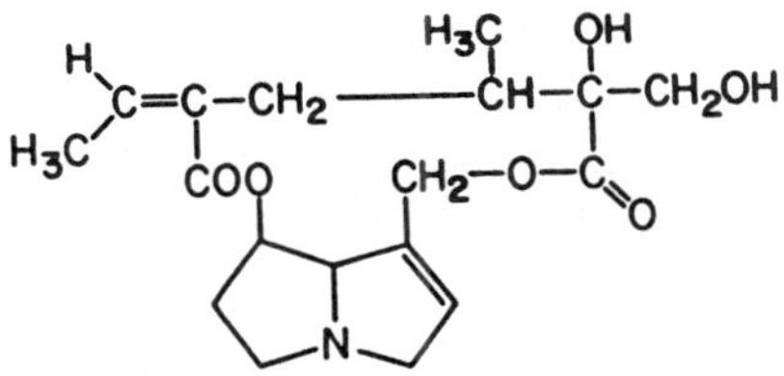

(a) Retrorsine

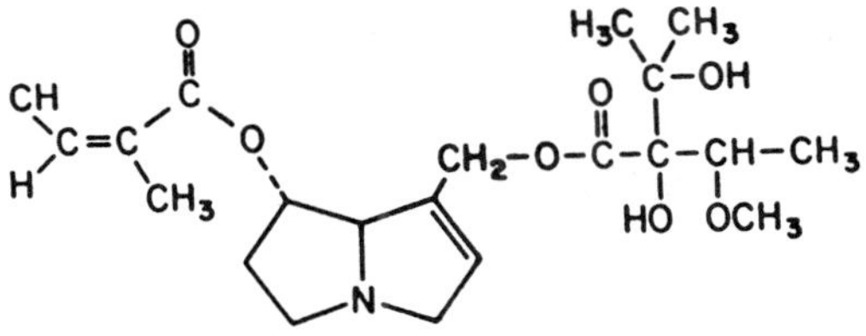

(b) Lasiocarpine

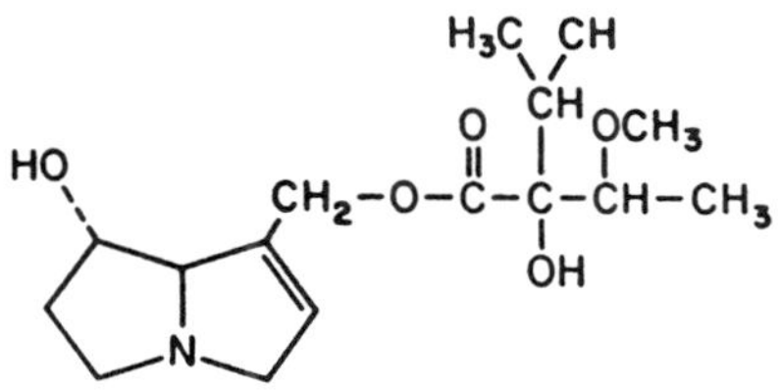

(c) Heliotrine

FIG. 5.11. Examples of types of toxic pyrrolizidine alkaloids. (a) Cyclic diesters, e.g., retrorsine; (b) noncyclic diesters, e.g., lasiocarpine; (c) monoesters, e.g., heliotrine.

Toxicity, Metabolism, and Metabolic Effects

The toxic effects of PAs are due to their bioactivation in the liver to toxic metabolites called pyrroles, or dihydropyrrolizine derivatives (Fig. 5.12). The pyrroles are chemically very reactive. Hepatotoxic PAs have a 1,2 double bond which facilitates aromatization of the B ring by the hepatic mixed function oxidases (Culvenor et al. 1976). The pyrroles are powerful alkylating agents and react with tissue components. Pyrroles from diester PAs may act as bifunctional alkylating agents and can cross-link DNA, explaining the antimitotic effects of PAs. Pyrroles may covalently bind with soluble nucleophiles such as glutathione and be excreted in the urine, or they may form covalent bonds with DNA or liver enzymes. Small amounts of pyrrole may enter the blood and be transported to the lungs, causing pulmonary lesions. Besides formation of pyrroles, other possible fates of PAs include urinary excretion (with or without conjugation), hydrolysis of the ester groups by esterases with excretion of the acid and amino alcohol products, and formation and excretion of highly water-soluble N-oxides. Factors which tend to increase hepatotoxicity of PAs are those that maximize pyrrole production and minimize the other pathways (Mattocks 1981). Factors favoring pyrrole production include a high degree of lipid solu-

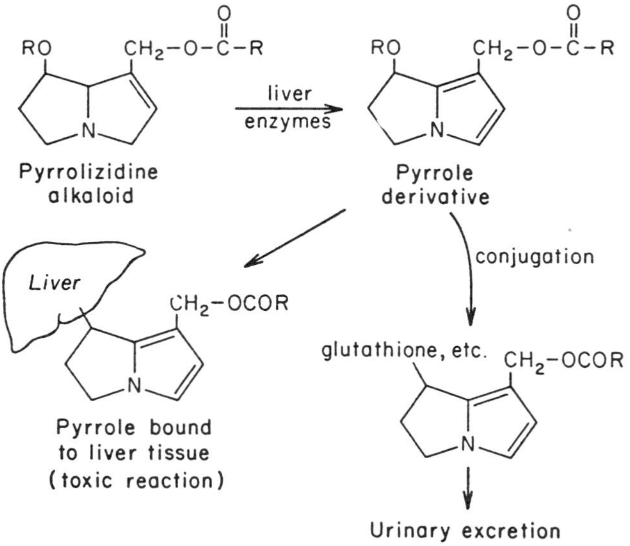

FIG. 5.12. Hepatic metabolism of pyrrolizidine alkaloids.

bility, a conformation favoring dehydrogenation rather than N-oxidation, and resistance to ester hydrolysis.

There may be additional metabolites produced in PA metabolism. Segall et al. (1984) demonstrated the production of two dihydropyrrolizine metabolites, methoxydehydroretronecine and hydroxydanaidal, from metabolism of senecionine by mouse microsomes. A further metabolite, trans-4-hydroxy-2-hexenal (t-4HH), has been isolated (H. J. Segall, 1984, personal communication, University of California, Davis). The 4-hydroxyalkenals are very reactive and may be the ultimate metabolites of PA metabolism. The origin of these metabolites is the pyrrolizidine ring:

Hydroxydanaidal $\longrightarrow$ $CH_3CH_2CH(OH)-CH=CH-CHO$ trans-4-hydroxy-2-hexenal

The hepatoxic effects of PAs are due to the reaction of pyrroles and/or other metabolites with tissue components, principally DNA. Typical hepatoxic signs are swelling of hepatocytes, centrilobular necrosis (the centrilobular region has microsomal enzyme activity, producing pyrroles), megalocytosis of the parenchymal cells (enlarged cells), karyomegaly (enlarged nuclei), fibrosis, bile duct proliferation, veno-occlusion, and loss of liver function. These effects are due to the alkylation of DNA with pyrroles, impairing cell division. The antimitotic effect results in a prevention of successful cell division. The specific stage of mitotic blockage is either in the latter half of the S phase or in early G_2 phase. The cells may be from 10 to 30 times normal size, and the DNA content may be 200 times normal. Megalocytosis is the result of an interaction between the antimitotic effect and the stimulus for regeneration. In most species affected by PA poisoning, the liver becomes extremely hard, fibrotic, and shrunken. In sheep, it tends to be mushy rather than hard. Liver weight as a percentage of body weight falls progressively in PA poisoning. Liver function is severely impaired. Because of decreased bile secretion, bilirubin levels in the blood rise, causing a typical jaundiced condition of the skin and mucous membranes. Other common clinical signs of PA toxicity include a rough unkempt appearance, diarrhea, prolapsed rectum, ascites, edema of tissues of the digestive tract, lassitude and dullness, photosensitization reactions, and abnormal behavior. In horses, neurological disturbances are seen, including "head pressing" against solid objects

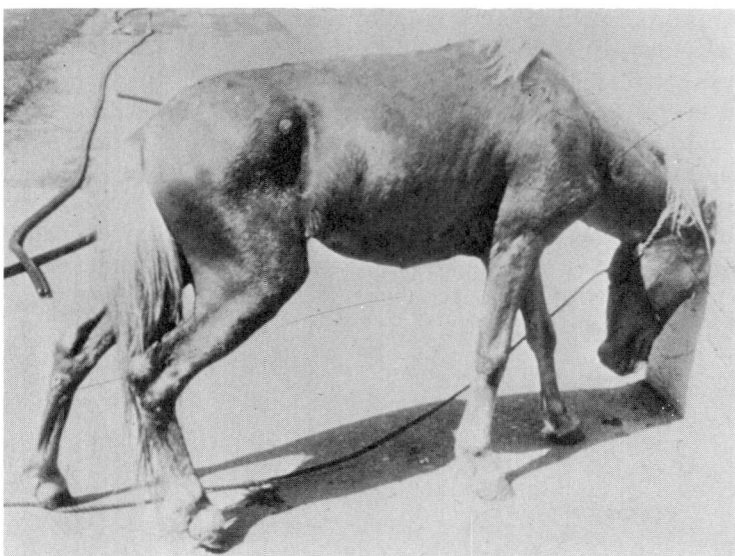

FIG. 5.13. A horse exhibiting head pressing, a neurological sign associated with pyrrolizidine alkaloid toxicosis. This animal had consumed feed contaminated with *Senecio vulgaris*.
Courtesy of Humphrey Knight.

(Fig. 5.13) and walking in a straight line, regardless of obstacles or dangers in the path. Many deaths of horses are due to misadventure as a result of these symptoms. The clinical signs can be explained in terms of the loss of liver function. Due to decreased liver synthesis of plasma proteins, osmotic relationships between blood and other body fluids are disturbed, causing fluid movement from the blood to interstitial spaces (edema) and body cavities (ascites) (see Fig. 5.14). Cattle poisoned by PAs may have a bloated appearance, and there may be several gallons of ascitic fluid in the abdominal cavity. Neurological signs in horses are due to elevated blood ammonia caused by decreased ability of the liver to convert ammonia to urea.

With chronic PA toxicity, animals may look normal even after they have consumed a lethal dose. The liver damage is progressive, and death may occur months or even years after exposure to toxic plants containing PAs. Several serum enzymes have been proposed for diagnostic purposes. These include γ-glutamyl transpeptidase, alkaline phosphatase, and glutamate dehydrogenase. While of some value for diagnostic purposes, changes in serum levels of these enzymes can be

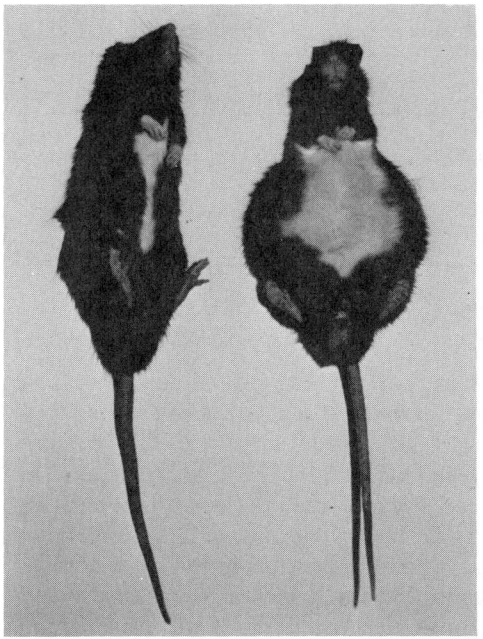

FIG. 5.14. Tansy ragwort toxicosis in rats. The intoxicated rat (right) shows pronounced ascites, compared to the control (left).

caused by other factors, and elevations in their levels due to PA poisoning are transitory.

Damage caused by pyrroles is not confined to the liver. Lesions in the lungs, heart, gastrointestinal tract, and kidneys have been observed. These may be due to extrahepatic metabolism of PAs to pyrroles, or a "spillover" of pyrroles from the liver. The damage from *Senecio, Heliotropium,* and *Echium* is largely confined to the liver, while *Crotalaria* intoxication involves significant lung damage.

Another effect of PA intoxication is anemia. This may be due to changes in copper and iron metabolism or may be a specific effect on hematopoesis. Swick *et al.* (1982A) demonstrated that rats poisoned with *S. jacobaea* had a severely impaired ability to incorporate ^{59}Fe into red blood cells (Table 5.2). The spleen is greatly enlarged in PA poisoning. Deposition of hemosiderin in various tissues occurs. Further studies are needed to conclusively identify the mechanisms by which PAs cause anemia.

A characteristic effect of PA poisoning in some animals is an increased liver copper concentration. In heliotrope poisoning of sheep, copper accumulates in the liver, even though forage copper is low, and

TABLE 5.2. Effect of Dietary Tansy Ragwort on Rat Tissue Levels of ^{59}Fe 3 Days after ^{59}Fe Administration[a]

Days tansy ragwort consumed prior to ^{59}Fe dose	Treatment	Percentage of injected ^{59}Fe dose in organ or tissue			
		Erythrocytes	Liver	Spleen	Tibia
5	Control	24.8	9.0	0.8	0.08
5	Tansy ragwort	25.5	12.8	1.0	0.13[b]
33	Control	23.2	11.0	0.6	0.17
33	Tansy ragwort	3.1[b]	4.1[b]	2.1[b]	0.05[b]

[a] Adapted from Swick et al. (1982A).
[b] Significantly different from control value.

leads to the hemolytic crisis of copper toxicity. In sheep, the liver copper content may be 4–10 times normal values, while in rats, a 50- to 100-fold increase may occur (Fig. 5.15). Swick et al. (1982C) demonstrated that the nuclear and debris fractions of the liver cells are the main intracellular sites of copper accumulation. Possibilities to explain the increased copper levels include an effect of PAs on copper absorption, a change in hepatic copper-binding proteins, or decreased ability to excrete copper. It is probable that cholestasis, or lack of bile secretion, accounts for copper retention in the liver. Other hepatoxins, such as phomopsins involved in lupinosis, also cause increased liver copper, suggesting that liver damage per se is the primary factor involved.

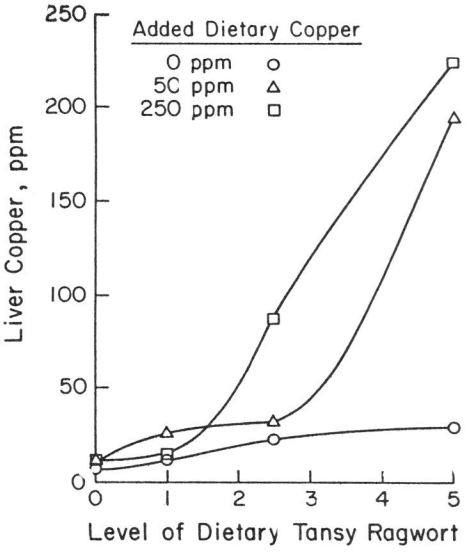

FIG. 5.15. Effect of various dietary levels of tansy ragwort on liver copper levels in rats. Adapted from Swick et al. (1982A).

The toxicity of PAs is influenced by their structure. Mattocks (1981) related the toxicity of PAs to the amount of covalently bound pyrrole in the liver. The liver pyrrole level 2 hr after adminstration of a dose of PAs is well correlated with the hepatoxic effect. The amino alcohols produce practically no pyrrole. Diesters produce more pyrrole than corresponding monoesters. The PAs that are more easily hydrolyzed by esterases produce less pyrrole. Branching in the side chains hinders hydrolysis and therefore leads to greater pyrrole production. The heliotridine esters are more toxic than similar retronecine esters.

Species Differences in Toxicity and Metabolism of PAs

Horses and cattle are the major livestock species poisoned by PAs, and they are of about equal susceptibility. In the case of *S. jacobaea* and *S. vulgaris* an intake of 5–10% of body weight is lethal. Sheep, on the other hand, are very resistant to the hepatoxic effects of PAs. They can tolerate an intake of several times their body weight of *S. jacobaea*. Goats are slightly more susceptible than sheep. Hooper (1978) ranked the relative intakes of *S. jacobaea* to produce mortality as pigs, 1; chickens, 5; cattle and horses, 14; rats, 50; mice, 150; sheep and goats, 200. Japanese quail, gerbils, rabbits, and guinea pigs are also very resistant to effects of dietary PAs.

There are several possibilities to account for species differences in susceptibility, and, in particular, resistance of some animals to PA toxicity. These include lack of PA absorption, degradation of PAs in the rumen, lack of hepatic formation of pyrroles, and conjugation and excretion of PAs and pyrroles. Shull *et al.* (1976) reported that, in general, susceptible species had a high rate of liver pyrrole production, while resistant species had a low rate (Table 5.3). There were some exceptions to this, such as the rabbit, which has a high rate of pyrrole production, but is very resistant to dietary PAs (Pierson *et al.* 1977). Evidence that hepatic metabolism is involved in some cases is that administration of inducing agents, such as phenobarbital, makes some resistant animals susceptible to PA poisoning. For example, guinea pigs are almost totally resistant to hepatoxic effects of monocrotaline. Prior adminstration of phenobarbital, which stimulates pyrrole production, results in guinea pigs becoming susceptible to monocrotaline toxicity. In contrast to their marked resistance to monocrotaline, guinea pigs are quite susceptible to *Senecio* PAs (Swick *et al.* 1982B).

The resistance of sheep to PA toxicity is of interest. Possibilities to account for their resistance are that PAs might be detoxified in the rumen, or that the sheep liver detoxifies pyrroles more efficiently than

TABLE 5.3. Characterization of Animal Species by Susceptibility to Pyrrolizidine Alkaloid Toxicity and in Vitro Hepatic Pyrrole Production Rate

Species	Susceptibility to PA toxicosis	In vitro pyrrole production rate[a]	Lethal dose (as % of body weight)[b]	Reference
Cow	High	High	3.6	Cheeke et al. (1985)
Horse	High	High	7.3	Garrett et al. (1984)
Sheep	Low	Low	302	White et al. (1984)
Goat	Low	?	205	Goeger et al. (1982A)
Rat	High	High	21	Goeger et al. (1983)
Mouse	Intermediate	High	?	Pierson et al. (1977)
Rabbit	Low	High	113	Cheeke and Pierson-Goeger (1983)
Guinea pig	Low	Low	119	Cheeke and Pierson-Goeger (1983)
Hamster	Low	High	338	Cheeke and Pierson-Goeger (1983)
Gerbil	Low	?	3640	Cheeke and Pierson-Goeger (1983)
Chicken	High	Low	39	Cheeke and Pierson-Goeger (1983)
Japanese quail	Low	Low	2450	Buckmaster et al. (1977)

[a] Adapted from Shull et al. (1976).
[b] Chronic lethal dose of Senecio jacobaea.

5. ALKALOIDS 113

does cattle liver. Lanigan (1976) has shown that a microorganism (*Peptococcus heliotrinreductans*) isolated from sheep rumen contents is capable of metabolizing heliotrope PAs to nontoxic 1-methylene and 1-methylpyrrolizidine derivatives by reduction of the 1,2 double bond in the nucleus and cleavage of the esters (Fig. 5.16). Culvenor *et al.* (1976) found that 7α-angelyloxy-1-methylene pyrrolizidine from in vitro metabolism of lasiocarpine did not produce liver necrosis in rats or yield pyrroles when incubated with rat liver microsomes. The formation of 1-methylene derivatives requires metabolic hydrogen. The addition of methane inhibitors to the diet appears to increase the rumen formation of the nontoxic metabolites from heliotrope PAs by providing more hydrogen (Lanigan *et al.* 1978). The resistance of sheep to *S. jacobaea* poisoning appears not to be due to rumen metabolism of PAs. Shull *et al.* (1976) and Swick *et al.* (1983A) found that incubation of *S. jacobaea* in sheep rumen fluid did not reduce its toxicity, and there was no evidence of formation of 1-methylene derivatives. The PAs in heliotrope are open esters of the base heliotridine, whereas those in *S. jacobaea* are all macrocyclic closed esters of retronecine. Because of steric hindrance, the closed esters are probably less readily converted to reduction metabolites. The resistance of sheep to *S. jacobaea* is probably due to a low rate of hepatic pyrrole production (Shull *et al.* 1976)

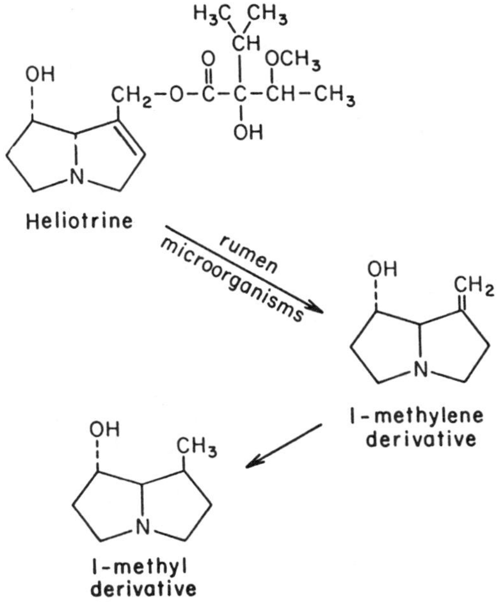

FIG. 5.16. Rumen metabolism of heliotrine.

and a high activity of detoxifying enzymes such as epoxide hydrolase (Swick *et al.* 1983B).

Milk Transfer of Pyrrolizidine Alkaloids

Dickinson and King (1978) demonstrated that PAs are found in the milk of dairy cattle and goats fed *S. jacobaea.* Goeger *et al.* (1982B) fed *S. jacobaea* to lactating goats and fed the milk to rats for a prolonged period. Very slight liver pathology was observed. They concluded that hazards to human health as a result of milk transfer of PAs were negligible.

Prevention and Treatment of Pyrrolizidine Alkaloid Poisoning.

Pyrrolizidine alkaloid poisoning of grazing animals in many areas can largely be eliminated through application of sound livestock management techniques. Plants containing PAs are unpalatable and not generally consumed in the presence of adequate feed. In the case of *S. jacobaea,* most poisonings occur when pastures are severly overgrazed and in early spring when other green forage is not available. The presence of *S. jacobaea* in a pasture is usually a reflection of poor management practices which have allowed invasion by the weed. In Australia, with harsh conditions and an extensive type of grazing system, exposure of livestock to PA-containing plants is often unavoidable. The potential toxicity problems associated with echium or heliotrope have to be balanced against the fact that often other feed is not available, hence the names Paterson's curse and Salvation Jane for *E. plantagineum.* Since it may take months or years for the lethal effects of PA ingestion to occur, grazing infested areas with animals destined for slaughter rather than with breeding stock has been recommended. Poisoning of swine and poultry generally is a result of contamination of grain with crotalaria seeds. Spraying of grain fields with broad-leaf herbicides can easily eliminate this problem.

Senecio jacobaea infestations of pastures can be controlled with herbicides. However, areas where the weed is a problem are often of low productivity, and it covers extensive areas of rough ground, making spraying economically unfeasible. Biological control measures have shown some success. The cinnabar moth (*T. jacobaeae*) larvae feed only on plants containing PAs. In the Pacific Northwest, the cinnabar moth has had a useful role in helping to control ragwort by defoliation and prevention of seed production. The flea beetle (*Longitarsus jacobaeae*) feeds on the roots of ragwort and, in western Oregon, has in conjunction with the cinnabar moth been an important management tool in controlling *S. jacobaea*. In Oregon and New Zealand, sheep have also been used as biological control agents. They are sufficiently resistant to PA toxicity that they can be used to control moderate ragwort infestations without likelihood of poisoning. Sheep are also used to control echium in Australia (St. George-Grambauer and Rac 1962), as their resistance to the plant's toxicity is sufficient to make poisoning unlikely.

An alternative to controlling the plant is to manipulate animal metabolism to make animals less sensitive to PAs. One approach is the feeding of sulfur-containing amino acids to supply sulfhydryl groups for conjugation with pyrroles. Buckmaster *et al.* (1976) found that feeding cysteine to rats increased their tolerance to *S. jacobaea*. The synthetic antioxidants butylated hydroxyanisole (BHA) and ethoxyquin have shown protective activity against PA poisoning in mice (Miranda *et al.* 1981A,B). A combination of cysteine and BHA as a dietary supplement has slightly increased the resistance of horses (Garrett *et al.* 1984) and beef cattle (Cheeke *et al.* 1985) to PA toxicity. Further evaluation of these supplements with cattle and horses is necessary to assess fully their potential as protective agents.

In Australia, copper toxicity in sheep grazing heliotrope is a major problem. The use of salt licks containing molybdenum, which increases copper excretion, should be examined as a potential protective regime. White *et al.* (1984) have studied interrelationships of copper and molybdenum with *S. jacobaea* toxicity and found that supplementary molybdenum tended to increase the susceptibility of sheep to PA toxicity.

The stability of PAs in hay and silage containing *S. alpinus* has been studied by Candrian *et al.* (1984). The PA content of hay remains constant over many months, while there is some degradation in silage. However, the reduction in PA levels during the ensilage process does not substantially reduce the toxicity of contaminated silage. Silage contaminated with more than about 5% *Senecio* plants is unsafe for cattle feeding. Clover and grass crops are sometimes contaminated with *S. jacobaea* and *S. vulgaris*, while corn may be contaminated with *S. vulgaris*.

PIPERIDINE ALKALOIDS

The principal piperidine alkaloids that cause livestock poisonings are those associated with poison hemlock (*Conium maculatum*). Poison hemlock is similar in appearance and name to water hemlock (*Cicuta maculata*), with which it is often confused.

Poison hemlock, a native of Europe, is distributed throughout the U.S., especially in the Pacific Northwest, and North-Central and Northeast regions. It is a vigorous plant, growing 6–10 ft high along roadsides, ditches, cattle yards, and the edges of grain fields. It resembles both wild carrot (*Daucus carota*) and water hemlock in appearance (Fig. 5.17). It has a fleshy, white tap root, whereas water hemlock has a thickened, branching root stalk much like a dahlia tuber in appearance. The lower portion of the stem of water hemlock is divided into chambers which contain the toxins. These features can be used to distinguish the two plants. Poison hemlock has a characteristic unpleasant mousy odor, which is detectable when one is near the plant or crushes a leaf or stem. The volatile alkaloids are toxic and have been shown to induce toxicities when inhaled. The conium alkaloids have two major effects: they cause acute effects on the nervous system, and they are teratogenic agents. According to Kingsbury (1964), an extract of poison hemlock was used to kill Socrates. The symptoms of acute poisoning, in order of their appearance, are nervousness, trembling, incoordination, dilation of the pupils, weakened heartbeat, coldness of

FIG. 5.17. Poison hemlock (*Conium maculatum*).
Courtesy of W. C. Krueger.

the extremities, coma, and death through respiratory failure. One of the major *Conium* alkaloids is coniine. Keeler *et al.* (1980) have reported the signs of sublethal coniine toxicity in cows, ewes, and mares. The alkaloid was given orally. Within 30–40 min after dosing in mares, and about 1.5–2 hr after dosing in cows and sheep, the animals became nervous, followed by tremors and ataxia. These signs lasted 4–5 hr in mares and 6–7 hr in cows and sheep. Cows were most sensitive, with horses intermediate, and sheep quite resistant to toxicity. Cows had severe toxic signs with a dose of 3.3 mg coniine per kilogram body weight, mares had severe signs at 15.5 mg/kg, while ewes showed only moderate toxic signs when given 44 mg/kg. Keeler *et al.* (1980) believe that the difference of about 10 times in susceptibility of cows and sheep to coniine is due to liver metabolism, because a similar difference in toxicity is seen when conium is administered either orally or by injection. They reported that when progressively larger daily doses of coniine were given by injection, about 16 mg/kg killed cows, whereas 240 mg/kg was the lethal dose for sheep. In the past, poisoning of swine and poultry by poison hemlock seeds in grain was fairly common. Concentration of the alkaloids is highest in the seeds. The use of herbicides in grain fields has virtually eliminated contamination of grain with conium seeds.

Panter *et al.* (1983) reported that a single oral dose of 1 g/kg body weight of *C. maculatum* seed or 8 g/kg of fresh plant (6- to 10-in. stage) were lethal to swine (gilts). The material was consumed avidly by the animals, and there was evidence of a craving and possible dependence on the plant. Daily sublethal doses were given to gilts for the study of teratogenic effects. About 140 g of seed and 1000 g of fresh plant per pig per day were used. About 15 min after dosing, the third eyelid began moving across the eye, causing complete blindness by 45 min. The blindness lasted about 15–30 min. The animals became weak and docile for about 1–2 hr. Trembling and weakness were observed. Both the seed and plant foliage were teratogenic. When fed in the 30- to 45-day period of gestation, *Conium* caused cleft palate. In the 43- to 53-day period, arthrogryposis occurred, along with severe flexure of fetlock joints, scoliosis, and hydrocephalus. When *Conium* was given in the 51- to 61-day period of gestation, limb malformations similar in type but less severe than in the 43- to 53-day group were observed. Edmonds *et al.* (1972) also noted that *C. maculatum* causes fetal deformities in swine.

In the western U.S., fetal deformities in cattle have been quite common. Crooked calf disease has been attributed primarily to maternal consumption of lupines at a particular stage of gestation. Implication of poison hemlock as a causative agent was reported by Keeler (1974).

Several crooked calves were found in herds that were not exposed to any known teratogenic plants. Poison hemlock was widely distributed in the pastures, suggesting it as a possibility. Because one of the conium alkaloids, coniine, was available commercially, Keeler and associates administered coniine to cows during the 55–75 days of gestation and observed skeletal deformities similar to those seen in the natural outbreak. Later studies (Keeler and Balls 1978) showed that administration of the whole plant to cows also caused crooked calf disease. Poison hemlock was given orally to cows daily during days 45–75 of gestation. Daily doses of green plant material of 410–840 g resulted in deformed calves. Cows forced to breathe fresh poison hemlock for about 20 hr/day had normal calves, but did show moderate sign of toxicity, demonstrating that the volatile alkaloids are toxic by inhalation. These workers further studied the structural features of the conium alkaloids that produced teratogenic effects. Five piperidine alkaloids in *C. maculatum* were identified (Fig. 5.18). Over 98% of the alkaloid in fresh plant was γ-coniceine, while it was less than 20% in dried plant material. A third of the alkaloid in dry plant was coniine. A variety of analogs of coniine were tested for teratogenic activity. Introduction of a double bond between C-1 and the nitrogen did not alter activity. Both the length of the side chain and the degree of unsaturation of the ring influenced activity. The fully unsaturated ring analog of coniine was not teratogenic. Compounds with a side chain of less than three carbons did not have activity. Piperidine alkaloids with a side chain α to the nitrogen, at least propyl in length, and with a saturated ring are candidates for having teratogenic activity.

In the study of Keeler *et al.* (1980) in which coniine was adminstered to pregnant cows, mares, and ewes, only the cows had deformed off-

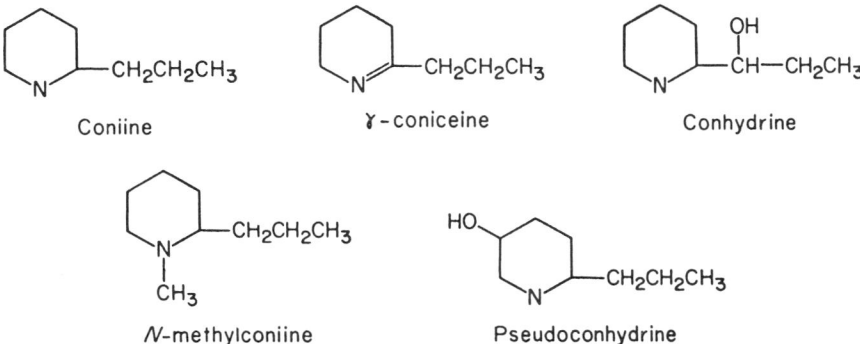

FIG. 5.18. Structure of the five naturally occurring conium alkaloids.

spring. The foal fetus may be resistant because the mares showed severe toxic signs and had normal foals. They concluded that the sheep fetus is definitely resistant to the teratogenic effects of coniine because the large doses given over a wide period (12–65 days) of gestation did not cause fetal defects.

PYRIDINE ALKALOIDS

The pyridine alkaloids are similar in structure to piperidine alkaloids except that the ring is unsaturated. An example of a pyridine alkaloid is nicotine:

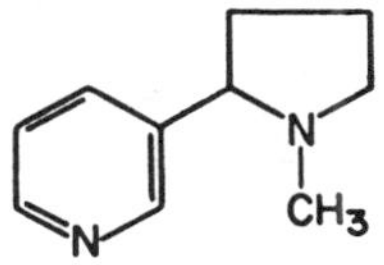

Nicotine is a highly toxic alkaloid produced by *Nicotiana* spp., the wild and cultivated tobaccos. Livestock are exposed to nicotine through the consumption of tobacco or through exposure to nicotine-containing insecticides such as Black Leaf 40. Wild tobaccos (*N. attenuata* and *N. trigonophylla*) are erect herbacious annuals found in dry sandy desert soils of the western U.S. Poisonings of sheep, cattle, and horses from consumption of these plants have been reported (Kingsbury 1964). Kingsbury (1964) also mentions a report of a family that used wild tobacco as a boiled green; one fatality resulted. Numerous cases of livestock poisonings from consumption of cultivated tobacco (*N. tabacum*) are known. Horses have died from consumption of tobacco leaves, while pigs have been fatally poisoned when they broke into a tobacco field (Kingsbury 1964). Nicotine is the presumed toxic agent. An unusual case is the death of a dog which consumed a package of cigarettes (Kaplan 1968).

Symptoms of nicotine toxicity include excitement, shaking and twitching, rapid respiration, staggering, weakness and prostration, followed by coma, paralysis, and death. Death is due to respiratory paralysis. Nicotine blocks autonomic ganglia and neuromuscular junctions. Large doses cause a descending paralysis of the central nervous system.

Nicotine has been implicated as a teratogen. Crowe (1978) reviewed a number of reports on congenital abnormalities in swine caused by consumption of tobacco stalks. A total of 15 farms in Kentucky experienced epidemics of congenital limb deformities (arthrogryposis) involving 246 sows and 1148 abnormal pigs. The pigs had a persistent flexure or contracture of a limb joint (Fig. 5.19). The affected bones were thickened, curved, or twisted, with a misalignment of joints, resulting in an inability of the pigs to flex or extend their limbs. The pregnant sows had access to freshly discarded tobacco stalks. The stalks were very large, with an abundance of succulent pith, resembling celery. The sows shredded the stalks and consumed the pith. Experimental feeding of the stalks confirmed the teratogenic effect, but it was not conclusively shown that nicotine was the active principle. Later studies (Keeler 1978) have shown that nicotine is not teratogenic and that the active compound is probably anabasine, a piperidine-pyridine alkaloid:

Nicotiana glauca (tree tobacco) is an evergreen shrub or small tree commonly found in low-elevation areas in California and Arizona. Consumption of the plant results in teratogenic effects in cattle (Keeler *et al.* 1981A), pigs (Keeler *et al.* 1981B), and sheep (Keeler and Crowe 1984). Calves from cows gavaged with dried *N. glauca* had arthrogryposis of the forelimbs or curvature of the spine, with the deformities clinically indistinguishable from those caused by maternal ingestion of lupine or *C. maculatum*. Fetal defects in sheep included carpal flexure, lordosis, and cleft palate. The arthrogrypotic defects in pigs given *N. glauca* were similar to those reported by Crowe (1978) in pigs from dams that consumed tobacco stalks, except that palate closure defects noted with *N. glauca* consumption (Keeler and Crowe 1983) are not seen following maternal consumption of tobacco. About 99% of the total alkaloid in *N. glauca* is anabasine (Keeler *et al.* 1981B); this alkaloid is also found in tobacco. It meets the criteria for teratogenicity of coniine analogs established by Keeler and Balls (1978) and is the probable teratogen in both *N. glauca* and tobacco. It has been shown to be teratogenic in swine (Keeler *et al.* 1984).

INDOLE ALKALOIDS

The most important livestock-poisoning alkaloids of this class are the ergot alkaloids produced by fungi which parasitize the seed heads of various grasses and grains.

5. ALKALOIDS 121

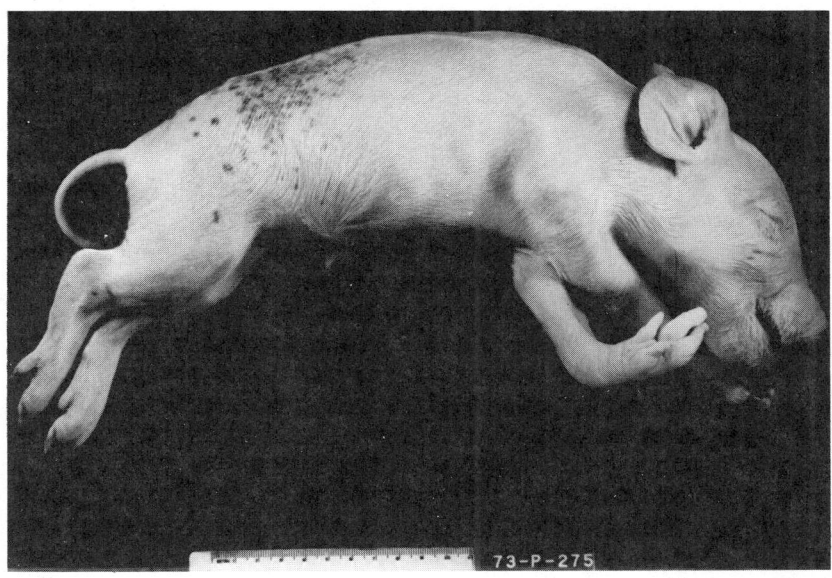

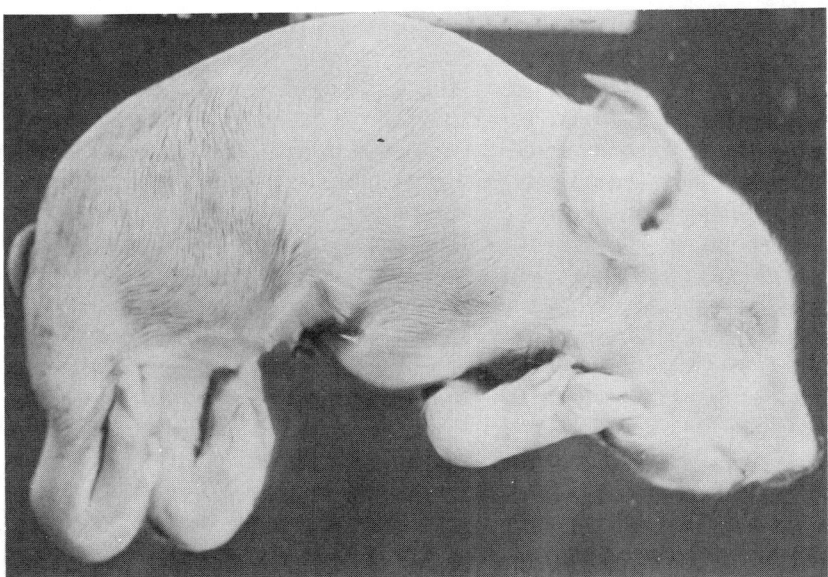

FIG. 5.19. Arthrogryposis in pigs from a sow that consumed tobacco stalks during gestation.
Courtesy of M. Ward Crowe.

The term "ergot" is used in general as a common name for species of the *Claviceps* fungi. The three most significant *Claviceps* are *C. purpurea, C. paspali,* and *C. cinerea. Claviceps purpurea* grows on rye, wheat, barley, and several wild grasses; *C. paspali* grows on grasses of the *Paspalum* species (e.g., Dallis and Bahia) and *C. cinerea* parasitizes several other grasses (e.g., tobasagrass). Ergot also specifically refers to the sclerotium formed by *C. purpurea* when it grows on rye (*Secale cereale*). Ergot is used for medicinal purposes such as controlling hemorrhaging at childbirth.

Consumption of ergotized grain was responsible for epidemics of human ergotism in Europe in the Middle Ages. Ergot poisoning of livestock from consumption of infected grain and grazing pastures containing grasses with ergotized seed heads is relatively common. Ergot infection tends to be greatest around the edges of a field because wild grasses near cultivated ground may harbor spores and infect the grain.

Historical Aspects of Ergotism

Ergotism in humans was a major problem in France from the ninth to the fourteenth centuries. Convulsive ergotism was manifested by itching, numbness, severe muscle cramps, sustained spasms and convulsions, and extreme pain. A foot, leg, or less commonly, an arm would be affected, with the victim experiencing feelings of cold alternating with severe burning sensations (St. Anthony's fire). Numbness and dry gangrene followed, with loss of fingers, hands, feet, and even entire limbs. The epidemics decreased due to changes in farming practices. Wheat replaced rye as the major grain crop and was much less susceptible to ergot infection. The advent of deep plowing resulted in the sclerotia being buried so they did not successfully germinate and form spores.

The general effects of ergot on livestock can be categorized as follows:

1. behavioral effects—convulsions, incoordination, lameness, difficulty in breathing, excessive salivation, and diarrhea;
2. dry gangrene of the extremities;
3. reproductive effects—abortion, high neonatal mortality, reduced lactation; and
4. reduced feed intake and weight gains.

These effects are not all seen in all types of livestock; they are fairly species specific and are modified by the ergot source, amount consumed, period of exposure, and age and stage of production of the animal.

Life Cycle of Ergot

The *Claviceps* spp. parasitize the ovary of the developing grass flower. The fungus sends filaments throughout the ovary tissue and prevents the development of the seed. At the tips of the filaments spores are formed which are shed in droplets of sticky exudate, or "honeydew." Insects spread the spores to infect additional grass heads. At the same time as spores are produced, the filaments harden into a pink or purple structure which replaces the grain or grass seed. This hard structure is called the sclerotium or ergot and is the poisonous stage of the *Claviceps* life cycle (Fig. 5.20). The sclerotia are either harvested with the seed head or are shed to overwinter and produce spores the following spring. These spores begin the infection of the crop again in the second season.

The extent to which grass or grain is infected by ergot depends on numerous factors, so the severity of infection is of high seasonal variability. A cool damp spring favors germination of sclerotia and, by delaying pollination, extends the period of plant susceptibility to infection. Rye and triticale are more susceptible than other grains because they are cross-pollinators, requiring a longer period to become polli-

FIG. 5.20. Ergot infestation of triticale seed heads.
Courtesy of R. J. Metzger.

nated than wheat, oats, and barley. Shallow cultivation and seeding leave the sclerotia near the soil surface where they germinate readily. "No-till" farming, and lack of crop rotation increase the likelihood of ergot infection of crops.

Structures of Ergot Alkaloids

Ergot alkaloids have lysergic acid as their basic component [lysergic acid diethylamide (LSD), the hallucinatory drug, is an ergot derivative]. The major toxic alkaloid groups in ergot are ergotamine, ergonovine, and ergotoxine. Ergotamine and ergotoxine are polypeptide derivatives of lysergic acid. Ergotoxine is a mixture of three alkaloids. The structures of lysergic acid, ergotamine, and ergonovine are shown in Fig. 5.21.

Effects of Ergot on Livestock

Ergot alkaloids have a direct stimulatory effect on smooth muscle, causing vasoconstriction and elevated blood pressure. During the third trimester of pregnancy, ergot has an oxytocin-like effect; the uterine muscle at this stage is more sensitive to ergot than other smooth muscles.

Two general forms of ergotism are the convulsive and gangrenous forms. The gangrenous form affects all types of livestock. The extremi-

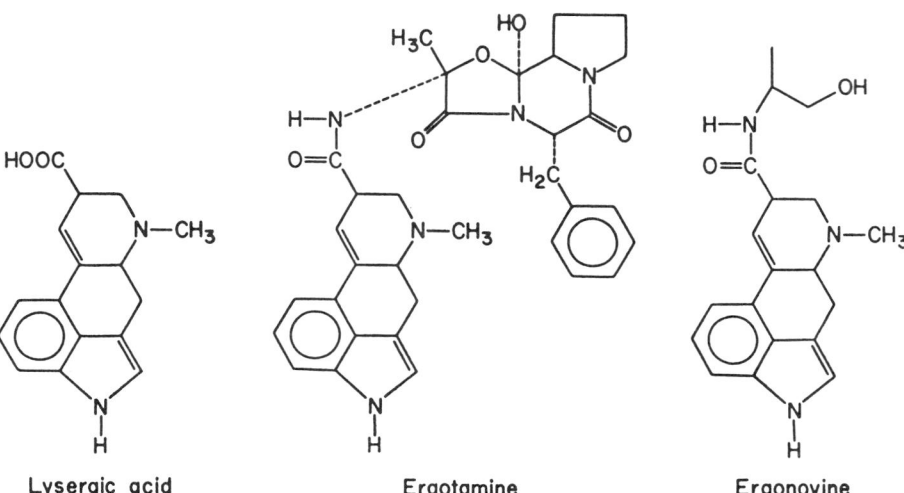

FIG. 5.21. Structures of ergot alkaloids.

ties (nose, ears, tail, and limbs) are affected due to vasoconstriction of arterioles. The early signs are manifested by evidence of pain, such as lameness, and stamping of the feet. The affected areas may feel cool. A sharply defined band encircles the limb, separating the normal tissue from the affected area. Dry gangrene follows, with the affected portion sloughing off, leaving a clean, rapidly healing surface. The digestive tract may also be affected, with inflammation, internal bleeding, vomiting, and constipation or diarrhea.

Cattle may exhibit both convulsive and gangrenous ergotism. The convulsive form is primarily with *C. paspali* infection of *Paspalum* spp. grasses (e.g., Dallis grass), and not with *C. purpurea* ergot. The most important ergot problem in livestock in the U.S. is due to cattle consumption of ergotized Dallis grass. The clinical signs are hyperexcitability, belligerency, incoordination, convulsions, and opisthotonus (star-gazing posture). Removal of cattle from the affected pasture usually results in recovery in 3–10 days. Gangrenous ergot in cattle is caused by ergot from both *C. paspali* and *C. purpurea*. Gangrene of the ear tips and tail may occur, but generally it is the feet that are affected. Signs include tenderness of the hind feet, with development of gangrene and sloughing of the hooves. There is little effect on reproduction, with abortion and agalactia (which are seen in swine) not observed.

Sheep which consume *C. purpurea* ergot show breathing difficulty, excessive salivation, diarrhea, and internal bleeding within the digestive tract. Sheep tend not to graze on grass flowers and therefore are less affected than cattle due to a difference in grazing behavior. Horses grazing grasses infected with *C. paspali* may develop symptoms of convulsive ergotism.

Classical signs of convulsive and gangrenous ergotism are usually not seen in swine. Abortions may occur, and newborn pigs have a high rate of mortality due to depressed lactation by affected sows. Swine are less sensitive than other livestock to ergot. Growing pigs fed ergotized grain may have reduced feed intake and lowered rate of gain, with some gut lesions. Poultry develop comb gangrene as a major symptom.

Ergot alkaloids are not transferred in the milk of cows consuming ergot.

Treatment and Prevention of Ergotism

Removal of animals from ergot-infected pastures or removal of contaminated grain from feeds is the only effective treatment. Ergot infestation of grain fields can be minimized by using clean seed, crop rotations, and deep cultivation. Growing ergot-resistant grains (wheat, barley, or oats) rather than rye or triticale may be advisable in areas where ergot is a problem. Sclerotia can be removed from grain by standard seed-cleaning techniques. Of course, the screenings from ergotized grain should not be used in feeds. Infected grain can be blended with clean grain to reduce the ergot concentration to a nontoxic level. The tolerance level for ergot in grain in the U.S. is 0.3% crude ergot alkaloid. Levels of 0.1% ergot in complete feeds may have adverse effects on livestock performances.

Ergotism Associated with *Balansia* Fungi

Porter *et al.* (1977) have reported that a group of parasite endophytic fungi (endophytes are parasitic organisms living within the body of the host plant) belonging to the Balansiae tribe of the Clavicipitaceae family may be responsible for livestock toxicities which in the past have been attributed to other sources. Some toxicities, such as fescue poisoning, Bermuda grass toxicity, Dallis grass poisoning, and ryegrass staggers, attributed to toxins in forage grasses, may in some cases be caused by *Balansia* infection of pasture weeds. There are 17 species of *Balansia,* including 13 endemic to the U.S. and other Western Hemisphere countries (Diehl 1950). *Balansia* infection of grasses is not obvious, as these fungi do not cause signs of infection such as the sclerotia which are characteristic of *Claviceps* infection. These endophytes inhabit the leaf tissues and may cause only minor symptoms, such as dwarfness, lack of flower and seed production, and deformation of the flag leaf. *Balansia*-infected grasses are sterile. About 125 grass species in the Western Hemisphere are subject to *Balansia* infections; most of these are weedy species of low forage quality, with the exception of Bermuda grass. In tall fescue pastures in the U.S., smut grass (*Sporobolus poiretii*), love grass, panic grass, and broomsedge are often parasitized by *B. epichloe.*

Balansia spp. produce ergot alkaloids identical to those produced by *Claviceps* spp.; these alkaloids are not localized in the seed head but rather are uniformly distributed in the leaves and stems. The production of the alkaloids is host-dependent because not all grass species infected by *Balansia* contain alkaloids. Bacon *et al.* (1979) have developed a test for the rapid evaluation of the presence of ergot alkaloids in *Balansia*-infected grasses to determine whether they have the potential of being toxic to livestock. Signs of *Balansia*-induced toxicity are identical to those caused by consumption of *Claviceps* alkaloids. Wallner *et al.* (1983) observed that *B. epichloe* alkaloids depressed serum prolactin levels in lactating cows and suggested that cows grazing on weeds infected with *B. epichloe* during gestation could have depressed milk production following parturition.

QUINOLIZIDINE ALKALOIDS

Quinolizidine alkaloids are found in lupines, *Cytisus* (Scotch broom), and *Laburnum* (golden chain tree). The basic quinolizidine nucleus is as follows:

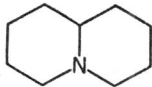

Lupines

Lupines (*Lupinus* spp.) (Fig. 5.22) are of interest as crop plants and as toxic range forages. In recent years, there has been extensive devel-

FIG. 5.22. An example of a range *Lupinus* sp.
Courtesy of R. F. Keeler.

opment of lupines as grain legumes, particularly in Australia. The lupine seed, containing about 35–40% crude protein, is used as a protein supplement for livestock and for direct human consumption. The crop is a winter annual, being sown in the fall and harvested in late spring or early summer. In the U.S., lupines are occasionally grown as a grain crop in areas where soybeans do not thrive.

The cultivated lupines such as *L. luteus* and *L. albus* are generally of European origin; *L. mutalilis,* another edible variety, originated in the Andean region of South America. The crop varieties have been selected for low alkaloid content. Two of the quinolizidine alkaloids found in these plants, lupanine and sparteine, are the most bitter of all alkaloids. Thus, the unselected types high in alkaloids are referred to as bitter lupines, while the selected varieties are called sweet lupines. The pharmacological effects of the alkaloids in humans include nausea, respiratory and visual disturbances, ataxia, progressive weakness, and coma. Because new varieties are of the sweet lupine type, the alkaloids in crop lupines are of minor concern as toxic agents. Another toxicological problem associated with the crop lupines is the condition of lupinosis in livestock grazing lupine stubble. This is caused by hepatoxic fungal metabolites produced by fungi that grow on the stalks and seed pods. Lupinosis will be discussed in detail in the section on mycotoxins (Chapter 12).

In the western range areas of the U.S., wild lupines have been responsible for great losses of range sheep. The principal toxic species are *L. leucophyllus* (woolly leaved lupine), *L. leucopsis* (big bend lupine), *L. argenteus* (silvery lupine), and *L. sericeus* (silky lupine). The greatest concentration of alkaloids is in the seeds. The preflowering plants are generally low in toxicity and provide good forage. Sheep are the main livestock affected; lupine poisoning causes greater losses of sheep than any other poisonous plant in Montana, Idaho, and Utah (Kingsbury 1964). Consumption of a large quantity of lupine in a short period of time disposes sheep to poisoning. A lethal dose is 0.25–0.5% of body weight for seeds and about 1.5% of body weight for pods and seeds. Symptoms appear within a few hours. The breathing is heavy and labored, often with snoring, and the animal becomes comatose and dies. Sometimes there may be trembling and convulsions. Death is from respiratory paralysis. Lupine poisoning can be controlled by good animal management, avoiding conditions under which large amounts of lupine would be consumed in a short period of time. These would include avoiding moving hungry sheep through heavy stands of lupine and avoiding unloading or bedding down sheep in areas dominated by lupines. Cattle and horses are less commonly poisoned, probably because they find lupine pods less palatable and because they are not herded and so are less likely than sheep to be exposed to and consume a large quantity at one time.

The specific alkaloids in wild lupines responsible for losses of range sheep have not been well identified. They may include lupanine and sparteine.

Crooked Calf Disease. In the western range areas of the U.S., a high incidence of skeletal deformities in calves has been observed in some years. This has been referred to as crooked calf disease (Fig. 5.23). The condition is characterized by twisted or bowed limbs (arthrogryposis), twisted neck, spinal curvature, cleft palate, or a combination of these conditions. Researchers at the U.S. Department of Agriculture (USDA) Poisonous Plants Laboratory in Logan, Utah identified consumption of lupine by pregnant cows as the causative factor in most cases. *Lupinus sericeus, L. caudatus,* and *L. laxiflorus* are high in the causative alkaloids. The teratogenic condition may develop if cows consume these lupines between days 40 and 70 of pregnancy. The principal alkaloid responsible for crooked calf disease is anagyrine:

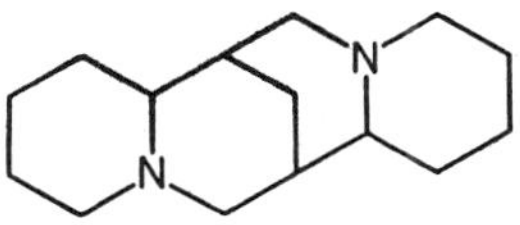

The concentration of anagyrine is high in the early stages of growth of the plant and is also high in mature seeds. A pregnant cow is at greatest risk when grazing teratogenic lupine early in the plant growth phase or when seed pods have formed. Crooked calf disease can be avoided by cattle and range management through altering breeding schedules or grazing rotations so that cows are not exposed to lupines high in anagyrine when they are from 40- to 70-days pregnant. Sheep are not affected by the teratogenic properties of anagyrine, even when high levels of lupine are fed to pregnant ewes (James 1977).

The U.S. Soil Conservation Service has developed a variety of nontoxic sickle-keeled lupine (Hederma) for use in reseeding roadsides, eroded areas, and rangelands to avoid toxicities in wildlife and grazing livestock.

Kilgore *et al.* (1981) reported an apparent case of human teratogenicity associated with lupine alkaloids in northern California. A baby boy with severe bone deformities in his arms and hands was born to a woman who had regularly consumed goat's milk during her pregnancy. The goats were grazing in an area where *Lupinus latifolius* formed the principal forage available. Goat kids from these goats were deformed, and puppies from dogs fed the goat's milk also showed de-

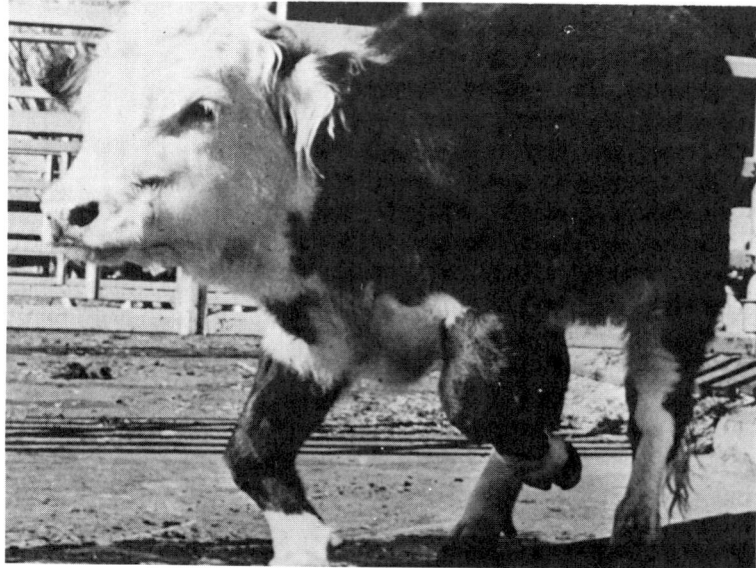

FIG. 5.23. Crooked calf disease caused by maternal consumption of lupine.
Courtesy of J. L. Shupe.

formities. Analysis of the lupine forage showed it to contain a very high concentration of anagyrine. When the lupines were fed to a lactating goat, anagyrine was detected in the milk. While it is impossible to prove that the deformities in the baby were due to lupine alkaloids transferred to the mother through consumption of goat's milk, the circumstantial evidence strongly implicated that etiology. The parents initially attributed the problem to the spraying of local forests with herbicides.

Other Plants with Quinolizidine Alkaloids

The golden chain tree or laburnum (*Laburnum anagyroides*) is a small ornamental tree grown in parks and gardens. It has long hanging racemes of yellow flowers followed by pods containing small seeds. The seeds contain the alkaloid cytisine. Livestock and humans have been fatally poisoned from consumption of the seeds. The tree is not recommended as an ornamental in areas such as public parks where children may play. Symptoms of poisoning include excitement, incoordination, convulsions, coma, and death.

Scotch broom (*Cytisus scoparius*) is a leguminous weedy shrub native to Europe. It has been widely introduced to various parts of the world such as the east and west coasts of North America, Australia, and New Zealand. It contains the toxic quinolizidine alkaloids sparteine and isosparteine. Livestock poisonings are very rare, and in many areas the plant provides good feed, particularly for sheep.

Mountain thermopsis (*Thermopsis montana*), also referred to as poison bean plant and false lupine, contains two principal alkaloids, anagyrine and thermopsine. It has been implicated in cattle poisoning (Chase and Keeler 1983). Signs of toxicity include loss of appetite, a humped-up appearance with drawn-up flanks, swollen eyelids, and depression. Thermopsis is a perennial legume that grows in high mountain meadows and along streambanks in the intermountain range areas of the western U.S. The young plants are high in alkaloid content but are unlikely to be hazardous because other forage is usually abundant. The seeds are high in alkaloids and are the major source of poisoning, since other desirable forage is often scarce during the seed-set period. Poisonings can be prevented by ensuring that cattle have access to sufficient good-quality forage.

STEROID ALKALOIDS

Solanum Type

Solanum spp. such as nightshades, potatoes, the Jerusalem cherry (a Christmas ornamental), as well as the closely related tomato, contain steroid alkaloids. Solanine, a glycoalkaloid in potatoes, was discovered in 1826. Solanine is a glycoside containing a steroid alkaloid nucleus (the aglycone) with a side chain of sugars. Thus, glycoalkaloids are glycosides of alkaloids. The aglycone is called solanidine. In 1954, another glycoalkaloid, chaconine, was discovered in potatoes (Jadhav *et al.* 1981). The structures of these compounds are shown in Fig. 5.24. Solanine and chaconine have the same aglycone (solanidine) but differ with respect to the carbohydrate side chain.

The two principal effects of solanum alkaloid poisoning are gastrointestinal tract irritation and impairment of the nervous system. The alkaloids are much more toxic when administered parenterally than when given orally because of relatively low gastrointestinal absorption (Nishie *et al.* 1971). The glycosides (e.g., solanine) are more toxic than the aglycones (e.g., solanidine). Apathy, drowsiness, salivation, dyspnea (difficult or labored breathing), trembling, weakness, paralysis, and loss of consciousness are manifestations of the effects on the nervous system. Solanum alkaloids are cholinesterase inhibitors (Jadhav,

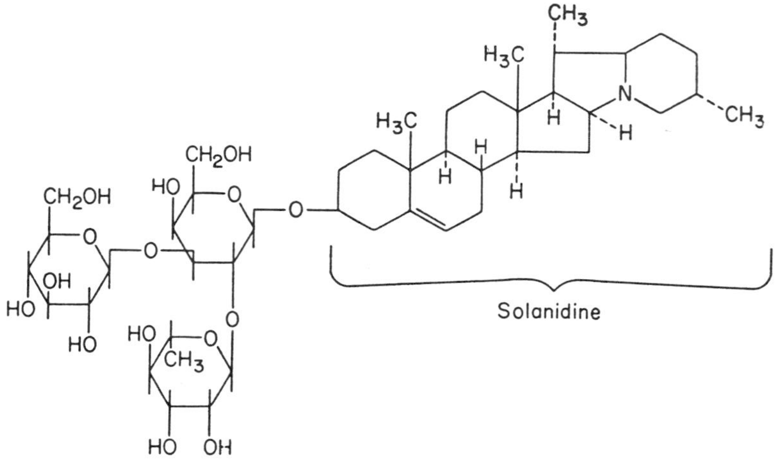

α-Solanine

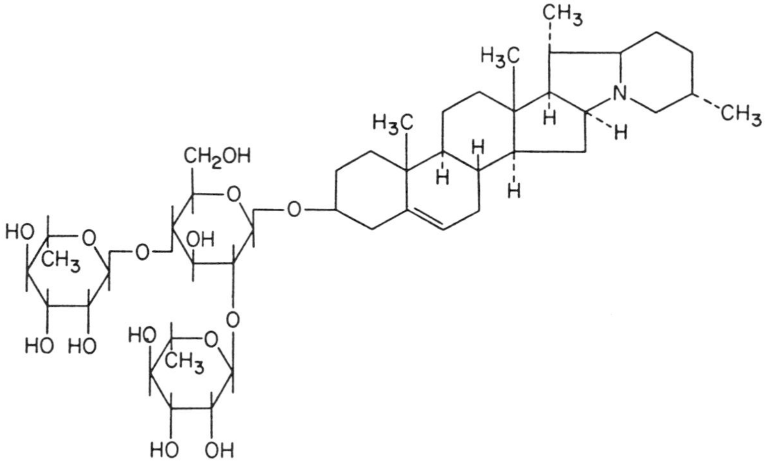

α-Chaconine

FIG. 5.24. Structures of solanidine, solanine, and chaconine.

et al. 1981), which explains their neural effects. Acetylcholine is the neurotransmitter formed at the terminals of all preganglionic nerve fibers and at the endings of the postganglionic fibers of the parasympathetic nerve system and mediates the transfer of impulses from somatic motor nerves to skeletal smooth muscle (neuromuscular junctions). Inhibition of acetylcholinesterase results in the accumulation of acetylcholine in nerve tissue and effector organs, with the result that neural function is impaired. Neurological signs, such as ataxia, convulsions, coma, muscle weakness, and involuntary urination are the consequence. Besides solanum alkaloids, common pesticides such as organophosphates are cholinesterase inhibitors.

Gastrointestinal tract effects of solanum alkaloids are manifested by inflammation of intestinal mucosa, hemorrhage or ulceration, abdominal pain, and constipation or diarrhea.

Teratogenic effects of potato alkaloids have been suggested, but a substantial body of evidence has indicated that the solanum alkaloids are probably not teratogens (Jadhav et al. 1981).

Poisoning of both livestock and humans from potatoes has occurred. The solanum alkaloids are in highest concentration in green sprouts and green potato skins (peels). The greening of potatoes occurs when the tubers are exposed to sunlight during growth or after harvest. The green pigment is chlorophyll; the increased concentration of solanum alkaloids in green potatoes is due to the fact that similar environmental conditions promote development of both chlorophyll and glycoalkaloids (Jadhav et al. 1981). Jadhav and his co-workers, as well as Kingsbury (1964), cite examples of human deaths from consumption of green potatoes. Livestock have been poisoned after being fed potato sprouts, peelings, and sunburned or spoiled potatoes. Potato vines have also caused toxicity, since the alkaloids are in highest concentration in green tissues. The potato alkaloids are not destroyed as a result of boiling, baking, frying, or drying at high temperatures (Jadhav et al. 1981). The relatively rare occurrence of solanine poisoning has been explained by Nishie et al. (1971) as the result of the following factors: (1) solanine is poorly absorbed; (2) solanine is hydrolyzed to a considerable extent in the gut to the less toxic solanidine; and (3) there is a rapid urinary and fecal excretion of metabolites.

In the rumen, potato glycoalkaloids are hydrolyzed to solanidine, which is further metabolized to a 5,6-dihydro analog of solanidine (King and McQueen 1981).

An interesting consequence of solanum alkaloids was the withdrawal of a new potato variety, Lenape, in the late 1960s. This variety had excellent characteristics for french fry production, but after commercial production began, it was found to have a toxic level of solanine. New potato varieties are now screened before release, and glycoalkaloid levels must be below 20 mg/100 g. Levels above about 14 mg/100 g are bitter, while levels above 20 cause a burning sensation of the mouth and throat. Lenape had a level of about 30 mg/100 g (Jadhav et al. 1981).

Nicholson et al. (1978) reported a study on the potential feeding value of potato vines. According to them, potato growers often wish to harvest their crop before the vines have died to minimize the transfer of viruses to the tubers. Conventional potato harvesters cannot handle a large bulk of vines; the usual practice is to remove the vines with mechanical beaters or to kill them with chemicals. Either operation is expensive and returns no revenue. Harvesting the vines for livestock feed would provide a use for them and utilize the nutrients they contain. Nicholson *et al.* (1978) concluded that potato vines harvested prior to senescence are nontoxic and provide a useful level of nutrients to ruminants. Possible pesticide residues are the primary concern, and a modified pest management system would be needed for vines intended for livestock feed.

Black nightshade (*Solanum nigrum*) (Fig. 5.25) is a common annual weed that grows from 1 to 3 ft in height. It has white flowers and shiny black berries. It is a common garden weed and also invades forage crops such as new seedings of alfalfa, and cereals. Mortality has been reported for cattle, sheep, swine, horses, chickens, and ducks (Kingsbury 1964) from consumption of the berries or from grazing the plant. Nightshade in stubble fields may be grazed in preference to the dry stubble.

FIG. 5.25. Examples of nightshades (*Solanum* spp.). Black nightshade (flowers and immature berries) on left and climbing nightshade (red berries) on right.

Climbing nightshade (*Solanum dulcamara*) (Fig. 5.25) is a climbing or trailing perennial reaching 6 ft in length. The flowers are blue or purple, followed by large red berries. Climbing nightshade has been implicated in poisoning of livestock and children (Kingsbury 1964). Silverleaf nightshade (*Solanum eleagnifolium*) is a common perennial weed in the U.S. southwestern states. Considerable losses of cattle from consumption of the plant have occurred (Kingsbury 1964). Ripe berries produce moderate to severe poisoning when ingested at 0.1–0.3% of body weight.

A cultivated vegetable plant, the garden huckleberry, is apparently a domesticated type of black nightshade. The berries are used for pies and are nontoxic. The wild and domestic plants do not interbreed.

Tomatoes contain a toxic solanum alkaloid called tomatine (aglycone is tomatidine). The structure of tomatine, first identified in 1948, is provided by Jadhav *et al.* (1981). The mature fruit is low in alkaloid, but green tomatoes and the vines contain appreciable quantities. Cattle and pigs have been poisoned when fed tomato vines. Tomatine is a cholinesterase inhibitor.

A number of *Solanum* spp. such as *S. fastigiatum, S. kwebense,* and *S. dimidiatum* cause cerebellar degeneration in cattle, characterized by progressive vacuolation and degeneration of Purkinje cells. The disorder is apparently a lysosomal storage disease similar to that caused by *Swainsona* and *Astragalus* spp. The *Solanum*-induced disorder is discussed in the section Indolizidine Alkaloids (this chapter).

Veratrum-Type Alkaloids

Veratrum californicum (false hellebore) (Fig. 5.26) is a tall, coarse, erect herb that grows in moist habitats in mountain valleys of the U.S. Pacific Northwest and Rocky Mountain range states. Investigators at the USDA Poisonous Plant Laboratory demonstrated that consumption of this plant by pregnant ewes on the fourteenth day of pregnancy resulted in production of "cyclops" or "monkey-face" lambs. This condition had occurred in western range areas, particularly in Idaho, in epidemic proportions. In southwest Idaho the incidence of the condition varied from less than 1% to nearly 25% of the lambs born in a given band of sheep. The lamb deformities vary from slight deformities of the upper jaw to a complete cyclops condition, with one centrally placed eye (Figs. 5.27 and 5.28).

A ewe carrying a deformed lamb(s) due to *Veratrum* may often have a prolonged gestation period. The deformed lamb continues to live and grow in utero and may reach a weight of 20–30 lb. Eventually both the lamb and ewe die.

FIG. 5.26. Leaves and flowers of *Veratrum californicum*.
Courtesy of M. E. Fowler.

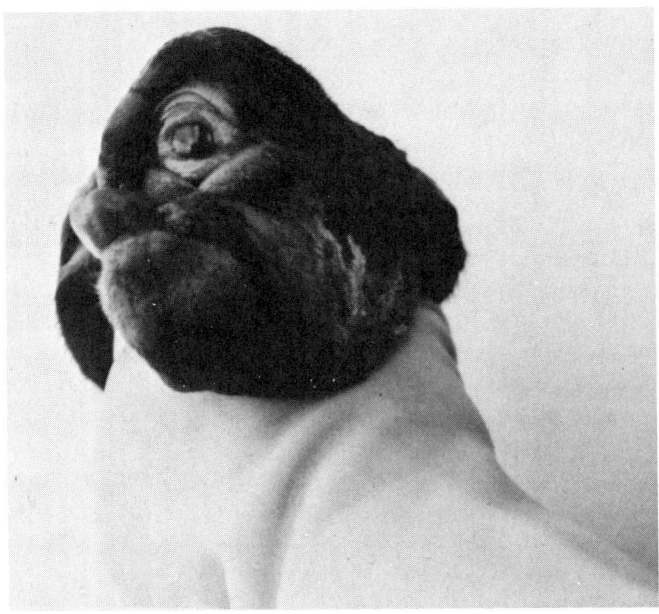

FIG. 5.27. A cyclops lamb from a ewe that had consumed *Veratrum californicum* in early gestation.
Courtesy of R. F. Keeler.

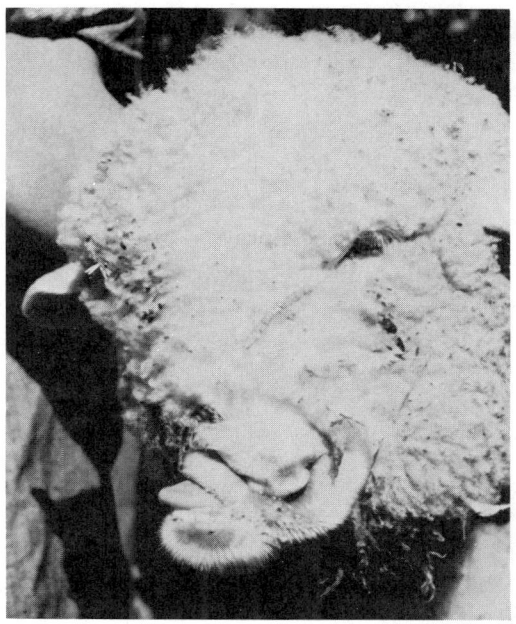

FIG. 5.28. Deformity in a sheep caused by maternal consumption of *Veratrum californicum.*
Courtesy of M. E. Fowler.

When the veratrum is consumed by the ewe on days 28–30 of pregnancy, the lamb will have shortened metatorsal and metacarpal bones, resulting in an abnormal squat appearance. A cleft lip and palate may also occur.

Other animals such as cattle and goats are affected by the teratogenic properties of veratrum when it is administered experimentally. Field cases are seen only with sheep.

Veratrum spp. contain over 50 individual steroid alkaloids. Veratramine is one found in high concentration. Five of the alkaloids found in largest concentrations in *V. californicum* are shown in Fig. 5.29.

The compounds that are active in producing teratogenicity are cylopamine, cycloposine, and jervine. Other veratrum alkaloids are not teratogenic. The fused rings with an oxide bridge appear necessary for activity. Prevention of cylops lambs can be accomplished by waiting until after the first frost before breeding ewes; veratrum is killed by frost and loses its toxicity.

The veratrum alkaloids have pharmacological effects with marked hypotensive properties. Lowering of blood pressure is caused by dilation of arterioles with constriction in venous vascular beds and a slowing of heart rate. The alkaloids have been used for medicinal purposes. Sheep sometimes consume toxic quantities of *Veratrum;* signs of intox-

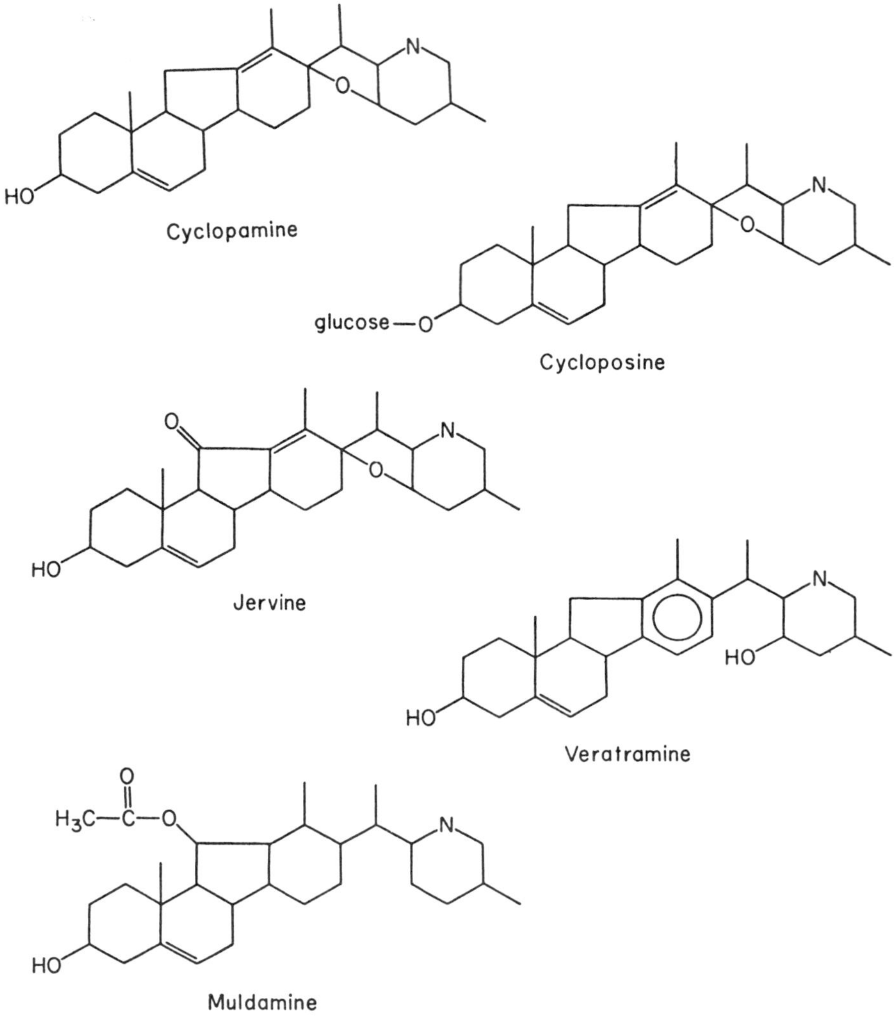

FIG. 5.29. Structures of five steroidal alkaloids from *Veratrum californicum*.

ication include salivation, prostration, depressed heart action, weakness, and dyspnea.

Death camas (*Zigadenus* spp.) (Fig. 5.30) contains veratrum-type alkaloids. Death camas resembles a wild onion in appearance, with a small bulb and grasslike leaves. Various spp. grow over the western U.S. Death camas is one of the major poisonous plants affecting range

FIG. 5.30. Death camas (*Zigadenus paniculatus*).
Courtesy of W. C. Krueger.

sheep. The plant begins growth in early spring and may be consumed because of the lack of other forage. All parts of the plant are toxic. The toxic bulbs may be pulled up and eaten when the ground is wet. Individual herd losses of 500–2000 sheep have occurred (Kingsbury 1964). The lethal dose varies from about 0.6 to 6% of body weight. Death camas contains several steroid alkaloids, including zygacine:

POLYCYCLIC DITERPENE ALKALOIDS

The polycyclic diterpene alkaloids occur in larkspurs (*Delphinium* spp.) and in *Aconitium* spp. (aconite or monkshood). These genera are similar in appearance, habitat preference, and toxic effects. Various species of *Aconitium* are found in western range areas of the U.S. They are much less common than larkspurs and do not appear to cause significant losses of livestock (Kinsbury 1964). Larkspurs, on the other hand, are responsible for large cattle losses. Since before 1900, and to the present, larkspurs have been considered to cause more cattle losses in the U.S. than any other poisonous plant. Cronin and Nielsen (1979) indicate that more cattle deaths are attributed to larkspurs than to all other poisonous plants combined. An intensive study has been made of the 20,000-acre Manti Canyon Cattle Allotment in the Wasatch Plateau of central Utah. Up to 870 head of cattle are allowed on the allotment each year. From 1956 to 1970, annual losses from larkspur poisoning averaged 36 cows, or 4% of the herd (Cronin and Nielsen 1979). Calf losses were about 1%. The average annual loss of cows and calves was $10,200. Addition of other economic factors, such as extra labor for cattle management in larkspur-infested areas, brought the average annual losses from larkspur to over $14,000 on an 870 cow herd. This is probably typical of the larkspur problem in much of the western U.S.

The larkspurs are divided arbitrarily into two groups: low larkspurs and tall larkspurs. This refers specifically to their growth habits; low larkspurs are usually under 3 ft in height, while tall larkspurs average from 3 to 6 ft or more. The low larkspurs are found at lower elevations, while tall larkspurs are found in high mountain regions (Fig. 5.31). *Delphinium barbeyi, D. occidentale, D. glaucum,* and *D. trolliifolium* are principal species of tall larkspur, while *D. andersonii* and *D. menziesii* are major low larkspur types. The low larkspurs tend to grow at lower elevations where moisture is quite limiting in the summer; they begin growth very early in the spring, sometimes growing through snow. Cattle losses occur at this time when animals are seeking out new green herbage. The tall larkspurs grow at high elevations in deep soils where moisture is abundant. They are found in open meadows or under dense tree cover such as aspen and conifer groves. They are climax species and increase with improving range conditions. They are quite palatable to cattle.

Cattle are moved from low elevations in the spring to high mountain ranges in the summer. They are exposed to a progression of various spp. of low and tall larkspurs as they make this annual migration. Calves are often poisoned in the mountain ranges. They tend to remain

FIG. 5.31. A single specimen of tall larkspur on a mountain range.
Courtesy of A. P. Appleby.

in groves of conifers or aspen with a "babysitting cow" while most of the cow herd is out grazing in meadows. The calves may graze on small patches of tall larkspur in the tree groves and because of their small size, can easily ingest a lethal dose.

Larkspur toxicity symptoms are rarely seen in range livestock because of the quick-acting effects of the alkaloids. Rapid bloating occurs as the animals often fall down with their head pointing downhill, resulting in an accumulation of rumen gases. Cattle have been experimentally poisoned (Olson 1978A), and the toxic signs are those of impairment of the nervous system. Signs include uneasiness, stiffness of gait, a straddled stance, with the hindlimbs far apart, followed by collapse. Signs of nausea, abdominal pain, weakness, and involuntary muscle twitching are observed. Death occurs from respiratory paralysis. Sheep are less susceptible than cattle; it takes about four times as much larkspur per kilogram body weight to kill sheep as cattle. The lethal dose in cattle is about 17 g of green plant per kilogram body weight (Olson 1978A), or about 0.5% of body weight.

The resistance of sheep to larkspur has led to suggestions that infested areas be grazed with sheep rather than cattle. According to Cronin and Nielsen (1978), this method is not generally effective.

Sheep normally graze larkspur only late in the season, select portions of the plant seldom used by cattle, and would not reduce larkspur populations without severe damage to other vegetation. In some areas of Utah where sheep have been used, some poisoning and losses of animals have been observed. Even under heavy sheep grazing, most larkspur plants survive and set seed each year.

The toxic alkaloids are of a complex diterpene structure:

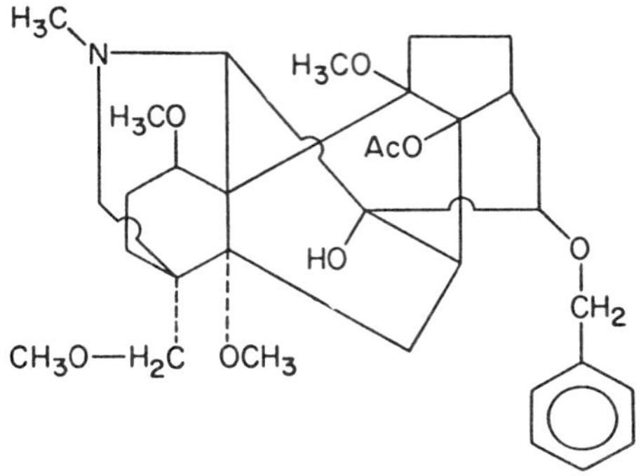

Studies with methyllycaconitine, an alkaloid in *Delphinium brownii*, have indicated that it functions as a potent neuromuscular blocking agent (Aiyar *et al.* 1979). Presumably the other diterpenes in larkspur act in this manner also. The alkaloids are found in highest concentrations in the early plant growth, in growing tips and new leaf growth. In general, the plants are most toxic in the spring and lose toxicity throughout the growing season, particularly after flowering. Losses can be reduced by managing cattle so that they are not allowed to graze larkspur-infested areas until after the plants have flowered. This is easier said than done, and also would result in overgrazing of lower elevation ranges. The most promising management technique appears to be spot spraying of larkspur patches with herbicides, followed by efforts to revegetate with other types of plants. Conversion of a larkspur-dominated plant community to one dominated by grass improves the range for livestock and wild animals and the quality of a watershed.

INDOLIZIDINE ALKALOIDS

Locoweed and *Swainsona*

Locoism, caused by the consumption of certain *Astragalus* and *Oxytropis* spp. by cattle, horses, and sheep on western rangelands, has been recognized for over 75 years as being very similar physiologically to *Swainsona* spp. poisoning of livestock in Western Australia. The toxic factor(s) in both cases remained elusive until 1979 when researchers at Murdoch University in Western Australia identified indolizidine alkaloids as the causative agents (Colegate et al. 1979).

There are over 50 species of *Swainsona* in Australia, with toxicity problems occurring mainly in Western Australia and the Northern Territory. These plants are found primarily in unimproved range areas where extensive livestock production with year-round grazing is conducted. The plants are perennials and following rains, are often the dominant forage available. Locoweeds in the U.S. are mainly *Astragalus* spp. and a few *Oxytropis* spp. (Fig. 5.32). There are at least 372 *Astragalus* spp. in the U.S., and most are toxic. Some contain poisonous nitro compounds (e.g., miserotoxin), while others accumulate toxic levels of selenium. A third group of about 13 species causes locoism. The symptoms of locoweed and *Swainsona* poisoning are similar and will be discussed collectively.

Consumption of the plants for a few weeks to a month is necessary before obvious signs of intoxication are seen. The principal signs are those of nervous system impairment. These include dullness and depression, excitement when disturbed (loco means crazy in Spanish), loss of sense of direction and of herding instinct in sheep, and habituation to the plants, or an apparent craving for them. Affected animals

FIG. 5.32. *Oxytropis serecia*, a locoweed.
Courtesy of L. F. James.

seek out locoweed. Up to 60% of ewes on locoweed-infested range may abort, while abortion is common in cows consuming *Swainsona*. With locoweed, congenital malformations in lambs and calves are seen. These include a permanent flexure of the carpal joints and contracted tendons (Fig. 5.33). These deformities are observed when ewes ingest locoweed at almost any period of gestation. When locoweed is consumed in the period of 90–120 days of pregnancy by ewes, fetal abnormalities such as enlarged heart, spleen, and thyroid may be observed. Fetal edema and lesions of the placenta may be seen. Limb deformities in foals from mares consuming locoweed have been reported (McIlwraith and James 1982).

Locoweed-poisoned animals seem to be susceptible to infections, such as pneumonia, footrot, and pink eye, suggesting possible impairment of the immune system (Sharma *et al.* 1984). Sharma and his associates reported that sheep fed locoweed had decreases in total leukocytes and lymphocytes, with an indication of a selective effect on cell-mediated immune responses.

Locoweed consumption may result in mortality of affected animals. Signs of toxicity in cattle and sheep are seen after consumption of

FIG. 5.33. Lamb from a ewe fed locoweed (*Astragalus lentiginosus*).
Courtesy of L. F. James.

about 90% of body weight of locoweed (Kingsbury 1964), or about 2 months of exposure. Death may occur after an intake of about 300% of body weight of locoweed. Horses are most susceptible, with intake of about 30% of body weight being a lethal dose. Affected horses are listless and unaware of their surroundings, but when excited, become wild and unmanageable. Horses in the early stages of loco intoxication are dangerous to ride. Deaths of affected livestock may occur from misadventure, such as tumbling over cliffs and banks.

Signs of toxicity from *Swainsona* poisoning are similar to those from North American locoweeds. Habituation to *Swainsona* is also seen. With both locoweeds and *Swainsona*, some differences in neurological signs between sheep and cattle are observed. Cattle have a low head carriage, a stiff clumsy gait, head shaking, and staring eyes. Sheep have a high head carriage, and a high-stepping stiff gait, and also show head tremors, staring eyes, disturbed vision and incoordination.

The predominant microscopic lesions are cytoplasmic vacuoles in organs and nerve tissue that have been identified as swollen lysosomes. Dorling *et al.* (1978) demonstrated that the lysosomes contained high levels of oligosaccharides composed largely of mannose. The lesions and composition of the lysosomes were strikingly similar to those observed in mannosidosis in humans and cattle, which is a lysosome storage disease due to a genetic deficiency of lysosomal α-mannosidase. Subsequent studies by these workers (Colegate *et al.* 1979) showed that *Swainsona* contains indolizidine alkaloids that inhibit α-mannosidase. The major alkaloid involved is swainsonine:

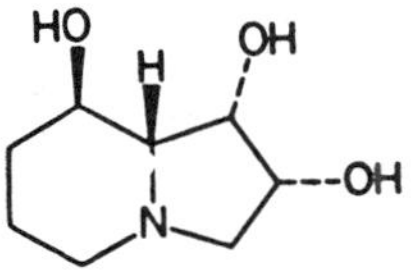

The α-mannosidase acts upon the mannose residues appearing at the terminal nonreducing end of a chain of sugars. Other glycosidases remove their respective sugars from oligosaccharides. If α-mannosidase is inhibited, other glycosidases remove their respective sugars until a mannose residue is reached, and then hydrolysis of the carbohydrate stops. The undigested oligosaccharides accumulate within the lysosomes. The number of lysosomes in the affected cells increases to accommodate the increasing quantity of oligosaccharide, which eventually disrupts cellular function (Huxtable and Dorling 1982). The effect of indolizidine alkaloid on α-mannosidase is reversible, and the accumulation of stored carbohydrate diminishes when alkaloid is no longer consumed. The vacuoles will largely disappear in 10–12 days, so that animals showing early clinical signs generally recover completely. Advanced clinical signs are irreversible. Axon degeneration may result from the crowding of perikaryon with storage vacuoles. Although vacuolation occurs in most tissues, it is the central nervous system that is most sensitive to mannose accumulation, with functional impairment and degenerative changes. After the Australian work, indolizidine alkaloids were shown to be the active compounds in locoweeds (Molyneux and James 1982).

The alkaloids are apparently not detoxified prior to excretion, as urine from sheep fed locoweed is toxic to rats (James and Van Kampen 1976) and milk from cows fed *Astragalus* produced locoism in kittens, lambs, and calves (James and Hartley 1977).

Cattle grazing high-elevation rangelands infested with the locoweed *Oxytropis sericea* have a high incidence of congestive right heart failure (high mountain disease). Calves are most susceptible. Swelling under the jaw and of brisket are observed. L.F. James (1984, personal communication, Logan, UT) observed that at high elevations, feeding locoweed to calves caused congestive right heart failure, while at lower elevations, locoweed poisoning occurred. Because swainsonine influences the vascular system, its effects are probably intensified by high elevations.

Several *Solanum* spp. cause cerebellar degeneration in cattle of a similar nature to that induced by *Swainsona* and *Astragalus* spp. Riet-Correa *et al.* (1983) described the intoxication of cattle on a number of farms in Brazil, which was associated with consumption of *Solanum fastigiatum*. The disorder was characterized by recurrent seizures with loss of equilibrium, extension of the head and thoracic limbs, opisthotonus, rapid eye movements, and falling to the side or backward. The main pathological signs were vacuolation, degeneration, and loss of Purkinje cells, with cytoplasmic inclusion bodies similar to those of induced lysosomal storage diseases. Similar signs and lesions were described for intoxication of cattle by *Solanum dimidatum* in the southwestern U.S. (Menzies *et al.* 1979) and by *Solanum kwebense* in South Africa (Pienaar *et al.* 1976). The toxic factor in these *Solanum* spp. has not been identified, but an inhibitor of a lysosomal hydrolase is a likely possibility.

Slaframine

For many years farmers in the U.S. midwest were aware that clover pasture and hay often induced excessive salivation or "slobbers" in livestock. Red clover hay was particularly implicated; white clover pastures also were a problem. When draft horses were important as work animals, slobbering was an unpleasant part of farming when the horses were on white clover pasture. When they were bridled for work, the farmer was drenched with a cup or more of saliva!

The factor in clover that causes slobbering is a metabolite of a red clover fungal pathogen, *Rhizoctonia leguminicola*. This compound, called slaframine, is an indolizidine alkaloid:

This fungus also produces swainsonine (Huxtable and Dorling 1982).

The problems associated with slaframine have been noted in Missouri, Illinois, Indiana, Ohio, Virginia, Pennsylvania, and Wisconsin. The offending red clover hay, while not visibly moldy, contains a dark brown fungal mycelium. The fungal infestation is known as black patch disease of red clover. It develops most rapidly in rainy weather and in periods of high humidity. The fungus contaminates clover seeds, is spread via the seeds, and may overwinter on clover stubble. In areas of endemic black patch infection, careful monitoring of livestock fed red clover is advisable. Prompt removal of the toxic forage from livestock generally alleviates all signs of intoxication. Degradation of the toxin in hay may occur; Hagler and Behlow (1981) noted that levels of 50–100 ppm slaframine in fresh red clover hay were reduced to about 7 ppm after 10 months of storage.

Biological Effects of Slaframine. Slobber or salivary syndrome which occurs primarily in cattle and horses is a cholinergic action induced by slaframine. The clinical signs in intoxicated cattle, in addition to the excessive salivation, are lacrimation (eye discharge), bloat, frequent urination, and watery diarrhea. Excessive amounts of saliva may appear soon after exposure to slaframine and possibly continue for several days after consumption of the toxin has ceased. Decreased milk production occurs in affected dairy cattle. Weight loss and abortion may occur. Guinea pigs have been used as the primary assay animals. In 10–20 min following ingestion of the compound, guinea pigs experience a salivation episode of 6–8 hr. In cattle, it may last up to 3 days.

Slaframine is extremely toxic; the acute oral LD_{50} in guinea pigs is less that 1 mg/kg of body weight. Toxic effects other than salivation observed in guinea pigs and other species include increased pancreatic flow, bile flow, and gastric acidity; decreased heart rate, cardiac output, respiration rate, body temperature, and metabolic rate; and uterine hemorrhage and fetal abortion. There are no distinguishable lesions or long-term effects resultant from slaframine intoxication. Spontaneous recovery is complete in 2–3 days. Atropine and certain antihistamines are quite effective in alleviating some of the clinical signs of the toxicosis.

Metabolism of the alkaloid is an important factor in slaframine toxicosis. The toxin possesses no biological activity itself; it is activated by enzymes in the liver. The active metabolite is thought to be a quaternary amine that resembles acetylcholine, thereby accounting for its cholinergic activity (Guengerich and Aust 1977).

Advantageous effects of slaframine have been suggested by Froetschel et al. (1984). They proposed that slaframine might have potential therapeutic effects in treating digestive disorders associated with the feeding of high-concentrate, low-roughage diets to ruminants, as a means of increasing salivary flow rate.

Castanospermine

Castanospermine is an indolizidine alkaloid similar in structure to swainsonine and slaframine. It occurs in the leaves, seeds, and bark of *Castanospermum australe* (Moreton Bay chesnut), a tree native to Eastern Australia (Everist, 1981). Livestock poisonings occur during dry periods when forage is scarce and animals consume large quantities of the seeds. Severe diarrhea is the most consistent sign of toxicity, accompanied by general debilitation. Severe inflammation of the digestive tract occurs.

TRYPTAMINE ALKALOIDS

Phalaris spp. such as *P. tuberosa* and *P. arundinacea* (reed canarygrass) contain tryptamine alkaloids. These have been most important with respect to *P. tuberosa* in Australia. Three distinct pathological conditions have been observed in sheep and cattle grazing this forage plant. These are peracute *Phalaris* poisoning (sudden death syndrome), acute *Phalaris* poisoning, and chronic *Phalaris* poisoning, or *Phalaris* staggers. The peracute and acute conditions have been demonstrated to be due to tryptamine alkaloids in the plant, while the cause of *Phalaris* staggers is not yet known.

The peracute and acute *Phalaris* poisoning are due to pharmacological properties of tryptamine alkaloids on the central nervous system and the heart (Gallagher et al. 1964). The most potent alkaloid is 5-

5. ALKALOIDS

methoxydimethyltryptamine, followed in potency by 5-hydroxydimethyltryptamine and dimethyltryptamine (Fig. 5.34). These are very similar in structure to serotonin. The alkaloids are competitive inhibitors of monoamine oxidases, which are important in brain metabolism. Phalaris poisoning results in a buildup of serotonin and catecholamines in brain tissue.

The peracute *Phalaris* poisoning is manifested by sudden collapse and death of affected animals, with death caused by acute heart failure. The acute condition is characterized by hyperexcitability, incoordination, spasms, head nodding, twitching, and salivation. Administration of the alkaloids causes the same symptoms as consumption of *Phalaris* forage. A dosage of 0.05–0.1 mg/kg body weight of 5-methoxydimethyltryptamine produces marked neurological signs; at 0.5 mg/kg the effects are severe, and at 1.5–2 mg/kg, acute heart failure and death occur.

Chronic *Phalaris* staggers has been reported in sheep and cattle in Australia, New Zealand, and South America. Most outbreaks in Australia occur on *P. tuberosa* pastures in late summer and early autumn, within a few weeks of autumn rains and new forage growth. The first clinical signs are incoordination, a stiff stilted gait, muscle tremors, head nodding, and hyperexcitability (Hartley 1978). As the disease progresses, and particularly following disturbance, an affected animal may show severe signs and convulsions. After a rest it may recover and walk away. As the symptoms progress, the animal may become recumbent and die. Gross lesions are seen in the nervous system, consisting

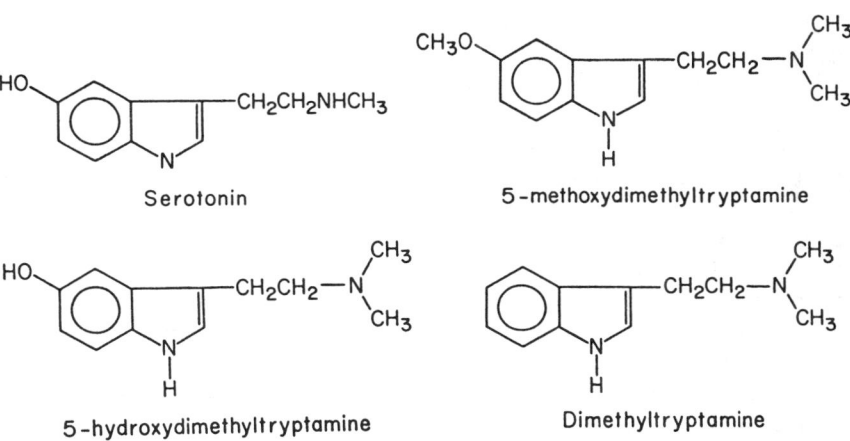

FIG. 5.34. Structures of serotonin and tryptamine alkaloids.

of gray to blue discoloration of the brain stem. The kidney and liver may also be pigmented. Microscopically, the pigment occurs throughout the brain as yellow brown granules in the cytoplasm of nerve cells. The pigment apparently destroys the affected neurons. It has been suggested that it may be a melanin pigment derived from tryptamine alkaloids (Jolly and Hartley 1977).

Administration of dietary cobalt protects against chronic *Phalaris* poisoning to some degree (Hartley 1978).

Reed canarygrass is an important forage grass in Canada and northern parts of the U.S. It thrives in poorly drained soils subject to prolonged flooding. It contains the same tryptamine alkaloids as *P. tuberosa* as well as some additional alkaloids. *Phalaris* staggers in animals grazing reed canarygrass has not been reported in North America, but has been seen in sheep in New Zealand (Hartley 1978). Reed canarygrass is somewhat unpalatable, and often gives animal performance lower than its nutrient composition would suggest. The concentration of tryptamine alkaloids is negatively correlated with palatability of the grass (Simons and Marten 1971), which may explain the effects on animal performance (Fig. 5.35). Reed canarygrass contains at least eight alkaloids. These include four derivatives of tryptamine, gramine (an indole alkaloid), hordenine, and two derivatives of β-carboline (carboline alkaloids). Some of the structures are shown in Fig. 5.36. The alkaloid content seems to peak in late summer, which may reflect moisture stress (Majak *et al.* 1979). Marten *et al.* (1976) observed that average daily gains were reduced and the incidence of diarrhea increased when sheep and cattle grazed high alkaloid (0.19–0.68% dry weight) genotypes of reed canarygrass. The effects of grazing high-alkaloid genotypes were reversible and no permanent disorders were observed (Marten *et al.* 1976), in contrast to the situation with *Phalaris* staggers. Marten *et al.* (1981) at Minnesota have developed a cultivar of reed canarygrass, MN-76, which is very low in tryptamine–carbolamine alkaloids. Comparison of lamb performance in grazing trials of MN-76 with two commercial cultivars, Rise and Vantage, is shown in Table 5.4. The results show improved lamb growth and a very low incidence of diarrhea with the low-alkaloid cultivar. Diarrhea, of unknown physiological cause, is associated with the tryptamine–carboline alkaloids. The indole alkaloid content above which lambs show reduced gains is about 0.2% of the dry weight of the plant. The mode of action of the alkaloid in reducing gain is not known, but could be related to both palatability and increased incidence of diarrhea. Coulman *et al.* (1977B) found no association between alkaloid content and *in vitro* digestibility, and suggested that the poor

FIG. 5.35. Plots of reed canarygrass cultivars at the University of Minnesota showing palatability differences. Sheep are walking through a plot of an unpalatable cultivar to graze a plot (foreground) that is palatable. The differences in acceptability are related to trypatamine alkaloid content.
Courtesy of G. C. Marten.

TABLE 5.4. Performance of Lambs Grazing Three Cultivars of Reed Canarygrass

Year	Cultivar	Total alkaoid content (%)	Type of alkaloid	Average daily gain (g)	Diarrhea incidence (%)
1	Rise	0.32	Gramine and tryptamines, carbolines	64	26
	Vantage	0.33	Gramine	67	1
	MN-76	0.12	Gramine	125	1
2	Rise	0.28	Gramine and tryptamines, carbolines	71	13
	Vantage	0.20	Gramine	94	3
	MN-76	0.09	Gramine	107	2

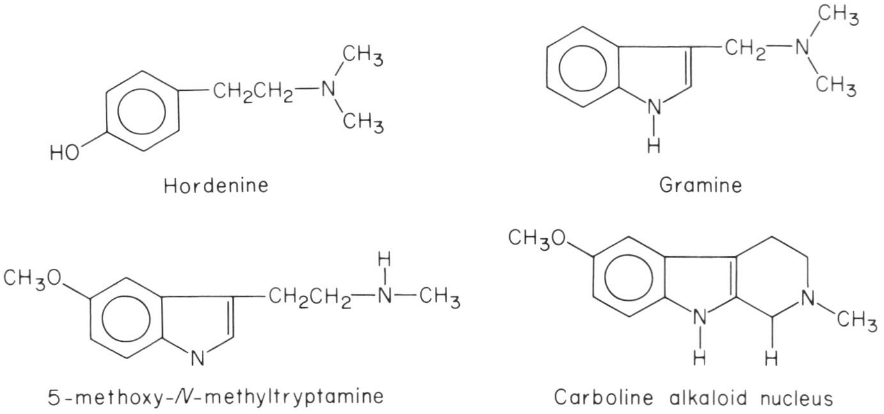

FIG. 5.36. Structures of alkaloids in reed canarygrass.

animal performance is probably due to factors other than an effect on rumen function.

It has been suggested (Majak *et al.* 1979; Parmar and Brink 1976) that the tryptamines in reed canarygrass might be implicated in a pasture-induced type of acute bovine pulmonary emphysema prevalent in British Columbia and the intermountain states of the U.S. Whether the canarygrass alkaloids are metabolized to 3-methylindole, the causative agent in bovine pulmonary emphysema, has not been demonstrated.

TROPANE ALKALOIDS

Jimsonweed, or thorn apple (*Datura stramonium*), is a common inhabitant of waste areas, barnyards, and edges of cultivated fields in much of the U.S. It is a large (3–5 ft tall) coarse annual. The fruit is an erect spiny capsule, while the flowers are erect, up to 4 in long, and white (Fig. 5.37). It contains tropane alkaloids, including atropine, which affect the central nervous system. It is unpalatable to livestock, although reports of poisoning of most classes of livestock are in the literature (Kingsbury 1964). Poisonings have resulted from grazing fresh material, from contamination of hay and silage, and from ingestion of the seeds. Symptoms in both humans and livestock include intense thirst, disturbed vision, delirium, and violent behavior. Human poisonings result from sucking the nectar from flowers or consuming the seeds, and from contamination of beans and grains. In 1676 a

FIG. 5.37. Jimsonweed (*Datura stramonium*) showing the spiny seed pods.

mass poisoning of soldiers sent to Jamestown, Virginia occurred, giving rise to the common names "Jamestown weed" and "Jimsonweed." The structure of atropine is as follows:

Leipold *et al.* (1973) reported an outbreak of arthrogryposis in newborn pigs that appeared to be associated with consumption of Jimsonweed by the pregnant sows. However, Keeler (1981) fed a high level (maintaining chronic signs of intoxication) of Jimsonweed to pregnant sows, and no arthrogryposis was observed in the offspring, suggesting that Jimsonweed is not teratogenic in pigs.

Worthington et al. (1981) investigated the toxicity of *Datura* seeds to pigs as a result of a legal action stemming from *Datura*-seed contamination of pig feed. They found that an alkaloid intake of 2.2 mg/kg body weight from seeds containing 0.2–0.6% alkaloid was tolerated with little or no effect. The seeds were very unpalatable, and it was not possible to produce severe toxicity signs because of feed rejection. A high proportion of the whole seed passed through the gut undigested. They concluded that it is difficult if not impossible to kill pigs by feeding *Datura* seeds because of rejection of levels high enough to cause toxicity.

In a study with cattle, Nelson et al. (1982) fed diets with various levels of Jimsonweed seed containing 0.26% atropine (hyoscyamine) and 0.55% scopolamine (hyoscine). They concluded that death of cattle from *Datura* poisoning is unlikely because rumen atony and anorexia limit intake of the contaminated feed below lethal levels. They suggested that the toxic dose is 2.49 mg atropine and 0.5 mg scopolamine per kilogram body weight, or about 107 seeds per kilogram body weight.

Williams and Scott (1984) noted that contamination of corn with 0.5% *Datura* seeds caused a variety of signs in horses, including anorexia, weight loss, rapid heart and respiration rates, dilation of the pupils, excessive thirst, diarrhea, and excessive urination. Removal of the offending feed resulted in recovery of the affected animals.

Day and Dilworth (1984) examined the toxicity of Jimsonweed seed to broiler chicks. About 1% Jimsonweed seed was the maximum level that could be used without detrimental effects. Higher levels (3 and 6%) caused pronounced growth depression. Thus, the major effects of Jimsonweed contamination of grain or grain screenings appears to be a depression of feed intake.

FESCUE ALKALOIDS

Tall fescue (*Festuca arundinacea*) is a vigorous, coarse perennial grass which grows in pronounced clumps. While somewhat unpalatable, it will be readily consumed by livestock in the absence of more palatable forage and in many areas is an excellent pasture species. It forms a sod that is particularly resistant to trampling, and it is quite drought resistant. In western Oregon, for instance, it is one of the few pasture grasses that forms a sufficiently dense sod to permit grazing during the wet winters, while it will continue to grow during the dry

5. ALKALOIDS

summers when other species become dormant without irrigation. It is grown extensively as a pasture and hay grass in the Pacific Northwest, Missouri, Kentucky, and throughout the southeastern U.S.

Two major problems have been associated with use of tall fescue pastures. These are summer fescue toxicosis, and fescue foot. An additional problem, fat necrosis, has been linked to the grazing of fescue pastures.

Fescue poisoning is widespread in the U.S. and has been reported in Argentina (Debanchero *et al.* 1983) and Australia and New Zealand (Seawright 1982). In Australia, fescue foot is the main problem observed, and sheep as well as cattle are affected (Seawright 1982).

Summer Fescue Toxicosis

The performance of cattle on tall fescue pastures often is poor during the summer. Cattle may show a pronounced loss of weight, a dull, rough hair coat caused by failure to shed the winter coat, a high respiration rate, elevated body temperature, and a susceptibility to heat stress. These characteristics seem to be associated with alkaloids contained in tall fescue.

Tall fescue contains several alkaloids, with perloline and perlolidine being the major ones. Perloline is a yellowish green fluorescent alkaloid that was first isolated from perennial ryegrass in 1943 by New Zealand researchers. The structures of these alkaloids are as follows:

Perlolidine   Perloline

Perloline has been demonstrated to have physiological effects on animals. After parenteral administration of the alkaloid, symptoms of convulsions, muscular incoordination, increased pulse and respiration rates, mild photosensitization, and coma have been seen in sheep. Boling *et al.* (1975) demonstrated that administration of 0.5% perloline to lambs reduced the digestibility of protein and cellulose and reduced nitrogen retention. Production of volatile fatty acids in the rumen was reduced. The body temperature of the perloline-fed lambs tended to be higher than for the control group. Inhibition of certain rumen cellulolytic organisms by perloline has also been demonstrated.

The perloline content of tall fescue increases gradually from early spring to the end of July, and then rises sharply during August. This coincides with the period of summer fescue toxicosis, which suggests a possible association between these two observations. On the basis of this apparent association, a group at Kentucky developed a plant-breeding program to produce both low- and high-perloline fescue. This objective was successfully achieved.

The Kentucky group has reported a series of studies comparing animal responses with the low- and high-perloline fescue lines. Surprising results were obtained. Typical results are shown in Table 5.5.

These results show that the signs of summer fescue toxicosis, including reduced gain, increased respiration rate, and increased rectal temperature, were most pronounced with the variety selected for low-perloline content. The cattle on the low-alkaloid variety were emaciated and had long, rough, dull hair coats. They seemed sensitive to heat stress and spent most of the day under shades, while cattle on the other pastures spent more time grazing and were less sensitive to heat. The cattle on the low-alkaloid grass pasture tended to congregate in muddy areas. In another study by the Kentucky group (Hemken *et al.* 1979), dairy cattle fed the same high- and low-perloline varieties showed similar results, with feed intake and milk production depressed in the cows fed the low-alkaloid fescue. In subsequent work (Hemken *et al.* 1981), it was demonstrated that the response to the low-alkaloid fescue depends on environmental temperature. At moderate (16°–18°C) temperatures, calf performance with the high- and low-

TABLE 5.5. Performance of Steers Grazing High- and Low-Perloline Tall Fescue and Two Commercial Varieties[a]

	Variety			
Item	Kentucky 31	High alkaloid	Low alkaloid	Kenhy
Perloline (μg/g)	172.9	432.0	66.5	164.2
Perlolidine (μg/g)	115.7	257.7	50.0	165.8
Average daily gain (kg)	0.42	0.38	0.29	0.40
Respiration rate (breaths/min)	66	63	82	65
Rectal temperature (°C)	38.7	38.7	39.1	38.8

[a] From Steen *et al.* (1979).

perloline varieties was similar. At high (above 31°C) temperature, calves fed the low-alkaloid fescue had elevated respiration rates and rectal temperatures, and reduced gains. These studies indicated that the low-perloline fescue contains another factor(s) which affects animal performance, and that these effects are manifested under conditions of high environmental temperature. Thus, summer fescue toxicosis is the result of an interaction between plant and environmental factors. Hurley et al. (1980) found that in cattle given toxic fescue, the normal rise in serum prolactin with increasing environmental temperature does not occur, suggesting an inhibition of prolactin secretion.

Bush et al. (1982) found that the low-perloline variety was high in N-acetyl loline and N-formyl loline alkaloids, which had increased with the selection for low perloline. These are pyrrolizidine alkaloids, but they do not contain the 1,2 double bond required for hepatotoxicity (see the section Pyrrolizidine Alkaloids). It was also shown that the concentration of these alkaloids was related to infestation of the plants with the endophytic fungus *Epichloe typhina* (also called *Acremonium coenophialum*). Treatment of the low-perloline fescue with the systemic fungicide benomyl reduced the N-acetyl- plus N-formyl loline concentrations from 875 μg/g in the untreated to 115 μg/g in the treated grass. Calves fed the treated forage had lower respiration rates and rectal temperatures than those fed the untreated fescue.

Schmidt et al. (1982) fed hay and seed to steers from pastures that were either fungus free or heavily invested with *E. typhina*. Presence of the fungus markedly decreased performance and produced signs of summer fescue toxicosis (Table 5.6). The toxicity was greatest for the seeds. The steers fed fungus-infested seed showed signs of severe heat stress during the hottest part of the day, and at these times had respiration rates so rapid and shallow that an accurate respiration rate

TABLE 5.6. Performance of Steers Fed Diets Containing Fungus-Infested Tall Fescue Hay and Seed[a]

Diet	Avg. daily gain (g)	Avg. daily feed intake (g)	Feed/gain	Rectal temperature (°C)	Respiration rate (breaths/min)
Fungus-free seed	960	641	6.7	39.0	56
Fungus-infested seed	200	414	20.7	39.6	71
Fungus-free hay	660	479	7.3	39.0	53
Fungus-infested hay	280	440	15.4	39.6	55

[a] Adapted from Schmidt et al. (1982).

count could not be made. Steers fed the infested seed were extremely nervous and excitable. Thus, there is little doubt that summer fescue toxicosis is associated with *E. typhina* infestation of the forage.

This work demonstrates that the performance of cattle grazing tall fescue can be affected by two classes of alkaloids. Perloline seems particularly to influence rumen fermentation. *N*-Acetyl- and *N*-formyl loline affect metabolism and are particularly detrimental under conditions of high environmental temperature. Their structures are as follows:

Loline   *N*-acetyl loline   *N*-formyl loline

It has not been conclusively established that *N*-acetyl- and *N*-formyl loline are the causative agents of summer fescue toxicosis. There is some evidence (R. C. Buckner, personal communication, University of Kentucky, Lexington 1983) that there may be other endophyte toxins involved.

Fairbourn (1982) noted that bromegrass (*Bromus inermus*) may contain perloline at levels sufficient to reduce in vitro dry-matter digestibility, particularly under conditions of high summer temperatures. Hence, perloline may have an adverse influence on the nutritional value of bromegrass for livestock.

Fescue Foot

Cattle grazing tall fescue pastures are subject to gangrene of the extremities, referred to as fescue foot (Fig. 5.38). It is similar in appearance to ergotism, but does not seem to be caused by a fungal toxin. Lameness, arched back, and diarrhea may occur from a few days to several months after cattle are put into tall fescue pastures. More severe signs include loss of body weight, emaciation, rough hair coat, and gangrene of the tail tip, rear hooves, and ears. The hooves and portions of the tail may slough off. Blood vessels in affected areas are congested, and some tissues contain perivascular hemorrhages. There is no particular seasonal distribution of the problem. Numerous unsucessful attempts have been made to isolate the toxin(s) involved. Williams *et al.* (1975) fractionated an ethanolic extract of fescue into cation, anion, and neutral fractions by ion-exchange chromatography.

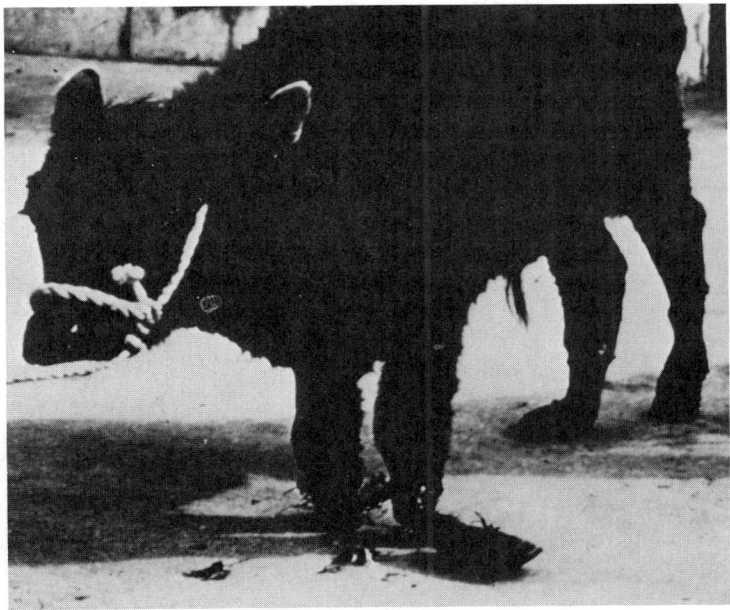

FIG. 5.38. A bovine with a severe case of fescue foot.
Courtesy of D. R. Jacobson.

The fractions were assayed for toxicity by intraperitoneal injection into calves. Clinical signs of fescue foot, including lameness, swelling and reddening of the rear coronary bands, and discoloration of the tip of the tail were observed in calves given the anion and crude ethanolic extracts. They concluded that the causative agent(s) has at least one acidic moiety and is moderately stable to heat and inorganic acids. Further analysis of the anion fraction did not reveal the identity of the causative agent(s) (Garner et al. 1982). More work is necessary for the identification of the toxic compound(s).

Fat Necrosis

Fat necrosis in the body fat of a variety of animals has been recognized for many years (Fig. 5.39). Rumsey et al. (1979) described fat necrosis in beef cattle grazing tall fescue pastures in Georgia. The necrotic fat lesions in these cattle were generally in the abdominal cavity and ranged in size from small nodules embedded in normal depot fat to large, irregularly shaped masses that surrounded and constricted the intestines and reproductive organs. The fat lumps were

FIG. 5.39. Cattle with fat necrosis. The animals had been grazing on tall fescue pasture for 5 years. Note the lumpy appearance of the back of the cow in the foreground.
Courtesy of J. A. Stuedemann.

hard and, when cut in a cross section, had a dry, hard, cheesy opaque appearance in contrast to the normal surrounding fat. Calcification was sometime noted. The necrotic fat was higher in ash, cholesterol, calcium, and magnesium than normal fat and much lower in ether extract (91.9 vs 47.7%). It is possible that fescue-induced fat necrosis is related to the vasoconstriction observed in fescue foot. Rumsey et al. (1979) suggest that vasoconstriction may be a direct causative factor or may lead to a general febrile (feverish) condition, which then may trigger a lipolytic process. The febrile condition associated with summer fescue toxicosis would also be consistent with this theory.

Mineral Metabolism

There is some evidence that tall fescue may affect mineral metabolism in cattle. Stoszek et al. (1979) reported that cattle on tall fescue pasture containing 6.6 ppm copper developed pronounced copper deficiency, while those on an adjacent quack grass (*Agropyron repens*) pasture containing a lower copper level (4.6 ppm) maintained normal blood copper and ceruloplasmin levels, and had increased liver copper stores. Cattle on tall fescue had a rapid decrease in plasma copper and ceruloplasmin levels and a progressive decline in liver copper level. Average daily gains were less on the tall fescue than on the quack grass pasture. The factor(s) involved in the apparent impairment of copper metabolism in animals grazing tall fescue was not identified. On the other hand, Hathaway *et al.* (1981) showed that the selenium in tall fescue was more available than selenium in quack grass. The studies of Stoszek, Hathaway, and colleagues were both conducted on similar pastures in muck soils in the Klamath Basin, Oregon.

Other Disorders Associated with Tall Fescue

Hemken *et al.* (1984), in a review of toxic factors in tall fescue, reported that agalactia (loss of milk production) and reproductive problems may be associated with tall fescue consumption. Reduced conception rates in cattle, horses, and sheep on tall fescue pastures have been noted, but since toxic fescues reduce feed intake, the cause of the lowered reproductive performance could be a lowered nutrient intake. Tall fescue has been implicated in prolonged gestation, abortion, a thickened placenta, and agalactia in horses (Hemken *et al.* 1984). Tall fescue extracts administered to pregnant rabbits caused an increased respiration rate, abortion, stillbirths, and agalactia (Daniels *et al.* 1984).

Fescue foot and summer fescue toxicosis are observed in the midwest and southeastern parts of the U.S. These problems do not occur in the Pacific Northwest, where tall fescue is a major pasture species. This is due to at least two factors: the endophyte fungus *E. typhina* does not appear to be found in this region, and the summer temperatures are generally quite moderate. Summer fescue toxicosis is associated with high environmental temperature. An additional factor is that the tall fescue cultivar "Fawn" is the most widely grown variety in the Pacific Northwest; Cornell *et al.* (1982) reported that Fawn fescue has a much lower fescue foot potential than other cultivars.

MISCELLANEOUS ALKALOIDS

Range sheep and goats that consume *Acacia berlandieri* develop a condition called "limberleg" or "guajillo wobbles," characterized by locomotor ataxia of the hindquarters. The toxic principle is a phenethylamine alkaloid, N-methyl-β-phenethylamine:

N-methyl-β-phenethylamine

The distribution of this compound and other phenethylamines in range plants of the U.S. Southwest has been discussed by Smith (1977) and Camp and Norvell (1966).

Sesbania spp. (*S. vesicaria, S. drummondii,* and *S. punicea*) growing along the coastal plains from the Carolinas to Texas are toxic to livestock, with signs of hemorrhagic diarrhea. The causative agent may be the cytotoxic alkaloid sesbanine (Powell *et al.* 1979):

Sesbanine

Various *Sesbania* spp. are tropical legume forage trees, of potential interest as livestock feed. They should be examined for possible toxicity problems.

REFERENCES

Pyrrolizidine Alkaloids

ALLEN, J. R., ROBERTSON, K. A., JOHNSON, W. D., and CARSTENS, L. A. 1979. Toxicological effects of monocrotaline and its metabolites. *In* Symposium on Pyrrolizidine (*Senecio*) Alkaloids: Toxicity, Metabolism, and Poisonous Plant Control Measures. P. R. Cheeke (editor), pp. 37–42. Nutrition Research Institute, Oregon State Univ., Corvallis.

BUCKMASTER, G. W., CHEEKE, P. R., and SHULL, L. R. 1976. Pyrrolizidine alkaloid poisoning in rats: Protective effects of dietary cysteine. J. Anim. Sci. *43,* 464–473.

BUCKMASTER, G. W., CHEEKE, P. R., ARSCOTT, G. H., DICKINSON, E. O., PIERSON, M. L., and SHULL, L. R. 1977. Response of Japanese quail to dietary and injected pyrrolizidine (*Senecio*) alkaloid. J. Anim. Sci. *45,* 1322–1325.

BULL, L. B., CULVENOR, C. C. J., and DICK, A. T. 1968. The Pyrrolizidine Alkaloids. Their Chemistry, Pathogenicity and Other Biological Properties. American Elsevier, New York.

BURNS, J. 1972. The heart and pulmonary arteries in rats fed on *Senecio jacobaea*. J. Pathol. *106*, 187–194.

CANDRIAN, U., LUTHY, J., SCHMID, P., SCHLATTER, C., and GALLASZ, E. 1984. Stability of pyrrolizidine alkaloids in hay and silage. J. Agric. Food Chem. *32*, 935–937.

CHEEKE, P. R. and PIERSON-GOEGER, M. L. 1983. Toxicity of *Senecio jacobaea* and pyrrolizidine alkaloids in various laboratory animals and avian species. Toxicol. Lett. *18*, 343–349.

CHEEKE, P. R., SCHMITZ, J. A., LASSEN, C. D., and PEARSON, E. G. 1985. Effects of dietary supplementation with ethoxyquin, magnesium oxide, methionine hydroxy analog and B vitamins on tansy ragwort (*Senecio jacobaea*) toxicosis in beef cattle. Am. J. Vet. Res. *46*, in press.

CULVENOR, C. C. J. 1978. Prevention of pyrrolizidine alkaloid poisoning. Animal adaptation or plant control? *In* Effects of Poisonous Plants on Livestock. R. F. Keeler, K. R. Van Kampen and L. F. James (Editors), pp. 189–200. Academic Press, NY.

CULVENOR, C. C. J. 1983. Estimated intakes of pyrrolizidine alkaloids by humans. A comparison with dose rates causing tumors in rats. J. Toxicol. Environ. Health *11*, 625–635.

CULVENOR, C. C. J., EDGAR, J. A., JAGO, M. V., OUTLERIDGE, A. PETERSON, J. E., and SMITH, L. W. 1976. Hepato- and pneumotoxicity of pyrrolizidine alkaloids and derivatives in relation to molecular structure. Chem.-Biol. Interact. *12*, 299–324.

CULVENOR, C. C. J., CLARKE, M., EDGAR, J. A., FRAHN, J. L., JAGO, M. V., PETERSON, J. E., and SMITH, L. W. 1980. Structure and toxicity of the alkaloids of Russian comfrey (*Symphytum* × *uplandicum* Nyman), a medicinal herb and item of the human diet. Experientia *36*, 377–379.

CULVENOR, C. C. J., JAGO, M. V., PETERSON, J. E., SMITH, L. W., PAYNE, A. L., CAMPBELL, D. G., EDGAR, J. A., and FRAHN, J. L. 1984. Toxicity of *Echium plantagineum* (Paterson's Curse). I. Marginal toxic effects in Merino wethers from long-term feeding. Aust. J. Agric. Res. *35*, 293–304.

DICKSON, J. O., and KING, R. R. 1978. The transfer of pyrrolizidine alkaloids from *Senecio jacobaea* into the milk of lactating cows and goats. *In* Effects of Poisonous Plants on Livestock. R. F. Keeler, K. R. Van Kampen, and L. F. James (Editors), pp. 201–208. Academic Press, NY.

FURMANOWA, M., GUZEWSKA, J., and BELDOWSKA, B. 1983. Mutagenic effects of aqueous extracts of *Symphytum officinale* L. and its alkaloidal fractions. J. Appl. Toxicol. *3*, 127–130.

GARRETT, B. J., and CHEEKE, P. R. 1984. Evaluation of amino acids, B vitamins and butylated hydroxyanisole as protective agents against pyrrolizidine alkaloid toxicity in rats. J. Anim. Sci. *58*, 138–144.

GARRETT, B. J., HOLTAN, D. W., CHEEKE, P. R., SCHMITZ, J. A., and ROGERS, Q. R. 1984. Effects of dietary supplementation with butylated hydroxyanisole, cysteine and B vitamins on tansy ragwort (*Senecio jacobaea*) toxicosis in ponies. Am. J. Vet. Res. *45*, 459–464.

GILES, C. J. 1983. Outbreak of ragwort (*Senecio jacobaea*) poisoning in horses. Equine Vet. J. *15*, 248–250.

GOEGER, D. E., CHEEKE, P. R., SCHMITZ, J. A., and BUHLER, D. R. 1982A. Toxicity of tansy ragwort (*Senecio jacobaea*) to goats. Am. J. Vet. Res. *43*, 252–254.

GOEGER, D. E., CHEEKE, P. R., SCHMITZ, J. A., and BUHLER, D. R. 1982B. Effect of feeding milk from goats fed tansy ragwort (*Senecio jacobaea*) to rats and calves. Am. J. Vet. Res. *43*, 1631–1633.

GOEGER, D. E., CHEEKE, P. R., RAMSDELL, H. S., NICHOLSON, S. S., and BUHLER, D. R. 1983. Comparison of the toxicities of *Senecio jacobaea, Senecio vulgaris* and *Senecio glabellus* in rats. Toxicol. Lett. *15*, 19–23.

HIRONO, I., MORI, H., and HAGA, M. 1978. Carcinogenic activity of *Symphytum officinale*. JNCI, J. Natl. Cancer Inst. *61*, 865–868.

HOOPER, P. T. 1978. Pyrrolizidine alkaloid poisoning—pathology with particular reference to differences in animal and plant species. *In* Effects of Poisonous Plants on Livestock. R. F. Keeler, K. R. Van Kampen, and L. F. James (Editors), Academic Press, NY.

HUXTABLE, R. J. 1979. Herbal teas and pyrrolizidine alkaloids. *In* Symposium on Pyrrolizidine (*Senecio*) Alkaloids: Toxicity, Metabolism and Poisonous Plant Control Measures. P. R. Cheeke (Editor), pp. 87–93. Nutrition Research Institute, Oregon State Univ., Corvallis.

JOHNSON, A. E. 1982. Failure of mineral-vitamin supplements to prevent tansy ragwort (*Senecio jacobaea*) toxicosis in cattle. Am. J. Vet. Res. *47*, 718–723.

KINGSBURY, J. M. 1964. Poisonous Plants of the United States and Canada. Prentice-Hall, Englewood Cliffs, NJ.

KNIGHT, A. P., KIMBERLING, C. V., STERMITZ, F. R., and ROBY, M. R. 1984. *Cynoglossum officinale* (hounds' tongue)—a cause of pyrrolizidine alkaloid poisoning in horses. J. Am. Vet. Med. Assoc. *185*, 647–650.

LANIGAN, G. W. 1976. *Peptococcus heliotrinreducans*, sp. nov., a cytochrome-producing anaerobe which metabolizes pyrrolizidine alkaloids. J. Gen. Microbiol. *94*, 1–10.

LANIGAN, G. W., PAYNE, A. L., and PETERSON, J. E. 1978. Anti-methanogenic drugs and *Heliotropium europaeum* poisoning in penned sheep. Aust. J. Agric. Res. *29*, 1281–1292.

LARSON, K. M., ROBY, M. R., and STERMITZ, F. R. 1984. Unsaturated pyrrolizidines from borage (*Borago officinalis*), a common garden herb. J. Nat. Prod. *47*, 747.

LUTHY, J., HEIM, T., and SCHLATTER, C. 1983. Transfer of [^{3}H]pyrrolizidine alkaloids from *Senecio vulgaris* L. and metabolites into rat milk and tissues. Toxicol. Lett. *17*, 283–288.

MATTOCKS, A. R. 1981. Relation of structural features to pyrrolic metabolites in livers of rats given pyrrolizidine alkaloids and derivatives. Chem.-Biol. Interact. *35*, 301–310.

MATTOCKS, A. R., and BIRD, I. 1983. Pyrrolic and N-oxide metabolites formed from pyrrolizidine alkaloids by hepatic microsomes in vitro: Relevance to in vivo hepatotoxicity, Chem.-Biol. Interact. *43*, 209–222.

MIRANDA, C. L., REED, R. L., CHEEKE, P. R., and BUHLER, D. R. 1981A. Protective effects of butylated hydroxyanisole against the acute toxicity of monocrotaline in mice. Toxicol. Appl. Pharmacol. *59*, 424–430.

MIRANDA, C. L., CARPENTER, H. M., CHEEKE, P. R., and BUHLER, D. R. 1981B. Effect of ethoxyquin on the toxicity of the pyrrolizidine alkaloid monocrotaline and on hepatic drug metabolism in mice. Chem.-Biol. Interact. *37*, 95–107.

MOHABBAT, O., SRIVASTAVA, R. N., YOUNOS, M. S., SEDIQ, G. G., MERZAD, A. A., and ARAM, G. N. 1976. An outbreak of hepatic veno-occlusive disease in North-Western Afghanistan. Lancet *2*, 269–271.

PETERSON, J. E., and JAGO, M. V. 1984. Toxicity of *Echium plantagineum* (Paterson's Curse). II. Pyrrolizidine alkaloid poisoning in rats. Aust. J. Agric. Res. *35*, 305–316.

PIERSON, M. L., CHEEKE, P. R., and DICKINSON, E. O. 1977. Resistance of the rabbit to dietary pyrrolizidine (*Senecio*) alkaloid. Res. Commun. Chem. Pathol. Pharmacol. *16*, 561–564.

ST. GEORGE-GRAMBAUER, T. D., and RAC, R. 1962. Hepatogenous chronic copper poisoning in sheep in South Australia due to the consumption of *Echium plantagineum* L. (Salvation Jane). Aust. Vet. J. *38*, 290–293.

SEGALL, H. J., DALLAS, J. L., and HADDON, W. F. 1984. Two dihydropyrrolizine alkaloid metabolites isolated from mouse hepatic microsomes in vitro. Drug Metabolism Disp. *12*, 68–71.

SHARROW, S. H., and MOSHER, W. D. 1982. Sheep as a biological control agent for tansy ragwort. J. Range Manage. *35*, 480–482.

SHULL, L. R., BUCKMASTER, G. W., and CHEEKE, P. R. 1976. Factors influencing pyrrolizidine (*Senecio*) alkaloid metabolism: Species, liver sulfhydryls and rumen fermentation. J. Anim. Sci. *43*, 1247–1253.

SWICK, R. A., CHEEKE, P. R., MIRANDA, C. L., and BUHLER, D. R. 1982A. The effect of consumption of the pyrrolizidine alkaloid-containing plant *Senecio jacobaea* on iron and copper metabolism in the rat. J. Toxicol. Environ. Health *10*, 757–768.

SWICK, R. A., CHEEKE, P. R., GOEGER, D. E., and BUHLER, D. R. 1982B. Effect of dietary *Senecio jacobaea* and injected *Senecio* alkaloids and monocrotaline on guinea pigs. J. Anim. Sci. *55*, 1411–1416.

SWICK, R. A., CHEEKE, P. R., and BUHLER, D. R. 1982C. Subcellular distribution of hepatic copper, zinc and iron and serum ceruloplasmin in rats intoxicated by oral pyrrolizidine (*Senecio*) alkaloids. J. Anim. Sci. *55*, 1425–1430.

SWICK, R. A., CHEEKE, P. R., RAMSDELL, H. S., and BUHLER, D. R. 1983A. Effect of sheep rumen fermentation and methane inhibition on the toxicity of *Senecio jacobaea*. J. Anim. Sci. *56*, 645–651.

SWICK, R. A., MIRANDA, C. L., CHEEKE, P. R., and BUHLER, D. R. 1983B. Effect of phenobarbital on toxicity of pyrrolizidine (*Senecio*) alkaloids in sheep. J. Anim. Sci. *56*, 887–894.

TANDON, B. N., TANDON, H. D., TANDON, R. K., NARNDRANATHAN, M., and JOSHI, Y. K. 1976. Epidemic of veno-occlusive disease of liver in Central India. Lancet *2*, 271–272.

WHITE, R. D., SWICK, R. A., and CHEEKE, P. R. 1983. Effects of microsomal enzyme induction on the toxicity of pyrrolizidine (*Senecio*) alkaloids. J. Toxicol. Environ. Health *12*, 633–640.

WHITE, R. D., SWICK, R. A., and CHEEKE, P. R. 1984. Effects of dietary copper and molybdenum on tansy ragwort (*Senecio jacobaea*) toxicity in sheep. Am. J. Vet. Res. *45*, 159–161.

Piperidine Alkaloids

EDMONDS, L. D., SELBY, L. A., and CASE, A. A. 1972. Poisoning and congenital malformation associated with consumption of poison hemlock by sows. J. Am. Vet. Med. Assoc. *160*, 1319–1324.

KEELER, R. F. 1974. Coniine, a teratogenic principle from *Conium maculatum*, producing congenital malformations in calves. Clin. Toxicol. *7*, 195–206.

KEELER, R. F. 1975. Toxins and teratogens of higher plants. Lloydia *38*, 56–86.

KEELER, R. F. 1978. Alkaloid teratogens from *Lupinus, Conium, Veratrum*, and related genera. *In* Effect of Poisonous Plants on Livestock. R. F. Keeler, K. R. Van Kampen, and L. F. James (Editors), pp. 397–408. Academic Press, NY.

KEELER, R. F., and BALLS, L. D. 1978. Teratogenic effects in cattle of *Conium maculatum* and Conium alkaloids and analogs. Clin. Toxicol. *12*, 49–64.
KEELER, R. F., BALLS, L. D., SHUPE, J. L., and CROWE, M. W. 1980. Teratogenicity and toxicity of coniine in cows, ewes and mares. Cornell Vet. *70*, 19–26.
KINGSBURY, J. M. 1964. Poisonous Plants of the United States and Canada. Prentice-Hall, Englewood Cliffs, NJ.
PANTER, K. E., KEELER, R. F., BUCK, W. B., and SHUPE, J. L., 1983. Toxicity and teratogenicity of *Conium maculatum* in swine. Toxicon, Suppl. *3*, 333–336.
WIDMER, W. R. 1984. Poison hemlock toxicosis in swine. VM/SAC, Vet. Med. Small. Anim. Clin. *79*, 405–408.

Pyridine Alkaloids

CROWE, M. W. 1978. Tobacco—a cause of congenital arthrogryposis. *In* Effects of Poisonous Plants on Livestock. R. F. Keeler, K. R. Van Kampen, and L. F. James (Editors), pp. 419–427. Academic Press, NY.
KAPLAN, B. 1968. Acute nicotine poisoning in a dog. VM/SAC, Vet. Med. Small Anim. Clin. *63*, 1033.
KEELER, R. F. 1978. Alkaloid teratogens from *Lupinus, Conium, Veratrum,* and related genera. *In* Effects of Poisonous Plants on Livestock. R. F. Keeler, K. R. Van Kampen, and L. F. James (Editors), pp. 397–408. Academic Press, NY.
KEELER, R. F., and BALLS, L. D. 1978. Teratogenic effects in cattle of *Conium maculatum* and Conium alkaloids and analogs. Clin. Toxicol. *12*, 49–64.
KEELER, R. F., and CROWE, M. W. 1983. Congenital deformities in swine induced by wild tree tobacco, *Nicotiana glauca*. Clin. Toxicol. *20*, 47–58.
KEELER, R. F., and CROWE, N. W. 1984. Teratogenicity and toxicity of wild tree tobacco, *Nicotiana glauca*, in sheep. Cornell Vet. *74*, 50–59.
KEELER, R. F., SHUPE, J. L., CROWE, M. W., OLSON, A., and BALLS, L. D. 1981A. *Nicotinia glauca*-induced congenital deformities in calves: Clinical and pathologic aspects. Am. J. Vet. Res. *42*, 1231–1234.
KEELER, R. F., BALLS, L. D., and PANTER, K. 1981B. Teratogenic effects of *Nicotinia glauca* and concentration of anabasine, the suspect teratogen in plant parts. Cornell Vet. *71*, 47–53.
KEELER, R. F., CROWE, M. W., and LAMBERT, E. A. 1984. Teratogenicity in swine of the tobacco alkaloid anabasine isolated from *Nicotiana glauca*. Teratology *30*, 61–69.
KINGSBURY, J. M. 1964. Poisonous Plants of the United States and Canada. Prentice-Hall, Englewood Cliffs, NJ.

Indole Alkaloids

BACON, C. W., PORTER, J. K., and ROBBINS, J. D. 1979. Laboratory production of ergot alkaloid by species of *Balansia*. J. Gen. Microbiol. *113*, 119–126.
BERDE, B., and SCHILD, H. O. (Editors) 1978. Ergot Alkaloids and Related Compounds. Springer-Verlag, NY.
BOVE, F. J. 1970. The Story of Ergot. S. Karger, NY.
DIEHL, W. W. 1950. *Balansia* and Balansiae in America. Agric. Monogr. *4*, 1–82.
LORENZ, K. 1979. Ergot on cereal grains. CRC Crit. Rev. Food Sci. Nutr. *11*, 311–354.

PORTER, J. K., BACON, C. W., ROBBINS, J. D., HIMMELSBACH, D. S., and HIGMAN, H. C. 1977. Indole alkaloids from *Balansia epichloe* (Weese). J. Agric. Food. Chem. 25, 88–93.

WALLNER, B. M., BOOTH, N. H., ROBBINS, J. D., BACON, C. W., PORTER, J. K., KISER, T. E., WILSON, R., and JOHNSON, B. 1983. Effects of an endophytic fungus isolated from toxic pasture grass on serum prolactin concentrations in the lactating cow. Am. J. Vet. Res. 44, 1317–1322.

WYLLIE, T. D., and MOREHOUSE, L. G. (Editors) 1978. Mycotoxic Fungi, Mycotoxins and Mycotoxicoses. An Encyclopedic Handbook. Marcel Dekker, NY.

YOUNG, J. C., and MARQUARDT, R. R. 1982. Effects of ergotamine tartrate on growing chickens. Can. J. Anim. Sci. 62, 1181–1191.

Quinolizidine Alkaloids

CHASE, R. L., and KEELER, R. F. 1983. Mountain thermopsis toxicity in cattle. Utah Sci. 44, 28–31.

DAVIS, A. M. 1982. The occurrence of anagyrine in a collection of western American lupines. J. Range Manage. 35, 81–83.

JAMES, L. F. 1977. Plant-induced congenital malformations in animals. World Rev. Nutr. Diet. 26, 208–224.

KEELER, R. F. 1978. Alkaloid teratogens from *Lupinus, Conium, Veratrum,* and related genera. *In* Effects of Poisonous Plants on Livestock. R. F. Keeler, K. R. Van Kampen, and L. F. James (Editors), pp. 397–408. Academic Press, NY.

KILGORE, W. W., CROSBY, D. G., CRAIGMILL, A. L., and POPPEN, N. K. 1981. Toxic plants as possible human teratogens. Calif. Agric. (Nov–Dec.), 6.

KINGSBURY, J. M. 1964. Poisonous Plants of the United States and Canada. Prentice-Hall, Englewood Cliffs, NJ.

SHUPE, J. L., BINNS, W., JAMES, L. F., and KEELER, R. F. 1967. Lupine. A cause of crooked calf disease. J. Am. Vet. Med. Assoc. 151, 198–203.

Steroid Alkaloids

Solanum Alkaloids

CLARINGBOLD, W. D. B., FEW, J. D., and RENWICK, J. H. 1982. Kinetics and retention of solanidine in man. Xenobiotica 12, 293–303.

DALVI, R. R., and BOWIE, W. C. 1983. Toxicity of solanine—an overview. Hum. Vet. Toxicol. 25, 13–15.

DAVIES, A. M. C., and BLINCOE, P. J. 1984. Glycoalkaloid content of potatoes and potato products sold in the UK. J. Sci. Food Agric. 35, 553–557.

JADHAV, S. J., SHARMA, R. P., and SALUNKHE, D. K. 1981. Naturally occurring toxic alkaloids in foods. CRC Crit. Rev. Toxicol. 9.1–104.

KING, R. R., and McQUEEN, R. E. 1981. Transformation of potato glycoalkaloids by rumen microorganisms. J. Agric. Food Chem. 29, 1101–1103.

KINGSBURY, J. M. 1964. Poisonous Plants of the United States and Canada. Prentice-Hall, Englewood Cliffs, NJ.

NICHOLSON, J. W. G., YOUNG, D. A., McQUEEN, R. E., DEJONG, H., and WOOD, F. A. 1978. The feeding value potential of potato vines. Can. J. Anim. Sci. 58, 559–569.

NISHIE, K., GUMBMANN, M. R., and KEYL, A. C. 1971. Pharmacology of solanine. Toxicol. Appl. Pharmacol. 19, 81–92.

OGG, A. G., ROGERS, B. S., and SCHILLING, E. E. 1981. Characterization of black nightshade (*Solanum nigrum*) and related species in the United States. Weed Sci. *29*, 27–32.

Veratrum Alkaloids

JAMES, L. F. 1977. Plant-induced congenital malformations in animals. World Rev. Nutr. Diet. *26*, 208–224.
KEELER, R. F. 1972. Known and suspected teratogenic hazards in range plants. Clin. Toxicol. *5*, 529–565.
KEELER, R. F. 1978. Alkaloid teratogens from *Lupinus, Conium, Veratrum*, and related genera. *In* Effects of Poisonous Plants on Livestock. R. F. Keeler, K. R. Van Kampen, and L. F. James (Editors), pp. 397–408. Academic Press, NY.
KINGSBURY, J. M. 1964. Poisonous Plants of the United States and Canada. Prentice-Hall, Englewood Cliffs, NJ.

Polycyclic Diterpene Alkaloids

AIYAR, V. N., BENN, M. H., HANNA, T., JACYNS, J., ROTH, S. H., and WILKENS, J. L. 1979. The principal toxin of *Delphinium brownii* Rydb., and its mode of action. Experientia *35*, 1367–1368.
CRONIN, E. H., and NIELSEN, D. B. 1978. Tall larkspur and cattle on high mountain ranges. *In* Effects of Poisonous Plants on Livestock. R. F. Keeler, K. R. Van Kampen, and L. F. James (Editors), pp. 521–534. Academic Press, NY.
CRONIN, E. H., and NIELSEN, D. B. 1979. The ecology and control of rangeland larkspurs. Bull.—Utah Agric. Exp. Stn. *499*, 1–34.
CRONIN, E. H., NIELSEN, D. B., and MADSON, N. 1976. Cattle losses, tall larkspur, and their control. J. Range Manage. *29*, 364–367.
KINGSBURY, J. M. 1964. Poisonous Plants of the United States and Canada. Prentice-Hall, Englewood Cliffs, NJ.
OLSON, J. D. 1978A. Tall larkspur poisoning in cattle and sheep. J. Am. Vet. Med. Assoc. *173*, 762–765.
OLSON, J. D. 1978B. Larkspur toxicosis. A review of current research *In* Effects of Poisonous Plants on Livestock. R. F. Keeler, K. R. Van Kampen, and L. F. James (Editors), pp. 535–543. Academic Press, NY.

Indolizidine Alkaloids

Locoweed and Swainsona

DORLING, P. R., HUXTABLE, C. R., and VOGEL, P. 1978. Lysosomal storage in *Swainsona* spp. toxicosis: An induced mannosidosis. Neuropathol. Appl. Neurobiol. *4*, 285–295.
DORLING, P. R., HUXTABLE, C. R., and COLEGATE, S. M. 1980. Inhibition of lysosomal α-mannosidase by swainsonine, an indolizidine alkaloid isolated from *Swainsona canescens*. Biochem. J. *191*, 649–651.
HUXTABLE, C. R., and DORLING, P. R. 1982. Poisoning of livestock by *Swainsona* spp.: Current Status. Aust. Vet. J. *59*, 50–53.
JAMES, L. F. 1981. Syndromes of *Astragalus* poisoning in livestock. J. Am. Vet. Med. Assoc. *158*, 614–618.
JAMES, L. F., and HARTLEY, W. J. 1977. Effects of milk from animals fed locoweed on kittens, calves and lambs. Am. J. Vet. Res. *38*, 1263–1265.

JAMES, L. F., and VAN KAMPEN, K. R. 1976. Effects of locoweed toxin on rats. Am. J. Vet. Res. *37,* 845–846.
JAMES, L. F., VAN KAMPEN, K. R., and HARTLEY, W. J. 1983A. *Astragalus bisulcatus*—a cause of selenium or locoweed poisoning? Vet. Hum. Toxicol. *25,* 86–89.
JAMES, L. F., HARTLEY, W. F., VAN KAMPEN, K. R., and NIELSEN, D. 1983B. Relationship between ingestion of the locoweed *Oxytropis sericea* and congestive right-sided heart failure in cattle. Am. J. Vet. Res. *44,* 254–259.
KINGSBURY, J. M. 1964. Poisonous Plants of the United States and Canada. Prentice-Hall, Englewood Cliffs, NJ.
McILWRAITH, C. W., and JAMES, L. F. 1982. Limb deformities in foals associated with ingestion of locoweed by mares. J. Am. Vet. Med. Assoc. *181,* 255–258.
MENZIES, J. S., BRIDGES, C. H., and BAILEY, E. M. 1979. A neurological disease of cattle associated with *Solanum dimidiatum.* Southwest. Vet. 32, 45–49.
MOLYNEUX, R. J., and JAMES, L. R. 1982. Loco intoxication: Indolizidine alkaloids of spotted locoweed (*Astragalus lentiginosus*). Science *216,* 190–191.
NELSON, B. K., JAMES, L. F., SHARMA, R. P., and CHENEY, C. D. 1980. Locoweed embryo toxicity in rats. Clin. Toxicol. *16,* 149–166.
PIENAAR, J. G., KELLERMAN, T. S., BASSON, P. A., JENKINS, W. L., and VAHRMEIJER, J. 1976. Maldronksiekte in cattle. A neuropathy caused by *Solanum kwebense.* Onderstepoort J. Vet. Res. *43,* 67–74.
RIET-CORREA, F., MENDEZ, M. D. C., SCHIELD, A. L., SUMMERS, B. A., and OLIVEIRA, J. A. 1983. Intoxication by *Solanum fastigiatum* var. *fastigiatum* as a cause of cerebellar degeneration in cattle. Cornell Vet. *73,* 240–256.
SHARMA, R. P., JAMES, L. F., and MOLYNEUX, R. J. 1984. Effect of repeated locoweed feeding on peripheral lymphocytic function and plasma proteins in sheep. Am. J. Vet. Res. *45,* 2090–2093.

Slaframine
CRUMP, M. H. 1973. Slaframine (slobber factor) toxicosis. J. Am. Vet. Med. Assoc. *163,* 1300–1302.
FROETSCHEL, M. A., CROOM, W. J., HAGLER, W. M., BROQUIST, H. P., and GASKINS, R. 1984. Effects of slaframine on salivary flow and rumen function. Can. J. Anim. Sci. *64,* (Suppl. 1), 64–65.
GUENGERICH, F. P., and AUST, S. O. 1977. Activation of the parasympathomimetic alkaloid slaframine by microsomal and photochemical oxidation. Mol. Pharmacol. *13,* 185–195.
HAGLER, W. M., and BEHLOW, R. F. 1981. Salivary syndrome in horses: Identification of slaframine in red clover hay. Appl. Environ. Microbiol. *42,* 1067–1073.
HUXTABLE, C. R., and DORLING, P. R. 1982. Poisoning of livestock by *Swainsona* spp.: Current status. Aust. Vet. J. *59,* 50–53.
SMALLEY, E. B. 1978A. Chemistry and physiology of slaframine. *In* Mycotoxic Fungi, Mycotoxins, Mycotoxicoses: An Encyclopedic Handbook. T. D. Wyllie and L. G. Morehouse (Editors), Vol. 1, pp. 449–457. Marcel Dekker, NY.
SMALLEY, E. B. 1978B. Salivary syndrome in cattle. *In* Mycotoxic Fungi, Mycotoxins, Mycotoxicoses: An Encyclopedic Handbook. T. D. Wyllie and L. G. Morehouse (Editors), Vol. 2, pp. 111–120. Marcel Dekker, NY.
SOCKETT, D. C., BAKER, J. C., and STOWE, C. M. 1982. Slaframine (*Rhizoctonia leguminicola*) intoxication in horses. J. Am. Vet. Med. Assoc. *181,* 606.

Castanospermine
EVERIST, S. L. 1981. Poisonous Plants of Australia. Angus & Robertson, Sydney, Australia.

Tryptamine Alkaloids

COULMAN, B. E., WOODS, D. L., and CLARK, K. W. 1977A. Distribution within the plant, variation with maturity, and heritability of gramine and hordenine in reed canarygrass. Can. J. Plant Sci. 57, 771–777.

COULMAN, B. E., CLARK, K. W., and WOODS, D. L. 1977B. Effects of selected reed canarygrass alkaloids on in vitro digestibility. Can. J. Plant Sci. 57, 779–785.

GALLAGHER, C. H., KOCK, J. H., MOORE, R. M., and STEEL, D. J. 1964. Toxicity of Phalaris tuberosa for sheep. Nature (London) 204, 542–545.

HARTLEY, W. J. 1978. Chronic phalaris poisoning or phalaris staggers. In Effects of Poisonous Plants on Livestock. R. F. Keeler, K. R. Van Kampen and L. F. James (Editors), pp. 391–393. Academic Press, NY.

JOLLY, R. D., and HARTLEY, W. J. 1977. Storage diseases of domestic animals. Aust. Vet. J. 53, 1–8.

KENDALL, W. A., and SHERWOOD, R. T. 1976. Palatability of leaves of tall fescue and reed canarygrass and of some of their alkaloids to meadow voles. Agron. J. 67, 667–671.

MAJAK, W., McDIARMID, R. E., VAN RYSWYK, A. L., BROERSMA, K., and BONIN, S. G. 1979. Alkaloid levels in reed canarygrass grown on wet meadows in British Columbia. J. Range Manage. 32, 322–326.

MARTEN, G. C., JORDAN, R. M., and HOVIN, A. W. 1976. Biological significance of reed canarygrass alkaloids and associated palatability to grazing sheep and cattle. Agron. J. 68, 909–914.

MARTEN, G. C., JORDAN, R. M., and HOVIN, A. W. 1981. Improved lamb performance associated with breeding for alkaloid reduction in reed canarygrass. Crop Sci. 21, 295–298.

PARMAR, S. S., and BRINK, V. C. 1976. Tryptamine levels in pasturage implicated in bovine pulmonary emphysema. Can. J. Plant Sci. 56, 175–184.

SIMONS, A. B., and MARTEN, G. C. 1971. Relationship of indole alkaloids to palatability of Pharlaris arundinacea L. Agron. J. 63, 915–919.

Tropane Alkaloids

DAY, E. J., and DILWORTH, B. C. 1984. Toxicity of Jimson weed seed and cocoa shell meal to broilers. Poult. Sci. 63, 466–468.

KEELER, R. F. 1981. Absence of arthrogryposis in newborn Hampshire pigs from sows ingesting toxic levels of Jimsonweed during gestation. Vet. Hum. Toxicol. 23, 413–415.

KINGSBURY, J. M. 1964. Poisonous Plants of the United States and Canada. Prentice-Hall, Englewood Cliffs, NJ.

LEIPOLD, H. W., OEHME, F. W., and COOK, J. E. 1973. Congenital arthrogryposis associated with ingestion of jimsonweed by pregnant sows. J. Am. Vet. Med. Assoc. 162, 1059–1060.

NELSON, P. D., MERCER, H. D., ESSIG, H. W., and MINYARD, J. P. 1982. Jimson weed toxicity in cattle. Vet. Hum. Toxicol. 24, 321–325.

WILLIAMS, S., and SCOTT, P. 1984. The toxicity of Datura stramonium (Thorn apple) to horses. N. Z. Vet. J. 32, 47.

WORTHINGTON, T. R., NELSON, E. P., and BRYANT, M. J. 1981. Toxicity of thornapple (Datura stramonium L.) seeds to the pig. Vet. Res. 108, 208–211.

Fescue Alkaloids

BOLING, J. A., BUSH, L. P., BUCKNER, R. C., PENDLUM, L. C., BURRUS, P. B., YATES, S. G., ROGOVIN, S. P., and TOOKEY, H. L. 1975. Nutrient digestibility and metabolism in lambs fed added perloline. J. Anim. Sci. 40, 972–976.

BUCKNER, R. C. 1983. University of Kentucky, Lexington. Personal communication.

BUSH, L. P., CORNELIUS, P. L., BUCKNER, R. C. VARNEY, D. R., CHAPMAN, R. A., BURRUS, P. B., KENNEDY, C. W., JONES, T. A., and SAUNDERS, M. J. 1982. Association of N-acetyl loline and N-formyl loline with Epichloetyphine in tall fescue. Crop Sci. 22, 941–943.

CORNELL, C. N., GARNER, G. B., YATES, S. G., and BELL, S. 1982. Comparative fescue foot potential of fescue varieties. J. Anim. Sci. 55, 180–184.

DANIELS, L. B., AHMED, A., NELSON, T. S., PIPER, E. L., and BEASLEY, J. N. 1984. Physiological responses in pregnant white rabbits given a chemical extract of toxic tall fescue. Nutr. Rep. Int. 29, 505–510.

DAVIS, C. B., CAMP, B. J., and READ, J. C. 1983. The vasoactive potential of halostachine, an alkaloid of tall fescue (Festuca arundinaceae Shreb) in cattle. Vet. Hum. Toxicol. 25, 480–411.

DEBANCHERO, E. P., DEMIGUEL, M. A. S., CORBELLINE, C. N., and MIQUET, J. M. 1983. Festuca arundinacea poisoning in the bovine. 2. Toxin-producing ability of Fusarium tricinctum strains isolated from Festuca arundinacea. In Proceedings of the International Symposium on Mycotoxins. K. Naguib, M. M. Naguib, D. L. Park, and A. E. Pohland (Editors), pp. 305–310. National Research Center, Dokki Cairo, Egypt.

FAIRBOURN, M. L. 1982. Alkaloid affects in vitro dry matter digestibility of Festuca and Bromus species. J. Range Manage. 35, 503–504.

GARNER, G. B., CORNELL, C. N., YATES, S. G., PLATTNER, R. D., ROTHFUS, J. A., and KWOLEK, W. F. 1982. Fescue foot: Assay of extracts and compounds identified in extracts of toxic tall fescue herbage. J. Anim. Sci. 55, 185–193.

HATHAWAY, R. L., OLDFIELD, J. E., and BUETTNER, M. 1981. Effect of selenium in a mineral-salt mixture on heifers grazing tall fescue and quackgrass pastures. Proc., Annu. Meet.—Am. Soc. Anim. Sci. West. Sect. 32, 32–33.

HEMKEN, R. W., BULL, L. S., BOLING, J. A., KANE, E., BUSH, L. P., and BUCKNER, R. C. 1979. Summer fescue toxicosis in lactating dairy cows and sheep fed experimental strains of ryegrass-tall fescue hybrids. J. Anim. Sci. 49, 641–646.

HEMKEN, R. W., BOLING, J. A., BULL, L. S., HATTON, R. H., BUCKNER, R. C., and BUSH, L. P. 1981. Interaction of environmental temperature and anti-quality factors on the severity of summer fescue toxicosis. J. Anim. Sci. 52, 710–714.

HEMKEN, R. W., JACKSON, J. A., and BOLING, J. A. 1984. Toxic factors in tall fescue. J. Anim. Sci. 58, 1011–1016.

HURLEY, W. L., CONVEY, E. M., LEUNG, K., EDGERTON, L. A., and HEMKEN, R. W. 1980. Bovine prolactin, TSH, T_4 and T_3 concentrations as affected by tall fescue summer toxicosis and temperature. J. Anim. Sci. 51, 374–379.

RUMSEY, T. S., STUEDEMANN, J. A., WILKINSON, S. R., and WILLIAMS, D. J. 1979. Chemical composition of necrotic fat lesions in beef cows grazing fertilized "Kentucky 31" tall fescue. J. Anim. Sci. 48, 673–682.

SCHMIDT, S. P., HOVELAND, C. S., CLARK, E. M., DAVIS, N. D., SMITH, L. A., GRIMES, H. W., and HOLLIMAN, J. L. 1982. Association of an endophytic fungus with fescue toxicity in steers fed Kentucky 31 tall fescue seed or hay. J. Anim. Sci. 55, 1259–1263.

SEAWRIGHT, A. A. 1982. Animal Health in Australia, Vol. 2. Chemical and Plant Poisons. Australian Government Publishing Service, Canberra.

STEEN, W. W., GAY, N., BOLING, J. A., BUCKNER, R. C., BUSH, L. P., and LACEFIELD, G. 1979. Evaluation of Kentucky 31, G1-306, and Kenhy Tall fescue as pasture for yearling steers. II. Growth, physiological response and plasma constituents of yearling steers. J. Anim. Sci. *48,* 618–623.

STOSZEK, M. J., OLDFIELD, J. E., CARTER, G. E., and WESWIG, P. H. 1979. Effect of tall fescue and quackgrass on copper metabolism and weight gains of beef cattle. J. Anim. Sci. *48,* 893–899.

WILLIAMS, M., SHAFFER, S. R., GARNER, G. B., YATES, S. G., TOOKEY, H. L., KINTNER, L. D., NELSON, S. L., and McGINITY, J. T. 1975. Induction of fescue foot syndrome in cattle by fractionated extracts of toxic fescue hay. Am. J. Vet. *36,* 1353–1357.

Miscellaneous Alkaloids

CAMP, B. J., and NORVELL, M. J. 1966. The phenylethylamine alkaloids of native range plants. Econ. Bot. *20,* 274–278.

FLORY, W., and HEBERT, C. D. 1984. Determination of the oral toxicity of *Sesbania drummondii* seeds in chickens. Am. J. Vet. Res. *45,* 955–958.

POWELL, R. G., WEISLEDER, D., MUTHARD, D. A., and CLARDY, J. 1979. Sesbanine, a novel cytotoxic alkaloid from *Sesbania drummondii.* J. Am. Chem. Soc. *101,* 2784–2785.

SMITH, T. A. 1977. Phenethylamine and related compounds in plants. Phytochemistry *16,* 9–18.

6

Glycosides

CYANOGENS

Cyanogens are glycosides of a sugar or sugars (usually glucose) and a cyanide-containing aglycone. At least 21 cyanogenic glycosides are known, distributed in over 250 plant genera and over 1000 species. They can be hydrolyzed by enzymatic action with the release of HCN (hydrogen cyanide, prussic acid). Hydrogen cyanide is a potent toxin, inhibiting the terminal respiratory enzyme cytochrome oxidase.

The major cyanogens of importance in animal nutrition are the following:

1. Amygdalin (laetrile). This glycoside is found in Rosaceae, such as chokecherries, wild cherries, mountain mahogany, saskatoon serviceberries, and the kernels of almonds, apricots, peaches, and apples. Prunasin is also found in these plants; it has the same structure as amygdalin except it has one glucose rather than two attached to the aglycone.
2. Dhurrin. This occurs in sorghum species such as grain sorghums, forage sorghums (Sudan grass, hay grazer), and Johnson grass.
3. Linamarin. This compound is found in white clover, flax (linseed), cassava, and lima beans.

Hydrogen cyanide is formed when the glycosides are hydrolyzed by plant enzymes in the following steps:

1. Cyanogenic glycoside $\xrightarrow{\beta\text{-glucosidase}}$ sugar + aglycone
2. Aglycone $\xrightarrow[\text{lyase}]{\text{hydroxynitrile}}$ HCN + aldehyde or ketone

The glycosides occur in vacuoles in plant tissue, while the enzymes are found in the cytosol. Damage to the plant from wilting, trampling, mastication, frost, drought, bruising (cassava), and so on results in the enzymes and glycosides coming together, causing HCN to be formed. These enzymes are also produced by rumen microorganisms. The optimal pH for the enzymes is near neutrality, so release of HCN is more rapid in the rumen than in the highly acid stomach of the nonruminant animal. For this reason ruminants are more sensitive to cyanogens than nonruminants. The specific reactions for each of the major types of cyanogens are illustrated.

1. Amygdalin

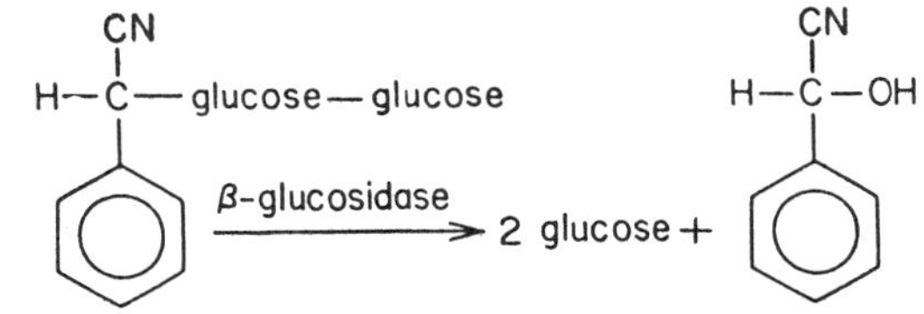

Amygdalin —(β-glucosidase)→ 2 glucose + Mandelonitrile

Mandelonitrile —(hydroxynitrile lyase)→ HCN + Benzaldehyde

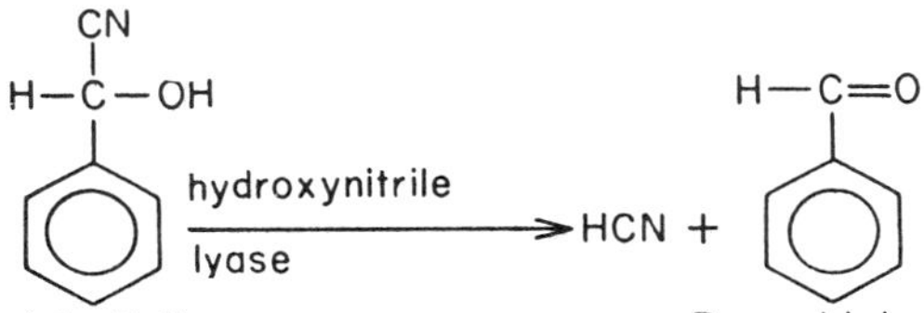

2. Dhurrin

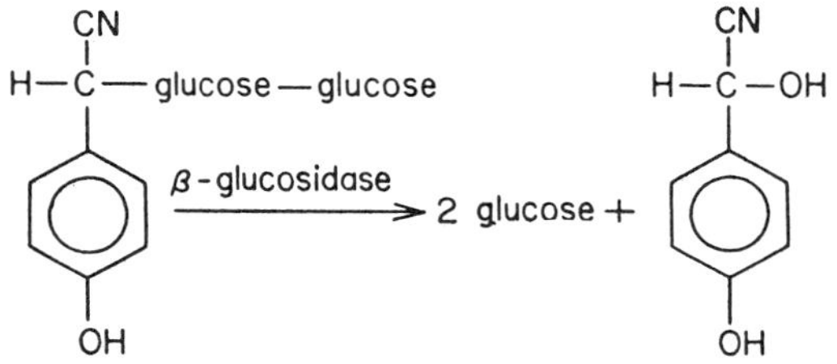

Dhurrin —(β-glucosidase)→ 2 glucose + p-hydroxymandelonitrile

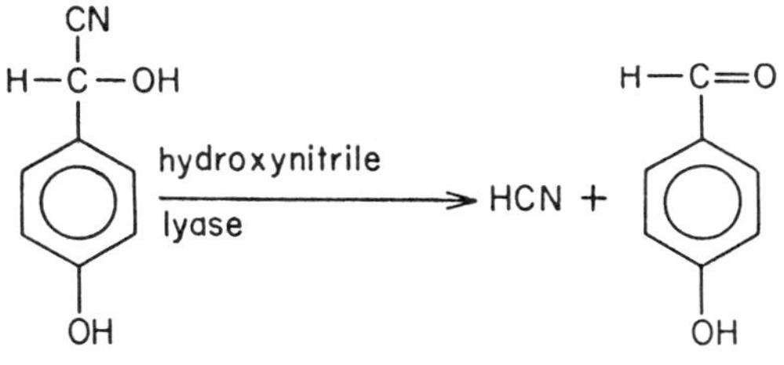

p-hydroxymandelonitrile —(hydroxynitrile lyase)→ HCN + p-hydroxybenzaldehyde

3. Linamarin

$$\text{glucose}-\underset{\underset{CH_3}{|}}{\overset{\overset{CH_3}{|}}{C}}-CN \xrightarrow[H_2O]{\beta\text{-glucosidase}} \text{glucose} + \underset{\underset{CH_3}{|}}{\overset{\overset{CH_3}{|}}{C}}=O + HCN$$

Linamarin Acetone

Metabolism of Hydrogen Cyanide and Signs of Cyanide Poisoning

Hydrogen cyanide is readily absorbed and enters individual tissue cells. It inhibits cytochrome oxidase, the terminal step in electron transport. When cytochrome oxidase is blocked, ATP formation ceases, and the tissues suffer energy deprivation. Death follows rapidly. Signs of cyanide poisoning are dyspnea (difficult or labored breathing), excitement, gasping, staggering, paralysis, convulsions, coma, and death. The odor of benzaldehyde or acetone may be detectable in the contents of the rumen if the dead animal is examined immediately.

Cyanide is readily detoxified, so acute toxicity occurs only if detoxification reactions are exceeded. Liver, kidney, and thyroid tissue contain an enzyme called rhodanese (thiosulfate sulfurtransferase), which catalyzes conversion of cyanide to thiocyanate:

$$S_2O_3^{2-} + CN^- \xrightarrow{\text{rhodanese}} SO_3^{2-} + SCN^-$$

Thiocyanate is excreted in the urine. This reaction is employed in the treatment of cyanide toxicity. Injection (IV) of sodium thiosulfate and sodium nitrate is used. Sodium thiosulfate participates in the above reaction, while nitrate converts hemoglobin to methemoglobin. Methemoglobin has a greater affinity for cyanide than does cytochrome oxidase, so it strips the cyanide from the enzyme. The dosage is critical, since there must be a balance between hemoglobin and methemoglobin sufficient to maintain oxygen transport. In most cases, treatment is not practical under farm or ranch conditions because of the rapidity of development of lethal cyanide poisoning.

Acute cyanide poisoning occurs in both humans and livestock. There are numerous examples of human deaths from consumption of apricot kernels, bitter almonds, and apple seeds. Major livestock poisoning problems involve the forage sorghums, chokecherries, and other cyanogen-containing plants on western ranges of the U.S. Forage sorghums such as Sudan grass (Fig. 6.1) have a high HCN potential when they have been stressed by drought or suffered frost damage. Very dark green plant growth tends to be higher in HCN. Since HCN is volatile, hay made from sorghums is generally safe.

FIG. 6.1. Sudan grass, a forage plant which may on occasion contain toxic levels of cyanogenic glycosides.
Courtesy of P. N. Drolsom.

Wild cherries (*Prunus* spp.) such as the chokecherry have caused extensive cattle losses from cyanide poisoning (Fig. 6.2). The chokecherry leaves are highest in cyanogen content in the spring and early summer. The hydrolysis of the glycosides requires water. It has been observed that mortality often occurs soon after drinking. The water promotes quick release of cyanide from previously ingested dry cherry leaves. Arrow grass (*Triglochin* spp.), which grows in damp areas, marshes, and sloughs over much of North America, has caused losses of livestock because it contains cyanogenic glycosides. Saskatoon serviceberry (*Amelanchier alnifolia*) is a cyanogen-containing shrub that is widely distributed on rangelands of western Canada and the northern states of the U.S. Frequently saskatoons and western chokecherries grow in the same areas and are important browse species for range livestock and wildlife. Majak *et al.* (1980B, 1981) have studied the toxicity of saskatoon serviceberries to cattle. They are much less toxic than western chokecherries, but could be hazardous during the bloom period.

FIG. 6.2. Leaves and flowers of chokecherry (*Prunus virginiana*).
Courtesy of B. R. LeaMaster and U.S. Sheep Experiment Station.

White clover (*Trifolium repens*) contains moderate amounts of linamarin. It does not generally cause livestock problems due to its cyanogenic activity. In midsummer, it often becomes unpalatable and animals will refuse to graze it. This may be due to bitterness associated with the cyanogens. Plant-breeding studies have shown that the cyanogen content of white clover can be easily modified by selection to produce low-cyanogen types. However, the cyanogens are not a problem of sufficient magnitude to warrant a serious effort to reduce them. The major livestock problem associated with white clover is bloat due to bloat-producing soluble proteins (see Chapter 7).

Bracken fern (*Pteridium aqualinum*) contains a number of toxins, including thiaminase (Chapter 7), carcinogens (Chapter 11), and a cyanogenic glycoside, prunasin. Prunasin appears to function primarily in deterrence of herbivory. Cooper-Driver and Swain (1976) noted that a small percentage of bracken plants lack prunasin or the enzyme which releases cyanide. These plants were heavily grazed by sheep and deer, whereas those which were cyanogenic were untouched, suggesting that cyanide was the deterrent.

Cassava (*Manihot*), also called tapioca and manioc, is a tropical tuber that is used for both human and animal feeding (Fig. 6.3). Cassava has traditionally been prepared for human consumption in such a way that its toxicity is largely eliminated. It can be grated and soaked in water to activate the conversion of glysoside to HCN and then washed to leach out the HCN. Alternatively, it can be cooked to destroy the enzymes, so the heat-stable glycoside cannot be converted to HCN. Plant breeders are selecting for low-cyanogen types of cassava.

Chronic exposure to moderate levels of cyanide occurs in some human populations. In tropical Africa, millions of people consume cassava as a staple dietary item and pulses (seed legumes) that contain cyanogens. A condition called tropical ataxic neuropathy occurs in West Africa and is apparently associated with the consumption of cassava. Symptoms are lesions of optic, auditory and peripheral nerves, elevated blood level of thiocyanate, and an increased incidence of goiter. Vitamin B_{12} and methionine supplementation have beneficial effects. The evidence is strong that chronic cyanogen ingestion can lead to neurological problems in humans, and that dietary vitamin B_{12} and

FIG. 6.3. Cassava tubers. Cassava is an important source of starch in human and livestock diets in the tropics. The tubers contain cyanogenic glycosides.
Courtesy of O. O. Tewe and S. K. Hahn.

sulfur amino acid status can have implications in these disturbances (Montgomery 1980).

The lethal dose of cyanide is from 0.5 to 3 mg/kg body weight. Examples of the cyanide yield reported from a number of cyanogen-containing plants (from Montgomery 1980, and Kingsbury 1964) are the following:

	mg CN/100 g plant tissue
Sorghum forage	250
Wild cherries (e.g., chokecherry)	140–370
Arrow grass	77
Bitter almonds	250
Lima beans	10–300
Bamboo tips	800
Linseed meal	53

Putting these figures in perspective, we find that wild cherry leaves with about 200 mg CN/100 g contain about 10 times the minimum level considered dangerous. Less than 100 g of leaves would be fatal to a 100-lb animal. Cyanogen-containing plants can have goitrogenic effects. For example, pregnant ewes fed linseed meal may produce lambs with enlarged thyroids. This is due to the thiocyanate produced during detoxification of cyanide by the rhodanese enzyme system.

The benzaldehyde produced during hydrolysis of cyanogenic glycosides is apparently not of sufficient magnitude to represent a toxicological problem.

It has been known for many years that linseed meal has protective effects against selenium toxicity (Halverson et al. 1955). Palmer et al. (1980) and Smith et al. (1980) have identified the protective factors as the cyanogenic glycosides linustatin and neolinustatin. Linustatin is a glycoside of linamarin and neolinustatin is a glycoside of lotaustralin. The protective activity of the cyanogens is due to the production of cyanide. Cyanide reduces the toxicity of selenium, while selenium reduces the toxicity of cyanide (Palmer 1981). The mechanism of the interaction could be via the rhodanese reaction, with a selenium-containing substrate used in place of the sulfur-containing compounds, such as thiosulfate, which are the traditional rhodanese substrates. There is some evidence (Palmer 1981) that dietary cyanide increases the severity of selenium deficiency.

There are suggestions that cyanogens may have teratogenic effects. Selby et al. (1971) reported an outbreak of swine malformations in Missouri which may have been associated with maternal consumption of wild black cherries during pregnancy. Deformed pigs were born with no tails, very small external sex organs, and limb deformities. Pritchard and Voss (1967) observed fetal deformities (ankylosis of the joints) in foals from mares grazing on Sudan grass pasture, while more recently, Seaman et al. (1981) reported limb deformities in calves from heifers grazing on Sudan grass. These three reports, indicating teratogenic effects when pregnant animals were exposed to cyanogen-containing plants, suggest a possible involvement of these glycosides as teratogens. Keeler (1984) hypothesized that this effect could be mediated through hypoxia induced by cyanide.

GLUCOSINOLATES

Glucosinolates are glycosides of β-D-thioglucose with an aglycone that yields an isothiocyanate, nitrile, thiocyanate, or similar structure upon hydrolysis. They occur widely in cultivated plants, particularly in the Cruciferae family. Most of the glucosinolate-containing crucifers that are important in human or animal nutrition are in the genus *Brassica*; examples include cabbage, broccoli, kale, rapeseed, mustard, and turnips.

The primary significance of glucosinolates in animal nutrition is that they occur in several oil meals used as protein supplements and have adverse effects on livestock consuming these supplements. Rapeseed is widely grown in Canada, the northern U.S., Europe, and Australia as a source of edible and industrial oils. The residue after oil extraction is used as a protein supplement. Other glucosinolate-containing oil meals of much less economic importance include crambe, mustard, and limnanthes (meadowfoam) meals.

The chemistry of glucosinolates is complex. The excellent review of Tookey et al. (1980) should be consulted for a more complete treatment of the structures of glucosinolate derivatives. In addition, see Chapter 1, this work.

Glucosinolates give rise to "hot" compounds with a biting taste, a property that has long been exploited in the use of condiments such as mustard and horseradish. A list of common vegetable and crop plants containing glucosinolates is given in Table 6.1. A comprehensive review of glucosinolates in food plants (Fenwick et al. 1983) provides detailed information on glucosinolate content of vegetables and the physiological effects of the compounds on humans.

The glucosinolates are hydrolyzed by an enzyme system (glucosinolase, or thioglucosidase) in a manner similar to the situation with cyanogenic glycosides. The enzyme is found in the plant and is released when the plant material is crushed. It is also produced by rumen mi-

TABLE 6.1. Examples of Glucosinolate-Containing Plants

Amoracia lapathifolia	Horseradish
Brassica oleracea	Cabbage, brussels sprouts, cauliflower, kohlrabi, broccoli, kale
Brassica chinensis	Pak-choi (Chinese white cabbage)
Brassica campestris	Turnips
Brassica napus	Rutabaga, rape
Brassica nigra	Black mustard
Crambe abyssinica	Crambe
Limnanthes alba	Meadowfoam
Nasturtium officinalis	Water cress
Raphanus sativus	Radish
Thlaspi arvense	Stinkweed

croorganisms. The products always include glucose and the acid sulfate ion. The organic aglycone may undergo various rearrangements, producing isothiocyanates, thiocyanates, or nitriles. Other products are sometimes produced. These include oxazolidine-2-thiones such as goitrin (Fig. 6.4). A variety of other compounds with complex structures may be produced and are discussed by Tookey *et al.* (1980). One of the main glucosinolates in rapeseed meal has the trival name progoitrin. It is converted to goitrin (Fig. 6.4), which has goitrogenic activity. Isothiocyanates, thiocyanates, and nitrites are also produced from rapeseed glucosinolates.

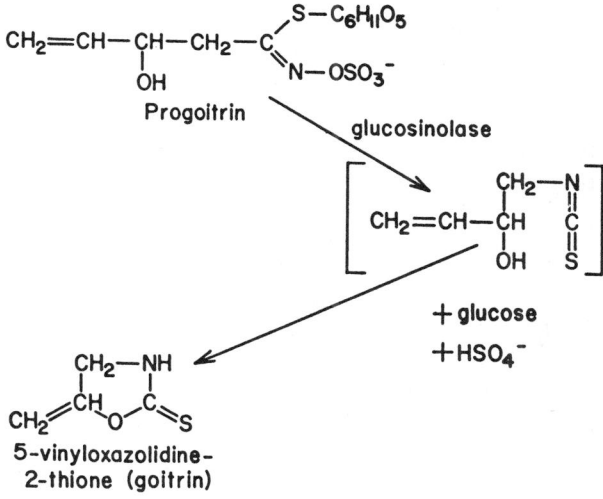

FIG. 6.4. Conversion of progoitrin to goitrin.

Metabolic Effects of Products from Glucosinolates

A major effect of the hydrolysis products of glucosinolates is inhibition of the function of the thyroid gland. The thyroid produces hormones such as thyroxine that are important in regulating the rate of cellular metabolism (metabolic rate). Antithyroid agents have four general effects. They may interfere with iodine uptake by the thyroid, interfere with iodination of tyrosine, suppress thyroxine secretion, or act as metabolic antagonists to thyroxine in the tissues. The general reactions involved in thyroid hormone synthesis are as follows:

$$HO-\underset{\text{Tyrosine}}{\underline{\bigcirc}}-CH_2-\underset{NH_2}{\overset{|}{C}H}-COOH \quad + \quad 2\ I^-$$

$$HO-\underset{I}{\underset{|}{\underline{\bigcirc}}}\underset{I}{\overset{|}{}}-CH_2-\underset{NH_2}{\overset{|}{C}H}-COOH$$

Diiodotyrosine

$$HO-\underset{I}{\underset{|}{\underline{\bigcirc}}}\underset{I}{\overset{|}{}}-O-\underset{I}{\underset{|}{\underline{\bigcirc}}}\underset{I}{\overset{|}{}}-CH_2-\underset{NH_2}{\overset{|}{C}H}-COOH$$

Thyroxine (Tetraiodothyronine)

The effects of some of the hydrolysis products of glucosinolates, such as goitrin, thiocyanates, isothiocyanates, and nitriles, are now discussed briefly.

1. **Goitrin.** Goitrin from rapeseed meal inhibits thyroid function. It causes reduced growth rate, and hyperplasia and hypertrophy of the thyroid. The oxazolidine-2-thiones act by inhibiting the incorporation of iodine into precursors of thyroxine and by interfering with thyroxine secretion. The antithyroid effect is not overcome by adding increased iodine to the diet.

2. **Thiocyanates and Isothiocyanates.** Thiocyanate inhibits iodine uptake by the thyroid. The effect is most pronounced when dietary iodine is low and can be overcome by increasing the iodine level of the

diet. Since isothiocyanates are irritating to mucous membranes, they are probably not consumed as such in toxic amounts. However, if they are consumed as the glucosinolate precursor, with the isothiocyanates released in the gut, they may act as antithyroid agents.

3. Nitriles. Nitriles are formed from glucosinolates in crambe meal (*Crambe abyssinica*) and rapeseed meal. Nitriles are toxic, causing poor growth and liver and kidney lesions, including bile duct hyperplasia, liver necrosis, and megalocytosis of the kidney tubular epithelium. The biochemical cause of these effects is not clear. Paik *et al.* (1980) suggested that since nitriles may be metabolized by mixed function oxidases, they may cause enzyme induction and a consequent increase in liver size.

Rapeseed meal nitriles may be partially converted to thiocyanate (Paik *et al.* 1980). Dietary thiosulfate overcomes part of the growth-depressing effects of high-nitrile rapeseed meals (Paik *et al.* 1980).

Effects of Glucosinolates on Humans and Livestock

1. Humans. Goiter in humans has been attributed to the consumption of large amounts of cabbage or other crucifers (Tookey *et al.* 1980). It is estimated that 96% of all human goiter is caused by uncomplicated iodine deficiency, so in reality the goitrogenic agents play a very minor role in human diseases. While it is plausible that in areas where iodine content of food is low, goiter may be accentuated by consumption of brassica vegetables, there is no direct proof that this is the case.

2. Livestock. Poultry and swine fed raw rapeseed meal exhibit enlarged thyroids and growth depression. Additional problems noted in poultry include perosis, lowered egg production, off-flavors in eggs, and liver damage. Liver hemorrhage in poultry fed rapeseed meal may be caused by nitriles. Swine may have enlarged livers when fed rapeseed meal. Levels of 5–10% rapeseed meal have been acceptable for nonruminants. Ruminants can tolerate at least 10% dietary rapeseed meal with no ill effect. They are less sensitive to glucosinolates and their derivatives than nonruminants, presumably because of rumen metabolism of these compounds.

Of major significance to the utilization of rapeseed meal by livestock has been the development in Canada of low-glucosinolate cultivars. These include so-called "double-zero" varieties, low in both glucosinolates and erucic acid. The meal from these new varieties can be used at a much higher level without reducing animal performance than for previous varieties. The difference in feeding value is so apparent that

Canadian producers have coined the term "canola meal" for the low-glucosinolate meal. Canola meal can be used as a total replacement for soybean meal for some classes and types of livestock, such as finishing pigs. Further improvements expected are a reduction in the fiber content and the introduction of genes for yellow seed color into commercial varieties. This will differentiate canola meal from rapeseed meal, which is brown in color. Also, the brown hulls are virtually indigestible, while the energy and protein in the yellow hulls are amenable to digestion. The yellow hulled strains have a lower hull content, and the hull contains less fiber than brown hulls (Bell 1984). Torch and Candle are yellow hulled cultivars, while Tower and Midas are brown hulled. The yellow varieties are derived from *Brassica campestris* while the brown types are derived from *B. napus*. Torch and Midas are high-glucosinolate types, while Tower and Candle are low-glucosinolate varieties.

In the development of low-glucosinolate cultivars, the glucosinolates having butenyl and pentenyl side chains have been markedly reduced (Table 6.2). The indolylmethyl types were not reduced, so their relative concentration is increased in the low-glucosinolate types (Table 6.2).

The development of improved varieties of rapeseed, reviewed by Bell

TABLE 6.2 Glucosinolate Content[a] of Selected High- and Low-Glucosinolate Rapeseed Cultivars[b]

Glucosinolate type	Side chain (R)	Cultivars	
		Torch	Candle
3-Butenyl	$CH_2=CH(CH_2)_2-$	31.2	4.5
4-Pentenyl	$CH_2=CH(CH_2)_3-$	22.9	3.9
2-OH-3-butenyl	$CH_2=CH-\underset{\underset{OH}{\vert}}{CH}-CH_2-$	22.5	5.2
2-OH-4-pentenyl	$CH_2=CH-CH_2-\underset{\underset{OH}{\vert}}{CH}-CH_2-$	3.8	1.3
3-Indolylmethyl	indole-CH_2- (N-H)	0.4	0.3
1-Methyl-3-indolylmethyl	indole-CH_2- (N-OCH_3)	12.3	12.5
Total glucosinolate		93.1	27.7

[a] μmol glucosinolate per gram meal.
[b] Adapted from Bell (1984).

FIG. 6.5. A single plant of rape, grown in western Canada for its oil-containing seeds.
Courtesy of J. M. Bell and R. K. Downey.

(1984), is a remarkable achievement of plant breeding. With our present knowledge of toxicity, rapeseed would in retrospect appear to be an unlikely candidate for crop development, as both the oil and the meal contain toxic factors (Fig. 6.5). The rate of return on investment from rapeseed research has exceeded that from hybrid corn and poultry research in the U.S. (Bell 1984).

Glucosinolates in Animal Products

Glucosinolates and their derivatives can be transferred in the milk of dairy animals, causing thyroid enlargement, increased ^{131}I uptake by the thyroid, and decreased blood thyroid hormones in experimental animals fed the milk. Placental transfer can also occur. Throckmorton *et al.* (1981) noted goiter and altered serum thyroid hormones in lambs from ewes fed raw meadowfoam meal (*Limnanthes alba*), while White and Cheeke (1983) noted evidence of thyroid changes in rabbits and goat kids fed milk from goats fed raw meadowfoam meal. Tissue residues of glucosinolates have been detected in cattle fed crambe meal (Van Etten *et al.* 1977). Embryonic thyroid enlargement in chicks from eggs from hens fed rapeseed meal was attributed to a low iodide content of the eggs due to increased iodide trapping by the enlarged maternal thyroid (March and Leung 1976).

Beneficial Effects of Glucosinolates and Their Derivatives

Consumption of cruciferous vegetables, which contain glucosinolates, has been suggested to have protective effects against colon and rectal cancer in humans (Fenwick et al. 1983). Benzyl isothiocyanate and thiocyanate inhibit tumor development in laboratory animals administered carcinogens (Fenwick et al. 1983). Indole-3-carbinol has been implicated as a compound in cruciferous vegetables which may have protective activity against cancer. Hendricks et al. (1982) found that indole-3-carbinol prevented hepatoma development in trout adminstered aflatoxin, a carcinogenic mycotoxin. Feeding cruciferous vegetables to rats enhanced the detoxification of the carcinogens aflatoxin B_1 and polybromobiphenyl (Fenwick et al. 1983). The activity of hepatic enzymes involved in detoxification, such as aryl hydrocarbon hydroxylase, mixed function oxidases, glutathione S-transferase, and 3,4-benzpyrene hydroxylase is enhanced by feeding crucifers (Fenwick et al. 1983), as is epoxide hydrolase activity (Hendrich and Bjeldanes 1983). Indole-3-carbinol also induces hepatic mixed function oxidase activity (Shertzer 1982) and reduces covalent binding of toxicants to hepatic macromolecules (Shertzer 1983). Thus, there are suggestions that glucosinolates and their derivatives may have beneficial effects on human health.

COUMARIN

Sweet Clover Poisoning

Sweet clover (*Melilotus alba* and *Melilotus officinalis*) contains a glycoside called melilotoside, an ether of glucose and coumarin (Fig. 6.6). Coumarin is metabolized by various molds, such as *Penicillium nigricans* and *Penicillium jensi,* producing dicoumarol. Dicoumarol is an inhibitor of vitamin K and induces a vitamin K deficiency. Vitamin K functions in blood clotting, so a deficiency is characterized by susceptibility to hemorrhaging.

Sweet clover has been widely grown in much of the northern U.S. and the prairie provinces of Canada as a forage and a soil-building (green manure) crop. In the early 1920s, widespread incidence of "bleeding disease" occurred in Ontario and the midwest. It was found to be associated with the consumption of moldy sweet clover hay. The condition was subsequently referred to as sweet clover poisoning. In-

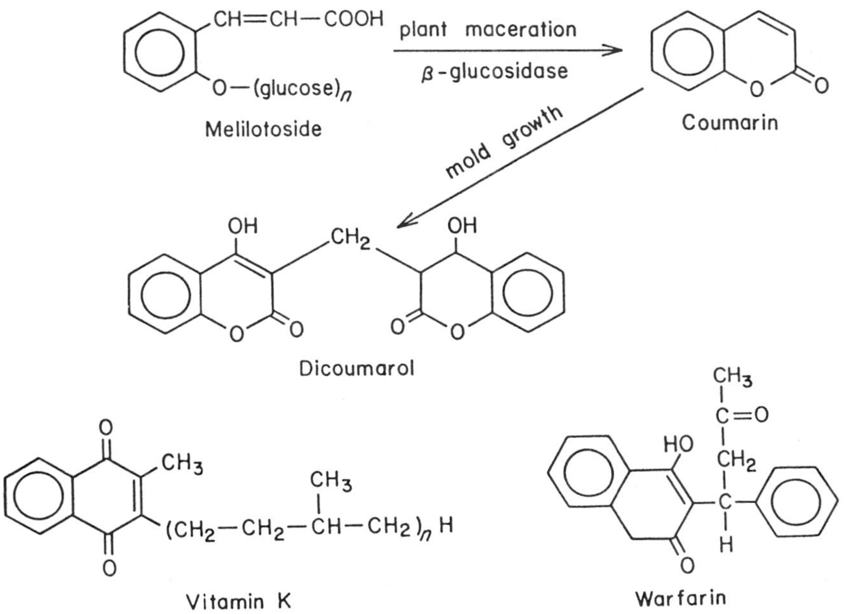

FIG. 6.6. Formation of dicoumarol and its structural relationship to vitamin K and Warfarin.

vestigators at The University of Wisconsin eventually elucidated the situation by demonstrating that a species of sweet clover which was not bitter did not cause sweet clover poisoning when moldy. It was shown that this species lacked coumarin. Addition of coumarin to alfalfa would result in poisoning only if the mixture was allowed to mold. The toxin was extracted, identified, and given the name dicoumarol. In association with these investigations, another vitamin K antagonist was produced. It was given the name Warfarin (for Wisconsin Alumni Research Foundation), and has been extensively used as a rat poison.

The widespread occurrence of sweet clover poisoning in the early 1920s was due to a combination of circumstances. The acreage of sweet clover for hay was rapidly expanding, and the summers were particularly wet, leading to a lot of moldy hay. Because of its succulent stem, sweet clover is difficult to cure without some molding.

Sweet clover poisoning occurs almost exclusively in cattle. The predominant sign is hemorrhaging, either external or internal. Internal hemorrhaging results in obvious subcutaneous swellings caused by pooling of blood. The mucous membranes are pale, and the animal becomes progressively weaker and dies without a struggle. Before internal hemorrhaging occurs, cattle fed moldy sweet clover hay have a prolonged blood-clotting time. Minor surgery such as dehorning or castration may lead to profuse hemorrhaging and death. Kingsbury (1964) cites an instance in which 21 of 22 cattle that were dehorned died from hemorrhage.

Vitamin K is involved in the activation of prothrombin. Dicoumarol inhibits this effect, so there is a deficiency of prothrombin in the blood of affected animals. The basic reactions involved in blood clotting are as follows:

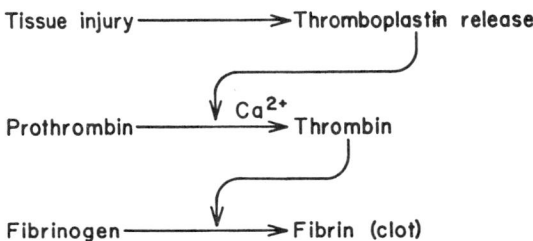

Thrombin acts as an enzyme to split one or more peptides from fibrinogen, altering its solubility and causing it to precipitate as a clot.

Levels of above 10 ppm dicoumarol in a sweet clover hay are suggestive of possible toxicity problems. Dicoumarol levels tend to be highest in small and in round bales, where opportunity for mold growth is greater than in stacks (Benson et al. 1981). Crimping or conditioning of sweet clover hay should speed up drying and might reduce the amount of mold growth, although Benson et al. (1981) found no difference in dicoumarol level of crimped vs noncrimped hay.

Sweet clover poisoning can be treated with injections of vitamin K and also by whole blood tranfusion (Radostits et al, 1980). It is a relatively minor problem now as the causative factors are known, and the plant is grown to a lesser extent than in years past. Low-coumarin varieties of sweet clover, such as Polara, have been developed (Goplen 1971).

Dicoumarol problems in livestock are not exclusively caused by moldy sweet clover. Pritchard et al. (1983) in England reported a hemorrhagic syndrome in cattle that was caused by dicoumarol in sweet vernal grass (Anthoxanthum odoratum) hay. Sweet vernal grass contains coumarin and o-coumaric acid, both of which are converted to dicoumarol by Aspergillus spp. fungi. Both sweet clover and sweet vernal grass have a pleasant or sweet odor caused by the presence of coumarins.

Coumarin compounds have commercial applications. Because of their pleasant and persistent vanilla-like odor, they have been used in perfumes and cosmetics (Scheel 1978), and in condiments. Coumarin is used as a fixative and enhancing agent for the odor of essential oils in perfumes, soap, toothpaste, and hair preparations, and is used in tobacco products to enhance the natural taste (Cohen 1979). Coumarin additions to food have not been permitted in the U.S. since 1954; it is permitted at 5 ppm in liquors (Cohen 1979). Hepatoxic effects have been reported in laboratory animals fed high levels of coumarin.

Furocoumarins

Furocoumarins are compounds with a furan ring fused with a coumarin nucleus. Most are derivatives of psoralen:

Psoralen

Furocoumarins are photosensitizing agents. They have been used for several thousand years in India and Egypt to treat skin depigmentation (leukoderma) by application to the skin with subsequent exposure to sunlight. Besides being natural components of many plants, they are in some cases phytoalexins; that is, they are produced when some plants are infested with fungal pathogens. Celery and parsnips are plants which may elaborate furocoumarins as phytoalexins.

In livestock the major toxicological problem associated with furocoumarins is photosensitization. These compounds are primary photodynamic agents, meaning that they react with light at the surface of the skin and cause cellular damage. They absorb long-wave ultraviolet light and become photodynamic. The photoactive compounds bind with pyrimidine bases and nucleic acids, thus inhibiting DNA synthesis (Ivie 1978).

Two furocoumarin-containing plants in the U.S. that cause livestock problems are bishop's weed (*Ammi majus*) and spring parsley (*Cymopterus watsonii*). Bishop's weed grows in coastal regions of the southern U.S. and has caused severe periodic outbreaks of photosensitization in Texas. In cattle, severe blistering and peeling of light-skinned animals occurs as well as clouding of the cornea, which may produce blindness. Photosensitization with erythema and blistering on the beak, feet, and eye lesions have occurred in poultry fed grain containing *Ammi* seeds.

Spring parsley grows on rangelands in Utah and Nevada; it is one of the first plants to begin growing in early spring. Sheep grazing on spring parsley suffer severe photosensitization, with the udder and teats so sensitive from erythema and blistering that the ewes may refuse to nurse their lambs. Losses of up to 25% of the lambs from starvation have occurred in some range sheep flocks exposed to the plant. Similar effects are seen in cattle. The main losses are of lambs and calves that the dams refuse to nurse. Losses are avoided mainly by keeping sheep and cattle off infested ranges in early spring until other plants begin growth. The furocoumarins responsible for the condition are xanthotoxin and bergapten.

Xanthotoxin   Bergapten   Trisoralen

Dutchman's breeches (*Thamnosma texana*) and *Thamnosma montana* are perennial weeds of the western and southwestern U.S. Oertli et al. (1983) identified at least nine psoralens, including xanthotoxin and bergapten, in *T. texana*. These workers induced photosensitization reactions in sheep by administering the plant (9–12 g/kg body weight/ day) and reported an incident of photosensitization in cattle grazing a *T. texana*-infested pasture.

Another condition due to furocoumarins is celery dermatitis. Outbreaks of this condition sometimes occur in workers handling the plant, with development of erythema and blistering on the hands and forearms. The causative agents are xanthotoxin and trisoralen, phytoalexins which celery produces in response to "pink rot" infections with *Sclerotinia sclerotiorum*.

STEROIDS AND TRITERPENOIDS

Cardiac Glycosides

Cardiac glycosides occur in the common garden foxglove (*Digitalis purpurea*), a native of Europe (Fig. 6.7). Foxglove has naturalized in the U.S. Pacific Northwest and is found along roadsides, in logged-off areas, and in pastures. Foxgloves contain a number of cardiac glycosides which strengthen the force of contraction of the heart and prolong the duration of the diastolic phase. Digitalis drugs such as digitonin are derived from *D. purpurea,* and are used extensively in human medicine.

FIG. 6.7. Foxglove (*Digitalis purpurea*) contains cardiac glycosides. It is a common roadside weed of the Pacific Northwest.

Foxglove poisoning of livestock is rare because the plant is not normally consumed. Signs of toxicity include gastric disturbances, drowsiness, irregular heartbeat and pulse, tremors, and convulsions. Kingsbury (1964) cites early work showing fresh foxglove leaves to be highly toxic with only a few hundredths percent of body weight being a lethal dose. Rabbits have been accidentally poisoned by being fed foxglove leaves as a forage. Two people who thought they were collecting an herb called comfrey but who in fact collected foxgloves died after drinking a tea prepared from the leaves (Huxtable 1979). An incidence of suspected foxglove tea poisoning was reported by Dickstein and Kunkel (1980).

Oleander (*Nerium oleander*) is an introduced ornamental shrub widely grown in California and other southerly areas of the U.S. It has large showy blossoms. Several cardiac glycosides from oleander have been isolated (Kingsbury 1964) and have similar pharmacological effects as digitalis glycosides. Kingsbury (1964) has reviewed studies on the toxicity of oleander. It is extremely toxic to livestock and humans, with all plant parts, green or dry, being toxic. A single leaf can be lethal to humans. People have been poisoned when meat or frankfurters have been skewered on oleander branches for cooking. An intake of leaves at 0.005% of body weight is lethal to cattle and horses. Symptoms of toxicity are similar to those of foxglove poisoning.

Cardenolides

Various species of milkweed (*Asclepias* spp.) grow throughout North America (Fig. 6.8). They are perennial herbs with a thick milky latex juice. Some are being evaluated as potential sources of latex and hydrocarbons (Nielsen *et al.* 1977). They are of two major groups, narrow leaved and broad leaved. Some, such as the woolly pod or California milkweed (*Asclepias eriocarpa*) and *Asclepias labriformis,* which occurs in Utah, are hazards to range animals. As little as 10–20 g of these milkweeds will kill a sheep. They are unpalatable and are consumed only in times of drought when the range is overgrazed or when contaminated hay is fed. Losses of domestic rabbits have occurred when milkweed-contaminated hay was fed (Cheeke *et al.* 1982).

Signs of milkweed toxicity in livestock include a profuse depression

FIG. 6.8. Plants of the broad-leaved milkweed *Asclepias syriaca.*

accompanied by staggering. Following collapse, there is labored respiration, elevated temperature, and dilation of the pupils. Pathological lesions include congestion of the lungs and kidneys and irritation of the intestinal mucosa.

The toxic principles of milkweeds are cardiac glycosides called cardenolides. A cardenolide contains a lactone ring attached to a steroid:

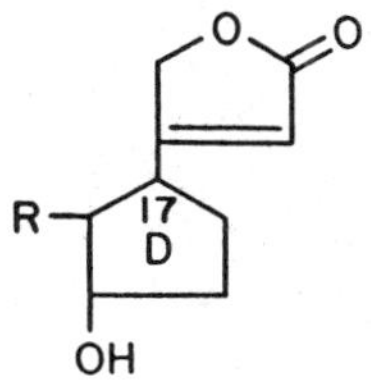

The milkweed cardenolides are structurally similar to oubain, a drug used in clinical medicine. Milkweed cardenolides have structures similar to the general structure shown below. Some of them contain one sugar moiety attached to the steroid nucleus through bonding with hydroxyl groups on both C-2 and C-3 positions of the aglycone, while others have one site of glycosidation, generally at C-3.

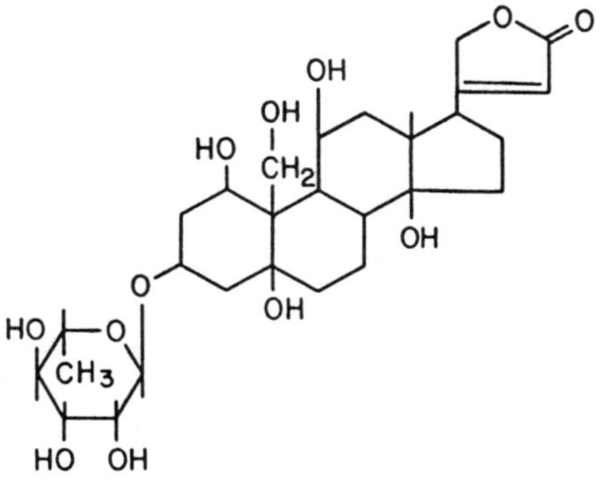

Oubain

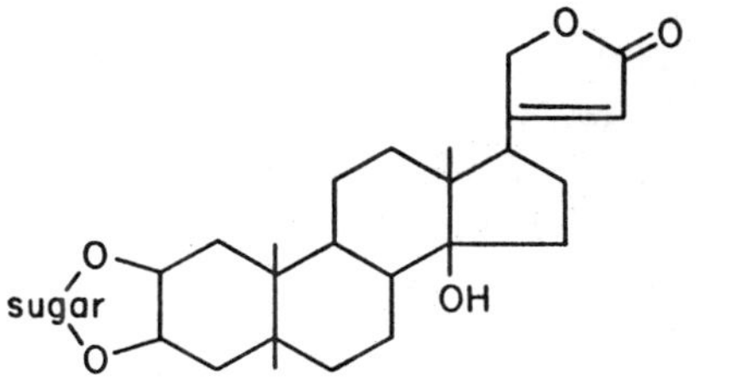

Milkweed Cardenolide

Benson et al. (1978) studied the effect of milkweed cardenolides on sheep. *Asclepias labriformis* and *A. eriocarpa* were acutely toxic at doses of 30–50 g of dry plant per sheep. The mode of action of the milkweed cardenolides appeared very similar to that of oubain, namely, an inhibition of Na^+, K^+-ATPase in cardiac muscle. Thus, their toxicity is due to their ability to inhibit the functioning of the heart muscle.

An interesting relationship is that of milkweeds and the monarch butterfly (Duffey 1980). The larvae of this butterfly feed on milkweed

FIG. 6.9. A monarch butterfly caterpillar feeding on leaves of *Asclepias*. The larvae accumulate cardiac glycosides, making them unpalatable to predators.
Courtesy of J. N. Seiber.

and accumulate the cardenolides (Fig. 6.9). If the level of toxin is sufficiently high, it causes birds which consume the caterpillar, pupa, or adult to vomit. This response may be learned by the predators, and they subsequently avoid the monarch butterfly. An extension of this relationship is that other butterflies which do not feed on milkweed have a nearly identical color pattern, so predators avoid these nontoxic insects also.

The diversity of the human palate is exemplified by the opinion of some that milkweed shoots are an asparagus-like delicacy (Gibbons 1972).

Bufadienolides

Bufadienolides are cardiac glycosides very similar to cardenolides, differing only in the structure of the C-17 substituent on the D ring. They are found in a variety of South African plant species (Naude 1977). In South Africa, *Morea polystachya* or blue tulp (Ker Gawl) causes livestock poisoning due to its cardiac glycosides (Joubert and

Schultz 1982). Clinical signs include gastrointestinal, cardiac, and neuromuscular disturbances. Gastrointestinal signs include rumen stasis, bloat, and diarrhea; cardiac effects involve bradycardia, tachycardia, and arrhythmia, while neuromuscular effects are hypersensitivity, paresis, ataxia, paralysis, dyspnea, and drooling. Joubert and Schultz (1982) reported that an oral administration of a mixture of activated charcoal (5 g/kg body weight) and potassium chloride (1 g/kg body weight) protected against blue tulp toxicity in sheep and cattle.

South African cattle raisers have for many years followed a management practice of predosing cattle orally with blue tulp before turning them into tulp-infested pastures, believing that the prior exposure lessened the likelihood of poisoning. However, Strydom and Joubert (1983) found no beneficial effects of predosing cattle with the plant in protecting against subsequent toxicity.

Anderson et al. (1983) in South Africa reported that *Kalanchoe lanceolata* contains three bufanienolides, causing a disease in sheep (krimpsiekte) with neurological signs typical of cardiac glycoside poisoning. This is the first report of bufanienolides in *Kalanchoe* spp. Williams and Smith (1984) found that several *Kalanchoe* spp. are toxic to chicks, causing neurological signs. They also cite an incident involving a rabbit which consumed three-fourths of a leaf of *K. daigremontiana* and was severely poisoned. The widespread use of *Kalanchoe* spp. as garden and houseplants and their apparent toxicity indicate that care should be taken to ensure that the plants are not accessible to children or herbivorous pets.

Saponins

Saponins are glycosides widely distributed among plants of economic importance. Many forage legumes grown in temperate areas contain saponins; they occur infrequently among the many tropical legumes being evaluated as human and animal foods. Saponins have a wide variety of biological effects, with both positive and negative implications. They show potential as dietary additives in lowering serum cholesterol levels in humans, thus possibly reducing the risk of atherosclerosis. On the negative side, they inhibit the productive performance of nonruminant animals (swine, poultry) fed diets containing alfalfa.

Saponins are characterized by a bitter taste and foaming properties (honeycomb foam in water). They are well known for their in vitro hemolytic effects, a property that has been used in analysis of plant material for saponin content. Saponins have had industrial and commercial applications, including use in soft drinks, shampoo, fire extinguishers, soap, and the synthesis of steriod hormones (e.g., birth control pills). Because the aglycone is nonpolar, while the carbohydrate side chain is water soluble, they have strong detergent properties.

The saponins in alfalfa are probably of the most interest in animal production. The principal saponins in common alfalfa cultivars are medicagenic acid, soyasapogenol A and B, and lucernic acid (Fig. 6.10). They differ in the hydroxyl and carboxyl groups attached to the sapogenin (aglycone). Medicagenic acid appears to be the major saponin in alfalfa with antinutritional effects.

The saponin content of alfalfa forage is influenced by environmental factors and follows a seasonal cycle, being high in midsummer and low in spring and fall. Consequently, the saponin content of second-cutting alfalfa is usually higher than that of first or third cuttings, which may contribute to the generally poorer animal performance observed with second-cutting alfalfa. Saponin in the roots is transferred to the foliage in response to flowering and various environmental factors. Plant breeders have developed strains or cultivars of alfalfa with both high and low saponin content. The low-saponin selections show potential as improved types of alfalfa for animal feeding. Among the effects of saponins on animals are growth inhibition of swine and poultry, reduced palatability, and increased excretion of cholesterol.

Low levels of dietary alfalfa meal reduce the growth rate of poultry and swine, primarily due to effects on palatability and feed intake (Cheeke 1976; Cheeke *et al.* 1977, 1978) rather than to metabolic effects. Use of low-saponin alfalfa strains or cultivars increases the level of alfalfa that can be fed to nonruminants without reducing growth performance. Saponins are bitter and irritate mucous membranes of the mouth and digestive tract. These effects depress voluntary feed intake.

Saponins are found in a variety of human feedstuffs (Oakenfull 1981; Fenwick and Oakenfull 1983). The food plants highest in saponin content include chick-peas (*Cicer arietinum*), soybeans, alfalfa sprouts, and common beans.

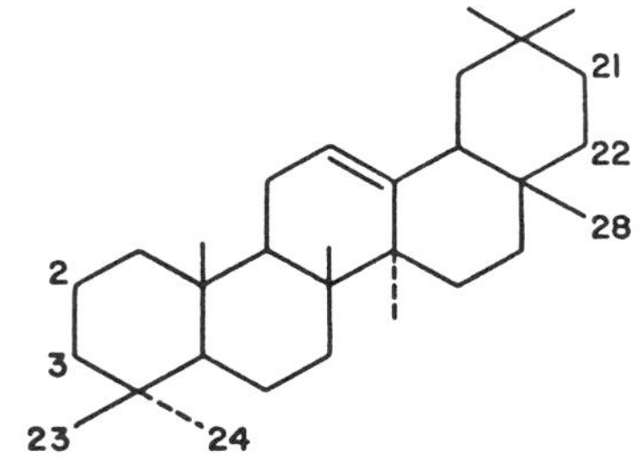

Saponin	C position of groups	
	OH	COOH
Medicagenic acid	2	2
Soyasapogenol A	4	—
Soyasapogenol B	3	—
Lucernic acid	3	1

FIG. 6.10. Structures of alfalfa saponins.

TABLE 6.3. Growth and Feed Intake of Turkey Poults Fed Alfalfa Protein Concentrate (APC) Prepared from High- and Low-Saponin Alfalfa[a]

Treatment	Average 5-week body weight (g)	Average daily feed intake (g)
Control (soybean meal)	995	38.4
High-saponin APC	427	15.3
Low-saponin APC	735	24.3

[a] From Cheeke (1983).

Alfalfa protein concentrate (APC) prepared from low-saponin alfalfa gives better growth performance of rats than APC prepared from high-saponin alfalfa (Hegsted and Linkswiler 1980). The addition of methionine was required for adequate growth. The reduced gains with high-saponin APC were due to lower feed intake and lower protein quality (Hegsted and Linkswiler 1980). Similar results (Table 6.3) were obtained with turkey poults (Cheeke, 1983) and chicks (Ameenuddin et al. 1983) fed alfalfa protein concentrate from high- and low-saponin alfalfa.

Saponins form insoluble complexes with cholesterol. Dietary alfalfa or alfalfa saponins cause a reduction in tissue and serum cholesterol, reduced cholesterol absorption, and increased cholesterol fecal excretion. Saponins may bind with bile salts that are needed for cholesterol absorption or, because of their surface-active properties, may cause bile salts to bind to polysaccharides in fiber. Dietary saponins increase the fecal excretion of bile acids and neutral sterols. The effect of saponins in lowering serum cholesterol is to prevent its reabsorption after it has been excreted in the bile. Data reported by Malinow et al. (1979) illustrate the effects of saponins on cholesterol absorption and show that the effect of alfalfa in reducing cholesterol absorption is due to its saponin content (Table 6.4).

TABLE 6.4. Effect of Alfalfa Meal and Alfalfa Saponins on Cholesterol Absorption in Rats[a]

Treatment	Percent absorption of oral dose of cholesterol	
	Trial 1	Trial 2
Control	77	76
Control + alfalfa meal	48	46
Control + alfalfa meal (saponins extracted)	81	80
Control + extracted alfalfa meal + saponin extract	55	48

[a] From Malinow et al. (1979).

The feeding of alfalfa meal to laying hens as a method of reducing egg cholesterol levels has been examined. Any effect of dietary alfalfa or alfalfa saponins on egg cholesterol levels is slight. Nakaue et al. (1980) found no difference in egg cholesterol from hens fed diets containing high- or low-saponin alfalfa meal.

For many years, alfalfa saponins were thought to be involved in bloat in ruminants. Evidence for such an involvement was that saponins are surface-active agents producing stable foams, they are found in legumes that cause bloat and are not found in nonbloating legumes, and experimental administration of alfalfa saponins has caused bloat. However, more recent work (Majak et al. 1980) has shown that there is no difference in the incidence of bloat in cattle receiving high- or low-saponin alfalfa. Other factors, such as cytoplasmic protein fractions, appear to be those responsible for bloat.

Saponins in several pasture and range weeds have been implicated in toxicological problems. Alfombrilla (*Drymaria arenaroides*) is a perennial weed native to northern Mexico. It contains about 3% saponin and is acutely poisonous to cattle. Symptoms include anorexia, diarrhea, arched back, depression, coma, and death. Enteritis is observed. There is concern (Williams 1978) that alfombrilla may invade the U.S. Southwest. Alfombrilla is a member of the Caryophyllaceae family. Saponins are the principal toxins in other members of this family, including corn cockle (*Agrostemma githago*), bouncing bet or soapwort (*Saponaria officinalis*), and cow cockle (*Saponaria vaccaria*). Broomweed or snakeweed (*Gutierrezia sarothrae*) is a perennial resinous shrub found in desert ranges of the southwestern U.S. Its toxicity is believed to be due to its saponin content. Symptoms include listlessness, anorexia and weight loss, a rough hair coat, and gastroenteritis. In addition, it causes abortions in cattle. Molyneux et al. (1980) suggested that the abortifacient effects of broomweed are very similar to those of ponderosa pine needles and presented evidence that this was not due to the essential oil fraction. There is some evidence that saponins in broomweed are the abortifacient fraction.

Steroid saponins (referred to as sarsaponin) occur in yucca (*Yucca shidegria*) (Fig. 6.11). Yucca is sometimes grazed by cattle in the U.S. Southwest, particularly in times of drought and feed shortage. Sarsaponins have been reported to have favorable effects upon ruminant digestion and performance of feedlot cattle (Goodall 1979). There is also some evidence that sarsaponins may influence microbial nitrogen metabolism; there is interest in the use of sarsaponin to control ammonia levels in poultry houses (Johnston et al. 1981) and in using it in sewage treatment processes.

FIG. 6.11. Yucca plants on rangeland near Las Cruces, New Mexico. Yucca contains saponins and is a commercial source of these compounds for industrial and agricultural applications.

NITRO-CONTAINING GLYCOSIDES

Simple aliphatic compounds containing a nitrite moiety are found in several *Leguminosae* of agricultural importance. These include various *Astragalus* spp., crown vetch (*Coronilla varia*), and *Indigofera spicata*, a tropical forage. The nitrocompounds in *Astragalus* spp. are of the most significance in livestock production.

Of the at least 372 *Astragalus* spp. in North America, about 263 are poisonous to livestock because of the nitro-containing glycosides they contain. About 13 species cause locoism, while about 25 species may accumulate toxic levels of selenium. Examples of some of these are listed in Table 6.5. A number of *Astragalus* spp. are excellent forages containing no toxic entities. An example is cicer milk vetch (*A. cicer*) which is grown as an irrigated and dryland hay and pasture crop in the

TABLE 6.5. Classification of Common *Astragalus* spp. According to Type of Toxin

Selenium accumulators	Indolizidine alkaloids (locoweeds)	Nitro-containing glycosides[a]
A. bisulcatus	A. argillophilus	A. atropubescens (3-NPOH)
A. racemosus	A. bisulcatus	A. canadensis (3-NPA)
A. pectinatus	A. earlei	A. cibaria (3-NPA)
A. pattersonii	A. lentiginosus	A. convallarius (3-NPOH)
	A. mollissimus	A. diversifolius (3-NPOH)
	A. nothoxys	A. falcatus (3-NPA)
	A. pubentissumus	A. miser (3-NPOH)
	A. thurberi	A. pterocarpus (3-NPOH)
	A. wootonii	
	Oxytropis lambertii	
	Oxytropis sericea (= O. lambertii var. sericea)	

[a] The glycosides are metabolized in ruminants to 3-NPOH or 3-NPA as indicated for each species in parentheses.

Great Plains and western states of the U.S. *Astragalus* spp. are N-fixing legumes.

Timber milk vetch (*A. miser*) is a prominent example of a nitro-containing species (Fig. 6.12). The toxic principles are glycosides containing a simple aliphatic moiety with a nitrite group. The glycoside in milk vetch is called miserotoxin, which contains β-D-glucose and 3-nitropropanol (3-NPOH). Some *Astragalus* spp. contain glycosides with 3-NPOH, while three others contain 3-nitropropionic acid (3-NPA). The two nitro compounds do not occur together in the same species.

Signs of Astragalus Poisoning

Both acute and chronic toxicity occur in livestock that consume *Astragalus* spp. containing nitroglycosides. Acute poisoning is characterized by weakness, convulsions, frequent urination, rapid beating of the heart, labored and rasping breathing, coma, and death. Cyanosis of the oral cavity and nasal passages is evident, associated with methemoglobinemia.

Chronic toxicity is often referred to as "cracker heels" because the fetlocks knuckle over and general weakness in the hindquarters causes the hooves to rub against each other (Fig. 6.13). There is loss of neural control of the extremities, so the animal staggers or weaves when walking. Blindness or impaired vision may occur because of degeneration of the optic nerve. Livestock rarely recover from *Astragalus*

FIG. 6.12. A specimen of *Astragalus miser* (timber milk vetch).
Courtesy of M. C. Williams.

intoxication, but may linger for several months after being removed from the toxic forage.

James *et al.* (1980) performed feeding trials with various *Astragalus* spp. and provide a good summary of toxic doses and signs of toxicity. Williams *et al.* (1979) described in detail an outbreak of emory milk vetch poisoning of cattle and sheep in New Mexico.

Physiological Basis for Astragalus Toxicity

The toxic principles are glycosides of 3-NPOH or 3-NPA. Glycosides containing 3-NPOH, such as miserotoxin, are hydrolyzed to release 3-NPOH in the rumen, whereas in nonruminants, 3-NPA is produced (see Chapter 1). The 3-NPOH is rapidly absorbed from the rumen and is toxic, whereas 3-NPA is absorbed more slowly and appears to be degraded in the rumen to nontoxic compounds. Majak and Clark (1980) have studied the ruminal metabolism of aliphatic nitrocompounds.

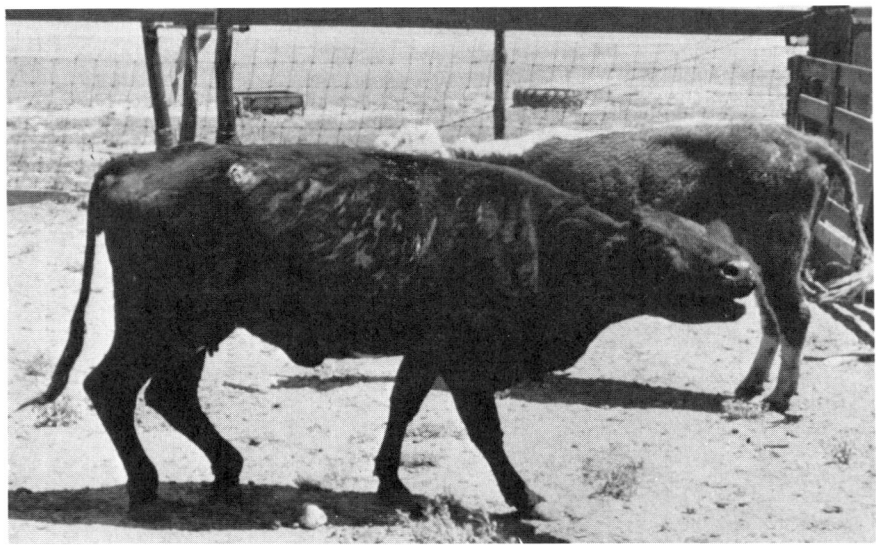

FIG. 6.13. Chronic toxicity signs due to ingestion of *Astragalus emoryanus* (Emory milk vetch). The cow displays the knuckled-over fetlocks, or "cracker heels," and general weakness of the hindquarters.
Courtesy of M. C. Williams.

Miserotoxin is rapidly hydrolyzed by rumen microorganisms; the 3-NPOH released can be degraded slowly, but at a much slower rate than the hydrolysis of the glycosidic bond. Per unit of NO_2 absorbed, 3-NPOH is much more toxic. However, *Astragalus* spp. containing 3-NPA often have several times as much glycoside as those with 3-NPOH, so the toxicity of the plants is often similar.

The metabolic basis of the toxic effect is not entirely clear. The acute toxicity is related to, but not exclusively caused by, methemoglobinemia in which hemoglobin is oxidized by nitrite:

$$\text{Hemoglobin Fe}^{2+} \text{ (ferrous state)} + NO_2^- \longrightarrow \text{Methemoglobin Fe}^{3+} \text{ (ferric state)}$$

Methemoglobin is incapable of carrying oxygen. Death usually occurs in acute astragalus toxicity when methemoglobin exceeds 20% of the total hemoglobin. This degree of methemoglobinemia is not lethal when induced by nitrite administration alone. Therefore, an additional factor, such as a metabolite formed when NO_2 is released from 3-NPOH or 3-NPA, seems to be involved. Majak *et al.* (1981A) studied the metabolism of intravenously administered 3-NPOH in cattle. They obtained evidence that 3-NPOH is metabolized at a site other than the circulatory system to yield NO_2^- and an unidentified metabolite(s) that may be involved in the intoxication. Nitrite is released into the blood, oxidizing hemoglobin to methemoglobin. Hemoglobin is subsequently regenerated by methemoglobin reductase, and the NO_2^- is converted to nitrate and excreted.

In chronic toxicity, levels of methemoglobin of 4–6% of total hemoglobin are observed. The mechanisms by which the nitrocompounds cause chronic toxicity are not conclusively known. Alston et al. (1977) have shown that 3-NPA is an inhibitor of the tricarboxylic acid cycle enzyme succinic dehydrogenase. The 3-NPA may bind to FAD, a cofactor for succinic dehydrogenase, making the FAD unreactive. In the process, nitrite is given off. The 3-NPA is structurally similar to succinate and may be acting as a competitive inhibitor. This reaction with FAD could explain both the metabolic effects and the occurrence of methemoglobinemia.

The astragalus nitrocompounds occur primarily in the leaves and reach their highest concentrations during pod formation. The levels drop rapidly when the leaves begin to dry. Herbicides cause the leaves to bleach and lose their activity. The toxins are stable in dried green specimens; nitro compounds have been measured in herbarium specimens collected over 100 years ago (Williams and James 1978).

The toxic dose of *A. miser* is about 4.8 g/kg body weight in cattle, which is equivalent to about 25 mg NO_2 per kilogram (about 100 mg NO_2 per kilogram is lethal). Sheep can tolerate 3–4 times as much per kilogram as cattle.

Williams (1981B) has examined 1690 spp. of Old World and South American *Astragalus* for alipatic nitrocompounds, using small (20 mg) samples of herbarium specimens. Nitrocompounds were detected in 12% of the Old World and 45% of the South American species, including some specimens collected between 1822 and 1836. Because *Astragalus* spp. are often considered for introduction on mine spoils, and reclamation of disturbed sites following oil, gas, and coal extraction on public grazing lands in the western states, it is important that introduced species be screened to be certain they are nontoxic. Sicklepod milk vetch (*A. falcatus*) was introduced to western ranges and subsequently was found to be highly toxic to sheep and cattle (Williams et al. 1976).

Bees have been poisoned by foraging on *Astragalus* spp. James et al. (1978) demonstrated that the pollen of *A. lentiginosus* was toxic to mice, so presumably the toxins would be present to affect bees. Majak et al. (1980) noted several outbreaks of high mortality of bees foraging on timber milk vetch (*A. miser*). The nectar was found to contain miserotoxin. Feeding trials with caged bees demonstrated the toxicity of the glycoside, with signs of incoordination, weakness, and inability to fly, followed by death. The possibility of toxicity of the honey, produced by bees feeding on *Astragalus* spp., to other organisms was not examined. The toxicity of *Astragalus* to bees is unusual in that comparatively few plants produce nectar or pollen which is poisonous to honeybees (Majak et al. 1980).

Crown Vetch Toxicity

Crown vetch (*C. varia*) is well adapted to the U.S. Northeast and Midwest (Fig. 6.14). It has considerable potential as a forage crop. It contains nitroglycosides such as coronarian:

$$\begin{array}{c} CH_2-O-\overset{O}{\underset{\|}{C}}-CH_2CH_2NO_2 \\ \text{(sugar ring with OH, OH, HO)} \\ O-\underset{\|}{\overset{}{C}}-CH_2CH_2NO_2 \\ O \end{array}$$

Crown vetch is not toxic to ruminants. Gustine et al. (1977) demonstrated that 3-NPA is degraded in rumen fluid to nontoxic metabolites. They established that 3-NPA can be completely detoxified in the rumen if its concentration does not exceed 1 mg/ml of rumen fluid. Such a level is unlikely to be surpassed as a result of consumption of crown vetch.

Nonruminants can be poisoned by crown vetch. Symptoms of 3-NPA toxicity include growth depression, ataxia, posterior paralysis, and death (Shenk et al. 1976). The toxicity may be due to a combination of methemoglobinemia and inhibition of succinate dehydrogenase (Alston et al. 1977).

VICINE (FAVISM)

The fava bean (*Vicia faba*), also called faba bean, horse bean, and broad bean, is an important protein source in the human diet (Fig. 6.15). It is grown extensively in Europe, particularly in Italy, Spain, Greece, and other countries of the Mediterranean region. It is also

FIG. 6.14. Crown vetch (*Coronilla varia*) is a forage plant grown in the U.S. Northeast and Midwest. It contains nitroglycosides, such as coronarian.

being evaluated as a protein supplement for livestock in Canada. This includes its use for poultry, swine, and as silage for cattle.

Consumption of fava beans or inhalation of the pollen of the plant sometimes has a deleterious effect, causing a condition called favism. It is characterized by acute hemolytic anemia. Symptoms, which may appear within minutes after inhalation of pollen, or 5–24 hr after consumption of beans, include headache, dizziness, nausea, yawning, vomiting, abdominal pain, and elevation of temperature. These symptoms may subside spontaneously or, in severe cases, acute hemolytic anemia with hemoglobinuria and icterus occur. Children are most affected, with a mortality rate of 6–8% reported in the past. Blood transfusion therapy has greatly reduced fatalities. Two peaks of incidence of favism are noted, one when the plant blooms, and the other in summer when the fresh beans appear on the market. As a matter of interest, the mathematician Pythagoras is said to have met his death at the hands of Greek soldiers rather than cross a field of fava beans. Pythagorus had founded a cult, and forbade his followers to eat fava beans or even walk among them (Crosby 1969; Marcus and Cohen 1967).

FIG. 6.15. Fava bean leaves and seed pods.
Courtesy of Manitoba Agriculture.

It is estimated that over 100 million people in the world are susceptible to favism (Mager *et al.* 1980). This is due to their genetic deficiency of a red blood cell enzyme, glucose-6-phosphate dehydrogenase (G6PD). The geographic distribution of G6PD deficiency parallels that of malaria, presumably because the enzyme deficiency increases the resistance of the red cells to the causative organism (Mager *et al.* 1980). Certain racial groups, such as Oriental Jews, Mediterranean Europeans, Arabs, Asians, and Blacks have a high incidence (5–50% of the population) of a low activity of G6PD. Northern Europeans, European Jews, American Indians, and Eskimos have virtually no incidence of the enzyme deficiency. In individuals susceptible to favism, G6PD activity is only 0–6% of normal.

The causative factor(s) is not completely identified, but appears to be the aglycones of glycosides in the fava beans. Two of the major glycosides are vicine and convicine; their respective aglycones are divicine and isouramil:

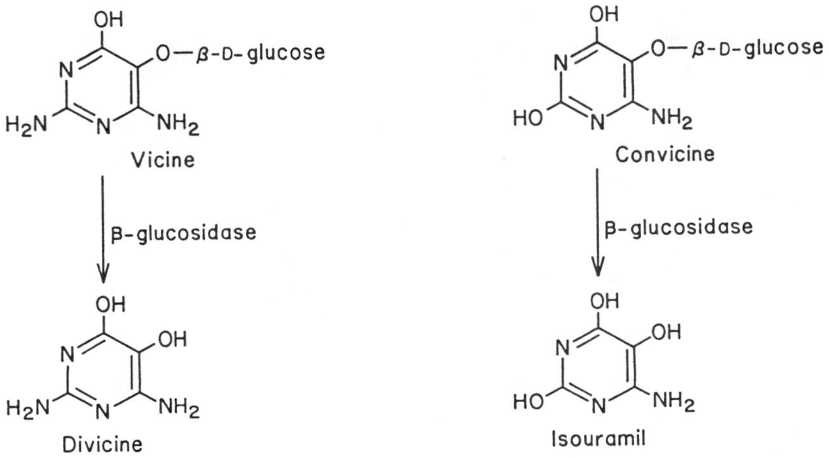

6. GLYCOSIDES

Vicine → (β-glucosidase) → **Divicine**

Convicine → (β-glucosidase) → **Isouramil**

The aglycones (divicine, isouramil) may either react with the red cell membrane directly or produce hydrogen peroxide, causing breakdown of the red cell membrane and hemolysis. In normal individuals, this is prevented by reduction of the oxidants by reaction with reduced glutathione (GSH). The supply of GSH is maintained by reactions of the pentose phosphate pathway in which G6PD functions. The major reactions are as follows:

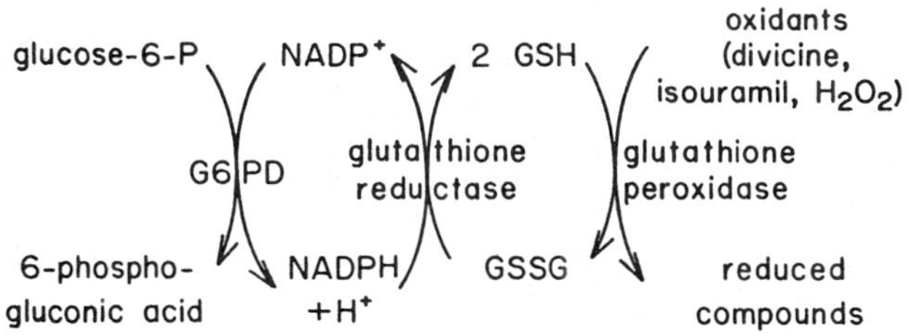

```
glucose-6-P      NADP⁺       2 GSH      oxidants
         \     /      \     /           (divicine,
          X           X                  isouramil, H₂O₂)
         G6PD   glutathione    glutathione
                reductase      peroxidase
6-phospho- /   \ NADPH /   \ GSSG      reduced
gluconic acid    +H⁺                    compounds
```

With a deficiency of G6PD, formation of reduced NADP cannot be increased by increased oxidation of glucose. As a result, reduced GSH cannot be regenerated fast enough, so the oxidants are not destroyed, but are available to attack the red cell membrane.

Of interest is that selenium functions as an essential nutrient as a part of glutathione peroxidase, the terminal enzyme in the above scheme. The essential metabolic role of selenium was discovered by Rotruck et al. (1973) when they were investigating why dietary selenium would protect against in vitro red cell hemolysis only when the incubation mixture contained glucose.

A consequence of the oxidant action of the fava bean compounds is the formation of methemoglobin in the red cell and the appearance of Heinz bodies. These are clumps of denatured hemoglobin resulting from the oxidation of its SH groups. The oxidant compounds may react with oxyhemoglobin to produce hydrogen peroxide, which may be the active membrane-rupturing factor. Hydrogen peroxide is converted to water by glutathione peroxidase.

It appears that individuals with a genetic deficiency of red cell G6PD activity can maintain sufficient GSH to cope with normal metabolic requirements, but when exposed to the stress of various oxidants, such as the fava bean glycosides or various drugs (e.g., the antimalarial primaquine), the regeneration of GSH is not rapid enough to dispose of the oxidants, and destruction of the red cell membrane occurs.

Utilization of Fava Beans by Livestock

Unprocessed fava beans contain factors that lower chick growth rate and feed efficiency, alter liver size, and increase pancreas size (Marquardt et al. 1974). In laying hens, dietary fava beans reduce feed efficiency, egg weight, and egg production rate (Campbell et al. 1980). Both thermostable and thermolabile factors are involved. The thermolabile factors include tannins, protease inhibitors, and lectins. Tannins are the major thermolabile antinutritional factor and account for over 50% of the chick growth depression (Marquardt et al. 1977). The thermostabile factors, vicine and convicine, are the other major antinutritional factors. Muduuli et al. (1981) studied the effects of dietary vicine on laying chickens. Feeding vicine caused a reduction in the number of developing ova, egg, and yolk weights, and reduced the fertility and hatchability of the eggs. Vicine consumption also elevated plasma lipid and peroxide levels, increased erthyrocyte hemolysis, and the birds had heavier livers with higher lipid peroxide and reduced glutathione levels. Vicine from fava beans therefore has a marked influence on the metabolism of the laying hen.

Whole crop fava beans have potential as a silage crop. Thorlacius and Beacom (1981) compared silage from fava beans, oats, corn, and field peas, in trials with lambs. The dry matter intake and rate of gain were greater for lambs fed fava bean silage than for those fed corn or oat silage, while the digestibility of dry matter and protein was generally higher for fava bean silage than for oats and corn silage. The field pea and fava bean silages were similar in most respects. Promising results with fava bean silage were also obtained by McKnight and MacLeod (1977) and Ingalls et al. (1979).

CALCINOGENIC GLYCOSIDES

Calcinosis is the deposition of calcium in the soft tissues. Some plants contain glycosides of 1,25-dihydroxycholecalciferol (1,25-OHD$_3$), the active metabolite of vitamin D. Consumption of these plants by grazing animals causes an induced vitamin D toxicity manifested by calcinosis.

Three plants have been implicated in calcinosis. These are *Solanum malacoxylon, Cestrum diurnum,* and *Trisetum flavescens,* which have caused calcinosis in grazing animals in South America (Argentina and Brazil), Florida, and the alpine region of Germany and Austria, respectively. The mode of action of the calcinogenic glycosides can best be perceived after a brief consideration of the metabolism and mode of action of vitamin D.

The primary function of vitamin D is the regulation of calcium (and secondarily, phosphorus) absorption. A metabolite of vitamin D (1,25-OHD$_3$) regulates the synthesis and activity of calcium-binding protein

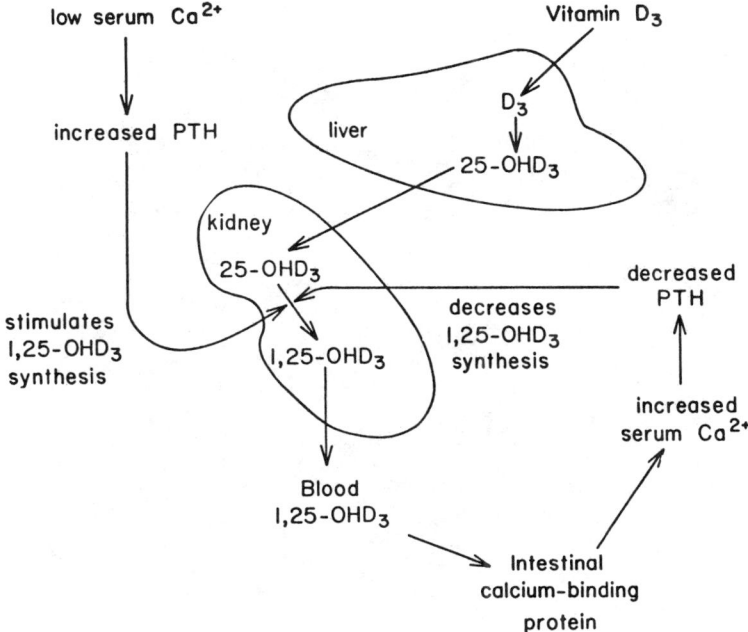

FIG. 6.16. Metabolism of vitamin D to produce the active metabolite, 1,25-dihydroxy vitamin D$_3$.

(CaBP) in the intestinal mucosa, which transports calcium from the intestine to the blood.

The level of 1,25-OHD_3 is governed by the serum calcium level. If the serum calcium falls, the parathyroid hormone (PTH) secretion is increased, which increases formation of 1,25-OHD_3, stimulating calcium absorption to bring the serum calcium level back to normal. These relationships are shown in Fig. 6.16.

Researchers at Cornell University (Wasserman 1978) have demonstrated the presence of glycosides of 1,25-OHD_3 in *S. malacoxylon* and *C. diurnum*. The plant compounds seem to act in an identical manner with the vitamin D metabolite formed in animal tissue. Thus, consumption of the plants overrides the regulatory role of 1,25-OHD_3 in animals in which its synthesis is governed by PTH, which in turn is controlled by the serum Ca level. The result is excessive calcium absorption and the deposition of the extra calcium in the soft tissues. Symptoms observed in grazing animals consuming these calcinogenic

FIG. 6.17. A 3-year-old Angus bull showing signs of *Cestrum diurnum* intoxication. A solid stand of *C. diurnum* is in the background.
Courtesy of L. Krook and The Cornell Veterinarian.

plants include progressive weight loss, lameness and stiffness of limbs, arching of the back, hypercalcemia and hyperphosphatemia, and calcification of the cardiovascular system, tendons, ligaments, lungs, diaphragm, and kidney (Fig. 6.17).

The activity of *S. malocoxylon* is about 1.3×10^5 IU vitamin D_3 equivalent per kilogram of dried leaf, while for *C. diurnum* the activity is about 3×10^3 IU D_3 per kilogram dried leaf, or about one-tenth that of *S. malacoxylon* (Wasserman 1978). A level of 1.5–3% dietary *S. malacoxylon* will induce calcinosis.

AZOXYGLYCOSIDES

Cycads are ancient palm like plants that were widely distributed in the Mesozoic period (Fig. 6.18). They are found in tropical and subtropical areas. Livestock poisonings from consumption of cycads are important in Australia, where *Cycas* and *Macrozamia* spp. are found in tropical grazing areas. Cycads survive adverse conditions, such as drought and fire. The roots, seeds, and stems contain high levels of starch. They are consumed by grazing animals and by humans in some tropical areas.

The cycads contain glycosides that cause hepatic and gastrointestinal diseases in livestock (Hooper 1978) and that have been shown to be

FIG. 6.18. A specimen of *Cycas media*, a cycad common in Western Australia.
Courtesy of P. R. Dorling.

carcinogenic (Wogan and Busby 1980). One of the main glycosides is cycasin, which contains methylazoxymethanol (MAM) as its aglycone. The structures are as follows:

$$\text{Cycasin} \xrightarrow{\beta\text{-glucosidase}} \text{glucose} + \text{HO-CH}_2\text{-N=N-CH}_3 \text{ (Methylazoxymethanol, MAM)}$$

Extensive losses of sheep have occurred in Australia as a result of consumption of *Macrozamia* and *Cycas* spp. (Hooper 1978). The seeds (nuts) as well as the leaves are eaten. Symptoms include liver cirrhosis, occlusion of central and hepatic veins, and gastroenteritis. The causative agent is MAM, released from the glycoside by β-glucosidase activity in the gastrointestinal tract.

Cattle consuming these plants develop a neural condition in which the hind legs become paralyzed because of axon degeneration in the central nervous system. The causative agent(s) has not been identified, but does not seem to be MAM (Hooper 1978).

Although MAM has been shown to be carcinogenic (Wogan and Busby 1980), there is no direct evidence that human populations which consume cycads have a higher cancer incidence than populations which do not consume it. Traditional methods of preparing cycad flour include fermentation, heating, water extraction, and sun-drying, which seem to destroy the carcinogenic activity.

An interesting historical note is that Captain Cook's crew consumed seeds of *Cycas media* when their ship ran aground at Cooktown, Australia. Cook wrote in his diary that the "sailors were violently ill, both upwards and downwards"! (A. E. Bell, Kew Gardens, personal communication, May, 1984).

Reddy *et al.* (1982) reported that butylated hydroxyanisole (BHA) when administered to mice protected against the acute toxicity of MAM. This protection was associated with increased hepatic levels of cytochromes P_{450} and b_5, and a reduction in necrotic changes in the liver. The MAM is a colon-specific carcinogen in mice.

CARBOXYATRACTYLOSIDE

Cocklebur (*Xanthium* spp.) is a coarse herbaceous annual weed found in many parts of the world. In the U.S., *X. strumarium* is the species involved in livestock poisoning. Significant losses of animals occur in the U.S., Australia, and South Africa. The cocklebur has a fruit containing two seeds surrounded by a spiny capsule. One seed germinates the first growing season and the other the following year. Cocklebur often grows in areas under water for extensive periods and that dry out during the summer. These conditions are found along streams or along the shores of shallow farm ponds. It also grows in pastures and fields. Only the seedlings in the cotyledon stage are poisonous (Fig. 6.19). As the first true leaves develop, toxicity is rapidly lost.

In the U.S., pigs seem to be the livestock most frequently poisoned. Signs of toxicity include depression, reluctance to move, a hunched posture, nausea, vomiting, weakness and prostration, dyspnea, opisthotonus, paddling of the limbs and convulsions when recumbent, coma, and death. Severe hypoglycemia occurs, with blood glucose levels going from a normal of about 100 mg/100 ml to levels as low as 16 mg/100 ml. The principal gross lesions seen are related to increased

FIG. 6.19. Cocklebur (*Xanthium strumarium*) seedlings which are the source of toxicity problems with cocklebur.
Courtesy of M. E. Fowler.

vascular permeability. These include edema of the gallbladder wall and ascites of the peritoneal cavity. There is evidence of gastrointestinal tract irritation. Acute centrilobular liver necrosis occurs.

For many years, the toxic agent in cocklebur was thought to be hydroquinone (Kingsbury 1964). Cole et al. (1980) and Stuart et al. (1981) demonstrated conclusively that the toxic agent is carboxyatractyloside, a glycoside which had previously been isolated from cocklebur and shown to be hypoglycemic (Kupiecki et al. 1974; Craig et al. 1976). Carboxyatractyloside causes uncoupling of oxidative phosphorylation, which probably contributes to its hypoglycemic effect. The structure of carboxyatractyloside is as follows:

Carboxyatractyloside

Hatch et al. (1982) have used various enzyme inducers and inhibitors to attempt to elucidate the mechanisms of metabolism of the cocklebur glycoside. The use of compounds such as phenobarbital to stimulate cytochrome P_{450}-dependent enzymes did not alter toxicity. Phenylbutazone, which apparently induces synthesis of a non-cytochrome P_{450}-dependent detoxification enzyme, did reduce the toxic effects. Glutathione precursors or blockers did not affect toxicity. Further studies are necessary to completely identify the metabolic pathways involved in carboxyatractyloside metabolism.

Cutler and Cole (1983) demonstrated that carboxyatractyloside is a plant growth inhibitor, and hypothesized that a higher content of the compound in one of the two seeds in the cocklebur seed bur may account for the delayed dormancy of one of the seeds.

ISOFLAVONES AND COUMESTANS

Subterranean clover (*Trifolium subterran*) has been widely sown for sheep pasture in many parts of Australia. It is also extensively grown on hill pasture lands in the U.S. Pacific Northwest. It is a winter annual, sprouting with the autumn rains and providing forage during the winter and spring. In late spring, it produces seeds and then dries up. The seeds are in burs that are pushed into the ground at maturity, so the plant reseeds itself. This characteristic provides the origin of its name. Sub clover, as it is commonly known, has greatly increased pasture productivity in regions where it has been adapted.

In the early 1940s, as sub clover became an important pasture species in Western Australia, a dramatic decrease in the fertility of sheep to a level of about 30% fertility was noted. The infertility was expressed as a failure to conceive and was accompanied by a cystic glandular hyperplasia of the cervix and uterus. Lactation in nonpregnant ewes and wethers suggested that a plant estrogen was involved in the so-called "clover disease." Australian researchers in the early 1950s extracted nearly 5 tons of fresh clover, from which they were able to isolate and identify two isoflavones, genistein and formononetin, that had estrogenic activity. These and other plant estrogens are referred to as phytoestrogens. Since that time, a great deal of Australian research has helped to identify the modes of action of phytoestrogens. North American research groups in this area have included Bickoff and colleagues at the United States Department of Agriculture Western Regional Research Laboratory in Berkeley, California, and Kitts and coworkers at the University of British Columbia, Vancouver.

Pasture species that cause livestock problems because of their phytoestrogen content include sub clover, red clover (*Trifolium pratense*), and alfalfa (*Medicago sativa*). The estrogens in clovers are usually isoflavones, while alfalfa contains coumestans. Structures of some common phytoestrogens are shown in Fig. 6.20. Their resemblance to estradiol can be readily seen. The phytoestrogens occur in plant tissue as water-soluble glycosides. The isoflavones are synthesized by plants from phenylalanine, while the coumestans are synthesized from cinnamic acid.

The mouse uterine weight bioassay has been extensively used in studies of phytoestrogens. Plant extracts or the isolated estrogens are injected into immature female mice, and 24 hr later the uterine weight is measured. Estrogens cause an increase in uterine weight (Fig. 6.21). Examples of some typical dose responses are shown in Table 6.6.

The equivalent potencies at a dosage required to produce a 25-mg uterus were estrone, 6900; coumestrol, 35; genistein, 1; daidzein, 0.75; biochanin A, 0.46; and formononetin, 0.26.

These results show that the phytoestrogens have an exceedingly low potency as compared to estrone. However, they can produce significant biological effects through an additive action with endogenous estrogen,

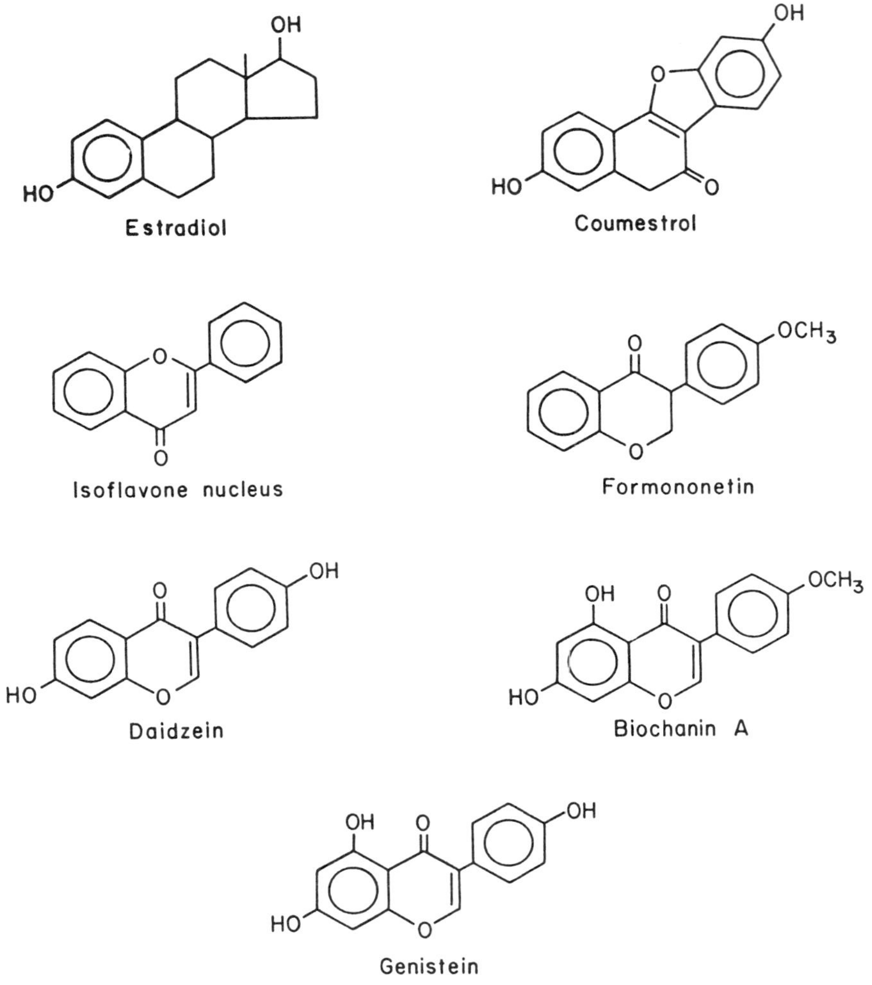

FIG. 6.20. Structures of some common phytoestrogens and estradiol.

and they may occur in plants at very high levels. The isoflavone content of sub clover may reach 5% of the dry weight.

A puzzling observation was that the estrogenic activity of sub clover pastures, as assessed by teat length of wethers, was correlated with the formononetin content of the pasture. Formononetin has a very low estrogenic activity (Table 6.6). The explanation resides in rumen me-

TABLE 6.6. Dose Response with Phytoestrogens Using Mouse Uterine Weight Bioassay[a]

Compound	μg/mouse	Uterine weight (mg)
Control	0	9.6
Estrone	0.5	14.7
	1	23.8
	2	45.3
Coumestrol	100	13.8
	200	24.2
	400	40.7
Genistein	5,000	19.4
	8,000	27.0
	12,000	32.4
Daidzein	5,000	17.3
	10,000	24.8
	15,000	31.2
Biochanin A	10,000	20.3
	20,000	27.9
	40,000	45.5
Formononetin	15,000	16.8
	25,000	23.2
	40,000	26.1

[a] Adapted from Livingston (1978).

tabolism. In the sheep rumen, biochanin A and genistein are degraded to p-ethylphenol and a phenolic acid, whereas formononetin is demethylated to daidzein and then metabolized to equol (Fig. 6.22). Equol is estrogenic. Hence, formononetin is bioactivated by rumen microorganisms to a more potent estrogen. The same metabolism occurs in the rumen of cows; the absorbed equol is excreted more rapidly in cattle, so they are less susceptible than sheep to the estrogenic effects of clover isoflavones.

Physiological Effects of Phytoestrogens

After sheep have grazed estrogenic pasture for several years, the fertility of the flock becomes depressed. A typical example is shown in Table 6.7, indicating that with a high-estrogen cultivar of sub clover, fertility of the flock eventually declined to zero. This condition of permanent infertility is known as clover disease. The main cause of the infertility is a failure of fertilization associated with poor sperm penetration to the oviduct. The cervical mucus has an altered consistency which impairs sperm storage in the cervix. Sperm are stored in the cervix after mating; in clover-affected ewes, the number of sperm present after 24 hr is less than 5% of that expected. Adams (1981)

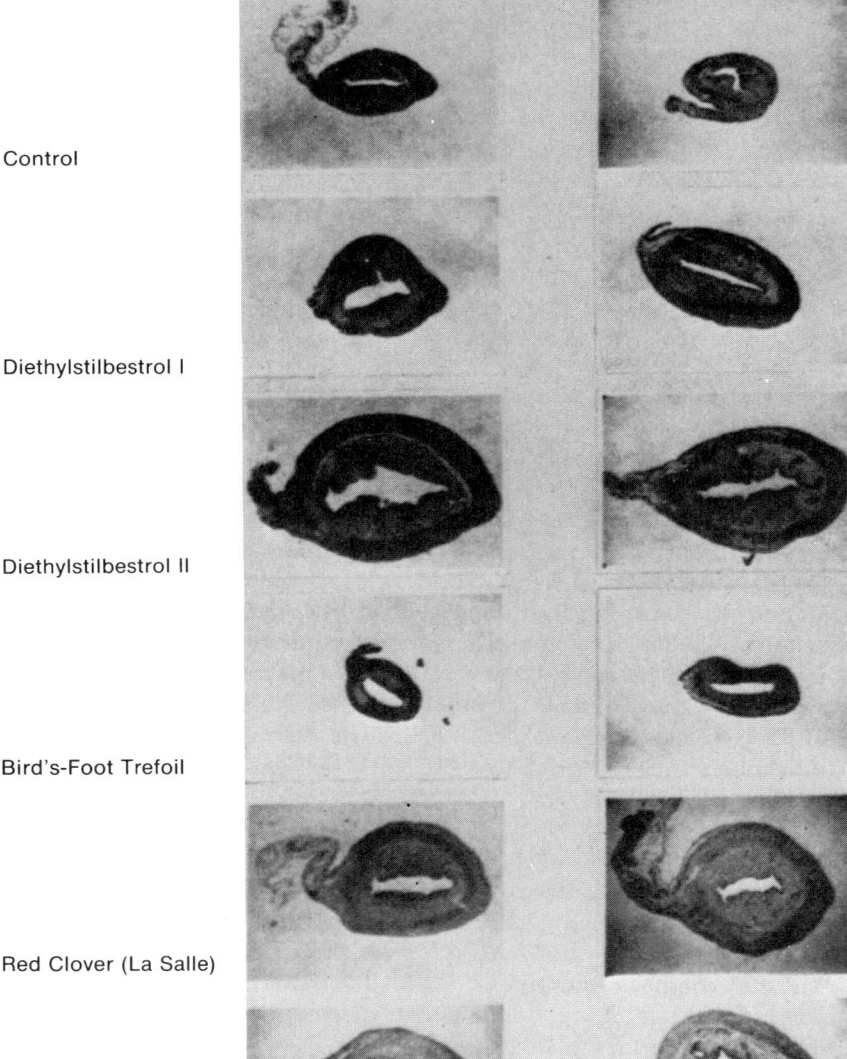

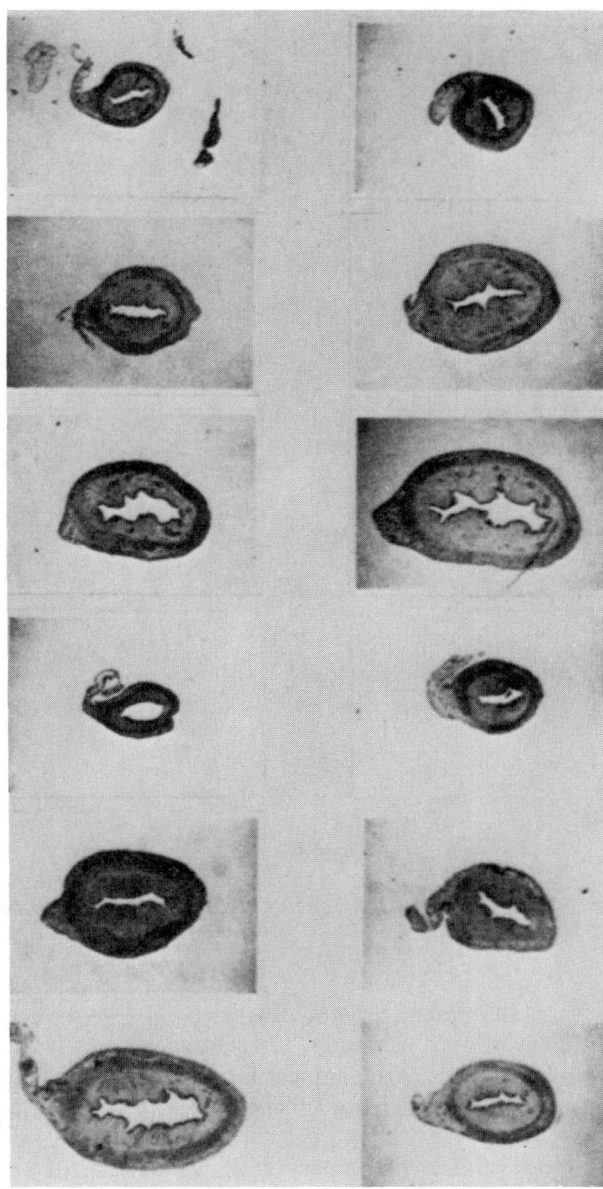

FIG. 6.21. Photomicrographs of cross sections of uteri from rats receiving phytoestrogens. Note the apparent "anti-estrogen" effect of birdsfoot trefoil extract. From Ostrovsky and Kitts (1963).
Courtesy of W. D. Kitts.

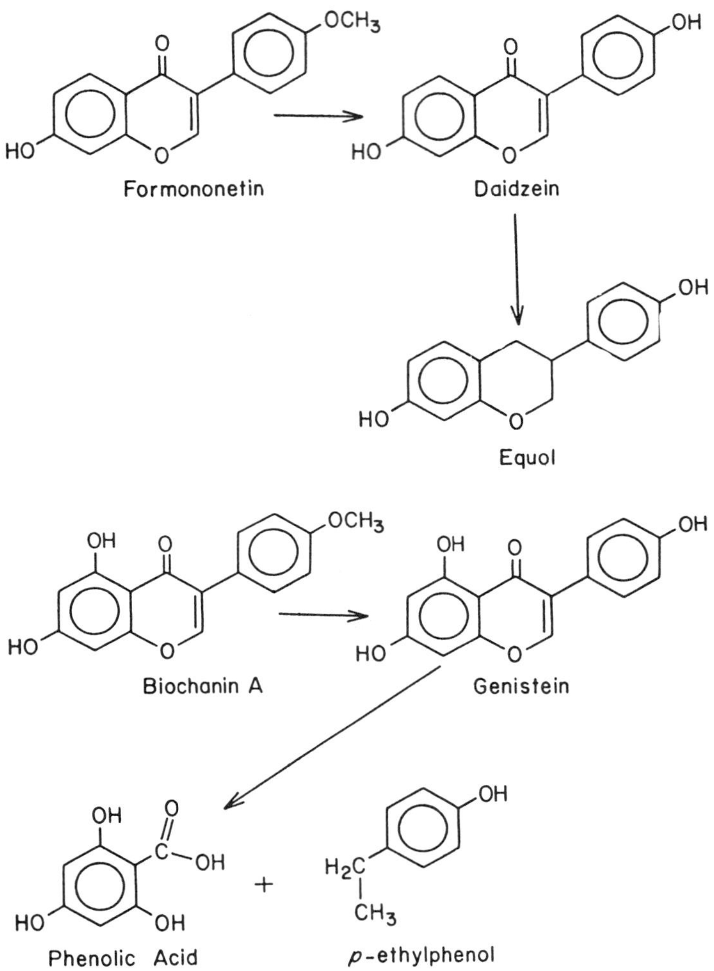

FIG. 6.22. Metabolism of subterranean clover isoflavone in the sheep rumen.

suggests that the change in mucus consistency is due to an altered responsiveness to stimulation with estrogen. Therefore, the cervix and vagina of ewes with clover disease fail to respond normally to endogenous estrogen stimulation to "prime" the cervix during the breeding season.

In ewes affected by clover disease, the cervix shows structural and functional changes. The cervical tissue changes in morphology to look more like uterine tissue than a cervix. The normal cervix has folds

TABLE 6.7. Effect of Formononetin Content of Subterranean Clover on Percentage of Ewes Lambing[a]

		Pasture type			
		Subterranean clover			
Year	Non-estrogenic control	Woogenellup (0.15)[b]	Geraldton (0.79)[b]	Dinninup (1.19)[b]	Dwalganup (1.30)[b]
1967	91	76	87	78	89
1968	73	84	78	72	56
1969	86	69	56	53	30
1970	59	41	42	35	6
1971	84	63	53	41	8
1972	85	67	52	38	0

[a] From Neil et al. (1974).
[b] Percentage formononetin.

(Fig. 6.23); in clover disease, the folds of the cervix fuse together, trapping epithelial tissue to make it look like glands (Fig. 6.24).

If sheep are bred while grazing estrogenic pasture, fertility depression can occur. The infertility does not persist if they are subsequently maintained on nonestrogenic pastures. This temporary infertility is especially pertinent to the coumestans. Problems in breeding have

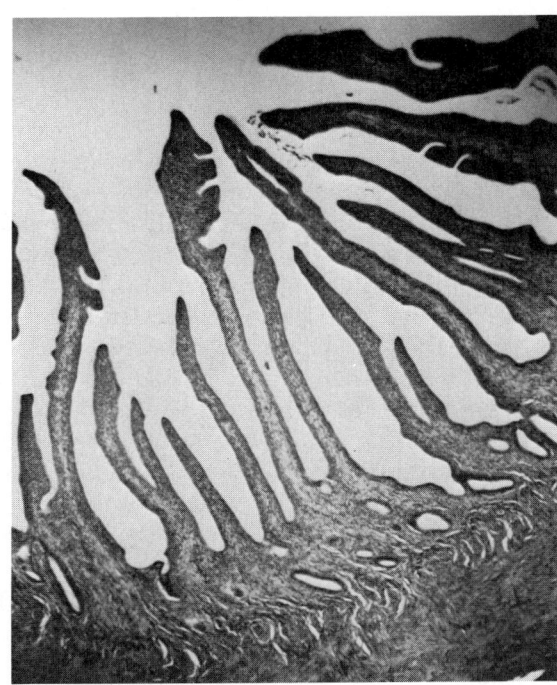

FIG. 6.23. The cross section of a normal cervix of a ewe.
Courtesy of N. R. Adams.

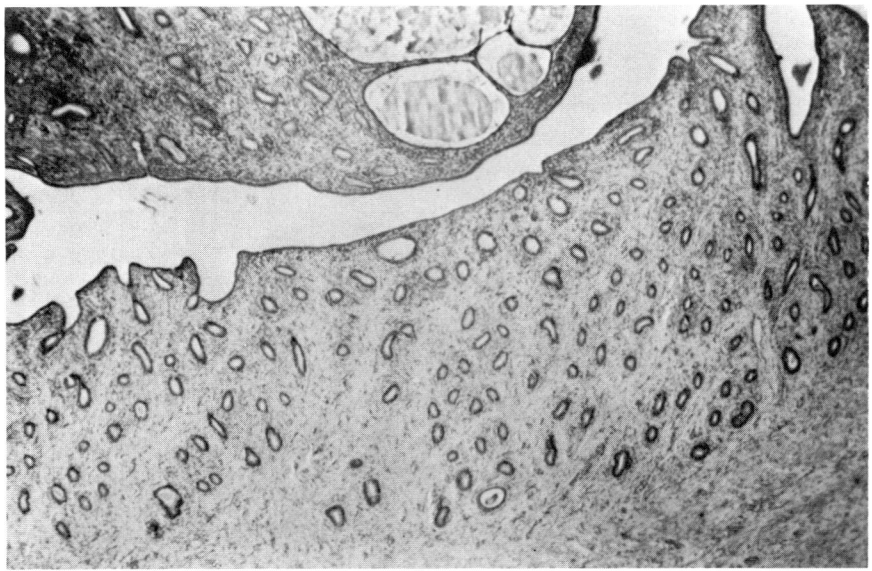

FIG. 6.24. A cross section of the cervix of a ewe with clover disease. The animal had grazed estrogenic subterranean clover for 3 years. The folds of the cervix tissue have folded together, trapping epithelial tissue to give the appearance of glands.
Courtesy of N. R. Adams.

occurred with dairy cattle fed alfalfa. These included decreased fertility because of cystic ovaries and irregular estrous cycles, as well as precocious mammary and genital development in heifers. Coumestans suppress estrus and inhibit ovulation, probably by lowering ovarian estrogen secretion. Sheep grazing estrogenic clover do not show the normal seasonal changes in serum LH (Chamley *et al.* 1981).

The fertility and sperm production of rams does not seem to be affected by their grazing on estrogenic pastures. Wethers may develop enlarged teats and begin lactating. Teat enlargement of wethers has been used as a sensitivity index of the potency of pastures.

An interesting situation is that phytoestrogens may be involved in the regulation of reproduction of California quail. Leopold *et al.* (1976) reported that during dry years, stunted desert plants produced high levels of estrogenic isoflavones that inhibited quail reproduction. In normal or wet years, the plants grow abundantly and the levels of phytoestrogens are low, resulting in higher reproduction rates in the quail.

Phytoestrogens may act as "antiestrogens." A high blood level of phytoestrogens may inhibit the release of gonadotropic hormones from the pituitary and may compete with endogenous estrogens for receptor sites in target tissues such as the uterus and cervix.

Problems associated with phytoestrogens and ewe fertility in Australia are much less than they once were. A major reason has been the development of low-formononetin cultivars of sub clover. Also, animal management to limit estrogenic exposure is practiced. However, moderate depression of fertility is still observed, and it has been estimated that about a million ewes fail to lamb each year in Australia because of phytoestrogen exposure (Cox 1978). The problem will persist for many years because of the difficulty of eliminating the high-estrogen cultivars from existing pastures. Sub clover has a high percentage of hard seeds which resist germination, and therefore a reservoir of seeds of the original cultivars persists in pastures that have been newly seeded to the low-formononetin types.

Laminitis or founder is a common problem in horses grazing both grass and clover. Steroidal hormones increase in the blood at the onset of founder (J. R. Coffman, Manhattan, Kansas, personal communication). Therefore, phytoestrogens might play a role in the etiology of clover founder.

Alfalfa tablets have become a common dietary supplement for humans, available in health food stores. Elakovich and Hampton (1983) analyzed three brands of commercial alfalfa tablets for their phytoestrogen content, which ranged from 20 to 190 ppm. They warned that this level of intake, in conjunction with other extraneous estrogen sources such as birth control pills and estrogen replacement therapy, could be potentially harmful. The benefits, if any, of consuming alfalfa tablets are unclear.

JOJOBA GLYCOSIDES

Jojoba (*Simmondsia californica*) is native to the desert areas of the U.S. Southwest. Its seeds contain a liquid wax, with properties similar to those of sperm whale oil, that is used in various applications such as cosmetics manufacture. The jojoba meal remaining after the oil has been extracted is a potential protein source for livestock. It gives very poor feeding results. The meal can be detoxified by treatment with ammonia (Liener 1980). Its toxicity has been attributed to a glycoside, called simmondsin:

$$\beta\text{-glucose}-O-\underset{\underset{OCH_3}{|}}{\overset{\overset{CN}{|}}{\underset{}{C}}}\begin{matrix}\\ OH \\ OCH_3\end{matrix}$$

Verbiscar et al. (1981) found that selected strains of *Lactobacillus acidophilus* and *L. bulgaricus* grew well on jojoba meal and decreased the level of simmondsin and related compounds by 95–98% after 21 days of incubation. As well as rendering the jojoba meal nontoxic to mice, poultry, sheep, and cattle, the treatment increased palatability.

RANUNCULIN

Buttercups (*Ranunculus* spp.) are a common pasture weed in North America, Europe, South Africa, and Australia. They generally grow in wet soils and marshy areas. Pastures heavily infested with buttercups are often characterized by acid soil, and poor fertilization and pasture management practices. Common species in North America are *Ranunculus acris* (tall field buttercup) and *Ranunculus repens* (creeping buttercup). Both are common across the northern U.S. and southern Canada.

Buttercups contain a glycoside, ranunculin, which upon crushing of the plant tissue is enzymatically converted to a yellow volatile oil, protoanemonin. Protoanemonin is unstable and either polymerizes to nontoxic anemonin or is volatilized. Therefore, the dried plant, as in hay, is nontoxic. Protoanemonin is an irritant and may cause blisters on the lips and irritation of the mouth and digestive tract, producing salivation, abdominal pain, and diarrhea. Buttercups are unpalatable and will not be consumed unless other feed is sparse. While a number of poisonings in North America have occurred in the past (Kingsbury 1964), buttercup is not a significant problem now, probably due to improved pasture management, including liming, fertilization, and seeding with improved pasture species. Therrien et al. (1962) and Hidiroglou and Knutti (1963) conducted feeding trials with sheep and cattle and found that even when high levels of buttercup (*R. acris*) were fed, there was no indication of toxicological problems. It appears that any hazards associated with consumption of common buttercup are slight.

However, in northwestern Oregon, some outbreaks of buttercup poisoning have been noted in cattle on dikeland pastures (J. Plummer, Ranier, Oregon, personal communication, June, 1984). Photosensitization and mild liver dysfunction are seen.

The structures of ranunculin and its breakdown products are as follows:

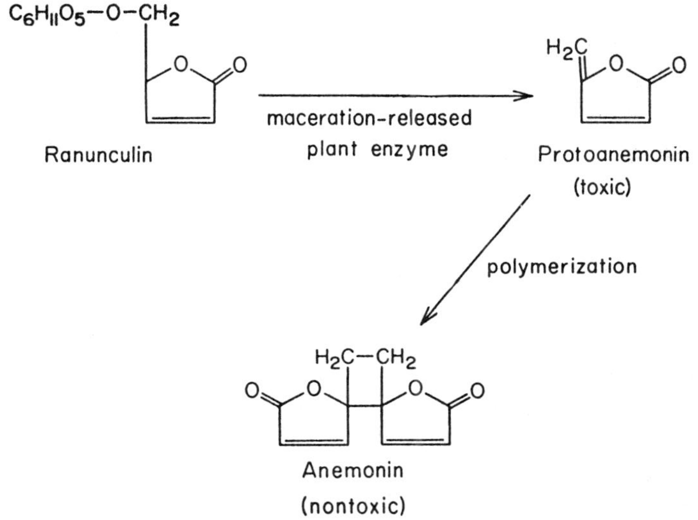

$C_6H_{11}O_5$—O—CH_2

Ranunculin

maceration-released plant enzyme

H_2C

Protoanemonin (toxic)

polymerization

H_2C—CH_2

Anemonin (nontoxic)

Bur buttercup (*Ceratocephalus testiculatus*) is an annual weed that infests large areas of several western states. It is of Mediterranean origin and was first identified in the U.S. in Utah in 1932. It is not a true buttercup (*Ranunculus* spp.), but closely resembles the buttercups in appearance. Bur buttercup frequents roadsides, heavily grazed areas, old sheep bed grounds, and other disturbed areas. It grows from 1 to 5 in. in height and forms dense stands. It has small, yellow buttercup-like flowers, with characteristic burlike clusters of fruit. The plant has been shown to be toxic to sheep. Olsen *et al.* (1983) reported an incident in Utah in which about 150 of a band of 800 ewes brought into lambing pastures died from bur buttercup poisoning. The sheep were hungry when unloaded into the pastures, which contained about 50% bur buttercup and 50% cheatgrass (*Bromus tectorum*). A few hours after grazing the plant, the sheep developed watery diarrhea, weakness, and labored breathing. Necropsy revealed edema of the rumen wall, hemorrhage on the inside of the left ventricle of the heart, congestion of the lungs, liver, and kidneys, and fluid accumulation in the thoracic and abdominal cavities. Toxicity of the bur buttercup was confirmed by an LD_{50} test with sheep gavaged with green plant material. About 10.9 g of green bur buttercup per kilogram of body weight was the lethal dose, with identical toxicity signs as observed in the field outbreak.

Bur buttercup is unpalatable and a lethal dose is not likely to be consumed under normal grazing conditions. When hungry sheep are put into an area heavily infested with the plant, significant stock losses can be expected. Nachman and Olsen (1983) demonstrated that bur buttercup contains ranunculin. The early flower stage had the highest concentration of the toxin.

REFERENCES

Cyanogens

BEILSTEIN, M. A., and WHANGER, P. D. 1984. Effects of cyanide on selenium metabolism in rats. J. Nutr. *114*, 929–937.

COOPER-DRIVER, G. A., and SWAIN, T. 1976. Cyanogenic polymorphism in bracken in relation to herbivore predation. Nature (London) *260*, 604.

HALVERSON, A. W., HENDRICK, C. M., and OLSON, O. E. 1955. Observation on the protective effect of linseed oil meal and some extracts against chronic selenium poisoning in rats. J. Nutr. *56*, 51–60.

KEELER, R. F. 1984. Teratogens in plants. J. Anim. Sci. *58*, 1029–1039.

KINGSBURY, M. 1964. Poisonous Plants of the United States and Canada. Prentice-Hall, Englewood Cliffs, NJ.

MAJAK, W., McDIARMID, R. E., HALL, J. W., and VAN RYSWYK, A. L. 1980A. Seasonal variation in the cyanide potential of arrowgrass (*Triglochin maritima*). Can. J. Plant Sci. *60*, 1235–1241.

MAJAK, W., UBENBERG, T., CLARK, L. J., and McLEAN, A. 1980B. Toxicity of saskatoon serviceberry to cattle. Can. Vet. J. *21*, 74–76.

MAJAK, W., McDIARMID, R. E., and HALL, J. W. 1981. The cyanide potential of saskatoon serviceberry (*Amelanchier alnifolia* and Chokecherry (*Prunus virginiana*). Can. J. Anim. Sci. *61*, 681–686.

MONTGOMERY, R. D. 1980. Cyanogens. *In* Toxic Constituents of Plant Foodstuffs. I. E. Liener (Editor), pp. 143–160. Academic Press, NY.

PALMER, I. S. 1981. Interactions between selenium and cyanogenic glycosides or cyanide. *In* Selenium in Biology and Medicine. J. E. Spallholz, J. L. Martin, and H. E. Ganther (Editors), pp. 336–342. AVI Publishing Co., Westport, CT.

PALMER, I. S., OLSON, O. E., HALVERSON, A. W., MILLER, R., and SMITH, C. 1980. Isolation of factors in linseed oil meal protective against chronic selenosis in rats. J. Nutr. *110*, 145–150.

PRITCHARD, J. T., and VOSS, J. L. 1967. Fetal ankylosis in horses associated with hybrid Sudangrass pasture. J. Am. Vet. Med. Assoc. *150*, 871–873.

SEAMAN, J. T., SMEAL, M. G., and WRIGHT, J. C. 1981. The possible association of a sorghum (*Sorghum sudanese*) hybrid as a cause of developmental defects in calves. Aust. Vet. J. *57*, 351–352.

SELBY, L. A., MENGES, R. W., HOUSER, E. C., FLATT, R. E., and CASE, A. C. 1971. An outbreak of swine malformations associated with wild black cherry, *Prunus serotina*. Arch. Environ. Health *22*, 496–501.

SMITH, C. R., WEISLEDER, D., MILLER, R., PALMER, I. S., and OLSON, O. E. 1980. Linustatin and neolinustatin: Cyanogenic glycosides of linseed meal that protect animals against selenium toxicity J. Org. Chem. *45*, 507–510.

Glucosinolates

BELL, J. M. 1984. Nutrients and toxicants in rapeseed meal: A review. J. Anim. Sci. *58*, 996–1010.
FENWICK, G. R., HEANEY, R. K., and MULLIN, W. J. 1983. Glucosinolates and their breakdown products in food and food plants. CRC Crit. Rev. Food Sci. Nutr. *18*, 123–201.
HENDRICH, S., and BJELDANES, L. F. 1983. Effects of dietary cabbage, Brussels sprouts, *Illicium verum, Schizandra chinesis,* and alfalfa on the benzo[a]pyrene metabolic system in mouse liver. Food and Chem. Toxicol. *21*, 479–486.
HENDRICKS, J. D., NIXON, J. E., BAILEY, G. S., and SINNHUBER, R. O. 1982. Inhibition of aflatoxin B_1 carcinogenesis in rainbow trout by dietary β-naphthoflavone (6-NF) and indole-3-carbinol (1-3-C). Toxicologist *2*, 102.
MARCH, B. E., and LEUNG, P. 1976. Effects of alterations in maternal thyroid metabolism on embryonic thyroid development in the chick. Can. J. Physiol. Pharmacol. *54*, 249–253.
PAIK, I. K., ROBBLEE, A. R., and CLANDININ, D. R. 1980A. Products of the hydrolysis of rapeseed glucosinolates. Can. J. Anim. Sci. *60*, 481–493.
PEARSON, A. W., GREENWOOD, N. M., BUTLER, E. J., CURL, C. L., and FENWICK, G. R. 1983. The involvement of trimethylamine oxide in fish meal in the production of egg taint. Anim. Feed Sci. Technol. *8*, 119–128.
SHERTZER, H. G. 1982. Indole-3-carbinol and indole-3-acetonitrile influence on hepatic microsomal metabolism. Toxicol. Appl. Pharmacol. *64*, 353–361.
SHERTZER, H. G. 1983. Protection by indole-3-carbinol against covalent binding of benzo[a]pyrene metabolites to mouse liver DNA and protein. Food Chem. Toxicol. *21*, 31–35.
THROCKMORTON, J. C., CHEEKE, P. R., PATTON, N. M., ARSCOTT, G. H., and JOLLIFF, G. D. 1981. Evaluation of meadowfoam (*Limnanthes alba*) meal as a feedstuff for broiler chicks and weanling rabbits. Can. J. Anim. Sci. *61*, 735–742.
TOOKEY, H. L., VAN ETTEN, C. H., and DAXENBICHLER, M. E. 1980. Glucosinolates. *In* Toxic Constituents of Plant Foodstuffs. L. E. Liener (Editor), pp. 103–142. Academic Press, NY.
VAN ETTEN, C. H., DAXENBICHLER, M. E., SCHROEDER, W., PRINCEN, L. H., and PERRY, T. W. 1977. Tests for epiprogoitrin, derived nitriles and goitrin in body tissues from cattle fed crambe meal. Can. J. Anim. Sci. *57*, 75–80.
WHITE, R. D., and CHEEKE, P. R. 1983. Meadowfoam (*Limnanthes alba*) meal as a feedstuff for dairy goats and toxicologic activity of the milk. Can. J. Anim. Sci. *63*, 391–398.

Coumarin

BENSON, M. E., CASPER, H. H., and JOHNSON, L. J. 1981. Occurrence and range of dicoumerol concentration in sweet clover. Am. J. Vet. Res. *42*, 2014–2015.
COHEN, A. J. 1979. Critical review of the toxicology of coumarin with special reference to interspecies differences in metabolism and hepatoxic response and their significance to man. Food Cosmet. Toxicol. *17*, 277–289.

DOLLAHITE, J. W., YOUNGER, R. L., and HOFFMAN, G. O. 1978. Photosensitization in cattle and sheep caused by feeding *Ammi majus* (Greater *Ammi*, Bishop's weed). Am. J. Vet. Res. *39*, 193–197.

GOPLEN, B. P. 1971. Polara, a low coumarin cultivar of sweetclover. Can. J. Plant Sci. *51*, 249–251.

IVIE, G. W. 1978. Toxicological significance of plant furocoumarins. In Effects of Poisonous Plants on Livestock. R. F. Keeler, K. R. Van Kampen, and L. F. James (Editors), pp. 475–485. Academic Press, NY.

KINGSBURY, J. M. 1964. Poisonous Plants of the United States and Canada. Prentice-Hall, Englewood Cliffs, NJ.

OERTLI, E. H., ROWE, L. D., SLOVERING, S. L., IVIE, G. W., and BAILEY, E. M. 1983. Phototoxic effect of *Thamnosma texana* (Dutchman's breeches) in sheep. Am. J. Vet. Res. *44*, 1126–1129.

PRITCHARD, D. G., MARKSON, L. M., BRUSH, P. J., SAWTELL, J. A. A., and BLOXHAM, P. A. 1983. Haemorrhagic syndrome of cattle associated with the feeding of sweet vernal (*Anthoxanthum odoratum*) hay containing dicoumarol. Vet. Rec. *113*, 78–84.

RADOSTITS, O. M., SEARCY, G. P., and MITCHELL, K. 1980. Moldy sweet-clover poisoning in cattle. Can. Vet. J. *21*, 155–158.

SCHEEL, L. D. 1978. The toxicology of sweet clover and coumarin anticoagulants. In Mycotoxic Fungi, Mycotoxins and Mycotoxicosis: An Encyclopedia Handbook. T. D. Wyllie and L. G. Morehouse (Editors), pp. 121–142. Marcel Dekker, NY.

WITZEL, D. A., DOLLAHITE, J. W., and JONES, L. P. 1978. Photosensitization in sheep feed *Ammi majus* (Biship's weed) seed. Am. J. Vet. Res. *39*, 319–320.

Steroids and Triterpenoids

Cardiac Glycosides

DICKSTEIN, E. S., and KUNKEL, F. W. 1980. Foxglove tea poisoning. Am. J. Med. *69*, 167–169.

HUXTABLE, R. J. 1979. Herbal teas and pyrrolizidine alkaloids. In Symposium on Pyrrolizidine (Senecio) Alkaloids. Toxicity, Metabolism and Poisonous Plant Control Measures. P. R. Cheeke (Editor), pp. 87–93. Nutrition Research Institute, Oregon State Univ., Corvallis.

KINGSBURY, J. M. 1964. Poisonous Plants of the United States and Canada. Prentice-Hall, Englewood Cliffs, NJ.

Cardenolides

BENSON, J. M., SEIBER, J. N., KEELER, R. F., and JOHNSON, A. E. 1978. Studies on the toxic principle of *Asclepias eriocarpa* and *Asclepias labriformis*. In Effects of Poisonous Plants on Livestock. R. F. Keeler, K. R. Van Kampen, and L. F. James (Editors), pp. 273–284. Academic Press, NY.

BENSON, J. M., SEIBER, J. N., BAGLEY, C. V., KEELER, R. F., JOHNSON, A. E., and YOUNG, S. 1979. Effects on sheep of the milkweeds *Asclepias eriocarpa* and *A. labriformis* and of cardiac glycoside-containing derivative material. Toxicon *17*, 155–165.

CHEEKE, P. R., PATTON, N. M., and TEMPLETON, G. S. 1982. Rabbit Production. Interstate Printers and Publishers, Danville, IL.

CLARK, J. C. 1979. Whorled milkweed poisoning. Vet. Hum. Toxicol. *21*, 431.

DUFFEY, S. S. 1980. Sequestration of plant secondary products by insects. Annu. Rev. Entomol. 25, 447–477.
GIBBONS, E. 1972. How to milk a milkweed. Org. Gard. Farming (Jan.), pp. 148–153.
NIELSEN, D. E., NISHIMURA, H., OTOS, J. W., and CALVIN, M. 1977. Plant crops as a source of fuel and hydrocarbon-like materials. Science 198, 942–944.

Bufadienolides
ANDERSON, L. A. P., SCHULTZ, R. A., JOUBERT, L. P. J., PROZESKY, L., KELLERMAN, T. S., ERASMUS, G. L., and PROCOS, J. 1983. Krimpiekte and acute cardiac glycoside poisoning in sheep caused by bufadienolides from the plant *Kalanchoe lanceolata* Forsk. Onderstepoort J. Vet. Res. 50, 295–300.
JOUBERT, J. P. J. 1983. Attempted prevention and treatment of *Geigeria filifolia* Mattf. poisoning (vermeersiekte) in sheep. J. S. Afr. Vet. Assoc. 54, 255–258.
JOUBERT, J. P. J., and SCHULTZ, R. A. 1982. The treatment of *Morea polystachya* (Thunb.) Ker-Gawl (cardiac glycoside) poisoning in sheep and cattle with activated charcoal and potassium chloride. J. S. Afr. Vet. Assoc. 53, 249–253.
NAUDE, T. W. 1977. The occurrence and significance of South African cardiac glycosides. J. S. Afr. Biol. Sci. 18, 7–20.
STRYDOM, J. A., and JOUBERT, J. P. J. 1983. The effect of predosing *Homeria pallida* Bak. to cattle to prevent tulp poisoning. J. S. Afr. Vet. Assoc. 54, 201–203.
WILLIAMS, M. C., and SMITH, M. C. 1984. Toxicity of *Kalanchoe* spp. to chicks. Am. J. Vet. Res. 45, 543–546.

Saponins
AMEENNUDDIN, S., BIRD, H. R., PRINGLE, D. J., and SUNDE, M. L. 1983. Studies on the utilization of leaf protein concentrates as a protein source in poultry nutrition. Poult. Sci. 62, 505–511.
CHEEKE, P. R. 1976. Nutritional and physiological properties of saponins. Nutr. Rep. Int. 13, 315–324.
CHEEKE, P. R. 1983. Biological properties and nutritional significance of legume saponins. *In* Leaf Protein Concentrates. L. Telek and H. D. Graham (Editors), pp. 396–414. AVI Publishing Co., Westport, CT.
CHEEKE, P. R., KINZELL, J. H., and PEDERSEN, M. W. 1977. Influence of saponins on alfalfa utilization by rats, rabbits and swine. J. Anim. Sci. 45, 476–481.
CHEEKE, P. R., PEDERSEN, M. W., and ENGLAND, D. C. 1978. Responses of rats and swine to alfalfa saponins. Can. J. Anim. Sci. 58, 783–789.
CHEEKE, P. R., POWLEY, J. S., NAKAUE, H. S., and ARSCOTT, G. H. 1983. Feed preference responses of several avian species fed alfalfa meal, high and low saponin alfalfa, and quinine sulfate. Can. J. Anim. Sci. 63, 707–710.
FENWICK, D. E., and OAKENFULL, D. 1983. Saponin content of food plants and some prepared foods. J. Sci. Food Agric. 34, 186–191.
GOODALL, S. R. 1979. Sarsaponin effect upon ruminant digestion and feedlot performance. Ph.D. Thesis. Colorado State Univ., Ft. Collins.
HEGSTED, M., and LINKSWILER, H. M. 1980. Protein quality of high and low saponin alfalfa protein concentrate. J. Sci. Food Agric. 31, 777–781.
JOHNSTON, N. L., QUARLES, C. L., FAGERBERG, D. J., and CAVENY, D. D. 1981. Evaluation of yucca saponin on broiler performance and ammonia suppression. Poult. Sci. 60, 2289–2292.
LEAMASTER, B. R., and CHEEKE, P. R. 1979. Feed preferences of swine: Alfalfa

meal, high and low saponin alfalfa, and quinine sulfate. Can. J. Anim. Sci. *59,* 467–469.

MAJAK, W., HOWARTH, R. E., FESSER, A. C., GOPLEN, B. P., and PEDERSEN, M. W. 1980. Relationships between ruminant bloat and the composition of alfalfa herbage. II. Saponins. Can. J. Anim. Sci. *60,* 699–708.

MALINOW, M. R., McLAUGHLIN, P., STAFFORD, C., LIVINGSTON, A. L., KOHLER, G. O., and CHEEKE, P. R. 1979. Comparative effects of alfalfa saponins and alfalfa fiber on cholesterol absorption in rats. Am. J. Clin. Nutr. *32,* 1810–1812.

MOLYNEUX, R. J., STEVENS, K. L., and JAMES, L. F. 1980. Chemistry of toxic range plants. Volatile constituents of broomweed (*Gutierreza sarothrae*). J. Agric. Food Chem. *28,* 1332–1333.

NAKAUE, H. S., LOWRY, R. R., CHEEKE, P. R., and ARSCOTT, G. H. 1980. The effect of dietary alfalfa of varying saponin content on yolk cholesterol level and layer performance. Poult. Sci. *59,* 2744–2748.

OAKENFULL, D. 1981. Saponins in foods—a review. Food Chem. *7,* 19–40.

WILLIAMS, M. C. 1978. Toxicity of saponins in alfombrilla (*Drymaria arenariodes*). J. Range Manage. *31,* 182–184.

WILLIAMS, M. C., JAMES, L. F., and LUIS, C. 1980. Seasonal concentration and toxicity of saponins in alfombrilla. J. Range Manage. *32,* 157–158.

Nitro-Containing Glycosides

ALSTON, T. A., MELA, L., and BRIGHT, H. J. 1977. 3-Nitro-propionate, the toxic substance of *Indigofera*, is a suicide inactivator of succinate dehydrogenase. Proc. Natl. Acad. Sci. U.S.A. *74,* 3767–3771.

GUSTINE, D. L. 1979. Aliphatic nitro compounds in crown vetch: A review. Crop Sci. *19,* 197–203.

GUSTINE, D. L., MOYER, B. G., WANGNESS, P. J., and SHENK, J. S. 1977. Ruminal metabolism of 3-nitropropan 1-D-glucopyranoses from crown vetch. J. Anim. Sci. *44,* 1107–1111.

JAMES, L. F., FOOTE, W., NYE, W., and HARTLEY, W. J. 1978. Effects of feeding *Oxytropis* and *Astragalus* pollen to mice and *Astragalus* seeds to rats. Am. J. Vet. Res. *39,* 711–712.

JAMES, L. F., HARTLEY, W. J., WILLIAMS, M. C., and VAN KAMPEN, K. R. 1980. Field and experimental studies in cattle and sheep poisoned by nitro-bearing *Astragalus* or their toxins. Am. J. Vet. Res. *41,* 377–382.

MAJAK, W., and CHENG, K. J. 1983. Recent studies on ruminal metabolism of 3-nitropropanol in cattle. Toxicon, Suppl. *3,* 265–268.

MAJAK, W., and CLARK, L. J. 1980. Metabolism of aliphatic nitro compounds in bovine rumen fluid. Can. J. Anim. Sci. *60,* 699–708.

MAJAK, W., NEUFELD, R., and CORNER, J. 1980. Toxicity of *Astragalus miser* V. *serotinus* to the honeybee. J. Apicult. Res. *19,* 196–199.

MAJAK, W., UDENBERG, T., McDIARMID, R. E., and DOUWES, H. 1981A. Toxicity and metabolic effects of intravenously administered 3-nitropropanol in cattle. Can. J. Anim. Sci. *61,* 639–647.

MAJAK, W., CHENG, K. J., and HALL, J. W. 1981B. The effect of cattle diet on the metabolism of 3-nitropropanol by ruminal microorganisms. Can. J. Anim. Sci. *62,* 855–860.

MUIR, A. D., MAJAK, W. PASS, M. A., and YOST, G. S. 1984. Conversion of 3-nitropropanol (miserotoxin aglycone) to 3-nitropropionic acid in cattle and sheep. Toxicol. Lett. *20,* 137–141.
SHENK, J. S., WANGANESS, P. J., LEACH, R. M., GUSTINE, D. L., GOBBLE, J. L., and BARNES, R. F. 1976. Relationship between β-nitropropionic acid content of crown vetch and toxicity in nonruminant animals. J. Anim. Sci. *42,* 616–621.
WILLIAMS, M. C. 1981A. Nitro compounds in *Indigofera* species. Agron. J. *73,* 434–436.
WILLIAMS, M. C. 1981B. Nitro compounds in foreign species of *Astragalus.* Weed Sci. *29,* 261–269.
WILLIAMS, M. C. 1983. Toxic nitro compounds in lotus. Agron. J. *75,* 520–522.
WILLIAMS, M. C., and JAMES, L. F. 1978. Livestock poisoning from nitro-bearing *Astragalus. In* Effect of Poisonous Plants on Livestock. R. F. Keeler, K. R. Van Kampen, and L. F. James (Editors), pp. 379–389. Academic Press, NY.
WILLIAMS, M. C., JAMES, L. F., and BLEAK, A. T. 1976. Toxicity of introduced nitro-containing *Astragalus* to sheep, cattle and chicks. J. Range Manage. *29,* 30–32.
WILLIAMS, M. C., JAMES, L. F., and BOND, B. O. 1979. Emory milkvetch (*Astragalus emoryanus*) poisoning in chicks, sheep and cattle. Am. J. Vet. Res. *40,* 403–406.

Vicine (Favism)

AHERNE, F. X., LEWIS, A. J., and HARDIN, R. T. 1977. An evaluation of faba beans (*Vicia faba*) as a protein supplement for swine. Can. J. Anim. Sci. *57,* 321–328.
CAMPBELL, L. D., OLABORO, G., MARQUARDT, R. R., and WADDELL, D. 1980. Use of fababeans in diets for laying hens. Can. J. Anim. Sci. *60,* 395–405.
CROSBY, D. G. 1969. Natural toxic background in the food of man and his animals. J. Agric. Food Chem. *17,* 532–538.
INGALLS, J. R., SHARMA, H. R., DEVLIN, T., BAREEBA, F. B., and CLARK, K. W. 1979. Evaluation of whole plant fababean forage in ruminant rations. Can. J. Anim. Sci. *59,* 291–301.
MAGER, J., CHEVION, M., and GLASER, G. 1980. Favism. *In* Toxic Constituents of Plant Foodstuffs. I. E. Liener (Editor), pp. 266–294. Academic Press, NY.
MARCUS, J. R., and COHEN, G. 1967. The riddle of the dangerous bean. Harpers Monthly Mag. *234*(1405), 98–102.
MARQUARDT, R. R., CAMPBELL, L. D., STOTHERS, S. C., and McKIRDY, J. A. 1974. Growth response of chicks and rats fed diets containing four cultivars of raw or autoclaved fababeans. Can. J. Anim. Sci. *54,* 177–182.
MARQUARDT, R. R., WARD, A. T., CAMPBELL, L. D., and CANSFIELD, P. E. 1977. Purification, identification and characterization of a growth inhibitor in fababeans. (*Vicia faba* L. var. *minor*). J. Nutr. *107,* 1313–1324.
McNIGHT, D. R., and MacLEOD, G. K. 1977. Value of whole plant fababean silage as the sole forage for lactating cows. Can. J. Anim. Sci. *57,* 601–603.
MUDUULI, D. S., MARQUARDT, R. R., and GUENTER, W. 1981. Effect of dietary vicine on the productive performance of laying chickens. Can. J. Anim. Sci. *61,* 757–764.
OLABORO, G., MARQUARDT, R. R., and CAMPBELL, L. D. 1981. Isolation of the egg weight depressing factor in fababeans. (*Vicia faba* L. var. *minor*). J. Sci. Food Agric. *32,* 1074–1080.

ROTRUCK, J. T., POPE, A. L., GANTHER, M. E., SWANSON, A. B., HAFEMAN, D. G., and HOEKSTRA, W. G. 1973. Selenium: Biochemical role as a component of glutathione peroxidase. Science *179*, 588–590.
THORLACIUS, S. O., and BEACOM, S. E. 1981. Feeding value for lambs of fababean, field pea, corn and oat silages. Can. J. Anim. Sci. *61*, 663–668.

Calcinogenic Glycosides

HUGHES, M. R., McCAIN, T. A., CHANG, S. Y., HAUSSLER, M. R., VILLAREALE, M., and WASSERMAN, R. H. 1977. Presence of 1,25-dihydroxyvitamin D_3-glycoside in the calcinogenic plant, *Cestrum diurnum*. Nature (London) *268*, 347–349.
KROOK, L., WASSERMAN, R. H., SHIVELY, J. H., TASHJIAN, A. H., BROKKEN, T. D., and MORTON, J. F. 1975A. Hypercalcemia and calcinosis in Florida horses: Implication of the shrub, *Cestrum diurnum*, as the causative agent. Cornell Vet. *65*, 26–56.
KROOK, L., WASSERMAN, R. H., McENTEE, K., BROKKEN, T. D., and TEIGLAND, M. B. 1975B. *Cestrum diurnum* poisoning in Florida cattle. Cornell Vet. *65*, 557–575.
MORRIS, K. M. L., and LEVACK, V. M. 1982. Evidence for aqueous soluble vitamin D-like substances in the calcinogenic plant *Tristetum flavescens*. Life Sci. *30*, 1255–1262.
WASSERMAN, R. H. 1978. The nature and mechanism of action of the calcinogenic principle of *Solanum malacoxylon* and *Cestrum diurnum*, and a comment on *Trisetum flavescens*. *In* Effects of Poisonous Plants on Livestock. R. F. Keeler, K. R. Van Kampen, and L. F. James (Editors), pp. 545–553. Academic Press, NY.

Azoxyglycosides

HOOPER, P. T. 1978. Cycad poisoning in Australia—Etiology and pathology. *In* Effects of Poisonous Plants on Livestock. R. F. Keeler, K. R. Van Kampen, and L. F. James (Editors), pp. 337–347. Academic Press, NY.
REDDY, B. S., FURUYA, K., HANSON, D., DIBELLO, J., and BERKE, B. 1982. Effect of dietary butylated hydroxyanisole on methlazoxymethanolacetate-induced toxicity in mice. Food Chem. Toxicol. *20*, 853–860.
TUSTIN, R. C. 1983. Notes on the toxicity and carcinogenicity of some South African cycad species with special reference to that of *Encephalartos lanatus*. J. S. Afr. Vet. Assoc. *54*, 33–42.
WOGAN, G. N., and BUSBY, W. F. 1980. Cycasin. *In* Toxic Constituents of Plant Foodstuffs I. E. Liener (Editor), pp. 350–353. Academic Press, NY.

Carboxyatractyloside

COLE, R. J., STUART, B. P., LANSDEN, J. A., and COX, R. H. 1980. Isolation and redefinition of the toxic agent from cocklebur (*Xanthium strumarium*). J. Agric. Food Chem. *28*, 1330–1332.
CRAIG, J. C., MOLE, M. L., BILLETS, S., and EL-FERALY, F. 1976. Isolation and identification of the hypoglycemic agent, carboxyatractylate, from *Xanthium strumarium*. Phytochemistry *15*, 1178.
CUTLER, H. G., and COLE, R. J. 1983. Carboxytractyloside—a compound from *Xanthium strumarium* and *Atractylis gummifera* with plant growth inhibiting properties—the probable inhibitor A. J. Nat. Prod. *46*, 609–613.

HATCH, R. C., JAIN, A. V., WEISS, R., and CLARK, J. D. 1982. Toxicologic study of carboxyatractyloside (active principle in cocklebur, *Xanthium strumarium*) in rats treated with enzyme inducers and inhibitors and glutathione precursor and depletor. Am. J. Vet. Res. *43*, 111–116.

KINGSBURY, J. M. 1964. Poisonous Plants of the United States and Canada. Prentice-Hall, Englewood Cliffs, NJ.

KUPIECKI, F. P., OGZEWALLA, C. D., and SCHELL, F. M. 1974. Isolation and characterization of a hypoglycemic agent from *Xanthium strumarium*. J. Pharm. Sci. *63*, 1166–1167.

STUART, B. P., COLE, R. J., and GOSSER, H. S. 1981. Cocklebur (*Xanthium strumarium*, L. var. *strumarium*) intoxication in swine: Review and redefinition of the toxic principle. Vet. Pathol. *18*, 368–383.

Isoflavones and Coumestans

ADAMS, N. R. 1977. Cervical mucus and reproductive efficiency in ewes after exposure to oestrogenic pastures. Aust. J. Agric. Res. *28*, 481–489.

ADAMS, N. R. 1981. A changed responsiveness to oestrogen in ewes with clover disease. J. Reprod. Fertil. Suppl. *30*, 223–230.

ADAMS, N. R., HEARNSHAW, H., and OLDHAM, C. M. 1981. Abnormal function of the corpus luteum in some ewes with phytoestrogenic infertility. Aust. J. Biol. Sci. *34*, 61–65.

CHAMLEY, W. A., ADAMS, N. R., HOOLEY, R. D., and CARSON, R. 1981. Hypothalamic-pituitary function in normal ewes and ewes which grazed oestrogenic subterranean clover for several years. Aust. J. Biol. Sci. *34*, 239–244.

COX, R. I. 1978. Plant estrogens affecting livestock in Australia. *In* Effects of Poisonous Plants on Livestock. R. F. Keeler, K. R. Van Kampen, and L. F. James (Editors), pp. 451–464. Academic Press, NY.

ELAKOVICH, S. D., and HAMPTON, J. M. 1983. Analysis of coumestrol, a phytoestrogen, in alfalfa tablets sold for human consumption. Agric. Food Chem. *32*, 173–175.

HETTLE, J. A., and KITTS, W. D. 1983. Effects of phyto-estrogenic alfalfa consumption of plasma LH levels in cycling ewes. Anim. Reprod. Sci. *6*, 233–238.

LEOPOLD, A. S., ERWIN, M., OH, J., and BROWNING, B. 1976. Phytoestrogens. Adverse effects on reproduction in California quail. Science *191*, 98–99.

LIGHTFOOT, R. J., and ADAMS, N. R. 1979. Changes in cervical histology in ewes following prolonged grazing on oestrogenic subterranean clover. J. Comp. Pathol. *89*, 367–373.

LIVINGSTON, L. 1978. Forage plant estrogens. J. Toxicol. Environ. Health *4*, 301–324.

NEIL, H. G., LIGHTFOOT, R. J., and FELS, H. E. 1974. Effect of legume species on ewe fertility in South Western Australia. Proc. Aust. Soc. Anim. Prod. *10*, 136.

OSTROVSKY, D., and KITTS, W. D. 1963. The effect of estrogenic plant extracts on the uterus of the laboratory rat. Can. J. Anim. Sci. *43*, 106–112.

Jojoba Glycosides

LIENER, I. E. (Editor) 1980. Toxic Constituents of Plant Foodstuffs, 2nd Edition. Academic Press, NY.

NGOUPAYOU, J. D. N., MAIORINO, P. M., and BIRD, B. L. 1982. Jojoba meal in poultry diets. Poult. Sci. *61*, 1692–1696.

VERBISCAR, A. J., and BANIGAN, T. F. 1978. Composition of jojoba seeds and foliage. Agric. Food Chem. *26*, 1456–1459.
VERBISCAR, A. J., BANIGAN, T. F., WEBER, C. W., REID, B. L., SWINGLE, R. S., TRIE, J. E., and NELSON, E. A. 1981. Detoxification of jojoba meal by lactobacilli. J. Agric. Food Chem. *29*, 296–302.

Ranunculin

HIDIROGLOU, M., and KNUTTI, H. J. 1963. The effects of green tall buttercup in roughage on the growth and health of beef cattle and sheep. Can. J. Anim. Sci. *43*, 68–71.
HILL, R., and VAN HEYNINGEN, R. 1951. Ranunculin: The precursor of the vesicant substance of the buttercup. Biochem. J. *49*, 332–335.
KINGSBURY, J. M. 1964. Poisonous Plants of the United States and Canada. Prentice-Hall, Englewood Cliffs, NJ.
NACHMAN, R. J., and OLSON, J. D. 1983. Ranunculin. A toxic constituent of the poisonous range plant bur buttercup (*Ceratocephalus testiculatus*.) J. Agric. Food Chem. *31*, 1358–1360.
OLSON, J. D., ANDERSON, T. E., MURPHY, J. C., and MADSEN, G. 1983. Bur buttercup poisoning of sheep. J. Am. Vet. Med. Assoc. *183*, 538–543
THERRIEN, H. P., HIDIROGLOU, M., and CHARETTE, L. A. 1962. The toxicity of tall buttercup (*Ranunculus acris* L.) to cattle. Can. J. Anim. Sci. *42*, 123–124.

7

Proteins and Amino Acids

TRYPSIN (PROTEASE) INHIBITORS

A wide variety of plants contain protein fractions which inhibit protein digestion in the digestive tract of animals. The trypsin inhibitors of soybeans are the best known and most widely studied. Other plants containing trypsin inhibitors include most types of beans, potatoes, rye, triticale, barley, and alfalfa. Those in soybeans, common beans, and triticale are of particular significance in livestock feeding. Although these factors are commonly referred to as trypsin inhibitors, protease inhibitor is probably a better term, since other enzymes such as chymotrypsin are also affected.

In soybeans, the protease inhibitors are of two main categories: those having a molecular weight of 20,000–25,000 with few disulfide bonds and with specific activity toward trypsin, and those having a molecular weight of 6000–10,000 with a high proportion of disulfide bonds and with inhibitory activity to trypsin and chymotrypsin at independent binding sites. These two have been referred to as the Kunitz inhibitor and the Bowman–Birk inhibitor, respectively. The Kunitz inhibitor has been isolated and its amino acid sequence determined. It contains 181 amino acids, with 2 disulfide bonds and with the active site at amino acid 63. The Kunitz inhibitor combines with trypsin stoichiometrically: 1 mole of inhibitor inactivates 1 mole of trypsin. The reaction is almost instantaneous, and the complex formed is a very tight one. It appears that trypsin reacts with the inhibitor in the same way that it reacts with other proteins it digests, but a number of noncovalent bonds formed at the active site result in an irreversible complex. The Bowman–Birk inhibitor has two active sites, one that binds trypsin and one that complexes chymotrypsin. It is a single polypeptide chain with 71 amino acids, and 7 disulfide bonds.

Trypsin inhibitors occur mainly in seeds, although in some cases they are also found in leaves. Besides soybeans, other legume seeds with trypsin inhibitors include lima, kidney, navy, pinto and common garden beans, peanuts, cowpeas, fava beans, and peas. While all common cereal grains (barley, wheat, corn, rye, rice, and sorghum) contain trypsin inhibitors, they are of practical significance only in the case of triticale. Triticale-breeding programs include efforts to lower trypsin inhibitor activity (Erickson *et al.* 1979). The physiological role of trypsin inhibitors in plants is uncertain, but they may be involved in defense mechanisms. Many plant leaves, after wounding by insects or bacterial infection, show an accumulation of protease inhibitors at the site of damage. They may inhibit insect and bacterial proteases.

Nutritional Significance of Trypsin Inhibitors

Soybean meal is the major protein supplement used in swine and poultry diets in the U.S. and much of the rest of the world. It must be heat-treated to destroy trypsin inhibitors (and several other toxins). Raw soybeans grown for on-the-farm feeding must be heat-treated before being used for feeding. In spite of a great deal of study on the subject, the mode of action of trypsin inhibitors is not entirely clear. Liener and Kakade (1980) present an extensive review of this subject. Some of the effects of feeding raw soybeans, particularly in nonruminants, include poor growth, reduced feed intake, a reduction in protein digestibility, pancreatic hypertrophy, and a deficiency of sulfur amino acids. The poor performance in many cases is not due to a simple interference with protein digestion. Trypsin inhibitors cause growth depression in rats even when included in diets where the protein is provided as free amino acids.

Chickens fed raw soybeans develop an enlarged pancreas (Fig. 7.1).

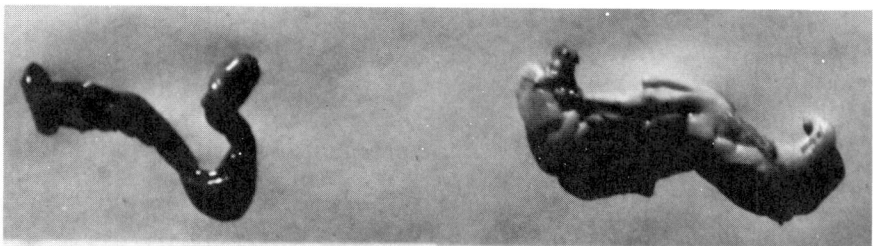

FIG. 7.1. The pancreas (5.70 g) from a chick (body weight 507 g) fed raw soybeans (right) is much larger than the pancreas (1.461 g) from the same sized bird (body weight 507 g) fed soybean meal (left).

Pancreatic hypertrophy has also been observed in rats and mice. However, it does not occur in the pig, dog, or calf. There is an interesting relationship between the normal size of the pancreas and the hypertrophic response to trypsin inhibitors (Liener and Kakade 1980). This is shown in Table 7.1. Species whose pancreas exceeds about 0.3% of body weight tend to develop an enlarged pancreas, whereas those with a smaller pancreatic size do not. Based on this relationship, it would be predicted that humans would not develop an enlarged pancreas. Liener and Kakade (1980) speculated that animals with a large pancreas have a greater secretion of trypsin and therefore are more sensitive to stimuli that cause increased trypsin secretion. The outpouring of trypsin from the enlarged pancreas causes a sulfur amino acid deficiency due to the fact that trypsin is high in cystine. Evidence in support of this theory is the fact that supplements of sulfur amino acids can overcome the growth depression caused by raw soybeans. Thus, in animals such as the rat, chicken, and mouse, growth depression is due to the high endogenous loss of sulfur amino acids from the excess trypsin produced. In animals such as the pig, calf, and young chick (in which the hypertrophic response of the pancreas is delayed), inhibition of protein digestion occurs because the amount of trypsin inhibitor exceeds the production of trypsin.

Trypsin inhibitors are readily destroyed by treatment of plant material with heat. Over 95% of the activity is destroyed in 15 min of 100°C heat treatment. For on-the-farm processing of soybeans, an extruder in which the beans are forced through a die is effective (Fig. 7.2). The friction in the die builds up sufficient heat to destroy the inhibitors. A simple test to determine if trypsin inhibitory activity is present involves incubation of treated soybeans with urea in the presence of an indicator dye. Soybeans contain urease, which is also heat sensitive. If heating has not been adequate to destroy urease, the urea is converted to ammonia, which changes the pH and causes color development. The

TABLE 7.1. Relation between Pancreatic Size and Hypertrophic Response to Trypsin Inhibitor[a]

Species	Pancreas weight as % of body weight	Pancreatic hypertrophy
Mouse	0.6–0.8	+
Rat	0.5–0.6	+
Chick	0.4–0.6	+
Dog	0.21–0.24	−
Human	0.09–0.12	?
Calf	0.06–0.08	−

[a] Adapted from Liener and Kakade (1980).

FIG. 7.2. An extruder, used to process soybeans and other feedstuffs which require heat treatment to destroy inhibitors. The feed is extruded through a die, with heat resulting from the friction involved.
Courtesy of Triple "F," Inc.

presence of active urease suggests that active trypsin inhibitors will also be present.

Alfalfa contains trypsin inhibitor activity; it appears to have little if any significant effect on the utilization of alfalfa meal by nonruminant animals (Mitchell and Parrish 1981).

The winged bean (*Psophocarpus tetragonolobus*) is a promising high-protein crop for the tropics, producing seeds, forage, and tubers which may be used for feed and food. The seeds of the winged bean contain trypsin inhibitors which are inactivated by heat (D'Mello et al. 1983). Leaf protein concentrate prepared from the foliage of the winged bean has a high nutritional value (Cheeke et al. 1980). When winged bean leaf meal was substituted for alfalfa meal in a diet for weanling rabbits, growth and protein digestibility were lowered (Harris et al. 1981), probably due to the effects of tannins.

AMYLASE INHIBITORS

A variety of foodstuffs contain amylase inhibitors, including wheat, oats, rye, beans, and potatoes (Liener 1980). The α-amylase inhibitors in wheat and other grains are effective against insect amylases, suggesting that they may be a defense mechanism of the seed against insect attack. Amylase inhibitors appear to be of little or no significance in animal nutrition. Amylase inhibitors have been marketed in the U.S. as "starch blockers" to reduce digestion of starch and thus help to reduce obesity. Their use has been discouraged by medical authorities. Prevention of starch digestion in the small intestine is likely to lead to its fermentation in the large intestine, resulting in flatus, diarrhea, and other digestive disturbances. In addition, trypsin inhibitor and lectin activities have been noted in starch blockers (Liener et al. 1984), representing a possible health risk. The starch blockers are crude extracts of *Phaseolus vulgaris*. In many cases, they contain sufficient amylase activity to overcome any amylase inhibitory activity they have (Liener et al. 1984).

CARBOXYPEPTIDASE INHIBITORS

Carboxypeptidases are a group of pancreatic enzymes that function in protein digestion. Specifically, they are exopeptidases which hydrolyze the terminal peptide bond at the carboxyl end of a polypeptide chain. Potatoes contain a heat-stable protein that is a carboxypeptidase inhibitor. Pearce et al. (1983) studied the effect of this inhibitor on the performance and metabolism of young chicks. They concluded that although some measurable effects on protein digestion and feed efficiency could be detected, the level of inhibitor that caused them was much higher than could be consumed by livestock or humans under normal circumstances. The level of carboxypeptidase inhibitor in potatoes (about 0.03% of fresh weight) is considered insignificant as an antinutrient.

ALLERGENIC PROTEINS

A wide variety of food items can induce allergenic responses in humans. Perlman (1980) has provided an in-depth review of food allergens, which should be consulted for further information on this subject.

In livestock feeding, the principal allergenic feedstuffs encountered in practical situations are soybean products containing undenatured proteins. Soybeans contain the storage globulins glycinin and β-conglycinin, which can cause gastrointestinal disturbances in young animals fed milk replacers containing soy products. Morphological changes in the intestinal mucosa occur, resulting in abnormal digesta movement and impaired nutrient absorption (Pedersen and Sissons 1984). Systemic immunological responses occur from the absorption of soybean antigens. Adequate heat treatment during soybean processing eliminates the allergenic potential.

HEMAGGLUTININS (LECTINS)

Hemagglutinins (phytohemagglutinins, lectins) cause the clumping or agglutination of red blood cells in vitro. They were first isolated from castor beans, which contain a potent lectin called ricin. Lectins are proteins that have a high affinity for certain sugar molecules. Many contain covalently bound sugars and so can be classed as glycoproteins. Their biological effects are probably due to their affinity for sugars. There are carbohydrate moieties in animal cell membranes which may bind to lectins and be altered in functional properties.

Lectins are found in soybeans and other field beans (*Phaseolus*) such as kidney, pinto, and navy beans (Table 7.2). The lectins in raw soybeans are relatively nontoxic, but feeding high levels of raw kidney beans to rats will kill them due to lectin activity. Hemagglutinins cause various adverse effects, including reduced growth, diarrhea, decreased nutrient absorption, and increased incidence of bacterial infection. The major effects seem to be on the intestinal mucosa. The hemagglutinins bind to cells in the intestinal wall and cause a nonspecific interference with nutrient absorption. In addition, there is evidence that lectins impair the immune system, leading to greater sensitivity to bacterial infection. Changes in gut permeability may lead to invasion by normally innocuous intestinal microflora. Pusztai *et al.* (1979) have reported that field bean lectins disrupt the brush borders of the cells lining the duodenum and jejunum (Fig. 7.3) and that an abnormally high rate of tissue-protein catabolism occurs in rats administered lectins. A dramatic proliferation of *Escherichia coli* occurs in the small intestine (Wilson *et al.* 1980).

Lectins are destroyed by moist heat. They are resistant to dry heat, so kidney and other beans should be soaked prior to cooking to ensure moisture penetrance of the entire seed.

Cull field beans are sometimes fed to livestock, so the potential adverse effects of lectins should be considered. Cull beans should be heat-

TABLE 7.2. Common Legume Seeds Classified as to Toxicity and Lectin Content[a]

Group A. Highly toxic, high lectin activity
Phaseolus cocceneus	Runner bean
P. vulgaris	Red or brown kidney bean
P. vulgaris	White or black kidney bean
P. acutifolius	Tepary bean

Group B. Growth inhibiting, moderate lectin activity
Psophocarpus tetragonolobus	Winged bean
Phaseolus lunatus	Baby lima bean

Group C. Nontoxic, low lectin activity
Lens esculentus	Lentils
Pisum sativum	Green peas
Cicer arietinum	Chick-peas
Vigna sinensis	Blackeyed peas
Cajanus cajan	Pigeon peas
Phaseolus aureus	Mung beans
Vicia faba	Broad beans
Phaseolus angularis	Aduki beans

Group D. Growth depression due to nonlectin factors
Glycine max	Soybeans
Phaseolus vulgaris	Pinto beans

[a] Adapted from Grant et al. (1983).

treated for swine feeding; raw beans even at dietary levels as low as 5% have adverse effects in pigs (Myer et al. 1982). The detrimental effects of raw beans on swine and poultry are reduced by the addition of antibiotics, perhaps due to reduced microbe colonization of the gut wall. Myer and Froseth (1983) found that extrusion of cull red beans was as effective as autoclaving in decreasing the growth-depressing effects in swine. Levels as high as 40% extruded beans gave satisfactory growth rates of pigs. Methionine supplementation of heat-treated beans also was effective in overcoming growth depression.

Kidney beans contain a factor(s) which impairs the utilization of vitamin E. Feeding raw kidney beans to sheep or chickens induces nutritional muscular dystrophy (Hintz and Hogue 1964). Autoclaving the beans partially overcomes the effect. The activity may be due to an α-tocopherol oxidase present in kidney beans, alfalfa, and soybeans (Liener 1980). Alfalfa contains a fat-soluble factor that reduces the availability of vitamin E to the chick (Pudelkiewicz and Matterson 1960; Olson et al. 1966).

Canavalia ensiformis (jack bean) is sometimes grown in tropical areas for livestock feed. An intake of jack bean seeds greater than 4% of body weight is toxic in cattle (Kingsbury 1964); signs include severe

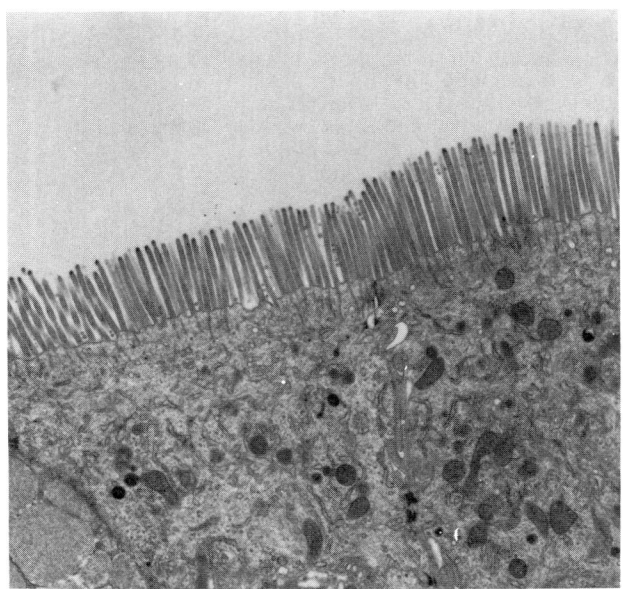

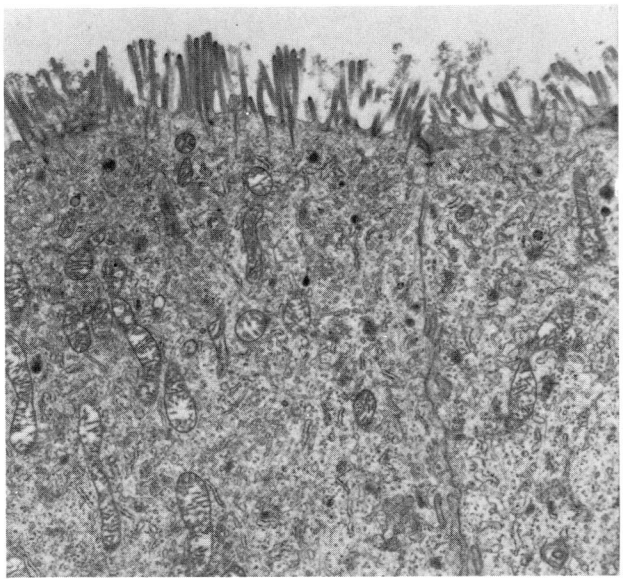

FIG. 7.3. Section of the jejunum of a pig fed a control diet (top) compared with a similar section from a pig fed raw kidney beans (bottom), showing disruption of the microvilli (×6300).
Courtesy of R. Begbie and T. P. King.

diarrhea, weakness, enteritis, and nephritis. Jack beans contain a lectin, concanavalin A. The precatory bean or rosary pea (*Abrus precatorius*) contains a lectin, abrin, that is one of the most deadly toxins known. It is a common tropical vine and occurs as a weed of fence rows, citrus groves, and waste areas in central and southern Florida. The seeds are sometimes made into rosaries, necklaces, or other decorations. Less than one seed, thoroughly masticated, is sufficient to kill a person. Abrin is a potent irritant of mucous membranes; signs of intoxication include severe gastrointestinal irritation and violent vomition. Abrin consists of two polypeptide chains joined by disulfide bonds.

The castor bean, *Ricinus communis*, is grown commercially as a source of castor bean oil which is used as an industrial lubricant and for medicinal purposes (Fig. 7.4). The seed, press cake, and foliage are poisonous; the oil is nontoxic. The poisonous principle is a lectin called ricin, one of the most toxic substances known. The minimum lethal dose by injection is about $1 \times 10^{-8}\%$ of body weight, which makes it about 1000 times more toxic than any other bean lectin (except for abrin). Ricin and abrin both consist of two polypeptide chains joined by disulfide bonds. The lighter chain contains the carbohydrate-binding site. In the U.S., castor bean poisoning has occurred in man and all classes of livestock (Kingsbury 1964). Livestock poisonings have been

FIG. 7.4. Leaves and developing seed pods of the castor bean (*Ricinus communis*).

due to castor bean seeds as a contaminant of grain. Ricin can be destroyed by moderate heat. Signs of toxicity are those of severe gastrointestinal tract irritation.

The black locust (*Robinia pseudoacacia*) is a small leguminous rapidly growing tree. Kingsbury (1964) reviewed several studies suggesting that the plant contains a toxic factor, a lectin called robin. Robin consists of two distinct polypeptide chains, but they are not joined by disulfide bonds (McPherson and Hoover 1979). It has been reported to cause symptoms of anorexia, lassitude, weakness, posterior paralysis in cattle and horses, nausea in humans, coldness of extremities, dilation of pupils, a weak irregular pulse, dyspnea, diarrhea, and death. Pathological lesions are mainly irritation and edema of the digestive tract. Poisoning occurs when horses strip and eat the bark of black locust trees to which they have been tied, and cattle and horses have been poisoned from eating black locust sprouts on cutover lands.

There is interest in the black locust as a potential forage plant, since it grows rapidly, is a nitrogen-fixing legume, its leaves are easily harvested by hand, and they are high in protein. These attributes would be useful under conditions of small-scale farming (e.g., backyard harvesting of the leaves for rabbit feed) or in developing countries where labor is inexpensive. Black locust tree leaf meal is being produced commercially in China. Liu and Jung (undated) described Chinese production of the meal and studies to ascertain its nutritional value. The locust leaves are collected by hand and sun-cured. They are taken to a feed mill where they are cleaned pneumatically, ground, and pelleted. The leaf meal contains 20–24% crude protein. In a trial with broiler chicks, Liu and Jung found that with 3% locust leaf meal or 3% alfalfa meal in a corn–soy diet, performance was consistently better with the alfalfa meal diet. There was a depression in both gain and feed conversion with the locust meal. Cheeke *et al.* (1983) demonstrated that 20% black locust meal in the diet of chicks caused a 30% reduction in growth rate; autoclaving the leaf meal partially overcame the growth-depressing effects. In rabbits, replacement of alfalfa by locust leaf meal resulted in growth depression and reduced crude protein digestibility (Harris *et al.* 1984). Horton and Christensen (1981) compared locust leaf meal with alfalfa meal in a lamb trial. Either meal was offered as the sole diet. They concluded that black locust leaf meal was not a satisfactory alternative to alfalfa meal in ruminant diets, as the digestibility of protein, organic matter, and phosphorus was much lower in the locust meal. The markedly reduced protein digestibility and the low palatability observed suggest that phenolic compounds could be present in black locust meal.

ENZYMES

Thiaminases

The thiaminase enzyme splits thiamine, a B vitamin, rendering it inactive. The enzyme is found in a variety of sources, including the viscera of certain fish such as carp, in bracken (*Pteridium aquilinum*), horsetail (*Equisetum arvense*), and nardoo (*Marsilea drummondii*), an Australian fern.

Bracken Fern Poisoning. Bracken fern (*P. aquilinum*) grows in humid temperate areas such as the west coast of North America, Great Britain, Western Europe, Japan, Australia, New Zealand, and South America (Fig. 7.5). Situations that lead to bracken consumption by livestock include early spring conditions when other green feed is short, grazing of succulent watery pasture so animals have a craving for coarse fibrous materials, use of bracken as bedding, contamination in hay, and exposure of rhizomes in plowed areas. Consumption of bracken can lead to several toxicity situations. Nonruminants (horses, pigs) may develop a thiamine deficiency, while cattle and sheep may

FIG. 7.5. Bracken fern (*Pteridium aquilinum*).

exhibit signs of poisoning manifested by a high body temperature, severe hematological disturbances, and bone marrow damage. Bracken also contains a carcinogen which may affect all species, including humans. These aspects are discussed in Chapter 11.

Evidence of an effect of bracken on thiamine activity was first reported by Weswig et al. (1946) at Oregon State University, who noted that rats fed a diet containing dried bracken fern produced classic signs of thiamine deficiency which were overcome with supplementary thiamine. Subsequent work demonstrated the presence of a thiaminase enzyme that splits the thiamine molecule, as shown in Fig. 7.6.

A cosubstrate, usually an amine or a sulfhydryl-containing compound such as proline or cysteine, is required. The pyrimidine analog can also be a thiamine antagonist, depending upon the structure of the cosubstrate, which increases the severity of the thiamine deficiency.

Thiamine deficiency due to bracken consumption has occurred primarily in horses. In the past, bracken poisoning was noted repeatedly in British Columbia, Washington, Oregon, and Great Britain. Ingestion of hay containing more than 20% bracken produces toxicity signs in about a month. Signs of thiamine deficiency include anorexia (loss of appetite), ataxia, opisthotonus (head retraction), convulsions, and death. Clinical signs are an elevated level of pyruvate in the blood, cardiac irregularity, and decreased blood thiamine. Response to thiamine administration is dramatic, with a complete reversal of symptoms within a short time.

Metabolically, thiamine is involved as a cofactor in decarboxylation reactions. These include the conversion of pyruvate to acetyl-CoA, and

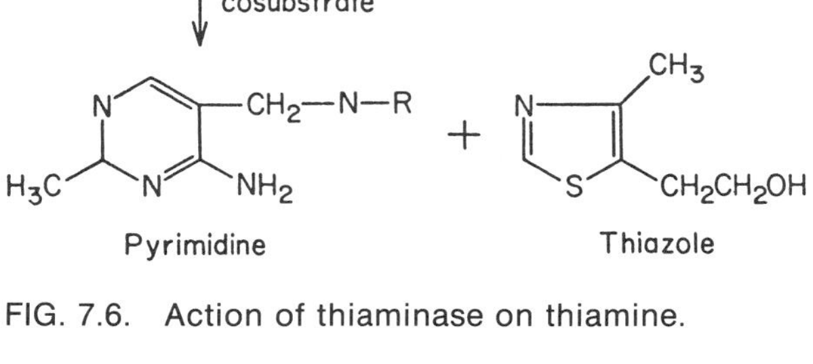

FIG. 7.6. Action of thiaminase on thiamine.

oxidation of α-ketoglutarate to succinyl-CoA in the citric acid cycle. These reactions are shown in Fig. 7.7.

As is apparent from Fig. 7.7, thiamine deficiency results in impaired pyruvate utilization. Therefore, as pyruvic acid is formed via glycolysis, it accumulates, and the blood pyruvate level rises.

The animal suffers from impaired energy metabolism and a cellular shortage of ATP. The elevated pyruvate may affect central nervous system function.

The thiaminase activity of bracken is highest in the rhizomes, and the activity shows a characteristic seasonal variation (Fig. 7.8).

Ruminant animals consuming bracken normally do not develop signs of thiamine deficiency because of the abundant synthesis of the vitamin in the rumen. Thiamine deficiency has been induced in sheep experimentally by feeding 15–25% dried rhizome, but such a situation is unlikely to occur under practical conditions.

Other Sources of Thiaminase. Horsetail (*E. arvense*) is a common weed in moist areas of the U.S. and Canada (Fig. 7.9). Horsetail contains thiaminase activity. Cases of poisoning of horses in North America have been documented; hay containing 20% or more horsetail may produce symptoms of thiamine deficiency in horses in 2–5 weeks.

Various fish such as carp contain thiaminase activity. These may produce thiamine deficiency in mink and other fur animals when large amounts of raw fish are used as feed. This condition is called Chastek's paralysis, named for the Minnesota fox farmer who first noted the problem. Mink develop classic signs of thiamine deficiency and recover dramatically when thiamine is injected.

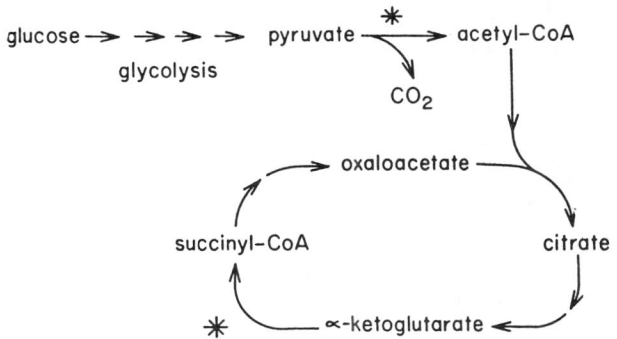

* thiamine-dependent reactions

FIG. 7.7. Role of thiamine in cellular metabolism.

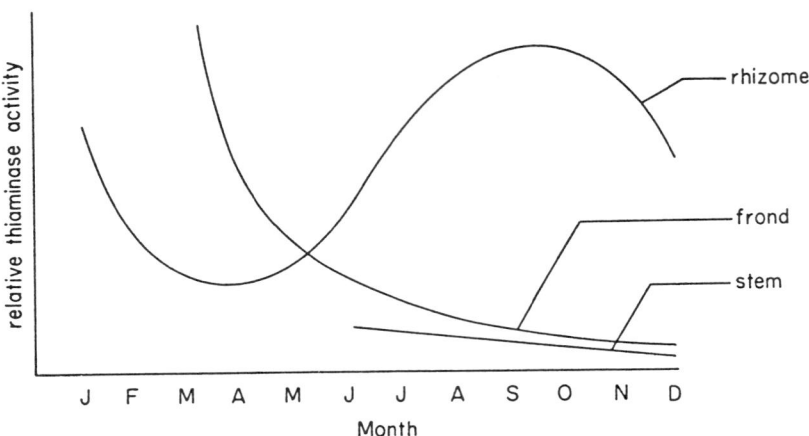

FIG. 7.8. Seasonal changes in the distribution of thiaminase activity in bracken.
Adapted from Evans (1976).

FIG. 7.9. Rush-like stems of an *Equisetum* sp.

7. PROTEINS AND AMINO ACIDS

FIG. 7.10. Nardoo, an Australian fern with high thiaminase activity.

Nardoo (*M. drummondii*) is an Australian fern that grows in damp areas such as water courses (Fig. 7.10). It often grows following periods of flooding or water logging. Extensive losses of sheep with typical symptoms of thiamine deficiency have occurred. Thiamine deficiency in ruminants is referred to as polioencephalomalacia, characterized by depression, incoordination, convulsions, and cerebrocortical necrosis (Fig. 7.11). Nardoo contains thiaminase levels up to 100 times those of bracken fern. Thiamine deficiency and polioencephalomalacia are widespread in New South Wales, Western Australia, Victoria, and South Australia.

As an interesting sidelight, the consumption of nardoo may have played a role in the deaths of the Australian explorers Robert Burke and William Wills, whose tragic expedition across Australia ended in their deaths at Cooper's Creek in 1861. Their epic journey has been described by Moorehead (1963) in his book *Cooper's Creek*. In their final days on the banks of Cooper's Creek, the explorers subsisted largely on the seeds and leaves of nardoo, which grew abundantly in the area and was a major food of the aborigines. In his diary, a few days before his death, Wills wrote

> my dear Father, these are probably the last lines you will ever get from me. We are on the point of starvation, not so much from absolute want of food, but want of

nutriment in what we can get—I am weaker than ever altho' I have a good appetite and relish the nardoo much but it seems to give us no nutriment—but starvation on nardoo is by no means very unpleasant, but for the weakness one feels, and the utter inability to move onself, for as far as appetite is concerned, it gives me the greatest satisfaction—.

These symptoms of weakness are consistent with a thiamine deficiency induced by the high thiaminase activity of nardoo. This conclusion was also reached by Bergin (1981) who recreated the Burke and Wills expedition. Burke and Wills exhausted their supply of pork, their main dietary source of thiamine, and so would have been low in thiamine. Symptoms described in their diary, including edema, wasting of muscles, altered sensitivity to cold and pain, and weakness in the legs, are indicative of beriberi. During April and May, when nardoo was being consumed, its toxicity is highest. Bergin (1981) concluded, "—nardoo would obviously have been lethal to men suffering from beri-beri!" The aborigines, who consumed large amounts of nardoo, avoided poisoning by cooking the plant, thus destroying thiaminase.

FIG. 7.11. A sheep with polioencephalomalacia, resulting from a thiamine deficiency induced from consumption of nardoo.
Courtesy of B. Chick.

Another Australian plant with a high thiaminase activity is rock fern (*Cheilanthes sieberi*), which is widely distributed in coastal areas and subtropical regions of Australia. It appears to contain a bone marrow toxin similar to that found in bracken (Everist 1981).

Kochia scoparia (kochia, summer cypress, burning bush, fireweed, Mexican fireweed) grows in arid lands of the U.S. West and Southwest and has been responsible for a number of poisonings of livestock. Signs of toxicosis include icterus and photosensitization, progressive central nervous system dysfunction, hepatic cirrhosis, gastrointestinal tract inflammation, and polioencephalomalacia. The occurrence of polioencephalomalacia suggests involvement of a thiaminase or a hepatotoxin which may lead to impaired thiamine utilization. Dickie and James (1983) found evidence of a thiamine-destroying principle but suggested that it is not the major factor in kochia toxicity. Hepatic necrosis and fibrosis, with photosensitization, suggests the presence of a hepatotoxin (Dickie and Berryman 1979). Kochia thrives on arid, infertile, high sodium soils where few other plants will grow. It can be used successfully as a forage plant. It is most toxic during periods of drought and at seed maturation. Supplemental feed should be provided at these times, especially if rain stimulates seed production (Dickie and James 1983). The high sulfate content of kochia generally results in the plant having a slight laxative effect. Other toxicants which might be associated with kochia are oxalates and nitrates.

Polioencephalomalacia is a serious disorder of feedlot cattle fed diets high in grain. It is associated with high levels of thiaminase activity in the rumen of affected animals (Edwin *et al.* 1982), leading to an induced thiamine deficiency. Several thiaminase-producing bacterial species, such as *Clostridium sporogenes* and *Bacillus* spp., have been implicated (Edwin *et al.* 1982; Haven *et al.* 1983). Haven *et al.* (1983) reported that the predominant gram-negative nonspore-forming microflora normally resident in the rumen may be the source of thiaminase, and that these organisms may for some reason develop metabolic alterations which lead to increased thiamine destruction.

Polioencephalomalacia is also known as cerebrocortical necrosis. Cattle and sheep are affected, with young animals between 2 and 7 months of age being the most susceptible (Edwin and Jackman 1982). Clinical signs of the disorder include aimless wandering, disorientation, blindness, recumbency, and opisthotonus. The brain is edematous, with yellowish discoloration of the cerebral cortex. Affected animals may respond dramatically to administration of thiamine, with doses of 200–500 mg for calves, 1000–2000 mg for adult cattle, and 100–500 mg for sheep, depending on body weight.

Lipoxidases (Lipoxygenases)

Lipoxidase is an enzyme that catalyzes the oxidation of lipids containing *cis-cis*-1,4-pentadiene systems to hydroperoxides:

$$R-CH=CH-CH_2-CH=CH-R_1$$
$$\text{cis} \qquad\qquad\qquad \text{cis}$$

$$\downarrow O_2 \text{ lipoxidase}$$

$$R-CH=CH-CH=CH-\overset{\overset{OOH}{|}}{CH}-R_1$$
$$\text{cis} \qquad \text{trans}$$

Because it is a specific catalyst for the *cis-cis*-1,4-pentadiene system, it will oxidize linoleic, linolenic, and arachidonic acids and their derivatives, but not those of oleic acid. Lipoxidases in grains, and particularly in legumes such as soybeans and alfalfa, readily destroy carotene and xanthophylls, reducing the vitamin A activity of feeds. In the baking industry lipoxidase is used to bleach carotene in dough. Sources of lipoxidase, in decreasing order of concentration, are soybeans, lentils, green peas, peanuts, field beans, wheat, barley, and sunflower seeds. The "beany" flavor of beans is due to the action of lipoxidases on free fatty acids in the seed, giving rise to ketones with undesirable flavors. Heat treatment of 80°C will denature the enzyme.

BLOAT-PRODUCING PROTEINS

Bloat is a distension of the rumen as a result of the inability of the animal to eructate gases produced in the normal processes of rumen fermentation (Fig. 7.12). The quantity of gas produced varies according to the amount of fermentable substrate and types of microorganisms in the rumen, but in general 30–50 liters of rumen gas per hour are produced in cattle in the period of 3–4 hr after feeding, with a daily production of about 400 liters in cattle and 50 liters in sheep. The principal gases are carbon dioxide and methane. In pasture bloat, these gases are trapped in the form of a stable foam. The eructation mechanism is inhibited by the presence of foam at the base of the esophagus; eructation of foam would result in it getting into the lungs. Bloat-producing plants, primarily legumes, contain substances which cause the production of a stable foam in the rumen. In addition, there are animal factors involved, as not all animals on a bloat-inciting

FIG. 7.12. An experimentally bloated animal, showing the high intraruminal pressures that may develop in bloat. The animal was fed lush alfalfa to cause bloat; when the cap of the rumen fistula was removed, the contents were forced out by the high pressure built up in the rumen.
Courtesy of G. Derrick and R. E. Howarth.

pasture develop the condition. Salivary secretions may have an effect. The rumen microorganisms have been postulated to be involved in the bloat complex.

Plant Factors Involved in Bloat

Bloat primarily occurs on legume pastures, but there are documented cases of the condition on lush grass pasture. In addition, it is a problem on wheat pastures. Some legumes are well known for their bloat-producing potential, while others are not bloat-producers. The most important bloating species in temperate regions are alfalfa (*Medicago sativa*), red clover (*Trifolium pratense*), and white clover (*Trifolium repens*). Subterranean clover (*T. subterraneum*) has occasionally been implicated in bloat, but generally does not cause problems. Bird's-foot trefoil (*Lotus corniculatus*), sainfoin (*Onobrychis viciifolia*), lespedeza (*Lespedeza stipulacea*), cicer milk vetch (*Astragalus cicer*), and crown vetch (*Coronilla varia*) do not cause bloat. Tropical legumes are not bloat-producers, with the occasional exception of lab lab (*Lablab purpureus,* also referred to as *Dolichos lablab*), a tropical leguminous

FIG. 7.13. Tropical legumes generally do not cause bloat in part because they often have tough fibrous leaves which reduce the rate of cell rupture. Here, cattle in Australia are grazing *Lablab purpureus,* which has occasionally caused bloat.
Courtesy of M. P. Hegarty.

vine used as a forage in Australia (Fig. 7.13). It appears that the primary plant factors involved in bloat production are cytoplasmic proteins that occur in the bloat-producing forages. These have been referred to as fraction I, fraction II, and 18 S proteins. Howarth *et al.* (1977) found a good relationship between bloat incidence and total soluble protein level. Other factors associated with the total soluble protein content in causing bloat include the rate at which cell rupture occurs in the rumen, the soluble carbohydrate level in the forage, and the presence or absence of protein precipitants. Rapid release of cell contents, allowing a high fermentation rate (and thus a high rate of rumen gas production), favors development of bloat. The soluble proteins form the membrane of the gas bubbles, producing a stable foam. Chloroplast fragments may serve as nucleation sites for bubble formation. The soluble carbohydrate content of the forage is important in determining the rate of fermentation and may also exert an influence on the rumen protozoa. In the presence of high-forage soluble carbohydrate, protozoa numbers may increase markedly and then suddenly die. They may store excessive starch and then burst, or they may die because of changes in rumen environment such as pH. The protozoa cell contents may contribute to the formation of a stable foam. A major reason for the lack of bloat with some legumes such as bird's-foot trefoil and sainfoin appears to be their content of protein precipitants, such as tannins. By denaturing soluble proteins, tannins prevent foam formation. Plant-breeding efforts to incorporate tannins into bloat-producing forages such as alfalfa are in progress. Howarth et al. (1977) also suggest that selection for a reduced ratio of soluble to total protein might be effective.

Lush legume pastures, especially immature plant tops, are more likely to cause bloat than more mature forage, or hay. This is because the immature foliage is high in soluble protein and carbohydrate and has a rapid rate of cell rupture in the rumen. Drying causes protein denaturation, thus reducing the level of soluble protein. Climatic factors can influence bloat. Warm days and cold nights lead to a high starch content of legume tissue, which may trigger bloat outbreaks. The lack of bloat with most tropical legumes may be due to the presence of tannins in some cases, and in others the tough fibrous nature of the leaves may retard the rate of cell rupture.

Animals likely to bloat have a vigorous microbial population with a high fermentation potential so that when an abundance of forage with high fermentable carbohydrate is consumed, rapid release of cell contents results in rapid fermentation (thus high rumen gas production) and viscous rumen contents, aiding in the trapping of gases in stable foam (Majak et al. 1983). The rumen bacteria colonize particulate matter such as chloroplast fragments. Majak et al. (1983) summarized the rumen conditions that lead to onset of bloat as the following: (1) colonization of chloroplast particles and other particulates by rumen bacteria; (2) accumulation of these particles in fluid of the dorsal sac of the rumen; (3) predisposition to frothiness by entrapment of gas bubbles among the suspended particles; (4) provision of an active inoculum for the rapid disintegration and release of mesophyll cell contents from ingested forage; (5) flotation of digesta because of microbial gas production; and (6) inability of the animal to clear fermentation gases because of entrapment by buoyant, frothy digesta.

Prevention of Bloat

In the U.S., the use of antifoaming agents such as poloxalene (Bloat Guard) in legume pastures has vastly reduced the bloat problem (Fig.

FIG. 7.14. A steer licking a Bloat Guard block. These blocks, when properly distributed in a legume pasture, are highly effective in preventing bloat.
Courtesy of SmithKline Corporation.

7.14). These blocks have a high molasses content and serve as the only source of salt, so cattle lick them frequently during the day, ensuring the continual presence of the antifoaming agent in the rumen. One block per five head of cattle should be used, and the blocks should be well distributed throughout the pasture. In areas of extensive grazing with scattered patches of bloat-producing legumes, the blocks may be less effective, or it may not be feasible to use them because of the nature of the terrain and plant distribution.

Other methods of minimizing bloat incidence include using a mixture of grasses and legumes rather than a pure stand of legume. The bloat potential is highest for lush, immature forage which has a rapid rate of cell rupture in the rumen, so it may be advisable to graze legumes only when they are approaching maturity. Strip-grazing to force consumption of the entire plant rather than selective grazing of tops is another useful management practice. In New Zealand, pastures have been strip-grazed with vegetable oil or fats sprayed on the daily allotment; the lipids serve as antifoaming agents. The ultimate solution will be the development through plant breeding of legume cultivars that have a minimal bloat potential. This might be accomplished through selection for reduced proportions or rate of release of soluble protein, or incorporation of tannins into the genetic structure of the plant. Since no species with tannins have been found in some plant genera that cause bloat, such as *Medicago* (alfalfa and medics), novel techniques such as mutagenesis or "gene-splicing" procedures are being explored as methods of introducing protein precipitants. In a few cases, it may be possible to plant bloat-producing legumes with other forages that do contain tannins, such as sainfoin. In parts of Australia, bloat in cattle grazing pastures dominated by white clover has been controlled by the concurrent ingestion of carpet grass, which is rich in tannins (McWilliam 1973).

Howarth *et al.* (1982) in Canada have a breeding program to develop a bloat-safe cultivar of alfalfa. Their program is based on observation that the major foaming agents involved are proteins, and the major cellular site of these proteins is in the mesophyll cells of leaves. The mesophyll cells of bloat-safe legumes are more resistant to rupture from chewing or rumen microbial digestion than those from bloat-causing legumes. For example, the release of proteins from alfalfa leaves through cell rupture is more rapid than from leaves of bird's-foot trefoil, Cicer milk vetch, and sainfoin, which are nonbloating legumes. Howarth *et al.* (1982) found differences in rate of cell rupture among alfalfa cultivars, indicating that selection for a bloat-free alfalfa may be possible. They measured the rate of cell rupture using nylon bag digestion of alfalfa in rumen contents. Rapid dry matter disappearance is associated with a rapid rate of cell rupture. Sant and Wilson (1982) have employed a cellulolytic enzyme technique to measure the rate of mesophyll cell rupture.

Results from the Canadian plant-breeding program to develop bloat-safe alfalfa based on rate of cell rupture indicate that a reduction in cell-rupture rate has been achieved, and the bloat-safe alfalfa has a low bloat potential (R. E. Howarth, 1984, personal communication, Agriculture Canada, Saskatoon, Sask.).

MIMOSINE

Leucaena leucocephala, commonly referred to as leucaena, ipil-ipil (Southeast Asia), and kao haole (Hawaii), is a tropical legume with great potential as a protein source for livestock (Fig. 7.15). It is vigorous, rapidly growing, drought-tolerant, palatable, high yielding, and its leaves contain from 25 to 35% crude protein. It can be grown in tropical pastures or can be harvested and dried, with the leaf meal

FIG. 7.15. *Leucaena leucocephala,* a tropical forage legume containing the toxic amino acid mimosine.

used in place of alfalfa meal in poultry and swine diets. It can be harvested in "cut and carry" feeding systems of small farmers and used for feeding cattle, goats, and rabbits. These potential attributes are presently limited by the occurrence of the toxic amino acid mimosine in leucaena.

Mimosine is structurally very similar to tyrosine, but does not seem to act as a tyrosine antagonist. In nonruminant animals, mimosine causes poor growth, alopecia (loss of hair), eye cataracts, and reproductive problems. Levels of leucaena meal above 5–10% of the diet for swine, poultry, and rabbits generally result in poor animal performance. The biochemical mode of action of mimosine in producing its toxic effects is not clear. Hegarty (1978), who has worked extensively on mimosine and leucaena in Australia, suggests that it may act as an amino acid antagonist, it may complex with pyridoxal phosphate, or it may complex metals such as zinc. Stunzi *et al.* (1979) have shown that under physiological conditions, mimosine binds copper and zinc ions more strongly than do most amino acids. Jones *et al.* (1978) in Queensland have reported that mineral supplements containing zinc reduce the toxicity of leucaena in cattle and suggested that some of the toxic signs (e.g., skin lesions) resemble zinc deficiency. Mimosine is an inhibitor of cystathionine synthetase and cystathionase, enzymes involved in the conversion of methionine to cystine. This inhibition could be a factor in mimosine-induced alopecia.

Ruminant animals grazing leucaena may show various symptoms such as poor growth, loss of hair, swollen and raw coronets above the hooves, lameness, mouth and esophageal lesions, depressed serum thyroxine levels, and goiter (Fig. 7.16). Some of these symptoms may be due to mimosine itself and others to a metabolite produced in the rumen. Hegarty et al. (1976) reported that in the rumen, mimosine is converted to 3,4-dihydroxypyridine (DHP).

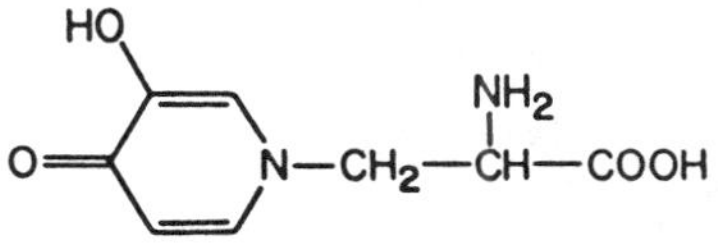

Mimosine

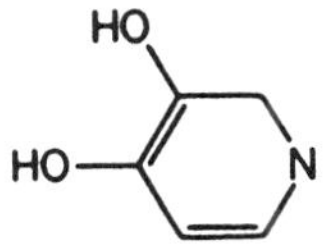

3,4-dihydroxypyridine

DHP is a goitrogen, impairing the incorporation of iodine into iodinated compounds in the thyroid gland.

In some parts of the world, such as Australia, consumption of leucaena by ruminants causes the problems previously discussed. In other places, such as Hawaii, cattle and goats can consume leucaena with impunity. Jones (1981) found that DHP is further metabolized in animals in Hawaii to unidentified nontoxic compounds. Mimosine and DHP rapidly disappear when incubated in vitro with rumen fluid from Hawaiian ruminants, but are much more stable when inoculum from Australian animals is used.

In subsequent work, it has been demonstrated that transfer of rumen fluid from animals in Hawaii to Australian ruminants resulted in a complete elimination of the toxic effects of leucaena (R. J. Jones 1984, personal communication, CSIRO, Townsville, Qld., Australia). The bacteria were gram-negative, strict anaerobe short rods. These organisms degrade mimosine and DHP to nontoxic metabolites in the

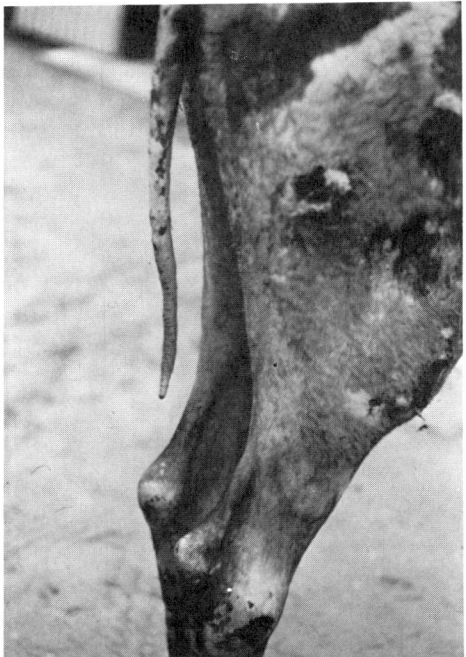

FIG. 7.16. Loss of hair from the switch and legs of a cow grazing *Leucaena leucocephala*.
Courtesy of R. J. Jones.

rumen. It is anticipated that a program will be established to introduce the organisms into cattle in all parts of Australia where leucaena is used, thereby having pronounced economic implications in Australian agriculture. It is likely that other plant toxicity problems might be overcome by transferring rumen microorganisms from animals where the plant is an indigenous species to animals in areas where the plant is not indigenous. For example, in Australia, ruminants are poisoned by indospicine in *Indigofera spicata,* whereas in Africa where the plant is indigenous, there are no reports of toxicity (R. J. Jones 1984, personal communication, CSIRO, Townsville, Qld., Australia), suggesting that indospicine-metabolizing rumen organisms might exist.

While the introduction of rumen microorganisms capable of mimosine–DHP degradation into susceptible ruminants appears to be a solution to the toxicity of leucaena to ruminants, the problem of mimosine toxicity to nonruminants remains. A solution to the mimosine problem could be the development of low-mimosine cultivars. This has not yet been successfully achieved; while there are some low-mimosine types, they are unproductive and of low vigor. Indonesian workers (Lowry *et al.* 1983) demonstrated enzymatic degradation of mimosine in leaf tissue. Wilting to allow enzymatic destruction of mimosine might be a means of reducing its toxicity to nonruminant animals such as rabbits. However, plant hydrolase activity may convert up to 30% of mimosine to DHP during the process of mastication (Lowry et al. 1983), so that reduction of mimosine levels might result in DHP toxicity.

Leucaena is a teratogenic plant, with mimosine the primary teratogenic agent. Deweede and Wayman (1970) fed mimosine to pregnant rats and noted a variety of fetal deformities. In swine fed leucaena during gestation, fetal resorptions and polypodia of the forelimbs of the fetuses were observed (Wayman et al. 1970).

Feeding leucaena meal to Japanese quail caused a reduction in egg production. Ross et al. (1980) fed leucaena meal and an equivalent amount of mimosine to quail and observed a much greater reduction in egg production with the leucaena meal, which suggests that leucaena may contain toxic factors other than mimosine.

The effects of leucaena and mimosine on nonruminants can be reduced to some extent by supplementation of the diet with ferrous sulfate. Mimosine forms a complex with iron which is excreted in the feces. El-Harith et al. (1979) found that addition of 2% ferrous sulfate to a diet containing 25% leucaena leaf meal prevented mimosine toxicity symptoms in rats. Dry heating the leaf meal at 90°C for 20 hr did not detoxify it.

TRYPTOPHAN

Acute bovine pulmonary emphysema (ABPE) is a respiratory disease of cattle caused primarily by metabolites of the amino acid tryptophan. It occurs after a sudden dietary change, generally from somewhat sparse feed to a lush, succulent pasture. In the western U.S., it occurs frequently in range cattle that are transferred from dry summer range to improved pastures or hay meadows in the late summer and autumn. In Britain, where the disease is called fog fever, it occurs after stock are turned into lush fields of kale, rape, or other succulent feeds. The term "fog fever" arises from the occurrence of the condition after cattle are given access to lush regrowth (foggage) pasture. In the U.S., there is a pronounced seasonal incidence of ABPE, with most cases in the autumn following a dry summer. The condition is rare in cattle less than 1-year old and is most prevalent in cattle over 2 to 3 years of age. This is apparently the cumulative result of repeated insults by the causative agent. The largest, heavily lactating cows are most likely to develop ABPE, perhaps because of their high intake of lush forage. The clinical signs are observed 2–10 days after an abrupt feed change.

They include an increased rate of respiration, with severe dyspnea (labored breathing). Cattle may stand with their head extended and lowered, tongue protruded, frothing at the mouth, and mouth breathing (Fig. 7.17). An audible expiratory grunt may be heard, giving rise to the common name "grunts" for the condition. Other local terms are cow asthma, green grass sickness, and summer pneumonia. The lungs are grossly enlarged, with considerable quantities of edematous fluid and interstitial emphysema (swelling due to air).

The causative agent is believed to be 3-methylindole (3-MI), a metabolite produced by rumen fermentation of tryptophan. This was a serendipitous discovery at Washington State University; calves were being dosed with different amino acids, and those given tryptophan developed emphysema. The 3-MI is absorbed and is either responsible for the lung damage itself or is metabolized by the pulmonary mixed function oxidase (MFO) system to other active metabolites. Postulated routes of metabolism are shown in Fig. 7.18. The toxic dose of tryptophan is about .25–35 g/kg body weight; .1 g 3-MI/kg has the same effect. It is not known for certain why a sudden diet change provokes increased production of 3-MI. It could be the combined result of an increase in available tryptophan and a diet-induced change in rumen microflora. Abrupt change from dry to lush forage is accompanied by pronounced changes in the composition and digestibility of the forage

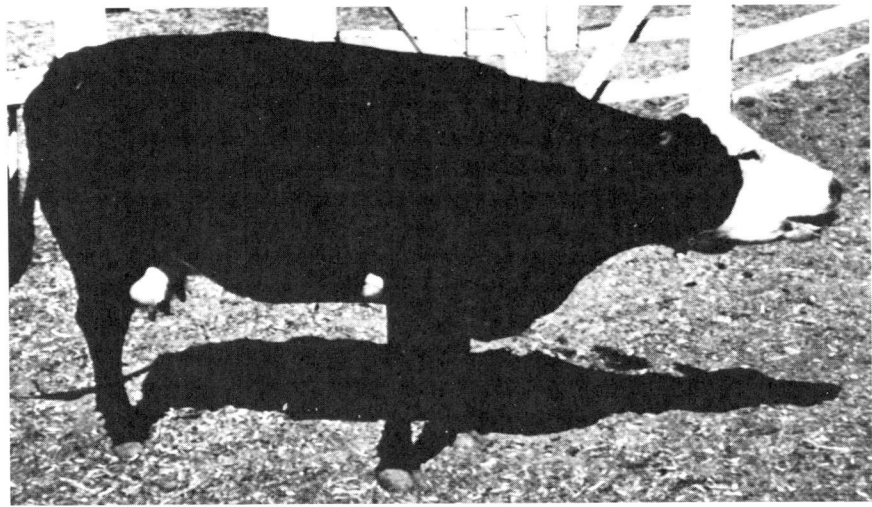

FIG. 7.17. A cow showing typical signs of acute bovine pulmonary emphysema.
Courtesy of J. R. Carlson.

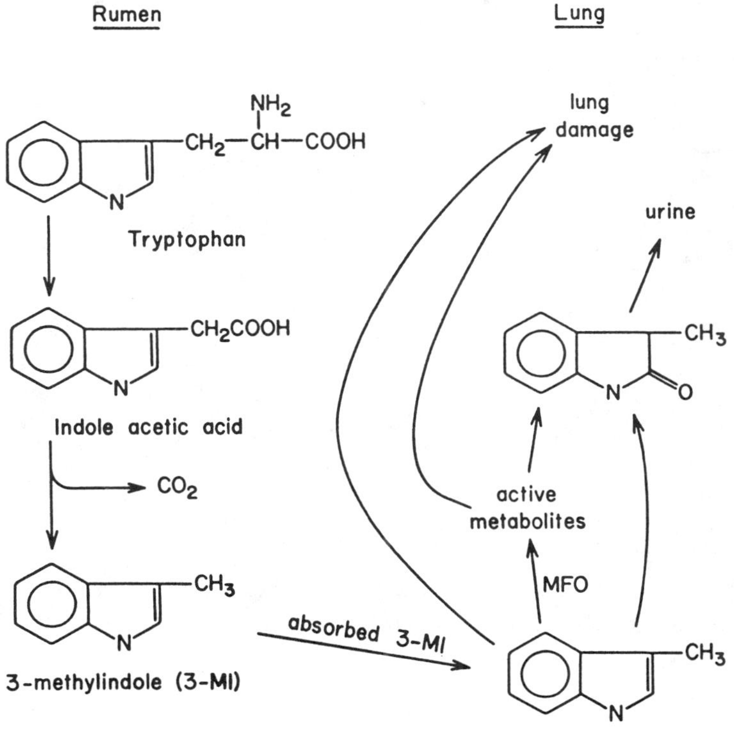

FIG. 7.18. Metabolism of tryptophan in acute bovine pulmonary emphysema.

consumed (Fig. 7.19). The fiber content may be halved and the crude protein content doubled in lush grass compared to mature forage of the same species. These changes, along with a large increase in feed intake, could markedly alter rumen fermentation.

The clinical signs of ABPE generally appear from 1 to 14 days after the pasture change, and death may occur in 2–4 days. Mortality is about 30% of affected animals. Animals surviving the episode of clinical signs may recover without serious permanent lung damage.

Evidence that dietary tryptophan is the initiating factor includes the observations that oral administration of tryptophan produces clinical signs and pulmonary lesions indistinguishable from those seen in field cases of ABPE, and that lush grass can contain sufficient tryptophan to produce toxic amounts of 3-MI. Possibly other indoles in forage could also contribute to 3-MI production. Rumen fermentation of tryptophan to 3-MI is a 2-step reaction, with deamination and decarboxylation

FIG. 7.19. A scene from Yorkshire, England showing typical conditions under which acute bovine pulmonary emphysema occurs. Cattle are brought from sparse pasture on the hillside to lush pastures in the lower fields.
Courtesy of J. R. Carlson.

producing indole acetic acid (IAA) and a subsequent decarboxylation producing 3-MI (Fig. 7.18). Carlson and co-workers at Washington State University have isolated a *Lactobacillus* sp. from rumen fluid which converts IAA to 3-MI (Carlson and Breeze 1984). There are a number of organisms which metabolize tryptophan to IAA. Dietary factors influence the rate of conversion of tryptophan to 3-MI, with the rate being more rapid on forage-based than on concentrate-based diets.

While 3-MI does have some direct effect on lung tissue, it appears that the mechanism of action is metabolism of 3-MI by pulmonary MFO to produce active metabolites. Pretreatment of animals with MFO inducers such as phenobarbital increases severity of lung injury, while MFO suppressors such as piperonyl butoxide inhibit 3-MI metabolism and prevent lung injury (Carlson and Breeze 1984). Pretreatment with cysteine to enhance lung glutathione levels reduces severity of lung damage, while administration of the glutathione depleter diethyl maleate increases lung injury, suggesting a role of glutathione in detoxification of 3-MI metabolites. The chemical identity of the reactive metabolite(s) has not yet been established.

There is no effective treatment of animals with ABPE. Prevention involves avoiding a sudden dietary change from a relatively poor pasture to a lush one. In the western range areas of the U.S., supplementation of cows on dry summer range may be advisable. Providing supplementary hay or other feed for the first 2 weeks after cattle are moved to lush grass and interrupting the grazing pattern by gathering and corralling the animals at intervals may be helpful management practices. Carlson and co-workers at Washington State have shown that modification of rumen metabolism may have potential as a means of preventing ABPE. The feeding of antibiotics, such as rumensin and chlortetracycline, reduced 3-MI production. The use of buffers to change the rumen microflora also shows potential.

Bovine pulmonary emphysema has also been encountered with feedstuffs containing 3-substituted furans. Moldy sweet potato poisoning of cattle results in lung disease that is caused by 3-substituted furans such as ipomeanol:

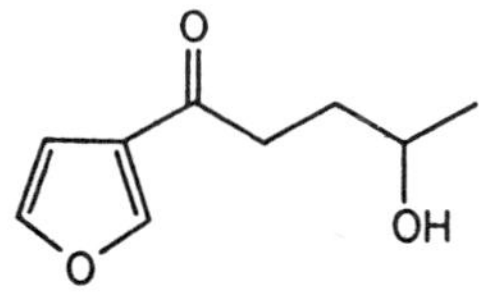

These compounds are stress metabolites of the sweet potato produced in response to fungal infections. The purple mint (*Perilla frutescens*), an herb introduced into the U.S. from the Orient, has caused outbreaks of pulmonary emphysema in Oklahoma and Arkansas (Wilson et al. 1978). It contains 3-substituted furans very similar in structure to those of moldy sweet potatoes.

SELENOAMINO ACIDS

Soils in many parts of the world contain high levels of selenium. Plants growing on these soils may accumulate toxic levels of the element, which may occur in the plant as selenoamino acids. In North America, high soil selenium levels are found in areas of the Great Plains, including parts of Montana, North and South Dakota, Nebraska, Kansas, Oklahoma, Texas, Wyoming, Colorado, New Mexico, Idaho, Utah, Arizona, Nevada, and the provinces of western Canada.

Selenium poisoning, or selenosis, may be of several types, including acute toxicity and chronic forms known as blind staggers and alkali disease. Acute toxicity results from the consumption of plants containing very high selenium levels (several hundred ppm). These types of plants are called selenium accumulators or selenium indicators, as their growth is restricted to areas of seleniferous soils. They may contain extremely high selenium levels, of as much as 10,000 ppm on a dry weight basis. Examples of indicator plants include about 25 species of *Astragalus* (e.g., *A. bisulcatus*) and *Xylorrhiza* spp., the woody asters. These plants are generally unpalatable because of the offensive odors of the selenium-containing compounds, but are eaten when forage is sparse. Signs of acute toxicity include staggering, diarrhea, prostration, hemorrhage of internal organs, and abdominal pain.

Chronic toxicity has been suggested to be of two types: alkali disease and blind staggers, although it is not completely accepted that blind staggers is a selenosis condition. The blind staggers syndrome has been associated with the consumption of indicator or secondary selenium absorber plants of moderate (less than 200 ppm) selenium content. Secondary absorber plants are those which accumulate moderate (several hundred ppm) levels of selenium when growing on seleniferous soils, and their growth is not restricted to such soils. The blind staggers syndrome involves loss of appetite, impairment of vision, and wandering in circles, followed by weakness, recumbency, paralysis of the tongue and swallowing mechanisms, labored respiration, abdominal pain, grating of the teeth, respiratory failure, and death. Consumption of *A. bisulcatus* has been associated with the condition. These symptoms resemble those of locoism and cannot be duplicated by feeding high levels of inorganic selenium. Van Kampen and James (1978) fed *A. bisulcatus* with a high (180 ppm) selenium content to ewes and produced signs of locoism rather than of blind staggers. They suggested that the condition previously described as blind staggers due to chronic selenium intoxication may in fact be locoweed poisoning. The situation is as yet not completely clarified.

Alkali disease is caused by the chronic consumption of plants containing 5–50 ppm of selenium. To put these and previous levels mentioned in perspective, the selenium requirement of animals is about 0.1–0.5 ppm, while frank deficiency symptoms (white muscle disease) occur when selenium levels are less than 0.05 ppm. Principal signs of chronic selenium toxicity include loss of the long hair from the mane and tail of horses (bobtail horses) (Fig. 7.20) and from the tail switch of cattle, loss of body hair from swine, sore feet with inflammation at the coronary band in horses, cattle, and swine, followed by hoof deformities (Fig. 7.21) and loss of condition, progressing to mortality. Decreased hatchability of chicken eggs due to embryonic malformation occurs.

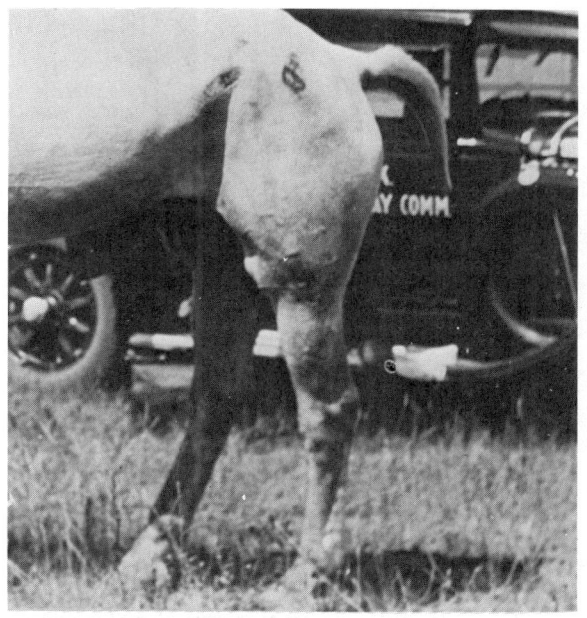

FIG. 7.20. Chronic selenium toxicosis in a horse, showing emaciation and loss of long hair from the tail.
Courtesy of O. E. Olson.

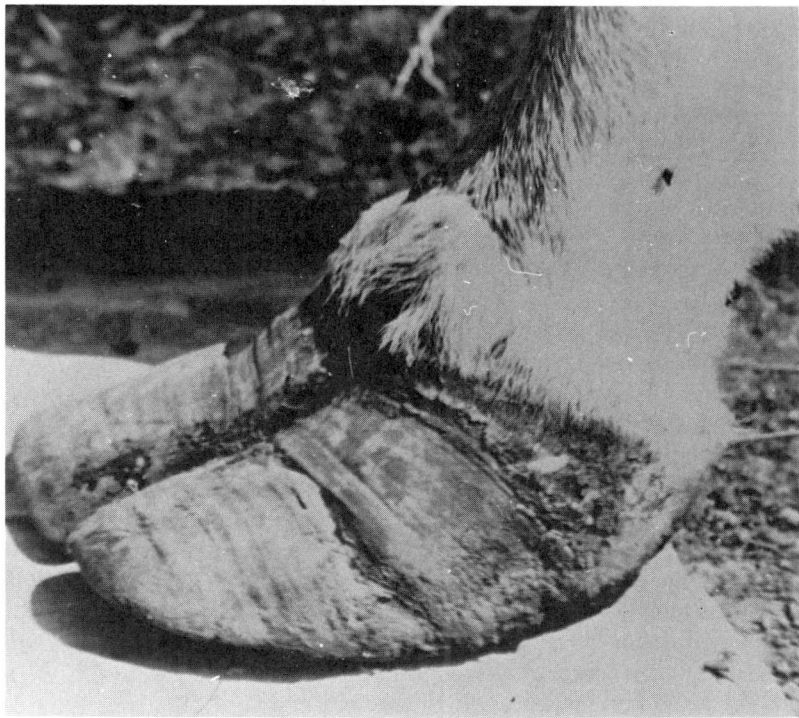

FIG. 7.21. Chronic selenium toxicosis in a bovine, showing elongated and cracked hooves.
Courtesy of O. E. Olson.

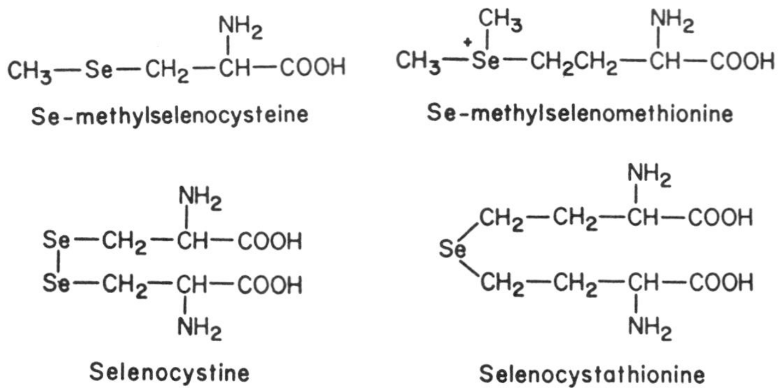

$$CH_3-Se-CH_2-\underset{\underset{NH_2}{|}}{CH}-COOH \qquad CH_3-\overset{\overset{CH_3}{|}}{\underset{+}{Se}}-CH_2CH_2-\underset{\underset{NH_2}{|}}{CH}-COOH$$

Se-methylselenocysteine   Se-methylselenomethionine

$$\begin{array}{c} Se-CH_2-\underset{\underset{}{|}}{\overset{\overset{NH_2}{|}}{CH}}-COOH \\ | \\ Se-CH_2-\underset{\underset{NH_2}{|}}{CH}-COOH \end{array} \qquad Se\begin{array}{c} CH_2-CH_2-\underset{\underset{}{|}}{\overset{\overset{NH_2}{|}}{CH}}-COOH \\ \\ CH_2-CH_2-\underset{\underset{NH_2}{|}}{CH}-COOH \end{array}$$

Selenocystine   Selenocystathionine

FIG. 7.22. Structures of some selenoamino acids.

Anemia, liver atrophy and cirrhosis, and hemosiderin deposits in tissues also occur. Alkali disease is associated with the consumption of normal crop and pasture species that have been grown in soils of moderate (<50 ppm) selenium content.

A wide variety of organic compounds containing selenium has been identified in plants. The accumulator or primary indicator plants synthesize largely Se-methylselenocysteine, while the nonindicators store Se-methylselenomethionine (Fig. 7.22). Other selenoamino acids found in plants include selenocystine and selenocystathionine (Fig. 7.22).

LATHYROGENS

Lathyrism is a crippling disease in humans caused by the consumption of seeds of *Lathyrus* spp., principally *L. sativus* (chick-pea). It has been a major public health problem in India and continues to afflict people in poorer sections of that country. *Lathyrus* spp. are hardy, vigorous legumes that grow in poor soil and under drought conditions. The crop is grown in India in spite of and with knowledge of its toxicity as insurance against the failure of other crops. In 1956–1957, 4 million acres of *L. sativus* were under cultivation in India (Padmanaban 1980). The seeds are used in making flour for bread and as a vegetable. At one time, landlords distributed wages to tenants in the form of *L. sativus* seeds. Several Indian states have now banned the sale of the seeds, but

7. PROTEINS AND AMINO ACIDS 269

FIG. 7.23. Seed pods of *Lathyrus latifolius*, the perennial sweet pea.

the crop is still widely grown. Epidemics of lathyrism occurred in 1958 and 1974. Lathyrism is of two types: neurolathyrism and osteolathyrism. Neurolathyrism, in which the nervous system is affected, is caused by neurotoxins in *L. sativus* and primarily affects humans. Osteolathyrism, in which skeletal deformities and aortic rupture occur, is seen mainly in livestock, particularly horses, consuming seeds of *L. odoratus*, the annual sweet pea. The perennial sweet pea, *L. latifolius* (Fig. 7.23), is also toxic and is widely naturalized in North America, as is the flat pea, *L. sylvestrus*. The rough pea, *L. hirsutus*, grows in the southern U.S. and contains lathyrogens. It was formerly grown as an annual winter forage or cover crop in southern areas of the U.S.

Osteolathyrism

This is the principal form of lathyrism that affects livestock. Consumption of seeds of *L. odoratus, L. sylvestris, L. hirsutus,* or related species of sweet pea causes skeletal deformity and aortic rupture due to defective synthesis of cartilage and connective tissue. Malformations of long bones are caused by irregular hyperplastic cartilage formed in the epiphysis (the area of the proliferative zone of cartilage at the end of a bone). Aortic rupture due to formation of aortic aneurysms (weakness of the artery wall) is due to defective collagen and elastin synthesis.

The lathyrogen in *L. odoratus* and related species is β-amino propionitrile (BAPN), an amino acid derivative:

$$NC-CH_2-CH_2-NH_2$$

It exists in plants in the form of β-(γ-L-glutamyl) amino propionitrile, but the glutamyl group is not necessary for lathyrogenic activity. BAPN interferes with the cross-linking of collagen and elastin molecules; consequently, the connective tissue loses its normal structure. Cross-linking of connective tissue fibers involves oxidation of the epsilon amino groups of lysine residues to hydroxylysine in the collagen or elastin proteins. The enzyme lysyl oxidase acts on the specific lysine residues involved in cross-linking; this enzyme also requires cupric ions for its activity. Lysyl oxidase is irreversibly inhibited by BAPN, thus accounting for the defective collagen and elastin formed in osteolathyrism. Copper deficiency produces very similar symptoms because of inhibited lysyl oxidase activity. BAPN toxicity is not counteracted by supplementary dietary copper. The reaction involving lysyl oxidase is as follows:

$$R-(CH_2)_4-NH_2 \xrightarrow[Cu^{2+}]{lysyl\ oxidase} R-(CH_2)_2-\underset{OH}{CH}-CH_2-NH_2$$

Neurolathyrism

Neurolathyrism is a paralysis of the legs due to nerve damage in the spinal cord caused by neurotoxins in *L. sativus*. The principal neurolathyrogen is β-N-oxalyl-L-α-β-diaminopropionic acid (ODAP):

$$HOOC-\overset{O}{\underset{}{C}}-NH-CH_2-\overset{NH_2}{\underset{}{CH}}-COOH$$

Neurolathyrism is a major public health problem in India. The disease is associated with both the consumption of seeds of *L. sativus* and environmental factors such as exposure to a wet environment and overwork. It generally appears whenever a diet containing *L. sativus* seeds is consumed for 3–6 months. The symptoms are muscular rigidity, weakness, and paralysis of the leg muscles, leading to death in extreme cases. The onset is unusually sudden, with development of stiffness or paralysis of the lower limbs. In mild cases, there is difficulty in walking. People with more advanced cases require a stick for support. This may advance to the two-stick stage, and finally the victims can only crawl (Figs. 7.24 and 7.25). Lathyrism affects primarily young men between 20 and 30 years of age. This may be interrelated with the environmental factors involved in the disease.

One of the major difficulties in studying neurolathyrism is that experimental animals do not generally respond to the neurotoxin. Day-old chicks have been used and exhibit head retraction and convulsions. Older chickens are not affected. The squirrel monkey also responds to the neurotoxins with muscle tremors and convulsions and has been used as an experimental model.

The specific mode of action of ODAP is not known, but it appears to have a direct effect on nerve cells. From studies with experimental animals it appears to be of low toxicity, which emphasizes the probable importance of other factors (preexisting disease, malnutrition, and stress of physical overexertion) in development of neurolathyrism in humans (*Parker et al.* 1979).

Lathyrus sativus seeds can be treated to eliminate the toxicity. Steeping and boiling in hot water removes most of the neurotoxin. They are often used in the raw form, e.g., as paste balls, so that the toxin is retained. Thus, by seemingly simple changes in preparation of the seeds for consumption, toxicity could be avoided. The ideal solution would be development of cultivars of *L. sativus* free of the toxin because the crop has agronomic features (drought resistance and hardiness) that would make it an excellent food plant if it were nontoxic.

Other Lathyrogens

Vicia sativa (common vetch) (Fig. 7.26) contains a neurolathyrogen, β-cyano-L-alanine:

$$NC-CH_2-\underset{\underset{NH_2}{|}}{CH}-COOH$$

Vetch causes neurolathyrism when fed to poultry. Vetch was at one time widely grown in western Oregon as a seed crop; contamination of grain with the seed resulted in a number of poultry losses. Chicks fed vetch develop convulsions, blindness, and a pronounced, plaintive

FIG. 7.24. The two-stick stage of lathyrism.
Courtesy of M. Mohan Ram.

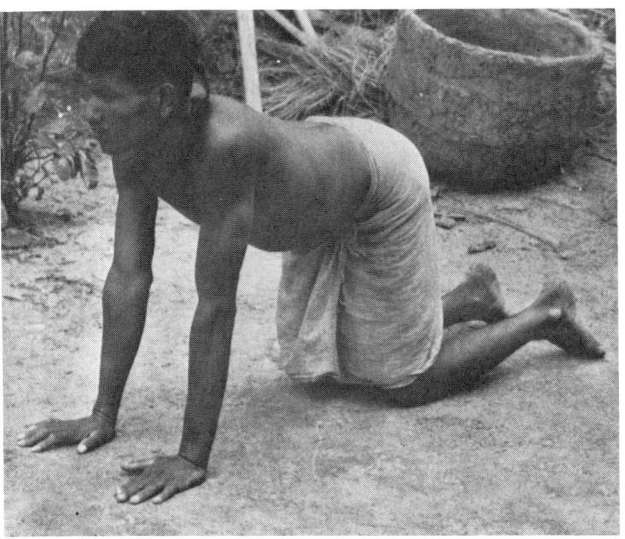

FIG. 7.25. Severe lathyrism in which the victim can no longer walk.
Courtesy of M. Mohan Ram.

FIG. 7.26. Leaves and stems of common vetch.

chirping, resembling a pyridoxine deficiency. Pyridoxine supplementation delays the onset of vetch toxicity symptoms.

β-Cyano-L-alanine may be a contributing factor to neurolathyrism in India, as vetch is sometimes a contaminant of lathyrus seeds.

Hairy vetch (*Vicia villosa*) has been implicated in mortality of cattle in Oklahoma (Panciera 1978; Kerr and Edwards 1982). Clinical signs in cattle grazing vetch include dermatitis, conjunctivitis with edema of the eyelids, and diarrhea. The disorder occurs sporadically, with a mortality rate of about 50% of affected animals. The morbidity in affected herds was 6–8%. Hairy vetch is also toxic to horses (Anderson and Divers 1983), leading to systemic granulomatous inflammation. Edema, particularly around the lips and eyes, is noted. Conjunctivitis and corneal ulceration occur. The causative agent of hairy vetch toxicity has not been identified. Most cases of poisoning have occurred in mid- to late spring when the vetch is approaching maturity.

Burroughs *et al.* (1983) described outbreaks of vetch poisoning in cattle in South Africa. These were characterized by a severe dermatitis, high morbidity, and mortality in older cows. Diarrhea was usually observed. The condition was virtually identical to the outbreaks in Oklahoma described by Panciera (1978).

LINATINE

Flax or linseed (*Linum usitatissimum*) meal contains an antagonist of the vitamin pyridoxine. The pyridoxine antagonist has been identified as the amino acid 1-amino-D-proline. It occurs in linseed meal as a dipeptide of 1-amino-D-proline and glutamic acid (Fig. 7.27), called linatine. The 1-amino-D-proline reacts with pyridoxal phosphate, forming a hydrazone, and thus impairing its function as a cofactor in amino acid metabolism. Pyridoxal phosphate is involved in transamination, decarboxylation, and other reactions of amino acid metabolism. Symptoms of pyridoxine deficiency, including depressed appetite, poor growth, and convulsions, may develop in chickens fed raw linseed meal (Fig. 7.28). Autoclaving and water extraction or pyridoxine supplementation of the meal will overcome the antipyridoxine effects.

INDOSPECINE

Indigofera spicata, or creeping indigo, is a tropical legume with potential value as a forage and soil improvement crop. It contains a toxic

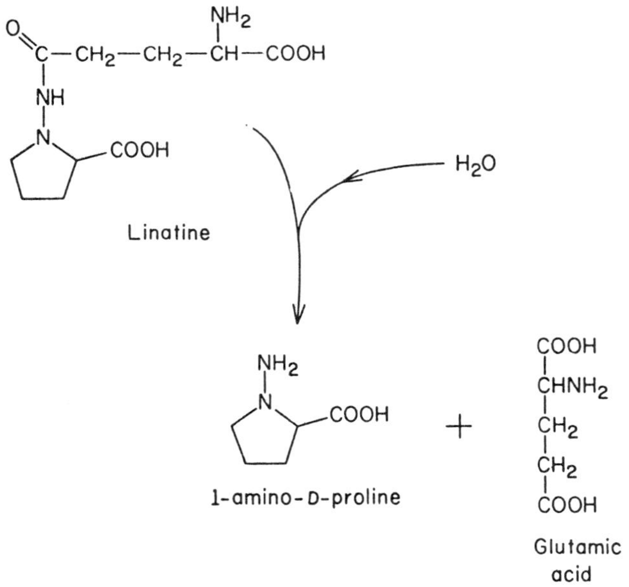

FIG. 7.27. Structure and hydrolysis products of linatine.

FIG. 7.28. Four-week-old broiler chicks fed diets in which the protein supplement was either soybean meal or raw linseed meal. A substantial part of the growth depression with the linseed meal is due to linatine and linamarin, while the remainder is a reflection of protein quality.

amino acid, indospecine, which is a structural antagonist of arginine (Fig. 7.29). It inhibits arginine incorporation into tissue proteins and causes liver damage in cattle and sheep consuming the plant, with necrosis and nodular cirrhosis. Plant breeding and selection efforts are under way in Australia to develop *Indigofera* selections that are nontoxic.

CANAVANINE

Canavanine, like indospecine, is an analog of arginine. It occurs in high concentrations (up to 5%) in the seeds of the jack bean (*Canavalia*

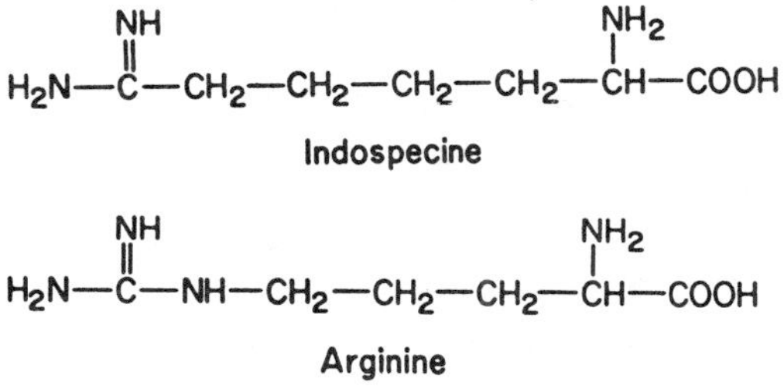

$$H_2N-\overset{NH}{\overset{\|}{C}}-CH_2-CH_2-CH_2-CH_2-\overset{NH_2}{\overset{|}{C}H}-COOH$$
Indospecine

$$H_2N-\overset{NH}{\overset{\|}{C}}-NH-CH_2-CH_2-CH_2-\overset{NH_2}{\overset{|}{C}H}-COOH$$
Arginine

FIG. 7.29. Structure of indospecine showing its similarity to arginine.

ensiformis) and a number of other legumes. Alfalfa sprouts contain about 1.5% of their dry weight of canavanine. A severe lupus erythematosus-like syndrome occurs when alfalfa sprouts are fed to monkeys (Malinow *et al.* 1982).

BRASSICA ANEMIA FACTOR

Plants in the *Brassica* genus, such as kale, rape, cabbage, cauliflower, and turnips, are important livestock feeds. Some, such as kale, forage rape, and turnips, are grown specifically for grazing animals, while with crops like cauliflower and brussels sprouts food-processing wastes and postharvest stubble may be used as feed for livestock. Kale and rape have been grown extensively in countries such as Britain and New Zealand, particularly as winter feeds for sheep. Turnip production has recently developed in the irrigated areas of Washington and Oregon to prevent soil erosion and provide winter feed for cattle. Turnip seed is applied by plane before wheat is harvested; after the wheat is combined, the stubble is irrigated to germinate the turnips. The turnips are grazed throughout the winter. Two types of sulfur-containing compounds limit the feeding value of these brassica crops. These are the glucosinolates and an amino acid, S-methylcysteine sulfoxide (SMCO). The growth performance of ruminants grazed on kale and other brassicas is lower than would be predicted from their nutrient content. It is generally believed that SMCO is the factor responsible for the poor performance.

When ruminants are fed mainly on a diet of kale, cabbage, or other brassicas, they may develop a severe hemolytic anemia after 3–4 weeks. The first clinical sign of the disease is the appearance of stainable granules within the red blood cells. These are termed Heinz–Ehrlich bodies (Fig. 7.30). The hemoglobin level falls from a normal level of about 11 g/100 ml to 8 g or lower. If the animals are removed from brassica pasture, the hemoglobin levels return to normal in 3–4 weeks and the Heinz–Ehrlich bodies disappear. If the animals are left on the brassica pasture, surviving animals make a spontaneous but incomplete recovery, followed by cycles of anemia and partial recovery. Other clinical signs are hemoglobinuria (red urine) when the hemoglobin level falls below 6 g/100 ml, loss of appetite, diarrhea, and jaundice. The liver becomes swollen and pale, with extensive hemosiderin deposits and necrosis. Kidneys and spleen also show massive hemosiderin deposits.

The SMCO is a fairly rare amino acid, found only in brassicas, garlic, and onion. In the brassicas it may occur at levels as high as 4–6% of

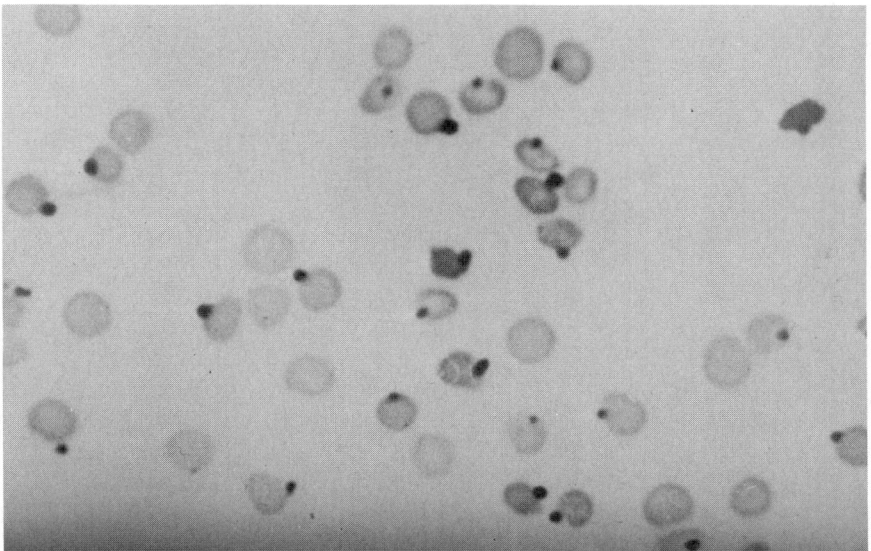

FIG. 7.30. Kale poisoning showing Heinz bodies in red blood cells caused by the brassica anemia factor.
Courtesy of R. H. Smith.

the dry matter. SMCO is probably not the primary hemolytic agent. It is metabolized in the rumen, producing dimethyl disulfide (Fig. 7.31).

Dimethyl disulfide is an oxidant that attacks the red cell membrane. It is inactivated by reacting with reduced glutathione (GSH), producing CH_3SH (methylmercaptan):

$$2\ GSH + CH_3-S-S-CH_3 \xrightarrow{\text{glutathione peroxidase}} G-S-S-G + 2\ CH_3SH$$

Glutathione peroxidase is a selenium-containing enzyme. This system is also involved in detoxifying the glycoside derivatives in favism, as previously discussed in Chapter 6.

In the normal red blood cell, hemoglobin is maintained in the reduced state by GSH. With low GSH or low activity of glucose-6-phosphate dehydrogenase (G6PD), hemoglobin under oxidative stress may suffer irreversible oxidative changes and be precipitated as granules—the Heinz–Ehrlich bodies.

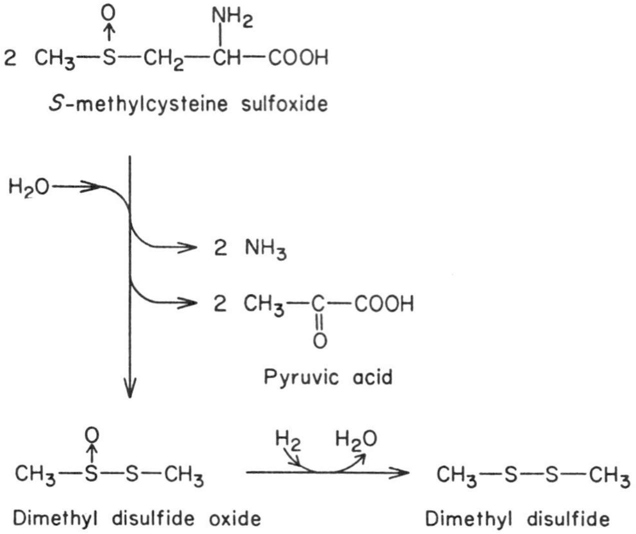

$$2\ CH_3 \overset{O}{\underset{\uparrow}{-}} S-CH_2-\underset{\underset{NH_2}{|}}{CH}-COOH$$

S-methylcysteine sulfoxide

$$H_2O \rightarrow \quad \searrow 2\ NH_3$$
$$\searrow 2\ CH_3-\underset{\underset{O}{\|}}{C}-COOH$$

Pyruvic acid

$$CH_3 \overset{O}{\underset{\uparrow}{-}} S-S-CH_3 \quad \underset{}{\overset{H_2\ \ H_2O}{\rightleftharpoons}} \quad CH_3-S-S-CH_3$$

Dimethyl disulfide oxide   Dimethyl disulfide

FIG. 7.31. Metabolism of S-methylcysteine sulfoxide in the rumen. The arrow bond (S → O) indicates a coordinate covalent bond in which one atom (sulfur) provides both electrons needed to form a covalent bond.

Typically in brassica poisoning there is a hemolytic crisis followed by a recovery phase. This may be due to an adaptation of the rumen microorganisms so less dimethyl disulfide is produced.

The SMCO content of brassicas tends to increase with increasing plant maturity. Through the winter, the toxicity of kale, cabbage, turnips, and other brassicas increases. It is advisable not to use these crops for pasture in late winter. Nitrogen fertilization tends to increase the SMCO content. McDonald et al. (1981) developed a fertilization program to produce kale with a low SMCO content by growing it on soils with low sulfate sulfur concentrations (5–10 mg sulfate sulfur/kg) and nitrogen fertilization to stimulate incorporation of the sulfur into plant protein. Barry et al. (1984) demonstrated that low SMCO kale produced in this manner was of lower toxicity to sheep than normal SMCO kale, but still produced signs of toxicity. Lambs fed kale showed large increases in serum thyroxin and growth hormone concentrations, suggesting that ruminants counteract the protein inactivation caused by dimethyl disulfide by increasing the activity of hormones which stimulate protein synthesis. The best long-term solution to the problem would be the development through plant breeding of cultivars of brassicas low in SMCO.

A good review of the brassica anemia factor is provided by Smith (1980).

SMCO has been demonstrated to have an antihypercholesterolemic effect (Fugiwara et al. 1972). The precursor of SMCO, S-methyl-L-cysteine, is found in appreciable quantities in legume seeds (Eyre et al. 1983). Eyre and co-workers demonstrated that this derivative of cysteine had no sparing effect on the methionine requirement of rats, but had no deleterious effects either.

Onion Poisoning of Livestock

Feeding onions to livestock may cause anemia of the same type as the brassica-induced anemia. The causative agent is not the same and is not an amino acid, but is discussed here as it is very similar to the brassica-poisoning situation.

Onions are commonly fed to sheep and cattle in the onion-producing areas of Oregon and Idaho. These may be cull onions or, at times, the main crop when overproduction prevents their marketing in normal channels. The onions are scattered on the ground in piles and are avidly consumed by livestock. Anemia may develop within a week of the onset of onion feeding. Clinical signs are hemoglobinuria, diarrhea, staggering, and collapse. Cattle are most susceptible, with horses intermediate, and sheep and goats somewhat resistant.

The toxic principle in onions is n-propyl disulfide. This compound is an oxidant that will cause red blood cell hemolysis and Heinz–Ehrlich bodies. It is reduced by the G6PD–glutathione peroxidase system previously described.

Onions also contain SMCO, which would presumably contribute to the anemia problem. Garlic contains S-allylcysteine sulfoxide, which is metabolized to allyl disulfide oxide. The reputed cholesterol-lowering properties of garlic may be due to a reaction of the disulfide group with the sulfhydryl group of CoA, leading to inhibition of lipid synthesis. A review of the effect of garlic consumption on blood lipids and atherosclerosis concludes that there are favorable effects (Lau et al. 1983).

Onion poisoning of livestock has been described by Kirk and Bulgin (1979) and Hutchinson (1977). Wild onions also cause hemolytic anemia in livestock (Van Kampen et al. 1970).

Maple Poisoning of Horses

The consumption of leaves and bark of the red maple (*Acer rubrum*) can cause hemolytic anemia and Heinz-body formation in horses (Tennant *et al.* 1981; George *et al.* 1982). The signs of toxicity are very similar to those seen with the brassica anemia factor. The condition occurs sporadically in the northeastern U.S. and is associated with the browsing of red maple leaves. The causative agent has not been identified.

HYPOGLYCIN

Blighia sapida is one of the plants that Captain Bligh brought to the West Indies on his second voyage, after he survived the mutiny on the Bounty. The plant is named after him. It is a small tree, native to Africa, that is cultivated in southern Florida and the tropics for its fruit, which is edible when ripe (Figs. 7.32 and 7.33). The fruit is called ackee in Jamaica and isin in Nigeria. Ackee is known as the national fruit of Jamaica and is popularized in calypso songs. The unripe fruit

FIG. 7.32. Fruit and leaves of ackee (*Blighia sapida*), a fruit widely used in Jamaica. Consumption of immature ackee fruit causes a condition called "vomiting sickness."
Courtesy of J. F. Morton.

FIG. 7.33. Ackee (*Blighia sapida*) trees in Coral Gables, Florida.
Courtesy of J. F. Morton.

has a high level of a toxic amino acid, β-methylenecyclopropyl alanine, known as hypoglycin A (Fig. 7.34). It is also found as a α-glutamyl dipeptide conjugate, hypoglycinin B.

When consumed by humans, ackee can cause a condition known as "vomiting sickness." It occurs in malnourished people who have consumed the unripe fruit. Its onset is sudden with violent vomiting, followed by convulsions, coma, and death, usually all in a period of 12 hr or less. The principal clinical sign is severe hypoglycemia, with a blood glucose level as low as 20 mg/100 ml blood, compared to a normal value of about 100 mg/100 ml. Hypoglycin is metabolized in a similar manner as the branched chain amino acids. It is deaminated to β-methylenecyclopropyl pyruvate, which then undergoes oxidative de-

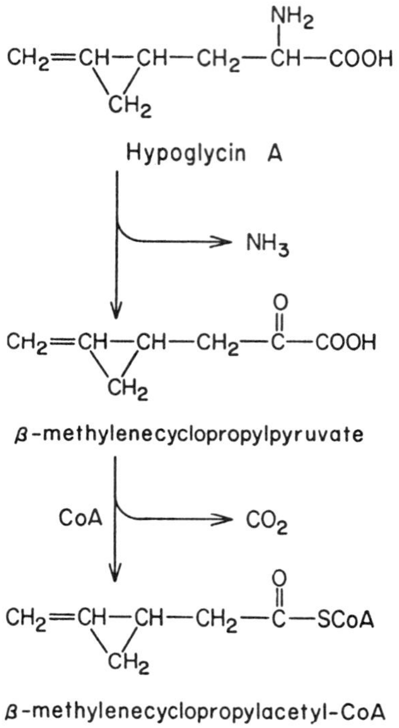

$$CH_2=CH-\underset{\underset{CH_2}{\diagdown\diagup}}{CH}-CH_2-\overset{NH_2}{\underset{|}{CH}}-COOH$$

Hypoglycin A

$\downarrow \rightarrow NH_3$

$$CH_2=CH-\underset{\underset{CH_2}{\diagdown\diagup}}{CH}-CH_2-\overset{O}{\underset{\|}{C}}-COOH$$

β-methylenecyclopropylpyruvate

CoA $\downarrow \rightarrow CO_2$

$$CH_2=CH-\underset{\underset{CH_2}{\diagdown\diagup}}{CH}-CH_2-\overset{O}{\underset{\|}{C}}-SCoA$$

β-methylenecyclopropylacetyl-CoA

FIG. 7.34. Metabolism of hypoglycin A.

carboxylation to β-methylenecyclopropyl acetyl-CoA (Fig. 7.34). This compound inhibits transfer of long-chain fatty acid CoA residues to carnitine, blocking the process of β-oxidation. This impairs gluconeogenesis, so total depletion of stored glycogen occurs and severe hypoglycemia follows. The vomiting syndrome may be a consequence of inhibition of leucine metabolism due to structural antagonism by the chemically similar hypoglycin. Isovalaric and α-methylbutyric acid are formed; since short-chain branched amino acids are depressants of the central nervous system, the vomiting syndrome could be a result of these compounds.

BIOGENIC (PRESSOR) AMINES

Biogenic amines are compounds present in foods that are potent vasoconstrictors and so cause elevated blood pressure (hence the name "pressor amines"). They are detoxified by the enzyme monoamine oxidase (MAO); therefore the administration of drugs that are MAO inhibitors, such as various antidepressants and amphetamines, may lead to hypertensive crises. Pressor amines occur in numerous common foodstuffs including pineapples, bananas, avocados, cheese, fish, and chocolate. Some individuals are quite sensitive to the presence of amines in the diet, perhaps due to a lower than average tissue level of MAO. People who have been treated with drugs that are MAO inhibitors are especially sensitive to some types of cheese that are high in tyramine. Migraine headaches and in severe cases intercranial bleeding may result (Blackwell *et al.* 1967). The consumption of large quantities of bananas or plantains, as in certain African countries, may lead to a high incidence of carcinoid heart disease. Bananas have a high content of serotonin, a biogenic amine.

Amines are detoxified by undergoing deamination catalyzed by MAO in the liver:

$$R-CH_2-NH_2 + H_2O + O_2 \xrightarrow{MAO} R-\overset{O}{\overset{\|}{C}}-H + NH_3 + H_2O_2$$

e.g.:

Serotonin $+ H_2O + O_2 \xrightarrow{MAO}$ aldehyde intermediate

5-hydroxyindole-3-acetic acid

Their lack of detoxification, due either to congenital inadequacy of MAO or the presence of MAO inhibitors, leads to an elevated circulating level of amines. Various amines (e.g., serotonin, norepinephrine, dopamine, and acetylcholine) function as neurotransmitters in the brain. Elevated blood amino levels may upset the neurotransmitter balance, leading to hypertension.

REFERENCES

Protease and Amylase Inhibitors

CHEEKE, P. R., TELEK, L., CARLSSON, R., and EVANS, J. J. 1980. Nutritional evaluation of leaf protein concentrates prepared from selected tropical plants. Nutr. Rep. Int. 22, 717–721.

D'MELLO, J. P. F., ACAMOVIC, T., and WALKER, A. G. 1983. Nutrient content and apparent metabolizable energy values of full-fat winged beans (Psophocarpus tetragonolobus) for young chicks. Trop. Agric. (Trinidad) 60, 290–293.

ERICKSON, J. P., MILLER, E. R., ELLIOTT, F. C., KU, P. K., and ULLREY, D. E. 1979. Nutritional evaluation of triticale in swine starter and grower diets. J. Anim. Sci. 48, 547–553.

HARRIS, D. J., CHEEKE, P. R., TELEK, L., and PATTON, N. M. 1981. Utilization of alfalfa meal and tropical forages by weanling rabbits. J. Appl. Rabbit Res. 4 (1), 4–9.

LIENER, I. E. 1980. Miscellaneous toxic factors. In Toxic Constituents of Plant Foodstuffs. I. E. Liener (Editor), 2nd Edition, pp. 429–467. Academic Press, NY.

LIENER, I. E., and KAKADE, M. L. 1980. Protease inhibitors. In Toxic Constituents of Plant Foodstuffs. I. E. Liener (Editor), 2nd Edition, pp. 7–71. Academic Press, NY.

LIENER, I. E., DONATUCCI, D. A., and TARCZA, J. C. 1984. Starch blockers: A potential source of trypsin inhibitors and lectins. Am. J. Clin. Nutr. 39, 196–200.

McNAUGHTON, J. L., REECE, F. N., and DEATON, J. W. 1981. Relationships between color, trypsin inhibitor content and urease index of soybean meal and effect on broiler performance. Poult. Sci. 60, 393–400.

MITCHELL, H. L., and PARRISH, D. B. 1981. Effect of trypsin inhibitor from alfalfa meal on rat growth. Nutr. Rep. Int. 24, 799–801.

PEARCE, G., McGINNIS, J., and RYAN, C. A. 1983. Effects of feeding a carboxypeptidase inhibitor from potatoes to newly hatched chicks. Proc. Soc. Exp. Biol. Med. 173, 447.

Allergenic Proteins

KILSHAW, P. J., and SISSONS, J. W. 1979. Gastrointestinal allergy to soyabean protein in preruminant calves. Allergenic constituents of soyabean products. Res. Vet. Sci. 27, 306–371.

PEDERSEN, H. E., and SISSONS, J. W. 1984. Effect of antigenic soyabean protein on the physiology and morphology of the gut in the preruminant calf. Can. J. Anim. Sci. 64 (Suppl. 1), 183–184.

PERLMAN, F. 1980. Allergens. In Toxic Constituents of Plant Foodstuffs. I. E. Liener (Editor), 2nd Edition, pp. 295–327. Academic Press, NY.

Hemagglutinins (Lectins)

BENDER, A. E., and REAIDI, G. B. 1982. Toxicity of kidney beans (Phaseolus vulgaris) with particular reference to lectins. J. Plant Foods 4, 15–22.

CHEEKE, P. R., GOEGER, M. P., and ARSCOTT, G. H. 1983. Utilization of black locust (*Robinia pseudoacacia*) leaf meal by chicks. Nitrogen Fixing Tree Res. Rep. *1*, 41.
GRANT, G., MORE, L. J., McKENZIE, N. H., STEWART, J. C., and PUSZTAI, A. 1983. A survey of the nutritional and haemagglutination properties of legume seeds generally available in the UK. Br. J. Nutr. *50*, 207–214.
HARRIS, D. J., CHEEKE, P. R., and PATTON, N. M. 1984. Evaluation of black locust leaves for growing rabbits. J. Appl. Rabbit Res. *7*, 7–9.
HINTZ, H. F., and HOGUE, D. E. 1964. Kidney beans (*Phaseolus vulgaris*) and the effectiveness of vitamin E for the prevention of nutritional muscular dystrophy in the chick. J. Nutr. *84*, 283–287.
HORTON, G. M. J., and CHRISTENSEN, D. A. 1981. Nutritional value of black locust tree leaf meal (*Robinia pseudoacacia*) and alfalfa meal. Can. J. Anim. Sci. *61*, 503–506.
JAFFE, W. G. 1980. Hemagglutinins (Lectins). *In* Toxic Constituents of Plant Foodstuffs. I. E. Liener (Editor), 2nd Edition, pp. 73–102. Academic Press, NY.
KING, T. P., BEGBIE, R., and CADENHEAD, A. 1983. Nutritional toxicity of raw kidney beans in pigs. Immunocytochemical and cytopathological studies on the gut and pancreas. J. Sci. Food Agric. *34*, 1404–1412.
KINGSBURY, J. M. 1964. Poisonous Plants of the United States and Canada. Prentice-Hall, Englewood Cliffs, NJ.
LIENER, I. E. 1980. Miscellaneous toxic factors. *In* Toxic Constituents of Plant Foodstuffs. I. E. Liener (Editor), 2nd Edition, pp. 429–467. Academic Press, NY.
LIS, H., and SHARON, N. 1973. The biochemistry of plant lectins (phytohemagglutinins). Annu. Rev. Biochem. *42*, 541–574.
LIU, C. C., and JUNG, Y. (Undated.) Nutritional value of Locust (*Robina pseudoacacia*) leaf meal. Unpublished report provided as a personal communication to P. R. Cheeke, from Peking Agricultural University, Peking, China.
McPHERSON, A., and HOOVER, S. 1979. Purification of mitogenic proteins from *Hura crepitans* and *Robinia pseudoacacia*. Biochem. Biophys. Res. Commun. *89*, 713–720.
MYER, R. O., and FROSETH, J. A. 1983. Heat-processed small red beans (*Phaseolus vulgaris*) in diets for young pigs. J. Anim. Sci. *56*, 1088–1096.
MYER, R. O., FROSETH, J. A., and COON, C. N. 1982. Protein utilization and toxic effects of raw beans (*Phaseolus vulgaris*) for pigs. J. Anim. Sci. *55*, 1087–1098.
NOAH, N. D., BENDER, A. E., REAIDI, G. B., and GILBERT, R. J. 1980. Food poisoning from raw kidney beans. Br. Med. J. *281*, 236–237.
OLOGHOBO, A. D., and FETUGA, B. L. 1983. Pathological observations on rats dosed with limabean and cowpea hemagglutinins. Toxicol. Lett. *18*, 301–306.
OLSON, G., PUDELKIEWICZ, W. J., and MATTERSON, L. D. 1966. Isolation of a compound from alfalfa lipids that inhibits tocopherol deposition in chick tissues. J. Nutr. *90*, 199–206.
PUDELKIEWICZ, W. J., and MATTERSON, L. D. 1960. A fat-soluble material in alfalfa that reduces the biological availability of tocopherol. J. Nutr. *71*, 143–148.
PUSZTAI, A., CLARKE, E. M. W., and KING, T. P. 1979. The nutritional toxicity of *Phaseolus vulgaris* lectins. Proc. Nutr. Soc. *38*, 115–120.
SANCHEZ, W. K., CHEEKE, P. R., and PATTON, N. M. 1983. Utilization of raw and heat-treated pinto beans by weanling rabbits. J. Appl. Rabbit. Res. *6*, 139–141.
TAKADA, K., NAKAZATO, T., ONO, K., HONDA, H., and YAMANE, T. 1980. Feeding value of leaf meal of acacia (*Robinia pseudoacacia*) in poultry feed. Jpn. Poult. Sci. *17*, 299–305.

WILSON, A. B., KING, T. P., CLARKE, E. M. W., and PUSZTAI, A. 1980. Kidney bean (*Phaseolus vulgaris*) lectin induced lesions in the rat small intestine. 2. Microbiological studies. J. Comp. Pathol. *90,* 597–602.

Thiaminases

BERGIN, T. 1981. In The Steps of Burke and Wills. Australian Broadcasting Commission, Sydney, N. S. W.

DICKIE, C. W., and BERRYMAN, J. R. 1979. Polioencephalomalacia and photosensitization associated with *Kochia scoparia* consumption in range cattle. J. Am. Vet. Med. Assoc. *175,* 463–465.

DICKIE, C. W., and JAMES, L. F. 1983. Kochia scoparia poisoning in cattle. J. Am. Vet. Med. Assoc. *183,* 765–768.

EDWIN, E. E., and JACKMAN, R. 1982. Ruminant thiamine requirement in perspective. Vet. Res. Commun. *5,* 237–250.

EDWIN, E. E., JACKMAN, R., and JONES, P. 1982. Some properties of thiaminases associated with cerebrocortical necrosis. J. Agric. Sci. *99,* 271–275.

EVANS, C. A., CARLSON, W. E., and GREEN, R. G., 1942. The pathology of Chastek paralysis in foxes. Am. J. Pathol. *18,* 79–90.

EVANS, W. C. 1976. Bracken thiaminase-mediated neurotoxic syndromes. Bot. J. Linn. Soc. *73,* 113–131.

EVANS, W. C., WIDDOP, B., and HARDING, J. D. 1972. Experimental poisoning by bracken rhizomes in pigs. Vet. Res. *90,* 471–475.

EVERIST, S. L. 1981. Poisonous Plants of Australia. Angus & Robertson Publishers, Sydney.

GALITZER, S. J., and OEHME, F. W. 1979. Studies on the comparative toxicity of *Kochia scoparia* (L.) Schrad (fireweed). Toxicol. Lett. *3,* 43–49.

HAVEN, T. R., CALDWELL, D. R., and JENSEN, R. 1983. Role of predominant rumen bacteria in the cause of polioencephalomalacia (cerebrocortical necrosis) in cattle. Am. J. Vet. Res. *44,* 1451–1455.

MOOREHEAD, A. 1963. Cooper's Creek. Harper & Row, NY.

ROBERTS, H. E., EVANS, E. T., and EVANS, W. C. 1949. The production of "bracken staggers" in the horse and its treatment with vitamin B_1 therapy. Vet. Rec. *61,* 549–550.

WESWIG, P. H., FREED, A. M., and HAAG, J. R. 1946. Antithiamine activity of plant materials. J. Biol. Chem. *165,* 737–738.

Bloat-Producing Proteins

HOWARTH, R. E., MAJAK, W., WALDERN, D. E., BRANDT, S. A., FESSER, A. C., GOPLEN, B. P., and SPURR, D. T. 1977. Relationships between ruminant bloat and the chemical composition of alfalfa herbage. I. Nitrogen and protein fractions. Can. J. Anim. Sci. *57,* 345–357.

HOWARTH, R. E., GOPLEN, B. P., BRANDT, S. A., and CHENG, K. J. 1982. Disruption of leaf tissues by rumen microorganisms: An approach to breeding bloat-safe forage legumes. Crop Sci. *22,* 564–568.

MAJAK, W., HOWARTH, R. E., FESSER, A. C., GOPLEN, B. P., and PEDERSEN, M. W. 1980. Relationships between ruminant bloat and the composition of alfalfa herbage II. Saponins. Can. J. Anim. Sci. *60,* 699–708.

MAJAK, W., HOWARTH, R. E., CHENG, K. J., and HALL, J. W. 1983. Rumen conditions that predispose cattle to pasture bloat. J. Dairy Sci. 66, 1683–1688.
McWILLIAM, J. R. 1973. Plant factors, environment and bloat. In Bloat. Proceedings of a Symposium. Reviews in Rural Science. R. A. Leng and J. R. William (Editors), No. 1, pp. 33–38. Univ. of New England Press, Armidale, N.S.W., Australia.
SANT, F. I., and WILSON, D. 1982. Use of a cellulolytic enzyme digestion technique to distinquish bloat-causing from non-bloat-causing legumes and to select for speed of mesophyll cell-wall disintegration in red clover (*Trifolium pratense* L.). J. Agric. Sci. 98, 99–102.

Mimosine

DEWEEDE, S., and WAYMAN, O. 1970. Effect of mimosine on the rat fetus. Teratology 3, 21–28.
D'MELLO, J. P. F., and ACAMOVIC, T. 1982. Growth performance of, and mimosine excretion by, young chicks fed on *Leucaena leucocephala*. Anim. Feed Sci. Technol. 7, 247–256.
EL-HARITH, E. A., SCHART, Y., and TERMEULEN, U. 1979. Reaction of rats fed on *Leucaena leucocephala*. Trop. Anim. Prod. 4, 162–167.
HEGARTY, M. P. 1978. Toxic amino acids of plant origin. In Effects of Poisonous Plants on Livestock. R. F. Keeler, K. R. Kampen, and L. R. James. (Editors). pp. 575–585. Academic Press, NY.
HEGARTY, M. P., COURT, R. D., CHRISTIE, G. S., and LEE, C. P. 1976. Mimosine in *Leucaena leucocephala* is metabolised to a goitrogen in ruminants. Aust. Vet. J. 52, 490–492.
HEGARTY, M. P., LEE, C. P., CHRISTIE, G. S., COURT, R. D., and HAYDOCK, K. P. 1979. The goitrogen 3-hydroxy-4(1*H*)-pyridone, a ruminal metabolite from *Leucaena leucocephala*: Effects in mice and rats. Aust. J. Biol. Sci. 32, 27–40.
HOLMES, J. H. G. 1981. Toxicity of *Leucaena leucocphala* for steers in the wet humid tropics. Trop. Anim. Health Prod. 13, 94–100.
HOLMES, J. H., HUMPHREY, J. D., WALTON, E. A., and O'SHEA, J. D. 1981. Cataracts, goitre and infertility in cattle grazed on an exclusive diet of *Leucaena leucocephala*. Aust. Vet. J. 57, 257–260.
JONES, R. J. 1981. Does ruminal metabolism of mimosine explain the absence of *Leucaena* toxicity in Hawaii? Aust. Vet. J. 57, 55.
JONES, R. J., and HEGARTY, M. P. 1984. The effect of different proportions of *Leucaena leucocephala* in the diet of cattle on growth, feed intake, thyroid function and urinary excretion of 3-hydroxy-4(1*H*)-pyridone, Aust. J. Agric. Res. 35, 317.
JONES, R. J., and MEGARRITY, R. G. 1983. Comparative toxicity response of goats fed on *Leucaena leucocephala* in Australia and Hawaii. Aust. J. Agric. Res. 34, 781–790.
JONES, R. J., BLUNT, C. G., and NURNBERG, B. I. 1978. Toxicity of *Leucaena leucocephala*: The effect of iodine and mineral supplements on penned steers fed a sole diet of *Leucaena*. Aust. Vet. J. 54, 387–392.
LOWRY, J. B., TANGENDJAJA, M., and TANGENDJAJA, B. 1983. Optimising autolysis of mimosine to 3-hydroxy-4(1*H*)-pyridone in green tissues of *Leucaena leucocephala*. J. Sci. Food Agric. 34, 529–533.
NATIONAL ACADEMY OF SCIENCES 1984. Leucaena. Promising Forage and Tree Crop for the Tropics. Washington, DC.

ROSS, E., WAYMAN, O., and OISHI, F. G. 1980. The use of Japanese quail to study the effect of *Leucaena leucocephala* and mimosine on egg production. Proc., Annu. Meet.—Am. Soc. Anim. Sci., West. Sect. *31,* 129–132.

STUNZI, H., PERRIN, D. A., TEITEI, T., and HARRIS, R. L. N. 1979. Stability constants of some metal complexes formed by mimosine and related compounds. Aust. J. Chem. *32,* 21–30.

TANGENDJAJA, G., HOGAN, J. P., and WILLS, R. B. M. 1983. Degradation of mimosine by rumen contents: Effects of feed composition and Leucaena substrate. Aust. J. Agric. Res. *34,* 289–293.

WAYMAN, O., IWANAGA, I. I., and HUGH, W. I. 1970. Fetal resorption in swine caused by *Leucaena leucocephala* (Lam.) De Wit in the diet. J. Anim. Sci. *30,* 583–588.

Tryptophan

BREEZE, R. G., and CARLSON, J. R. 1982. Chemical-induced lung injury in domestic animals. Adv. Vet. Sci. Comp. Med. *26,* 201–231.

BREEZE, R. G., LAEGREID, W. W., BAYLY, W. M., and WILSON, B. J. 1984. Perilla ketone toxicity: A chemical model for the study of equine restrictive lung disease. Equine Vet. J. *16,* 180–184.

CARLSON, J. R., and BREEZE, R. G. 1984. Ruminal metabolism of plant toxins with emphasis on indolic compounds. J. Anim. Sci. *58,* 1040–1049.

CARLSON, J. R., and DICKINSON, E. O. 1978. Tryptophan-induced pulmonary edema and emphysema in ruminants. *In* Effects of Poisonous Plants on Livestock. R. F. Keeler, K. R. Van Kampen, and L. F. James (Editors), pp. 261–272. Academic Press, NY.

CARLSON, J. R., HAMMOND, A. C., BREEZE, R. G., POTCHOIBA, M. J., and HEINEMANN, W. W. 1983. Effect of monensin on bovine ruminal 3-methylindole production after abrupt change to lush pasture. Am. J. Vet. Res. *44,* 118–122.

DERKSEN, F. J., ROBINSON, N. E., SLOCOMBE, R. F., and HILL, R. E. 1982. 3-Methylindole-induced pulmonary toxicosis in ponies. Am. J. Vet. Res. *47,* 603–607.

DICKINSON, E. O., and CARLSON, J. R. 1978. Acute respiratory distress of rangeland cattle. *In* Effects of Poisonous Plants on Livestock. R. F. Keeler, K. R. Van Kampen, and L. F. James (Editors), pp. 251–259. Academic Press, NY.

HAMMOND, A. C., CARLSON, J. R., and BREEZE, R. G. 1982. Effect of monensin pretreatment on tryptophan-induced acute bovine pulmonary edema and emphysema. Am. J. Vet. Res. *43,* 753–759.

POTCHOIBA, M. J., CARLSON, J. R., and BREEZE, R. G. 1982. Metabolism and pneumotoxicity of 3-methyloxindole, indole-3-carbinol, and 3-methylindole in goats. Am. J. Vet. Res. *43,* 1418–1423.

WILSON, B. J., GARST, J. E., LINNABARY, R. D., and DOSTER, A. R. 1978. Pulmonary toxicity of naturally occurring 3-substituted furans. *In* Effects of Poisonous Plants on Livestock. R. F. Keeler, K. R. Van Kampen and L. F. James (Editors), pp. 311–333. Academic Press, NY.

Selenoamino Acids

VAN KAMPEN, K. R., and JAMES, L. F. 1978. Manifestations of intoxication by selenium-accumulating plants. *In* Effects of Poisonous Plants on Livestock. R. F.

Keeler, K. R. Van Kampen, and L. F. James (Editors), pp. 135–138. Academic Press, NY.

VIRUPAKSHA, T. K., and SHRIFT, A. 1965. Biochemical differences between selenium accumulator and nonaccumulator *Astragalus* species. Biochim. Biophys. Acta *107*, 69–80.

Lathyrogens

ANDERSON, C. A., and DIVERS, T. J. 1983. Systemic granulomatous inflammation in a horse grazing hairy vetch. J. Am. Vet. Med. Assoc. *183*, 569–570.

BURROUGHS, G. W., NESER, J. A., KELLERMAN, T. S., and VANNIEKERK, F. A. 1983. Suspected hybrid vetch (*Vicia villosa* crossed with *Vicia dasycarpa*) poisoning of cattle in the Republic of South Africa. J. S. Afr. Vet. Assoc. *54*, 75–80.

KERR, L. A., and EDWARDS, W. C. 1982. Hairy vetch poisoning in cattle. VM/SAC. Vet. Med. Small Anim. Clin. *77*, 257–261.

PADMANABAN, G. 1980. Lathyrogens. *In* Toxic constituents of Plant Foodstuffs. I. E. Liener (Editor), 2nd Edition, pp. 239–263. Academic Press, NY.

PANCIERA, R. J. 1978. Hairy vetch (*Vicia villosa* Roth) poisoning in cattle. *In* Effects of Poisonous Plants on Livestock. R. F. Keeler, K. R. Van Kampen, and L. F. James (Editors), pp. 555–563. Academic Press, NY.

PANCIERA, R. J., JOHNSON, L., and OSBOURN, B. I. 1966. A disease of cattle grazing hairy vetch pasture. J. Am. Vet. Med. Assoc. *148*, 804–808.

PARKER, A. J., MEHTA, T., ZARGHAMI, N. S., CUSICK, P. K., and HASKELL, B. E. 1979. Acute toxicity of the *Lathyrus sativus* neurotoxin, L-3-oxalylamino-2-aminopropionic acid, in the squirrel monkey. Toxicol. Appl. Pharmacol. *47*, 135–143.

ROY, D. N. 1981. Toxic amino acids and proteins from *Lathyrus* plants and other leguminous species: A literature review. Nutr. Abstr. Rev., Ser. A: Hum. Exp. *51*, 691–707.

SIMPSON, C. F., and CARDEILHAC, P. T. 1983. Mortality, hemodynamics, and aortic properties among male and female turkeys fed β-aminopropionitrile. Proc. Soc. Exp. Biol. Med. *172*, 168–172.

Linatine

KLOSTERMAN, H. J. 1974. Vitamin B6 antagonists of natural origin. J. Agric. Food Chem. *22*, 13–19.

SASAOKA, K., OGAWA, T., MORITOKI, K., and KOMOTO, M. 1976. Antivitamin B-6 effect of 1-aminoproline on rats. Biochim. Biophys. Acta *428*, 396–402.

Indospecine

CHRISTIE, G. S., WILSON, M., and HEGARTY, M. P. 1975. Effects on the liver in the rat of ingestion of *Indigofera spicata*, a legume containing an inhibitor of arginine metabolism. J. Pathol. *117*, 195–205.

HEGARTY, M. P., and POUND, A. W. 1970. Indospecine, a hepatoxic amino acid from *Indigofera spicata:* Isolation, structure and biological studies. Aust. J. Biol. Sci. *23*, 831–842.

Canavanine

MALINOW, M. R., BARDANA, E. J., PIROFSKY, B., CRAIG, S., and McLAUGHLIN, P. 1982. Systemic lupus erythematosus-like syndrome in monkeys fed alfalfa sprouts: Role of a nonprotein amino acid. Science 216, 415-417.

Brassica, Onion, and Maple-Induced Anemia

BARRY, T. N., DUNCAN, S. J., SADLER, W. A., MILLAR, K. R., and SHEPPARD, A. D. 1983A. Iodine metabolism and thyroid hormone relationships in growing sheep fed on kale (Brassica oleracea) and ryegrass (Lolium perenne)-clover (Trifolium repens) fresh forage diets. Br. J. Nutr. 49, 241-254.
BARRY, T. N., MILLAR, K. R., BOND, G., and DUNCAN, S. J. 1983B. Copper metabolism in growing sheep given kale (Brassica oleracea) and ryegrass (Lolium perenne)-clover (Trifolium repens) fresh forage diets. Br. J. Nutr. 50, 281-290.
BARRY, T. N., MANLEY, T. R., MILLAR, K. R., and SMITH, R. H. 1984. The relative feeding value of kale (Brassica oleracea) containing normal and low concentrations of S-methyl-L-cysteine sulphoxide (SMCO). J. Agric. Sci. 102, 635-643.
BRADSHA, J. E., and BORZUCKI, R. 1982. Digestibility, S-methylcysteine sulfoxide content and thiocyanate ion content of cabbages for stockfeeding. J. Sci. Food Agric. 33, 1-5.
EYRE, M. D., PHILLIPS, D. E., EVANS, I. M., and THOMPSON, A. 1983. The nutritional role of S-methyl-L-cysteine. J. Sci. Food Agric. 34, 696-700.
FUGIWARA, M., ITOKAWA, Y., UCHINO, H., and INOUE, K. 1972. Antihypercholesterolemic effect of a sulphur containing amino acid, S-methyl-L-cysteine sulphoxide, isolated from cabbage. Experientia 28, 254-255.
GEORGE, L. W., DIVERS, T. J., MAHAFFEY, E. A., and SUAREZ, M. J. H. 1982. Heinz body anemia and methemoglobinemia in ponies given red maple (Acer rubrum L.) leaves. Vet. Pathol. 19, 521-533.
HUTCHINSON, T. W. S. 1977. Onions as a cause of Heinz body anemia and death in cattle. Can. Vet. J. 18, 358-360.
KIRK, J. H., and BULGIN, M. S. 1979. Effects of feeding cull domestic onions (Allium cepa) to sheep. Am. J. Vet. Res. 40, 397-399.
LAU, B. H. S., ADETUMBI, M. A., and SANCHEZ, A. 1983. Allium sativum (garlic) and atherosclerosis: A review. Nutr. Res. 3, 119-128.
LONG, P. H., and PAYNE, J. W. 1984. Red maple-associated pulmonary thrombosis in a horse. J. Am. Vet. Med. Assoc. 184, 977-978.
McDONALD, R. C., MANLEY, T. R., BARRY, T. N., FORSS, D. A., and SINCLAIR, A. G. 1981. Nutritional evaluation of kale (Brassica oleracea) diets. 3. Changes in composition induced by soil fertility practices, with special reference to SMCO and glucosinolate concentrations. J. Agric. Sci. 97, 13-23.
SMITH, R. H. 1980. Kale poisoning: The brassica anaemia factor. Vet. Rec. 107, 12-15.
TENNANT, T., GILL, S. G., GLICKMAN, L. T., MIRRO, E. J., KING, J. M., POLAK, D. M., SMITH, M. C., and KRADEL, D. C. 1981. Acute hemolytic anemia, methemoglobinemia, and Heinz body formation associated with ingestion of red maple leaves by horses. J. Am. Vet. Med. Assoc. 179, 143-150.
VAN KAMPEN, K. R., JAMES, L. F., and JOHNSON, A. E. 1970. Hemolytic anemia in sheep fed wild onions (Allium validum). J. Am. Vet. Med. Assoc. 156, 328-332.

Hypoglycin

MANCHESTER, K. L. 1974. Biochemistry of hypoglycin. FEBS Lett. *40*, S133–S139.
TANAKA, K., KEAN, E. A., and JOHNSON, B. 1976. Jamaican vomiting sickness. N. Engl. J. Med. *295*, 461–467.

Biogenic (Pressor) Amines

BLACKWELL, B., MARLEY, E., PRICE, J., and TAYLOR, D. 1967. Hypertensive interaction between monamine oxidase inhibitors and foodstuffs. Br. J. Psychiatry *113*, 349–353.
HURST, W. J., MARTIN, R. A., ZOUMAS, B. L., and TARKA, S. M. 1982. Biogenic amines in chocolate—a review. Nutr. Rep. Int. *26*, 1081–1086.

8

Carbohydrates, Lipids, and Conjugates

CARBOHYDRATES

Xylose

There are few toxicants that are carbohydrates, although numerous toxicants do affect carbohydrate metabolism. One simple sugar that has toxic properties is xylose. Xylose is a major component of the hemicellulose fraction of roughages; xylans in forages are essentially polymers of xylose. Xylose liberated from roughage digestion is fermented in ruminants, but may be absorbed in nonruminants. Wood molasses, sometimes available for feeding purposes, is high in xylose. Several studies with poultry and swine have shown that xylose produces toxic effects in these species. In pigs, dietary xylose causes depressed appetite, reduced gains and feed conversion, reduced nitrogen retention, a change in hair color (in red breeds) from deep red to yellowish red, and eye cataracts (Wise et al. 1954). In chickens, reduced growth, severe diarrhea, and mortality have been observed (Baker 1977). Little energy is derived from xylose by nonruminants, presumably because of low conversion to glucose. The eye cataracts may be a consequence of an inhibition of the phosphogluconate oxidative pathway.

Lactose

Other carbohydrates that may have antinutritional effects include lactose, pectins, and β-glucans. Lactose is a disaccharide of glucose and galactose. It is digested by the intestinal enzyme lactase, splitting it apart into the two sugars. The activity of carbohydrate-digesting enzymes changes with age (Fig. 8.1), as shown by Bailey et al. (1956) in the pig.

In postweaning individuals, the decline in lactase activity may cause difficulties in the utilization of high levels of dietary lactose (lactose intolerance). Because the lactose is not digested properly in the small intestine, it reaches the hindgut, where it is fermented. The result is irritation of the lining of the hindgut, gas production, and increased osmolality of the hindgut, causing dehydration and diarrhea. For individuals susceptible to lactose intolerance, lactose can be added to milk products to allow predigestion of the lactose. In livestock, feeding high levels of lactose, as, for example, dried whey, may cause diarrhea.

Oligosaccharides

Certain oligosaccharides may have negative nutritional properties. Raffinose and stachyose are oligosaccharides in beans that are poorly digested in nonruminants.

They undergo fermentation in the hindgut. In humans, this causes digestive problems associated with consumption of beans, including nausea, diarrhea, and flatulence. In swine and poultry, the presence of these carbohydrates in feeds (e.g., beans) lowers the digestible energy content of the feed from what it would be if a more digestible type of carbohydrate were present.

FIG. 8.1. Changes in activity of carbohydrate-digesting enzymes with age.

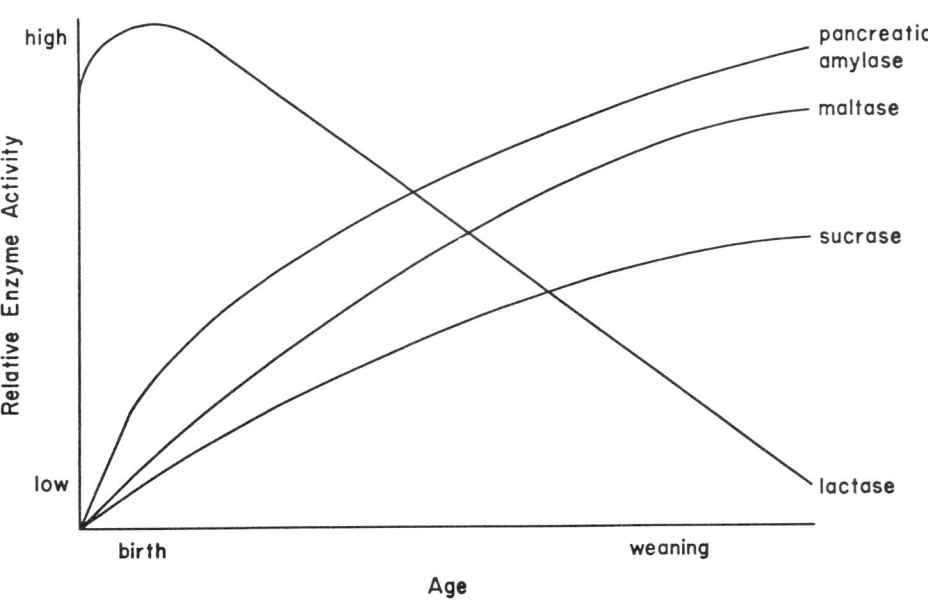

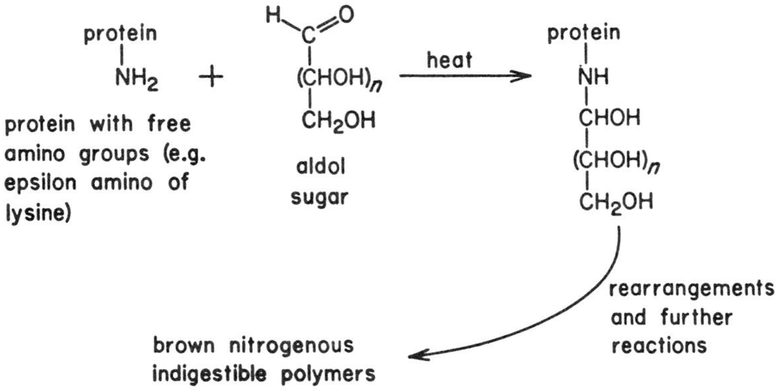

FIG. 8.2. The browning or Maillard reaction between free amino groups in proteins and reducing sugars.

Raffinose occurs in cottonseed meal and can reduce the availability of lysine by participating in the Maillard or browning reaction (Fig. 8.2). Soluble carbohydrates in soybean meal, such as raffinose and stachyose, may adversely affect the performance of baby pigs. Preparation of soy protein concentrates, removing a large proportion of the soluble carbohydrates, improves the value of soybean products for the young pig.

Pectins

Pectins are a group of polysaccharides that function as an intercellular cement in plant tissue. Pectin consists largely of unbranched chains of α-1,4-linked D-galacturonic acid units with many of the carboxyls esterified with methyl groups or neutralized with calcium or magnesium (Fig. 8.3). Pectins occur in the middle lamella and other membranes of the plant cell, while hemicellulose is a more complex fraction in the thickening of the secondary wall. Wagner and Thomas (1978) found that dietary levels of pectin above 1.5% decreased weight gains of chicks and caused sticky droppings. These effects are similar to results obtained with feeding high dietary levels of rye grain, which contains about 7.9% pectin, whereas barley, corn, oats, and wheat contain much lower pectin levels of 0.5, 0.2, 0.6, and 1.2%, respectively (McNab and Shannon 1974). The results of Wagner and Thomas (1978) show that pectin has growth-depressing properties in chicks, which may explain the poor performance obtained with rye in poultry diets. Misir and Marquardt (1978 A–D), in a series of studies, also found that rye reduced the performance of growing birds, and they attributed the sticky droppings and reduced feed conversion to a pectin fraction.

Swine are apparently not as sensitive as poultry to growth-depressing factors in rye. Honeyfield et al. (1983) found that dietary rye levels up to 68% of the diet did not adversely affect young pigs. Friend and MacIntyre (1969) reported similar results. Dietary antibiotics improve the utilization of rye by chickens but not by swine (Honeyfield et al. 1983).

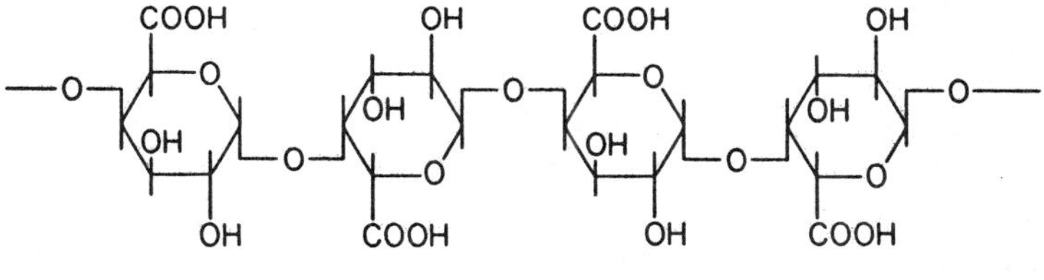

FIG. 8.3. Structure of pectin (polygalacturonic acid).

Pectins increase the viscosity of intestinal contents, impairing nutrient absorption and possibly stimulating proliferation of bacteria in the lower gut. Gamma irradiation of rye reduces its adverse effects, apparently by depolymerization and solubilization of polysaccharides such as pectins (Campbell et al. 1983).

trans-Aconitic Acid

Another carbohydrate that has been implicated in a livestock problem is *trans*-aconitic acid (*cis*-aconitic acid is a citric acid cycle intermediate). Stout et al. (1967) suggested that *trans*-aconitic acid, which is found in high concentration in forage in the spring, may chelate magnesium and reduce its availability. Therefore, it may be a causative agent in inducing grass tetany, or hypomagnesemia. Camp et al. (1968) found that administration of *trans*-aconitic acid to sheep reduced serum magnesium levels but did not cause grass tetany. Bohman et al. (1969) reported that *trans*-aconitic acid plus potassium chloride when administered orally to cattle would induce grass tetany. Bohman et al. (1983), in a study of grass tetany in cattle on wheat pasture, found a sharp and sudden rise in the *trans*-aconitic acid content of the wheat at the time when the cows developed tetany, providing further evidence of a role of aconitate in the etiology of the disease. *trans*-Aconitic acid is hydrogenated in the rumen to tricarballylic acid (J. B. Russell and N. E. Forsberg, personal communication, Cornell University, 1984). Tricarballylic acid is the form absorbed and may sequester magnesium or inhibit aconitase. Tricarballylate inhibits acetate oxidation in vitro, suggesting an aconitase-inhibitory effect.

β-Glucans

Barley is a major feed grain in parts of North America, including the U.S. Pacific Northwest, central and western Canada, and Alaska. It is the world's fourth most important grain, ranking after corn, wheat, and rice in tonnage produced. It is tolerant of greater extremes in climate, altitude, and soil fertility than most other grains.

Barley contains a number of carbohydrate fractions that are poorly utilized by nonruminant animals. The hull contains hemicellulose and indigestible gums; gums are also found in the aleurone and endosperm cell walls. The gums in the aleurone cell walls are polymers of pentosans, while those in endosperm cell walls are polyglucans consisting of β-linked D-glucose units, with β-1,3 and β-1,4 bonds, called β-glucans.

8. CARBOHYDRATES, LIPIDS, AND CONJUGATES

These water-soluble cereal gums are found at quite high levels in barley, especially that grown in the western U.S. under semiarid conditions. Premature ripening hastened by excessively hot and dry weather usually increases the β-glucan content. The β-glucan content of western barley ranges from 1.5 to 8.0%. The β-glucans cause some nutritional problems, particularly in poultry, including reduced growth, low metabolizable energy, and sticky wet droppings (Fig. 8.4). This is due to the inability of birds to digest the β-glucans and the high

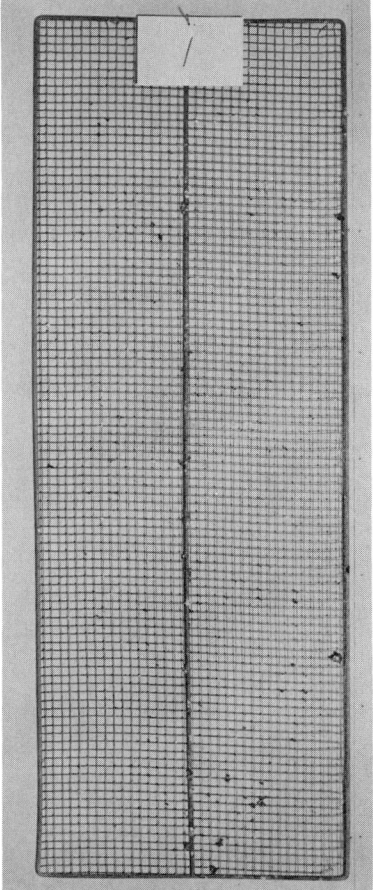 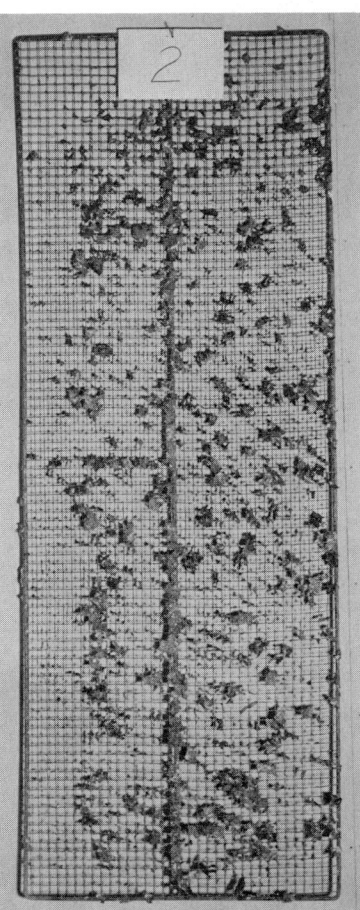

FIG. 8.4. Cage floors from cages of control birds (left) and from chicks fed western barley (right), showing the sticky droppings of the chicks fed barley. *Courtesy of G. H. Arscott.*

viscosity of the glucans. The growth-depressing properties of β-glucans and other mucilaginous types of fiber may be a result of shortened intestinal transit time, partitioning of intestinal contents, modification of intestinal mucosa, and changes in hormonal profiles caused by an altered rate of nutrient absorption (Vahouny 1982). A source of β-glucanase added to the feed will allow normal growth and eliminate the sticky droppings. This can be accomplished by soaking barley to allow growth of bacteria that produce β-glucanase or by mixing barley with corn, which contains the enzyme. Crude enzyme preparations of *Bacillus subtilis,* available commercially, are also effective. Swine also exhibit reduced growth rate and feed efficiency when fed barley due to its β-glucan content.

Sucrose

Mesquite (*Prosopis glandulosa*) is a leguminous shrub or small tree common on rangelands in the southwestern U.S. and Hawaii. It is considered an undesirable range weed, reducing forage production and in some areas forming impenetrable thickets. It has a large pod containing the mesquite seeds (beans). Ingestion of large amounts of mesquite pods by cattle over an extended period of time may cause rumen impaction and stasis, and eventual death. The pods contain about 30% sucrose, making them very palatable to livestock. It has been hypothesized (Adler 1949) that this amount of sugar represses bacterial cellulase activity so that the cellulose fraction of the pod is not digested, with the result that the pods may accumulate in and impact the rumen. A more plausible explanation is that the pod is refractory to bacterial enzymes and, because of its large size, cannot leave the rumen. Mesquite pods have been found in the rumen of cattle that have not had access to mesquite for as long as 9 months (Kingsbury 1964). Death from ingestion of mesquite pods is basically a result of starvation because of rumen stasis and impaction with indigestible pods.

There is interest in the potential use of mesquite as a crop plant in arid areas (Felker 1979). The pods contain about 30% sugar and 13% protein. They were a major food source for Indians in the southern California deserts. The trees are drought resistant and, being leguminous, require no nitrogen fertilization. Plant-breeding efforts are under way to develop improved mesquite varieties for orchard production in arid areas (Felker 1979).

Sucrose has been implicated as a cause of myocardial infarction and peripheral arterial disease in man (Yudkin 1964). Brooks *et al.* (1972)

in Hawaii found that feeding high (about 65%) levels of sucrose in the diet of growing pigs resulted in lesions in the left atrium of the heart. No lesions were found in controls fed similar diets with starch or corn in place of sucrose. Serum cholesterol levels were elevated in the pigs fed sucrose. No biochemical mechanism was advanced to explain the possible role of sucrose in the development of heart lesions.

Molasses toxicity has been reported in cattle on a liquid molasses–urea feeding system with restricted access to forage in numerous areas including Cuba, Kenya, Mexico, Australia, and the Dominican Republic (Edwin and Jackman 1982). The signs are similar to those of polioencephalomalacia, but molasses toxicity does not respond to thiamine administration. Presumably the condition is related to the sucrose in molasses. It may result from the change in rumen fermentation leading to a reduction in the molar proportion of propionic acid produced (Losada and Preston 1973), which in turn could result in reduced gluconeogenesis and low tissue glucose levels.

Gums

Plant gums are polymers of various types of monosaccharides, with uronic acid usually present. Gums are often produced in response to injury to a plant. They are generally water soluble or strongly hydrophilic. Often their glucuronic acid groups exist as sodium, potassium, or calcium salts.

Gums may cause a variety of problems in livestock, including growth depression, lack of feed palatability, and pasting of the beak in poultry. Feeds containing gums include rapeseed meal, leucaena, guar, and sesbania. Guar gum obtained from the Indian cluster bean (*Cyamopsis tetragonolobus*) is used extensively as an additive in food processing. Dietary levels of 2% guar gum can result in growth depression in chickens. Guar gum has hypocholesterolemic effects of interest in human nutrition.

In Australia, seeds of *Sesbania* spp. are often contaminants of sorghum grain, resulting in feed refusal and ill thrift of pigs fed the contaminated sorghum. Galactomannose gums may be the offending fraction of the *Sesbania* seeds (K. C. Williams and L. J. Daniels 1984, Personal communication, Animal Research Institute, Yeerongpilly, Qld., Australia).

Rapeseed gums are an aqueous emulsion of phospholipids and triglycerides produced during the refining of rapeseed oil (Leeson *et al.* 1977). Studies with poultry have shown that rapeseed gums can be used in layer diets with no detrimental effects (Leeson *et al.* 1977).

FATTY ACIDS

Erucic Acid

Erucic acid is one of the major fatty acids found in rapeseed oil. It has been implicated in myocardial lesions in rats and is therefore of toxicological interest.

Erucic acid ($22:1\omega 9$) has important industrial applications primarily because it is a longer-chain fatty acid than most fatty acids in plant oils. When cleaved at the ethylene bond, it yields brassylic acid, a dicarboxylic acid (Fig. 8.5) that is used in the manufacture of nylon and other polymers.

There is no other convenient commercial source for C_{13} dicarboxylic acid. Rapeseed oil is used in the lubrication of high-speed engines, as a flotation agent in mining, and for various other applications for which a long-chain fatty acid is useful. It also has a higher smoke point and vaporization temperature than other common oils. The erucic acid content of rapeseed oil varies from 25 to 35% in Canadian varieties up to 55% in some European varieties. While the high erucic acid content of rapeseed oil is advantageous for industrial applications it probably has negative implications in oil used for human consumption.

Feeding high erucic acid rapeseed oil to animals such as mice, rats, swine, guinea pigs, and poultry causes a reduction in growth rate, enlargement of the adrenals, increased mortality of the offspring of rats, moderate hepatic lipidosis, and degenerative changes in the kidney. In addition, myocardial lesions in rats occur. These develop in three stages: severe lipidosis, inflammation, and scar formation and

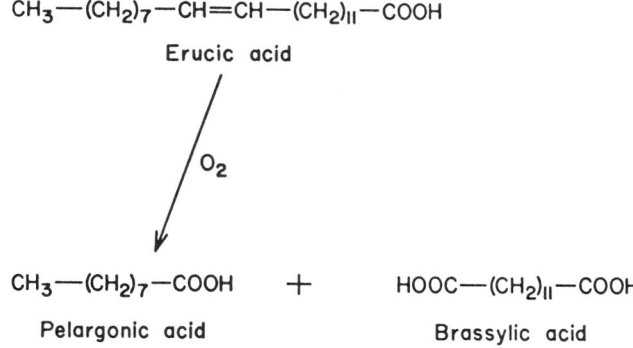

FIG. 8.5. Erucic acid and its conversion to brassylic acid.

fibrosis. The intracellular lipidosis starts within a few hours after rapeseed oil is fed and reaches a maximum in 3–6 days. Droplets of lipid are deposited in the myocardial fibers. After 4–8 weeks on a rapeseed oil-containing diet, foci of inflammation are found, associated with the presence of lipid-laden fragmented myofibers. At later periods, replacement of muscle fibers by scar tissue occurs, with hemosiderin deposits. The significance of these lesions reported in rats to human utilization of rapeseed oil is not certain. However, the question may become academic, since low erucic acid varieties of rapeseed have been developed. In general, the high erucic acid varieties are grown to produce industrial oil, while low erucic acid types are grown to produce edible oils for margarine, salad oil, and other food uses. The low erucic acid type of oil has been designated canbra oil, while the term rapeseed oil refers to the high erucic acid types.

Cyclopropenoid Fatty Acids

Two cyclopropene fatty acids, sterculic and malvalic acids, occur in cottonseed oil at levels of 1–2% of the crude oil.

$$H_3C-(CH_2)_7-\overset{\overset{\displaystyle CH_2}{\triangle}}{C=C}-(CH_2)_7-COOH$$

Sterculic acid

$$H_3C-(CH_2)_7-\overset{\overset{\displaystyle CH_2}{\triangle}}{C=C}-(CH_2)_6-COOH$$

Malvalic acid

Malvalic acid is present in a higher concentration than sterculic. The oil of the Java olive (*Sterculia foetida*) contains nearly 70% cyclopropene fatty acids and has been used as a source of the cyclopropenoids for experimental purposes. The cyclopropenoid fatty acids in cottonseed oil and meal intensify the effect of gossypol in causing olive green yolks of eggs and also cause the development of pink albumins during egg storage. These fatty acids have been shown to be cocarcinogens, increasing the tumor incidence and decreasing the latent period for tumor formation in trout fed carcinogens (Lee *et al.* 1971). The cyclopropenes in cottonseed oil have a pronounced effect on the mixed function oxidase system of trout liver (Loveland *et al.* 1979), resulting in reductions in microsomal protein, cytochrome P_{450}, NADPH-cytochrome c reductase, cytochrome b_5, and aldrin expoxidation, while there was an increase in benzo[a]pyrene hydroxylation activity. The conversion of aflatoxin B_1 to aflatoxicol in trout liver was inhibited, and production of aflatoxin M_1 was completely lacking in fish fed cyclopropenes. Similar but less pronounced changes were seen in rabbits fed cyclopropenoid fatty acids (Eisele et al. 1982).

Cyclopropenoid fatty acids inhibit fatty acid desaturation in numerous species, causing a rise in the stearate:oleate ratio of the body fat. Pigs fed diets containing cottonseed meal, which often contains residual lipids, produce fat with a high melting point. Dairy cattle fed cyclopropene fatty acids protected against biohydrogenation in the rumen have changes in the milk stearate:oleate ratio, probably due to a cyclopropene-mediated inhibition of mammary desaturase enzymes (Cook et al. 1976).

Cyclopropenoid fatty acids induce hypercholesterolemia and atherosclerosis in rabbits (Ferguson et al. 1976). The effects on plasma lipids and the degree of atherosclerosis are shown in Table 8.1.

Cyclopropenoid fatty acids may be of possible concern in the human diet. Glandless cottonseeds (lacking gossypol) have been approved by the Food and Drug Administration (FDA) as a nut substitute and snack item for human consumption. Since whole cottonseed kernels contain cyclopropenoid fatty acids, which are metabolically active and are cocarcinogens, potential exists for a human health risk. Hendricks et al. (1980) fed roasted cottonseed kernels to rainbow trout as part of a complete diet and found reduced growth and induction of liver carcinomas. They concluded that glandless cottonseed products could pose a possible health risk in the human diet.

Trans Fatty Acids

Unsaturated fatty acids of naturally occurring fats generally have a cis configuration of their double bonds. During partial hydrogenation of vegetable oil to produce margarine and partial hydrogenation in the rumen, trans isomers of unsaturated fatty acids may be formed.

TABLE 8.1. Effect of Dietary Cyclopropenoid Fatty Acids (CPFA) on Plasma Lipids and Aortic Lesions in Rabbits after 5 Weeks of Feeding[a]

Item	Control	Control + 0.27% CPFA
Total plasma cholesterol (mg/100 ml)	148.0	461.0
Free plasma cholesterol (%)	37.0	46.0
Plasma triglycerides (mg/100 ml)	43.0	64.0
Liver cholesterol (mg/g protein)	22.2	44.2
Incidence of aortic lesions	0/8	5/8

[a] Adapted from Ferguson et al. (1976).

Elaidic acid, a trans monoene of oleic acid, has been extensively used in studies of the biological effects of trans fatty acids. The potential adverse effects of dietary trans fatty acids have been of some concern in human nutrition. There is some evidence that they may behave more like saturated than unsaturated fatty acids in metabolism (Brisson 1982). They have been suggested as possible risk factors in atherosclerosis and cancer. Trans isomers may have hypercholesterolemic effects, and epidemiological studies have suggested a positive association between intake of trans fatty acids and cancer. The public health implications of these findings, if any, have not been resolved. There would appear to be virtually no significance of these compounds to livestock, except that they are produced in the rumen, are found in ruminant body fat, and thus are found in food products of animal origin. The significance of trans fatty acids has been reviewed by Applewhite (1981), Brisson (1982), and Kinsella *et al.* (1981).

Miscellaneous Fatty Acids

Hydroxylated fatty acids, such as ricinoleic acid in castor oil, cause damage to the intestinal membrane, and fluid and electrolyte loss to the intestine (Racusen and Binder 1979). This accounts for the cathartic effect of castor oil. Ricinoleic acid has also been identified in *Linum* seed oils (Green 1984). It is not a constituent of linseed (*Linum usitatissimum*) oil.

Ixiolaena brevicompta is a plant that has been implicated in sheep poisonings in New South Wales in Australia (Walker *et al.* 1980). The seeds contain an unusual C-18 acetylenic fatty acid, crepenynic acid (Ford *et al.* 1981). It has been postulated as a potential causative agent of the toxicity. Toxicity signs include acute myopathy.

Linolenic acid in spring pasture grass has been implicated in nutritional myopathy of grazing ruminants (Rice *et al.* 1981). These workers demonstrated that dietary linolenic acid, if protected from hydrogenation in the rumen, can induce muscle degeneration. It was suggested that with the change in diet when cattle are turned out to pasture in the spring, there may be changes in rumen microflora, with a period of time during which hydrogenation of linolenic acid is impaired.

Autoxidation Products of Lipids

Autoxidation of fats involves their oxidation at temperatures not exceeding 100°C, while thermal oxidation is oxidation of a fat at about 200°C in the presence of air. Rancidity generally refers to autoxidation processes occurring at ambient temperatures. Fats that have undergone these oxidation processes are of potential toxicity to animals. Feedstuffs such as corn and soybeans are susceptible to development of rancidity, particularly after grinding. Autoxidation involves oxidation of unsaturated fatty acids. Fats of animal origin are susceptible to development of rancidity because of the presence of heme compounds (myoglobin and hemoglobin) which are potent biological pro-oxidants. Plant feedstuffs, while having a high content of unsaturated fatty acids, often contain a high content of antioxidants such as the tocopherols and ascorbic acid, which confer protection against autoxidation. Trace elements such as copper and zinc are potent pro-oxidants.

Lipid peroxidation is associated with the addition of oxygen to a carbon atom adjacent to a double bond of a fatty acid, forming hydroperoxides. A discussion of the mechanisms is provided by Yannai (1980). The rate of formation of hydroperoxides increases dramatically with increases in unsaturation. Thus, linoleate is oxidized 10 times faster than oleate, with linolenate oxidized 20–30 times faster. Adverse effects of oxidized lipids on animal performance are related to unpalatability of feeds, destruction of other nutrients such as vitamins, and tissue damage. Hydroperoxides form adducts with cysteine, inhibiting enzyme activity. There is some suggestion that oxidized lipids have carcinogenic properties (Yannai 1980).

Fats used in animal feeding should be stabilized with antioxidants to prevent the adverse consequences of autoxidation.

GLYCOLIPIDS

Corynetoxins

Annual ryegrass toxicity (ARGT) is a disease of livestock caused by a group of highly toxic glycolipids, called corynetoxins, produced in annual ryegrass seed heads infected with a nematode-bacteria combination. ARGT is an important sheep-poisoning problem in South Australia and Western Australia and has been observed in South Africa and Oregon. It could occur in the future wherever annual ryegrass is grown as a pasture species. It occurs mainly in sheep, but has also been reported in elk (Mackintosh *et al*, 1982).

Signs of Toxicity in Livestock. The corynetoxins affect the nervous system, and their effects become most obvious when animals are stressed or excited. Signs of toxicity may appear as soon as 2 days or as late as 12 weeks after stock are introduced into toxic annual ryegrass pastures. If the animals are not inspected regularly, the first symptoms seen may be a number of mortalities. Close inspection of the herd or flock daily may reveal some animals with neurological signs following disturbance. The animal typically has a high stepping gait (the stair-climbing gait) with the head held high. There is a loss of coordination of the hind legs, and it may collapse, with convulsions and spasms (Fig. 8.6). It may appear to recover after a time, regain its feet, and return to the herd or flock. In more severe cases, it may regain its feet but remain standing only by propping itself on spread legs. In the terminal stages, animals remain lying on the ground, with spasms and convulsions, and with their feet in a paddling motion. Death usually occurs within about 24 hr. Pathological changes include diffuse fat deposits in the liver, hemorrhages in various organs, and vascular damage in the brain, particularly in the cerebellum.

The biological activities of corynetoxins are virtually identical to those of the closely related tunicamycin antibiotics (Jago *et al.* 1983). Both groups of compounds strongly inhibit UDP-N-acetylglucosamine: dilicholphosphate N-acetylglucosamine phosphate transferase, an enzyme essential for lipid-linked N-glycosylation of glycoproteins. Hence, annual ryegrass toxicity may be the result of the depletion or reduced activity of essential N-glycosylated glycoproteins.

Some aspects of corynetoxin poisoning resemble failure of the reticuloendothelial system. This function is largely determined by the blood levels of the N-glycosylated glycoprotein, fibronectin, an opsonic protein rendering bacteria susceptible to phagocytosis. Poisoning with either corynetoxin or tunicamycin reduces serum fibronectin levels

FIG. 8.6. Sheep suffering from annual ryegrass toxicity, showing the mounds of earth caused by the "paddling" action.
Courtesy of P. Vogel.

and reticuloendothelial function in a dose-related manner. Jago *et al.* (1983) hypothesized that depletion of fibronectin could lead to increased vascular permeability, accounting for the neurological signs in annual ryegrass toxicity.

Production of the Toxins. The infection of annual ryegrass resulting in toxin production involves a unique interrelationship between the grass, a nematode, and a bacterium. The nematode, *Anguina agrostis,* infects the ryegrass shortly after germination. The nematode larvae crawl up the plant to the growing tip. They remain passive at the growing point and are carried up as it develops. When the grass begins to flower, the larvae burrow into the developing flower where they mature into nematode worms. The flower does not develop a seed; the seed is replaced by a gall in which the adult nematodes lay eggs that hatch into larvae in the gall (Fig. 8.7). They remain dormant until the following season when they hatch out in the soil and begin the infestation anew. The nematode is nontoxic. However, if the nematode larvae carry a bacterium, *Corynebacterium rathayi,* the seed galls produce the toxins. It has not been experimentally possible to get cultures of *C. rathayi* to produce the toxins, except in the presence of ryegrass, suggesting that a plant factor is necessary to trigger toxin production. The bacteria produce a yellow slime on the seed heads. The slime can be seen as a yellowness in ryegrass fields; on close inspection it can be readily seen as a glistening sticky yellow mass oozing from the seed heads.

The toxins have been identified as glycolipids containing an amino sugar(s) with 3-hydroxy C-17 fatty acid residues (Vogel *et al.* 1981):

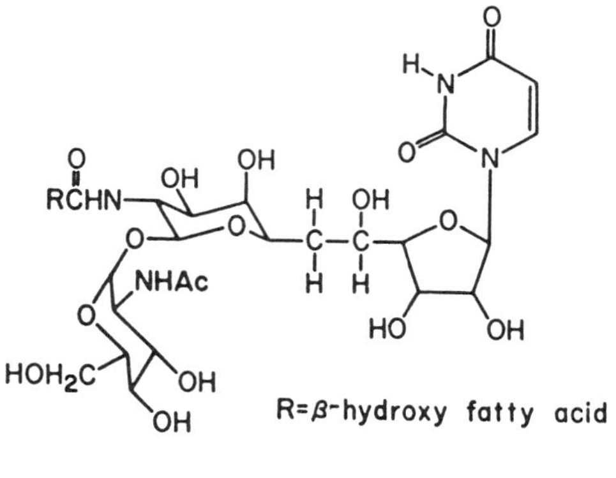

R=β-hydroxy fatty acid

Corynetoxin

FIG. 8.7. A parasitized seed head of annual ryegrass (left) that can produce annual ryegrass toxicity, contrasted with a normal seed head (right).
Courtesy of D. J. Schneider.

Control of Ryegrass Toxicity. Stock on ryegrass pastures in areas where ARGT has occurred should be inspected daily and removed at the first sign of neurological problems. Various pasture management schemes have been applied to eliminate the problem from pastures. These involve breaking the ryegrass-nematode-bacterium cycle at its weakest link—the nematode. Methods to eliminate nematode infestations include crop rotation, with chemical control of ryegrass in cereal crops grown in the rotation. Burning is effective in destroying seed galls in the stubble. In this regard, early reports of ARGT in Oregon (Shaw and Muth 1949; Galloway 1961) have not been followed up by further observations of the disease, probably because of the almost universal practice of burning ryegrass stubble in Oregon (much to the dismay of the urban population!). Clipping to destroy seed heads is also effective. Fallowing may be used to eliminate the nematode.

The condition is spreading throughout Australia, probably by transport of infected hay and grain seeds. Whirlwinds may transfer the nematode galls for short distances. Nematodes may be eliminated by the use of some of the above techniques, and then special efforts should be made to avoid reinfestation.

GLYCOPROTEINS

Examples of toxicants that are glycoproteins include lectins and avidin. Lectins have been discussed in Chapter 7.

Avidin is a glycoprotein with a molecular weight of about 43,500. It is secreted by the oviduct of birds into the egg white, and binds with the B vitamin biotin in a tight complex, resisting digestion and absorption. Biotin deficiency in many animals can be induced by feeding raw eggs, egg white, or avidin. Avidin is denatured by moist heat and is inactivated when eggs are cooked. A livestock problem due to avidin occurred in the U.S. Pacific Northwest. Mink fed the viscera from turkey-processing plants developed an achromatrichial condition in which the pelts were light in color (Fig. 8.8). This was called "turkey waste graying." Eventually this was traced by Oregon State University scientists (Stout and Adair 1969) to the presence of eggs from cull hen turkeys in the turkey viscera. This created a biotin deficiency that could be overcome by either cooking the turkey waste or supplementing the diet with biotin. When raw turkey eggs were fed, severe biotin deficiency characterized by fur graying, "spectacle eye," loss of fur, exudates from the eyes, nose, and mouth, and encrustation of paws was observed. In more recent work (Wehr et al. 1980), it was found that spray-dried eggs retained sufficient avidin to cause biotin deficiency in mink.

Turkey eggs contain about four times as much avidin as hen eggs. Since avidin occurs in the egg white and biotin is in the yolk, a biotin deficiency is most readily induced when egg whites rather than whole eggs are fed.

An interesting case (Anon. 1963) of avidin-induced biotin deficiency in humans was that of an Italian laborer who had eaten about 6 dozen raw eggs weekly since childhood, with little other food but 1 to 4 quarts

8. CARBOHYDRATES, LIPIDS, AND CONJUGATES

FIG. 8.8. A pelt from a biotin-deficiency mink (turkey waste graying) caused by the consumption of turkey viscera. The avidin in the eggs induced a biotin deficiency. Note the achromatrichia of the underfur.
Courtesy of J. Adair and J. E. Oldfield.

of wine daily! He experienced exfoliative dermatitis and conjunctivitis for many years; the symptoms disappeared after 2 weeks of hospitalization on a liberal diet and injections of biotin.

REFERENCES

Xylose and Lactose

BAILEY, C. B., KITTS, W. D., and WOOD, A. J. 1956. The development of the pig during its pre-weaning phase of growth. Can J. Agric. Sci. *36*, 51–58.

BAKER, D. H. 1977. Xylose and xylan utilization by the chick. Poult. Sci. *56*, 2105–2107.

WISE, M. B., BARRICK, E. R., WISE, G. H., and OSBORNE, J. C. 1954. Effects of substituting xylose for glucose in a purified diet for pigs. J. Anim. Sci. *13*, 365–373.

Pectins

CAMPBELL, G. L., CLASSEN, H. L., REICHERT, R. D., and CAMPBELL, L. D. 1983. Improvement of the nutritive value of rye for broiler chickens by gamma irradiation-induced viscosity reduction. Br. Poult. Sci. *24*, 205–212.

FRIEND, D. W., and MACINTYRE, T. M. 1969. Digestibility of rye and its value in pelleted rations for pigs. Can. J. Anim. Sci. *49*, 375–381.

GRAMMER, J. C., McGINNIS, J., and PUBOLS, M. H. 1983. The rachitogenic effects of fractions of rye and certain polysaccharides. Poult. Sci. *62*, 103–109.

HONEYFIELD, D. C., FROSETH, J. A., and McGINNIS, J. 1983. Comparative feeding value of rye for poultry and swine. Nutr. Rep. Int. *28*, 1253–1260.

McNAB, J. M., and SHANNON, D. W. F. 1974. The nutritive value of barley, maize, oats and wheat for poultry. Br. Poult. Sci. *15*, 561–567.

MISIR, R., and MARQUARDT, R. R. 1978A. Factors affecting rye (*Secale cereale* L.) utilization in growing chicks. I. The influence of rye level, ergot, and penicillin supplementation. Can. J. Anim. Sci. *58*, 691–701.

MISIR, R., and MARQUARDT, R. R. 1978B. II. The influence of protein type, protein level and penicillin. Can. J. Anim. Sci. *58*, 703–715.

MISIR, R., and MARQUARDT, R. R. 1978C. III. The influence of milling fractions. Can. J. Anim. Sci. *58*, 717–730.

MISIR, R., and MARQUARDT, R. R. 1978D. IV. The influence of autoclave treatment; pelleting, water extraction and penicillin supplementation. Can. J. Anim. Sci. *58*, 731–742.

WAGNER, D. D., and THOMAS, O. P. 1978. Influence of diets containing rye or pectin on the intestinal flora of chicks. Poult. Sci. *57*, 971–975.

trans-Aconitic Acid

BOHMAN, V. R., LESPERANCE, A. L., HARDING, G. D., and GRUNES, D. L. 1969. Induction of experimental tetany in cattle. J. Anim. Sci. *29*, 99–102.

BOHMAN, V. R., HORN, F. P., STEWART, B. A., MATHERS, A. C., and GRUNES, D. L. 1983. Wheat pasture poisoning. 1. An evaluation of cereal pastures as related to tetany in beef cows. J. Anim. Sci. *57*, 1352–1363.

BURAU, R., and STOUT, P. R. 1965. *Trans*-aconitic acid in range grasses in early spring. Science *150*, 766–767.

CAMP, B. J., DOLLAHITE, J. W., and SCHWARTZ, W. L. 1968. Biochemical changes in sheep given *trans*-aconitic acid. Am. J. Vet. Res. *29*, 2009–2013.

STOUT, P. R., BROWNELL, J., and BURAU, R. G. 1967. Occurrences of *trans*-aconitate in range forage species. Agron. J. *59*, 21–24.

β-Glucans

GOHL, B., ALDEN, S., ELSWINGER, K., and THOMKE, S. 1978. Influence of beta-glucanase on feeding value of barley for poultry and moisture content of excreta. Br. Poult. Sci. *19*, 41–47.

HESSELMAN, K., ELWINGER, K., and THOMKE, S. 1982. Influence of increasing levels of β-glucanase on the productive value of barley diets for broiler chickens. Anim. Feed Sci. Technol. *7*, 351–358.

VAHOUNY, G. V. 1982. Dietary fiber, lipid metabolism and atherosclerosis. Fed. Proc. *41*, 2509–2511.

WHITE, W. B., BIRD, H. R., SUNDE, M. L., MARLETT, J. A., PRENTICE, N. A., and BURGER, W. C. 1983. Viscosity of β-D-glucan as a factor in the enzymatic improvement of barley for chicks. Poult. Sci. *62*, 853–862.

Sucrose

ADLER, A. E. 1949. Indigestion from an unbalanced kiawe (mesquite) bean diet. J. Am. Vet. Med. Assoc. *115*, 263.

BROOKS, C. C., MIYAHARA, A. Y., HUCK, D. W., and ISHIZAKI, S. M. 1972. Relationship of sugar-induced lesions in the heart of the pig to live weight, serum cholesterol and diet. J. Anim. Sci. *35,* 31–37.
DOLLAHITE, J. W. 1964. Management of the disease produced in cattle on an unbalanced diet of mesquite beans. Southwest Vet. *17,* 293–295.
EDWIN, E. E., and JACKMAN, R. 1982. Ruminant thiamine requirement in perspective. Vet. Res. Commun. *5,* 237–250.
FELKER, P. 1979. Mesquite: An all purpose leguminous arid-land tree. *In* New Agricultural Crops. G. A. Ritchie (Editor), pp. 89–132. Westview Press, Boulder, CO.
FELKER, P., and BRANDURSKI, R. S. 1979. Uses and potential uses of leguminous trees for minimal energy input agriculture. Econ. Bot. *33,* 172–184.
KINGSBURY, J. M. 1964. Poisonous Plants of the United States and Canada. Prentice-Hall, Englewood Cliffs, NJ.
LOSADA, H., and PRESTON, T. R. 1973. Molasses toxicity and cerebrocortical necrosis (CCN). Cuban J. Agric. Sci. *7,* 169–178.
YUDKIN, J. 1964. Levels of dietary sucrose in patients with occlusive atherosclerotic disease. Lancet *2,* 6.

Gums

LEESON, S., SLINGER, S. J., and SUMMERS, J. D. 1977. Performance of laying hens fed diets containing gums derived from Tower rapeseed. Can. J. Anim. Sci. *57,* 479–483.

Erucic Acid

KRAMER, J. K. G., SAUER, F. D., and PIGDEN, W. J. 1983. High and Low Erucic Acid Rapeseed Oils: Production, Usage, Chemistry and Toxicological Evaluation. Academic Press, NY.

Cyclopropenoid Fatty Acids

COOK, L. J., SCOTT, T. W., MILLS, S. C., and JOHNSON, A. R. 1976. Effects of protected cyclopropene fatty acids on the composition of ruminant milk fat. Lipids *11,* 705–711.
EISELE, T. A., LOVELAND, P. M., KRUK, D. L., MEYERS, T. R., NIXON, J. E., and SINNHUBER, R. O. 1982. Effect of cyclopropenoid fatty acids on the hepatic microsomal mixed-function-oxidase system and aflatoxin metabolism in rabbits. Food Chem. Toxicol. *20,* 407–412.
FERGUSON, T. L., WALES, J. H., SINNHUBER, R. O., and LEE, D. J. 1976. Cholesterol levels, atherosclerosis and liver morphology in rabbits fed cyclopropenoid fatty acids. Food Cosmet. Toxicol. *14,* 15–18.
HENDRICKS, J. D., SINNHUBER, R. O., LOVELAND, P. M., PAWLOWSKI, N. E., and NIXON, J. E. 1980. Hepatocarcinogenicity of glandless cottonseeds and cottonseed oil to rainbow trout (*Salmo gairdnerii*). Science *208,* 309–311.
LEE, D. J., WALES, J. H., and SINNHUBER, R. O. 1971. Promotion of aflatoxin-induced hepatoma growth in trout by methyl malvalate and sterculate. Cancer Res. *31,* 960–963.
LOVELAND, P. M., NIXON, J. E., PAWLOWSKI, N. E., EISELE, T. A., LIBBEY, L. M., and SINNHUBER, R. O. 1979. Aflatoxin B_1 and aflatoxicol metabolism in rain-

bow trout and the effects of dietary cyclopropene. J. Environ. Pathol. Toxicol. *2*, 707–718.

PHELPS, R. A., SHENSTONE, R. S., KEMMERER, A. R., and EVANS, R. J. 1965. A review of cyclopropenoid compounds: Biological effects of some derivatives. Poult. Sci. *44*, 358–395.

Trans Fatty Acids

APPLEWHITE, T. H. 1981. Nutritional effects of hydrogenated soya oil. J. Am. Oil Chem. Soc. *58*, 260–269.

BRISSON, G. J. 1982. Lipids in Human Nutrition. Burgess, Inc., Englewood, NJ.

KINSELLA, J. E., BRUCKNER, G., MAI, J., and SHIMP, J. 1981. Metabolism of trans fatty acid with emphasis on the effect of *trans,trans*-octadecadienoate on lipid composition, essential fatty acid, and prostaglandins: An overview. Am. J. Clin. Nutr. *34*, 2307–2318.

Miscellaneous Fatty Acids

FORD, G. L., WHITFIELD, F. B., and WALKER, K. H. 1981. Fatty acid composition of *Ixiolaena brevicompta* F. Muell. seed oil. Lipids *18*, 103–105.

GREEN, A. G. 1984. The occurrence of ricinoleic acid in *Linum* seed oils. J. Am. Oil Chem. Soc. *61*, 939–940.

RACUSEN, L. C., and BINDER, J. H. 1979. Ricinoleic acid stimulation of active anion secretion in the colonic mucosa of the rat. J. Clin. Invest. *63*, 743–749.

RICE, D. A., BLANCHFLOWER, W. J., and McMURRAY, C. H. 1981. Reproduction of nutritional degenerative myopathy in the postruminant calf. Vet. Rec. *109*, 161–162.

WALKER, K. H., THOMPSON, D. R., and SEAMAN, J. T. 1980. Suspected poisoning of sheep by *Ixiolaena brevicompta*. Aust. Vet. J. *56*, 64–66.

Autoxidation Products of Lipids

YANNAI, S. 1980. Toxic factors induced by processing. *In* Toxic Constituents of Plant Foodstuffs. I. E. Liener (Editor), pp. 371–427. Academic Press, NY.

Glycolipids

BERRY, P. H. and VOGEL, P. 1982. Toxicity studies of the toxins isolated from annual ryegrass (*Lolium rigidum*) by *Corynebacterium rathayi*. Aust. J. Exp. Biol. Med. Sci. *60*, 129–132.

BERRY, P. H., HOWELL, J. M., and COOK, R. D. 1980. Morphological changes in the central nervous system of sheep affected with experimental annual ryegrass (*Lolium rigidum*) toxicity. J. Comp. Pathol. *90*, 603–617.

BERRY, P. H., RICHARDS, R. B., HOWELL, J. M., and COOK, R. D. 1982. Hepatic damage in sheep fed annual ryegrass, *Lolium rigidum*, parasitised by *Anguina agrostis* and *Corynebacterium rathayi*. Res. Vet. Sci. *32*, 148–156.

EDGAR, J. A., FRAHN, J. L., COCKRUM, P. A., ANDERTON, N., JAGO, M. V., CULVENOR, C. C. J., JONES, A. J., MURRAY, K., and SHAW, K. J. 1982.

Corynetoxins, causative agents of annual ryegrass toxicity; their identification as tunicamycin group antibiotics. J. Chem. Soc., Chem. Commun. pp. 222–224.

FLETCHER, L. R., and HARVEY, I. C. 1981. An association of a lolium endophyte with ryegrass staggers. N. Z. Vet. J. *29*, 185.

GALLAGHER, R. T., WHITE, E. P., and MORTIMER, P. H. 1981. Ryegrass staggers—Isolation of potent neurotoxins lolitrem-A and lolitrem-B from staggers-producing pastures. N. Z. Vet. J. *29*, 189.

GALLOWAY, J. H. 1961. Grass seed nematode poisoning in livestock. J. Am. Vet. Med. Assoc. *139*, 1212–1214.

JAGO, M. V., PAYNE, A. L., PETERSON, J. E., and BAGUST, T. J. 1983. Inhibition of glycosylation by corynetoxin, the causative agent of annual ryegrass toxicity: A comparison with tunicamycin. Chem.-Biol. Interact. *45*, 223–234.

LANIGAN, G. W., PAYNE, A. L., and FRAHN, J. L. 1976. Origin of toxicity in parasitised annual ryegrass (*Lolium rigidum*). Aust. Vet. J. *56*, 244–246.

MACKINTOSH, C. G., ORR, M. B., GALLAGHER, R. T., and HARVEY, I. C. 1982. Ryegrass staggers in Canadian wapiti deer. N. Z. Vet. J. *30*, 106.

SHAW, J. N., and MUTH, O. H. 1949. Some types of forage poisoning in Oregon cattle and sheep. J. Am. Vet. Med. Assoc. *114*, 315–317.

STYNES, B. A., and WISE, J. L. 1980. The distribution and importance of annual ryegrass toxicity in Western Australia and its occurrence in relation to cropping rotations and cultural practices. Aust. J. Agric. Res. *31*, 557–569.

STYNES, B. A., PETTERSON, D. S., LLOYD, J., PAYNE, A. L., and LANIGAN, G. W. 1979. The production of toxin in annual ryegrass, *Lolium rigidum*, infected with a nematode, *Anguina* sp., and *Corynebacterium rathayi*. Aust. J. Agric. Res. *30*, 201–209.

VOGEL, P., PETTERSON, D. S., BERRY, P. H., FRAHN, J. L., ANDERTON, N., COCKRUM, P. A., EDGAR, J. A., JAGO, M. V., LANIGAN, G. W., PAYNE, A. L., and CULVENOR, C. C. J. 1981. Isolation of a group of glycolipid toxins from seed heads of annual ryegrass (*Lolium rigidum* Gaud.) infected by *Corynebacterium rathayi*. Aust. J. Exp. Biol. Med. Sci. *59*, 455–467.

Glycoproteins

ANON. 1963. Vitamin Manual. Upjohn Company, Kalamazoo, MI.

STOUT, F. M., and ADAIR, J. 1969. Biotin deficiency in mink fed poultry by-products. Am. Fur Breeder *42* (6), 10.

WEHR, N. B., ADAIR, J., and OLDFIELD, J. E. 1980. Biotin deficiency in mink fed spray-dried eggs. J. Anim. Sci. *50*, 877–855.

9

Metal-Binding Substances and Inorganic Toxicants

OXALATES

Various pasture and range plants are toxic because of their high oxalate content. In the U.S., the major oxalate-related livestock problem has been halogeton (*Halogeton glomeratus*) poisoning of range sheep. Large numbers of sheep have died, including numerous instances of hundreds and even a thousand or more sheep dying in a single outbreak. In Australia, soursob (*Oxalis pes-caprae*), an introduced plant from South Africa, causes extensive problems. In Australia and other parts of the tropics, certain tropical grasses such as setaria (*Setaria sphacelata*) and *Panicum* spp. (elephant grass, guinea grass) may contain toxic oxalate concentrations under some conditions and have been implicated in oxalate-induced problems in cattle and horses. Other oxalate-containing plants for which livestock problems have been suggested include redroot pigweed (*Amaranthus retroflexus,*) kochia (*Kochia scoparia,*) and greasewood (*Sarcobatus vermiculatus*).

Oxalate is found in plants in two major forms. Some plants, such as soursob, have a cell sap pH of about 2, and the oxalate exists as salts of acid oxalate ($H_2CO_4^-$), such as acid potassium oxalate (Fig. 9.1). Other plants, such as halogeton, have a cell sap pH of about 6, and the oxalate exists as soluble sodium and insoluble calcium and magnesium oxalates. With the acid oxalate salts, both acute and chronic toxicity occurs, while with halogeton, only acute toxicity is seen.

Halogeton Poisoning

Halogeton is a branched annual herbaceous plant native to arid alkaline soils of Russia. It was accidentally introduced into the U.S. as

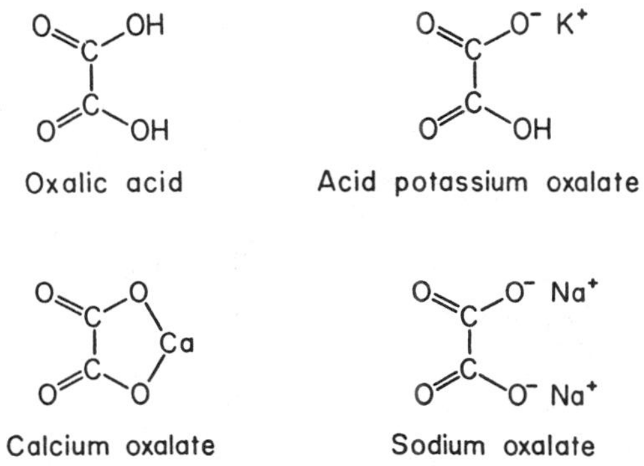

9. METAL-BINDING AND INORGANIC SUBSTANCES

FIG. 9.1. Structure of oxalates.

a contaminant of agricultural products and was first collected and identified in 1934 in Nevada. Since that time, it has spread over more than 10 million acres of western rangelands, particularly in Nevada, Utah, and Idaho. Losses of sheep from halogeton consumption were suspected in the 1930s; in 1942, severe losses in Nevada prompted experimental studies which proved that halogeton is toxic. In the 1940s, the range sheep industry was still quite large in the western states, and therefore with a large number of animals exposed to the plant, some spectacular losses occurred. Numerous cases are documented where 500–1500 sheep died in a single day when a band was herded through halogeton-infested areas. Since that time, losses have been few, both because of increased awareness on the part of sheep producers of the toxicity of the plant and because the decline of the western range sheep industry has put fewer animals at risk. In the 1970s, the death of hundreds of sheep on an army base led to charges that nerve gases had been released, but the sheep mortality was shown to be caused by ingestion of halogeton.

Halogeton cannot compete with established perennial plants and therefore is found primarily in disturbed or barren soils and many winter ranges. It grows along roadsides, railroad tracks, and other disturbed areas. Livestock losses, including some cattle poisonings (Lincoln and Black 1980), occur most frequently when hungry stock are unloaded in halogeton-infested areas or trailed along roads with stands of halogeton on the roadside (Fig. 9.2). The oxalate concentration is highest in fall and winter, and the plant is most likely to be consumed at this time after fall rains have softened the dry plant. Generally plants must contain at least 10% oxalate on a dry weight basis for toxicity to occur.

FIG. 9.2. Halogeton growing along a road on a western range. Sheep are frequently poisoned when being moved along roads in such areas.
Courtesy of L. F. James.

Signs of halogeton poisoning include labored breathing, depression, weakness, coma, and death. Some animals may have convulsions, and tetany may be seen. Animals with obvious symptoms may have serum calcium reduced by 20% or more, while severe hypocalcemia, with serum calcium only 20% of normal, occurs in animals that die (Fig. 9.3). Gross pathology includes hemorrhages and edema of the rumen wall, swollen kidneys, and calcium oxalate crystals found in the rumen wall, kidney tubules, and other tissues. Rumen stasis also occurs.

The mode of action of oxalate in causing toxicity is not totally clear. Acute toxicity may be due to hypocalcemia, while uremia from kidney damage may contribute to chronic toxicity. Complicating the situation are the observations (James 1978) that sheep made hypocalcemic by dialysis or EDTA infusion do not die, while blood calcium levels can be experimentally maintained in sheep poisoned by halogeton, and they still die. Hemorrhagic rumenitis and shock may be contributory factors. Rumen stasis may cause an increased pH of the rumen. Oxalate also inhibits several respiratory enzymes, including succinic dehydrogenase. Enzymes activated by calcium or magnesium may be inhibited by oxalate.

Dietary oxalate can be degraded by rumen microorganisms (Allison *et al.* 1981). Ruminants adapted to diets with high oxalate content can

FIG. 9.3. Sheep losses from halogeton poisoning. Typically the animals die in a sleeping position, characteristic of effects of hypocalcemia.
Courtesy of L. F. James.

tolerate oxalate levels that are lethal to nonadapted animals. Allison *et al.* (1981) demonstrated the presence of a rumen microorganism that degrades oxalate to CO_2 and formate and depends on oxalate for its growth. Since most vegetation contains a small amount of oxalate, residual populations of an oxalate-dependent organism could survive in the rumen. Alfalfa, for instance, has an appreciable portion of its calcium in the form of calcium oxalate (Ward *et al.* 1979).

Prevention of halogeton poisoning depends primarily on livestock management. Hungry animals should not be exposed to abundant stands of halogeton, particularly if they have not had prior exposure to it. At least 4 days of oxalate exposure are needed for development of oxalate-degrading capacity of the rumen. Such preconditioning results in about a 30% increase in the level of oxalate required to kill sheep. The provision of calcium supplements has had some beneficial effect.

Other Oxalate-Containing Plants

In the U.S., other plants besides halogeton have been implicated in livestock losses. Greasewood (*S. vermiculatus*) is an erect spiny shrub that grows in alkaline soils on western ranges. The oxalate content of the leaves varies between 10 and 20% of the dry weight, reaching a maximum in late summer. Heavy losses of sheep have occurred from oxalate poisoning due to consumption of greasewood. It is regarded as a useful forage, and toxicity can be avoided with good animal management.

Redroot pigweed (*A. retroflexus*) causes perirenal edema and nephrosis in cattle and swine. Oxalates (Marshall *et al.* 1967) have been suggested as one of the causative agents. Domesticated varieties of amaranthus are being developed as grain sources; they have growth-depressing properties which could be influenced by the fairly high oxalate content. Various *Chenopodiaceae* such as lamb's-quarters (*Chenopodium album*) accumulate oxalates. *Rumex* spp., the sorrels and docks, contain oxalates and sometimes cause problems in Australia and New Zealand; no *Rumex* poisonings in North America have been reported. Soursob (*O. pes-caprae*) causes significant sheep losses in Australia. It contains acid potassium oxalate and causes chronic poisoning. Kidney damage due to formation of oxalate crystals is the major problem.

An interesting situation is the development of nutritional secondary hypoparathyroidism, or osteodystrophy fibrosa, in horses consuming certain tropical grasses. This condition is also called "bighead." The frontal bones of the face undergo decalcification and fibrosis, and become enlarged (Fig. 9.4). Bighead develops as a result of a high ratio of absorbed phosphorus to calcium. Calcium is mobilized from the bone under the stimulus of parathyroid hormone to maintain serum calcium levels, which tend to drop because of the effects of hyperphosphatemia. Horses grazing grasses such as setaria in Queensland in Australia develop this condition. Setaria is moderately high in oxalate and low in calcium. The calcium in the diet tends to be precipitated in the gut as calcium oxalate, leading to a high phosphorus, low calcium situation. Administration of a dietary calcium supplement prevents the disease from developing. McKenzie *et al.* (1981B) reported that an effective supplement is a mixture of ground limestone and dicalcium phosphate; phosphorus is needed to replace the large urinary losses.

Nutritional secondary hyperparathyroidism occurs in horses on pastures that support beef cattle with no hypocalcemia problems. Oxalates have a much greater negative effect on calcium absorption in horses than in cattle, which can be explained in terms of their anatomical differences in site of oxalate degradation and calcium absorption.

Besides bighead, nutritional secondary hyperparathyroidism causes severe lameness of horses. In Australia, this interferes with normal cattle management, as stock owners need to keep large herds of horses

FIG. 9.4. Nutritional secondary hyperparathyroidism or osteodystrophy fibrosa ("bighead") in a horse that had been grazing on tropical grass pasture species which are high in oxalate and low in calcium.
Courtesy of R. A. McKenzie.

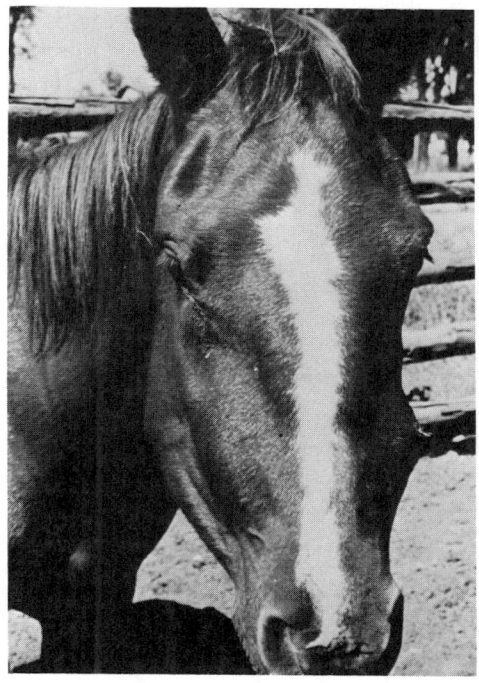

to ensure that a minimum number are fit for use. The disorder generally occurs when horses are grazed solely on pastures in which the calcium to oxalate ratio is less than 0.5; most offending grasses contain in excess of 0.5% total oxalate. The oxalate content of the grasses is highest under conditions of active growth.

Cattle mortalities have occurred on setaria pastures in Australia (Jones *et al.* 1970). Death follows acute signs of hypocalcemia and is accompanied by deposition of calcium oxalate crystals in the kidney (Seawright *et al.* 1970). Blaney *et al.* (1982) found that in tropical grasses such as *Setaria sphacelata,* which contain calcium oxalate, the availability of calcium to cattle is about 20% lower than in low-oxalate grasses. They suggest that the availability of calcium to cattle grazing tropical grasses should be considered to be a maximum of 50%. Neither magnesium nor phosphorus availability in tropical grasses was affected by oxalate.

Tropical grasses which contain sufficient levels of oxalate to impair mineral utilization include buffel (*Cenchrus ciliaris*), pangola (*Digitaria decumbens*), setaria (*S. sphacelata*), and kikuyu (*Pennisetum clandestinum*).

Oxalate is the toxic component of rhubarb (*Rheum rhaponeticum*) leaves. Rhubarb (pieplant) stems are a common item of the human diet, and it is frequently grown in home gardens. The leaves are poisonous to humans and livestock and have resulted in severe poisoning and death (Kingsbury 1964).

Ward *et al.* (1979) demonstrated that about 20–30% of the calcium in alfalfa is in the form of oxalate and is unavailable to ruminants. While oxalate toxicity from alfalfa is not likely, the low availability of about one third of its calcium content should be considered in diet formulation.

Dieffenbachia sequine

The popular houseplant *Dieffenbachia sequine* (Fig. 9.5), or dumb cane, contains crystals of calcium oxalate that cause severe irritation of the mouth and throat if the plant is consumed. Dieffenbachia is a native of the West Indies, and according to Woodhouse (1983), it was used in torturing slaves. Woodhouse (1983) gives an interesting account of the occurrence of calcium oxalate in this plant. Found throughout the plant tissues are cells called idioblasts, shaped like double-ended microscopic lemons, which contain raphides or needle-shaped crystals of calcium oxalate. The crystals are slender, sharp, and packed together embedded in a gelatinous substance. If the tip of the

FIG. 9.5. Dumb cane (*Dieffenbachia sequine*) causes intense irritation of the buccal cavity due to calcium oxalate crystals.

idioblast is broken, juice from the plant or saliva enters and causes the gelatinous material to swell, increasing the internal pressure and expelling the needles. As Woodhouse (1983) describes,

> They emerge like bullets one at a time, with sufficient force to cause the cell to recoil like a gun. This can be watched under a microscope, and goes on for many minutes. The result, then, of eating a piece of dieffenbachia leaf, is not a taste sensation, but simple pain.

In addition, dieffenbachia contains a toxic protein which causes swelling of the mucous membranes of the mouth and throat. Thus, if one is unfortunate enough to sample a dieffenbachia leaf, it is unlikely he would wish to repeat the experience.

PHYTATES

Plant seeds contain phytic acid, an organic compound containing six phosphates. This phosphate (phytin phosphate) is largely unavailable to nonruminant animals. In ruminants, bacterial phytases release the bound phosphate. Of further nutritional significance is that phytic acid can chelate with various minerals, producing phytates.

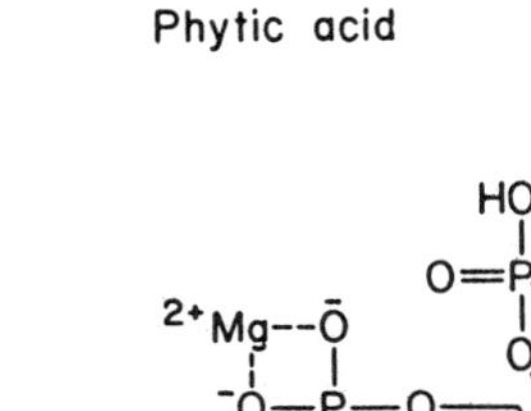

Phytic acid

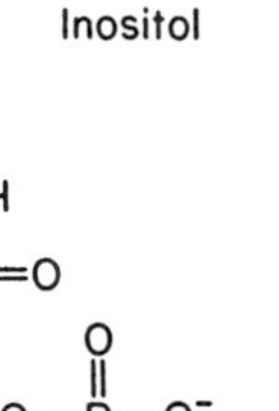

Inositol

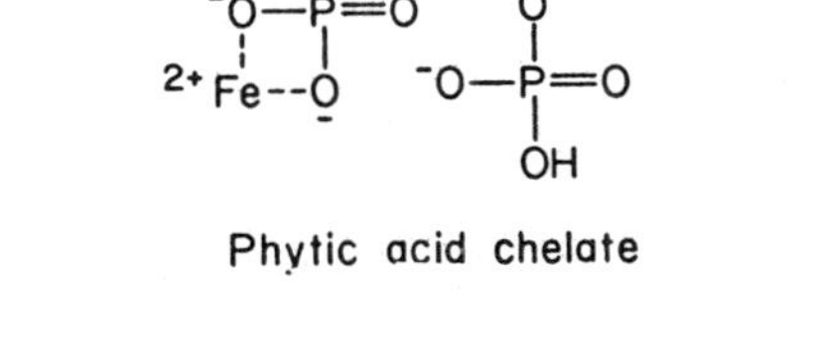

Phytic acid chelate

TABLE 9.1. Phytic Acid Content of Common Feedstuffs[a]

Feedstuff	Phytic acid (%)
Barley	0.97–1.08
Cottonseed meal	2.86–4.29
Oats	0.84–1.01
Rapeseed meal	3–5
Sesame meal	1.44–5.18
Soybean meal	1.00–1.47
Wheat	0.62–1.35

[a] Adapted from Maga (1982).

This is of particular nutritional significance with respect to zinc. When soybean meal was first used in large quantitites as a protein supplement for swine, a high incidence of parakeratosis was observed in swine herds. This is a classic zinc deficiency symptom of poor growth and feed conversion, and dermatitis. The tie-up of zinc can be overcome by supplementing with sufficient zinc to exceed the chelating capacity of the soybean meal. Other elements, such as copper, manganese, iron, calcium, and magnesium, can also be bound in phytate form and be rendered nutritionally unavailable. The decreasing order of stability of phytate–mineral complexes is zinc, copper, cobalt, magnesium, and calcium. Graf and Eaton (1984) found that phytate had no appreciable effect on the absorption of calcium and ferric iron in mice, and suggested that phytate may have a beneficial role in suppressing iron-mediated oxidative processes.

The phytic acid content of common feedstuffs is shown in Table 9.1. In mature cereal grains, 60–80% of the total phosphorus is bound as phytic acid, while 50–60% of the phosphorus in soybeans is as phytic acid. Phytate is not destroyed readily by heat or soaking, but fermentation does liberate phosphates from phytic acid.

Amino Acids

While numerous amino acids form chelates, the only negative nutritional implications are those associated with mimosine, an amino acid from *Leucaena leucocephala* (see Chapter 7). As previously discussed (Chapter 7), there is some evidence that mimosine may induce a zinc deficiency in livestock.

TRIMETHYLAMINE OXIDE AND FORMALDEHYDE

A condition in mink called "cotton fur" in which the underfur is not pigmented has caused large losses in the fur industry. The pelts from cotton fur mink are of no economic value. Researchers in Norway (Helgebostad and Martinsons 1958) and Oregon (Stout et al. 1960A) reported an association of this condition with the feeding of certain types of marine fish to mink. Cooking the fish overcame the problem, suggesting the presence of a heat-labile factor. Stout et al. (1960B) showed that the cotton fur condition was accompanied by a microcytic, hypochromic anemia which would respond to parenteral but not to oral iron administration. This suggested that the heat-labile factor in raw fish rendered dietary iron unavailable. Such a condition was confirmed in a study by Costley (1970), in which the absorption of radioactive iron in mink was reduced when administered with raw Pacific hake (*Merluccius productus*), one of the marine fish that induces cotton fur. These results are shown in Table 9.2. Both ferrous and ferric iron absorption were impaired with the raw fish. Further studies by Costley (1970) identified trimethylamine oxide and formaldehyde as two components of the fish that might account for its inhibition of iron metabolism.

Japanese scientists discovered that certain marine fish, mainly members of the cod family, develop formaldehyde in their tissues when frozen for a period of time (Amano and Yamada 1964). They reported that while trimethylamine oxide is widely distributed in marine fish, some fish, such as hake and cod, contain an enzyme in the pyloric ceca which during cold storage converts the trimethylamine oxide to formaldehyde and dimethylamine (Fig. 9.6). Costley (1970) found that fresh raw hake did not impair iron absorption, while the same raw fish after a period of frozen storage did inhibit iron uptake. This would be explained by the formation of formaldehyde during cold storage. Mink ranchers typically do freeze a supply of fresh fish for feeding, leading to outbreaks of cotton fur syndrome. Trimethylamine oxide inhibited iron absorption when added chemically, but not when as a natural component of raw hake, suggesting that it is bound in a nonreactive form in the fish tissue. On the other hand, formaldehyde did not inhibit iron absorption unless it was as a natural component of Pacific hake, or added to a hake-containing diet.

TABLE 9.2. Effect of Raw and Cooked Pacific Hake on the Absorption of Iron in Mink[a]

Type of iron	% Dose of Fe absorbed	
	Raw hake	Cooked hake
^{59}Ferric chloride	3.4	8.9
^{59}Ferrous citrate	1.3	9.3

[a] From Costley (1970)

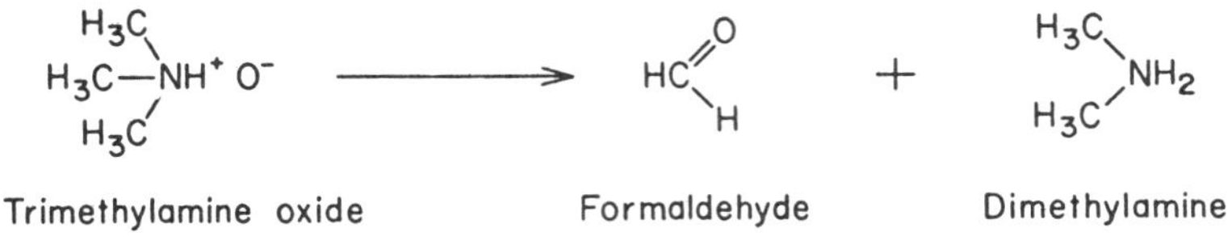

$$\underset{\text{Trimethylamine oxide}}{\overset{H_3C}{\underset{H_3C}{\overset{|}{C}}}-NH^+\ O^-} \longrightarrow \underset{\text{Formaldehyde}}{HC\overset{O}{\underset{H}{\diagdown}}} + \underset{\text{Dimethylamine}}{\overset{H_3C}{\underset{H_3C}{\diagdown}}NH_2}$$

FIG. 9.6. Conversion of trimethylamine oxide to formaldehyde and dimethylamine.

These studies indicate that certain raw marine fish subjected to cold storage contain a factor that inhibits iron absorption in animals consuming the fish, producing iron deficiency symptoms of reduced growth, anemia, and achromatrichea (cotton fur syndrome) (Figs. 9.7 and 9.8). Costley (1970) demonstrated these effects in mink and rats. The evidence suggests that the active compound is formaldehyde, but that it requires the presence of the fish tissue (either raw or cooked) for it to exert an iron-inhibiting effect. Levels of 100–500 ppm formaldehyde in the diet inhibited iron absorption.

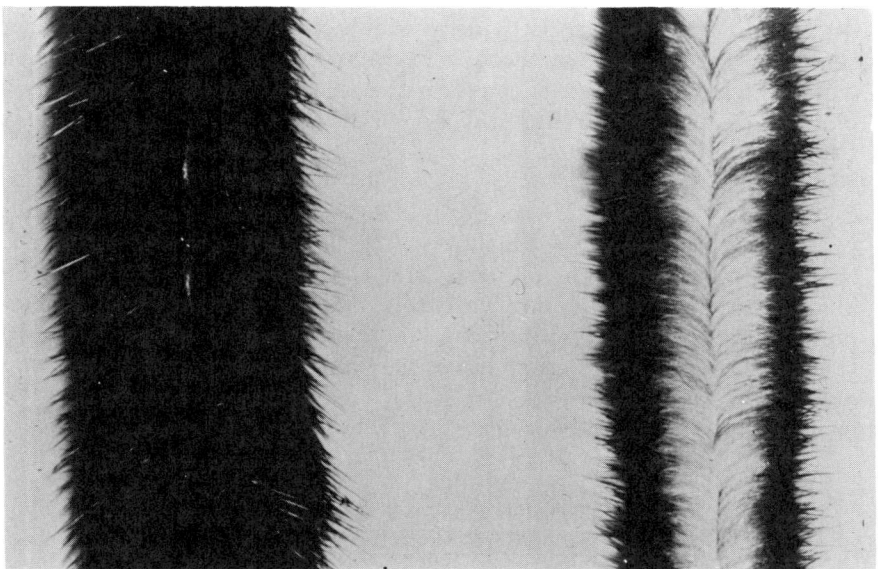

FIG. 9.7. A normal dark mink pelt (left) and a cotton fur pelt (right) in which the underfur shows lack of pigmentation. Cotton fur is caused by an induced iron deficiency due to iron-binding compounds in certain types of raw fish that are often fed to mink.
Courtesy of J. Adair and J. E. Oldfield.

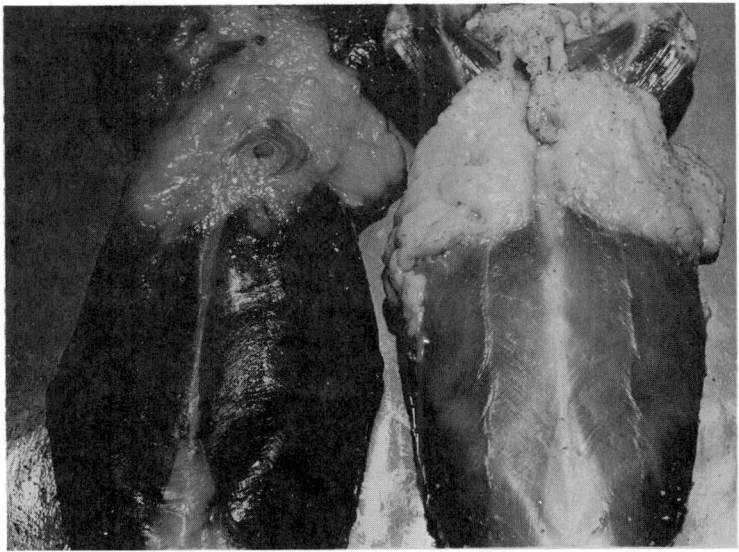

FIG. 9.8. The carcass from a normal mink after pelting (left) contrasted with the anemic carcass of a mink that had cotton fur (right).
Courtesy of J. Adair and J. E. Oldfield.

SILICA UROLITHIASIS (URINARY CALCULI)

Cattle and sheep on the semiarid rangelands of the northern Great Plains of North America are susceptible to the development of urinary calculi composed mainly of silica (Fig. 9.9). This condition is also a problem in parts of Australia and the U.S.S.R. Calculi are present in the urinary tracts of at least 50% of the cattle in North American range herds (Bailey 1981). Calculi form in both sexes but are only a problem in males. Displacement of large calculi from the bladder to the urethra causes an obstruction to the normal flow of urine. Within a few days of obstruction, urine pressure causes the bladder or urethra to rupture, and urine enters the abdominal cavity, causing distension of the abdomen, known as water belly. Obstruction and rupture of the bladder is fatal unless an opening is surgically constructed into the urethra to permit draining of the urine. This opening will usually stay open long enough to restore the animal to a condition fit for slaughter for meat. However, the obstruction generally occurs among range cattle when they are not under close observation, so mortality of affected animals is high. Obstruction does not occur in females because the

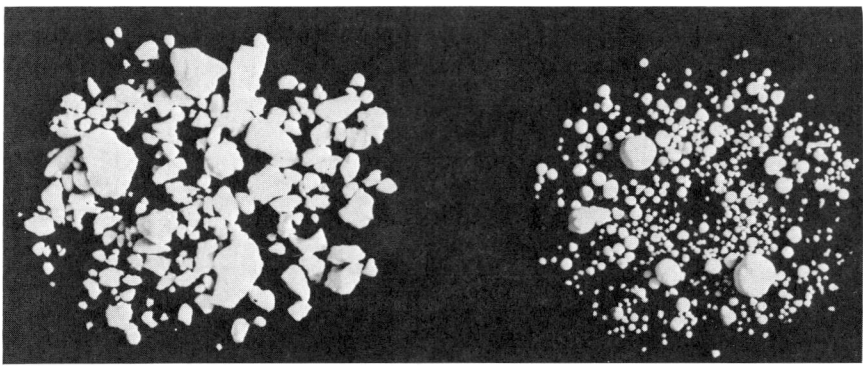

FIG. 9.9. Mixed selection of calculi removed from the kidneys (left) and bladders (right) of steers with subclinical urolithiasis. The largest bladder calculus is 3 mm in diameter.
From Bailey (1981). Courtesy of C. B. M. Bailey.

short, wide urethra and rapid urine flow causes passage of calculi before they reach obstructive size. Obstruction is more common in steers than in bulls, since castration retards urethral development. In North America, obstructive urolithiasis is most common in range calves during the winter, while in Australia, it occurs most frequently in the hottest months. This is probably due to climatic effects on water intake. In North America, cold temperatures reduce water intake, while in Australia, water is likely to be in short supply in the summer. A low output of water in the urine relative to the excretion of silicic acid, giving rise to a high concentration of silicic acid in the urine, is the primary cause of calculi formation (Bailey 1981).

The main component of the calculi is silica (silicon dioxide), comprising about 75% of the total weight. The organic component contains protein and some carbohydrate.

Bailey at Agriculture Canada, Lethbridge (Alberta) has conducted extensive studies on the causes and prevention of urinary calculi. An excellent review article (Bailey 1981) summarizes his work. His studies have shown that the silica content of range grasses is low at the start of the growing season and progressively increases, even after plant growth ceases. Plants exposed to weathering, which removes soluble components, have the highest silica content. The material may contain over 6.5% total silica. This is found as silica associated with plant cell walls, or as free silica, consisting of unpolymerized silicic acid. Calves on native range hay can ingest as much as 500 g of silica per day, while sheep in Australia can consume as much as 40 g/day. A portion of the ingested silica is solubilized in the rumen fluid to keep the rumen saturated with silicic acid. This acid is absorbed and is the form of silicon found in the blood. In range cattle, because of the low level of protein and minerals other than silica in the forage, water resorption in the kidney is high and urine output is low. This concentrates the silicic acid in the urine.

Bailey (1981) has shown that when water intake is below 3.5 kg per kilogram of grass intake, the urine is supersaturated with silicic acid and formation of calculi will occur. Water intakes below this level are frequent in preweaning calves, and calculi are often present by weaning. Mortality is most common from 6 to 8 months postweaning.

Bailey (1981) has shown that the most effective way of reducing the urinary calculi problem is to induce animals to drink more water to reduce the concentration of silicic acid in the urine. This can be accomplished by increasing the salt intake. However, when provided loose or as a lick, salt will not be consumed in sufficient amounts to influence water intake. Ingestion of salt at a rate of about 1 g/kg body weight increases water intake sufficiently to eliminate calculus formation. This can be accomplished by providing a palatable grain supplement, with 15% salt, in a creep feeder for calves beginning at about 4 months of age. This should be fed continuously through the first winter until the calves are about 1 year of age.

NITRATE–NITRITE TOXICITY

Common crop and pasture plants and numerous weeds may accumulate toxic levels of nitrates. In ruminants, nitrate is readily reduced to nitrite, which is absorbed and causes toxicosis. Nitrite ions oxidize ferrous iron in hemoglobin to produce methemoglobin (ferric iron). Methemoglobin cannot react with oxygen, so anoxia occurs. Clinical signs of toxicity may be seen when methemoglobin levels reach 30–40% of total hemoglobin, while death occurs at levels of 80–90% methemoglobin. Clinical signs of nitrate–nitrite toxicity include dyspnea (labored breathing), cyanotic mucous membranes, and evidence of abdominal pain. The outstanding feature is dark brown or chocolate-colored blood due to the methemoglobin. Common sources of toxic levels of nitrate are forage plants and water. Abnormal accumulation of nitrate in plants is provoked by high levels of nitrogen fertilization, drought conditions, and herbicide treatment with phenoxyacetic herbicides such as 2,4-D. Nitrate accumulates in the vegetative tissue, particularly in stalks with less in the leaves. Seeds (grains) do not generally contain toxic nitrate levels. Some plants are more likely than others to accumulate toxic levels of nitrate. Among weeds, pigweed (*Amaranthus* spp.), nightshades, and Johnson grass are known as nitrate accumulators. Sudan grass, oats, rape, wheat, barley, and corn accumulate nitrate. Heavy fertilization of grass pastures, especially with cool, cloudy weather, may result in toxic nitrate levels. Water sources may be contaminated from barn and feedlot runoff, silage juice, or nitrogen fertilizers. Forage levels of 0.5% nitrate and above are potentially dangerous, with acute poisoning likely to occur if the nitrate level exceeds 1%. Levels of 200 ppm nitrate in water are potentially hazardous, with 1500 ppm causing acute toxicity.

When silage is made from high-nitrate forage, toxic silo gas may be produced. Anaerobic silage fermentation results in reduction of nitrate to oxides of nitrogen such as nitrogen dioxide (NO_2) and nitrogen tetraoxide (N_2O_4). These gases are yellowish-brown and may collect in barns in sufficient concentration to kill animals.

Chronic nitrate toxicity has been suggested to cause reduced growth, vitamin A deficiency, abortion, infertility, goiter, and other nonspecific problems. In general, these effects are not documented with experimental evidence. The principal concern with nitrates is acute toxicity.

NITROSAMINES AND NITROSAMIDES

Nitrosamines and related compounds have been of concern as potential carcinogens in the human diet and have caused problems in livestock production as well. They are formed by the reaction of amines with nitrite and thus are likely to be formed when foods or feeds are preserved with nitrates or nitrites. Nitrates are readily reduced to nitrites. The major livestock poisoning problems have involved mink fed fish meal preserved with high levels of sodium nitrite. Massive liver cancer (hepatoma) occurred in mink fed the treated fish meal. Methylamines are abundant in fish and react with nitrate for form dimethylnitrosamine:

$$\begin{array}{c} H_3C \\ \diagdown \\ N-H \\ \diagup \\ H_3C \end{array} \xrightarrow{H^+NO_2^-} \begin{array}{c} H_3C \\ \diagdown \\ N-N=O \\ \diagup \\ H_3C \end{array}$$

Dimethylamine   Dimethylnitrosamine

Nitrosamides, which are less stable than nitrosamines, are also powerful carcinogens and may be formed in foods. An example of nitrosamide formation from methylurea is the following:

$$\underset{\text{Methylurea}}{H_3C-\underset{H}{N}-\overset{\overset{O}{\|}}{C}-NH_2} \xrightarrow{H^+NO_2^-} \underset{\text{Methylnitrosourea}}{H_3C-\underset{\underset{O}{\overset{\|}{N}}}{N}-\overset{\overset{O}{\|}}{C}-NH_2}$$

Nitrosamines are formed by nitrosation of secondary, tertiary, and some primary amines, and quaternary ammonium compounds. As these are widely distributed in foods, widespread occurrence of nitrosamines is possible. Nitrites and nitrates are widely used as preserving agents in the production of cured meats and fish. Nitrite is added to cured meats as a preservative, to form a pink color, and for beneficial effects on flavor. Considerable controversy surrounds its use, and efforts to reduce the employment of nitrates and nitrites in the food industry have been implemented. Nitrosamines have also been found in beer and cheese. In livestock feeds, fish meal preserved with nitrate or nitrite is the major source of nitrosamines. Losses from liver cancer of mink and other fur animals fed fish meal have occurred in Canada, the U.S., and Norway.

REFERENCES

Oxalates

ALLISON, M. J., COOK, H. M., and DAWSON, K. A. 1981. Selection of oxalate-degrading rumen bacteria in continuous cultures. J. Anim. Sci. 53, 810–816.

BLANEY, B. J., GARTNER, R. J. W., and McKENZIE, R. A. 1981A. The effects of oxalate in some tropical grasses on the availability to horses of calcium, phosphorus and magnesium. J. Agric. Sci. 97, 507–514.

BLANEY, B. J., GARTNER, R. J. W., and McKENZIE, R. A. 1981B. The inability of horses to absorb calcium from calcium oxalate. J. Agric. Sci. 97, 639–641.

BLANEY, B. J., GARTNER, R. J. W., and HEAD, T. A. 1982. The effects of oxalate in tropical grasses on calcium, phosphorus and magnesium availability to cattle. J. Agric. Sci. 99, 533–546.

GARTNER, R. J. W., BLANEY, B. J., and McKENZIE, R. A. 1981. Supplements to correct oxalate-induced negative calcium and phosphorus balances in horses fed tropical grass hays. J. Agric. Sci. 97, 581–589.

GROENENDYK, S., and SEAWRIGHT, A. A. 1974. Osteodystrophia fibrosa in horses grazing *Setaria sphacelata*. Aust. Vet. J. 50, 131–132.

HINTZ, H. F., SCHRYVER, H. F., DOTY, J., LAKIN, C., and ZIMMERMAN, R. A. 1984. Oxalic acid content of alfalfa hays and its influence on the availability of calcium, phosphorus and magnesium to ponies. J. Anim. Sci. 58, 939–942.

JAMES, L. F. 1978. Oxalate poisoning in livestock. *In* Effects of Poisonous Plants on Livestock. R. F. Keeler, K. R. Van Kampen, and L. R. James (Editors), pp. 139–145. Academic Press, NY.

JONES, R. J., SEAWRIGHT, A. A., and LITTLE, D. A. 1970. Oxalate poisoning in animals grazing the tropical grass *Setaria sphacelata*. J. Aust. Inst. Agric. Sci. *36,* 41–43.

KINGSBURY, J. M. 1964. Poisonous Plants of the United States and Canada. Prentice-Hall, Englewood Cliffs, NJ.

LINCOLN, S. D., and BLACK, B. 1980. Halogeton poisoning in range cattle. J. Am. Vet. Med. Assoc. *176,* 717–718.

MARSHALL, V. L., BUCK, W. B., and BELL, G. L. 1967. Pigweed (*Amaranthus retroflexus*): An oxalate-containing plant. Am. J. Vet. Res. *28,* 888–889.

McKENZIE, R. A., BLANEY, B. J., and GARTNER, R. J. W. 1981A. The effect of dietary oxalate on calcium, phosphorus and magnesium balances in horses. J. Agric, Sci. *97,* 69–74.

McKENZIE, R. A., GARTNER, R. J. W., BLANEY, B. J., and GLANVILLE, R. J. 1981B. Control of nutritional secondary hyperparathyroidism in grazing horses with calcium and phosphorus supplementation. Aust. Vet. J. *57,* 554–557.

SEAWRIGHT, A. A., GROENENDYK, S., and SILVA, K. I. 1970. An outbreak of oxalate poisoning in cattle grazing *Setaria sphacelata*. Aust. Vet. J. *46,* 293–296.

WARD, G. M., HARBERS, L. H., and BLAHA, J. J. 1979. Calcium-containing crystals in alfalfa: Their fate in cattle. J. Dairy Sci. *62,* 715–722.

WOODHOUSE, E. D. 1983. Talking of dumb cane. Pac. Hortic. *44* (1), 47–48.

Phytates

ERDMAN, J. W. 1979. Oilseed phytates: Nutritional implications. J. Am. Oil Chem. Soc. *56,* 736–741.

GRAF, E., and EATON, J. 1984. Effects of phytate on mineral availability in mice. J. Nutr. *114,* 1192–1198.

MAGA, J. A. 1982. Phytate: Its chemistry, occurrence, food interactions, nutritional significance, and methods of analysis. J. Agric. Food Chem. *30,* 1–9.

Trimethylamine Oxide and Formaldehyde

AMANO, K., and YAMADA, K. 1964. Formaldehyde formation from trimethylamine oxide by the action of pyloric caeca of cod. Bull. Jpn. Soc. Sci. Fish. *30,* 639–645.

COSTLEY, G. E. 1970. Involvement of formaldehyde in depressed iron absorption in mink and rats fed Pacific hake (*Merluccius productus*). Ph.D. Thesis. Oregon State Univ., Corvallis.

ENDER, F., and HELGEBOSTAD, A. 1968. Studies on the anemiogenic properties of trimethylamine oxide, an etiological factor in fish-induced anemia in mink. Acta Vet. Scand. *9,* 174–176.

GJONNES, B., and HELGEBOSTAD, A. 1965. Fish-induced anemia in rats. Acta Vet. Scand. *6,* 239–248.

HELGEBOSTAD, A., and MARTINSONS, E. 1958. Nutritional anemia in mink. Nature (London) *191,* 1660–1661.

STOUT, F. M., OLDFIELD, J. E., and ADAIR, J. 1960A. Nature and cause of the cotton fur abnormality in mink. J. Nutr. *70,* 421–426.

STOUT, F. M., OLDFIELD, J. E., and ADAIR, J. 1960B. Aberrant iron metabolism and the cotton fur abnormality in mink. J. Nutr. *72,* 46–52.

Urinary Calculi

BAILEY, C. B. 1975. Siliceous urinary calculi in bulls, steers, and partial castrates. Can. J. Anim. Sci. *55,* 187–191.
BAILEY, C. B. 1976A. Fate of the silica in prairie hay and alfalfa hay consumed by cattle. Can. J. Anim. Sci. *56,* 213–219.
BAILEY, C. B. 1976B. Relation of water turnover to formation of siliceous calculi in calves given high-salt supplements on range. Can. J. Anim. Sci. *56,* 745–751.
BAILEY, C. B. 1981. Silica metabolism and silica urolithiasis in ruminants: A review. Can. J. Anim. Sci. *61,* 219–235.

Nitrosamines

RAGELIS, E. P., EDWARDS, G. S., COOMBS, J. R., and REISCH, J. W. 1983. Determination of nitrate and volatile nitrosamines in animal diets. J. Agric. Food Chem. *31,* 1026–1029.
SCANLAN, R. A. 1983. Formation and occurrence of nitrosamines in food. Cancer Res. *43* (Suppl.), 2435s–2440s.

10

Tannins and Polyphenolic Compounds

Phenolics are water-soluble compounds in plants, containing one or more phenol groups:

Those with many phenolic hydroxyls are called polyphenolic compounds. Tannins are polyphenolic compounds which precipitate proteins from aqueous solution. The term "tannin" is derived from the use of plant extracts (e.g., oak leaves) to tan leather. Tannins can transform animal skins and hides that are soaked in tannin extracts into leather that is much more resistant to bacterial action, heat, and abrasion than the original material. Hydrogen bonds are formed between the phenolic hydroxyls and the peptide groups of collagen fibrils to form cross-links between adjacent protein chains. Oxidation of phenolic groups in the tannins to quinones

Phenol ⇌ Quinone

may give rise to covalent bonds with the epsilon amino groups of lysine and arginine, further increasing the durability of the leather.

Tannins have the ability to precipitate proteins because they contain a number of functional groups which form strong complexes with protein molecules, producing a large cross-linked protein–tannin complex. For this, the presence of o-dihydroxyphenols is required, forming hydrogen bonds with the protein, with hydrophobic binding also contributing to the stability of the protein–tannin complex.

This property of tannins to form bonds with soluble proteins accounts for their adverse effects on animals. They can bind with digestive enzymes to inhibit their activity and can bind with dietary proteins to prevent their digestion. They may bind with proteins in the intestinal mucosa and affect nutrient absorption. The interaction of tannins with the salivary proteins and glycoproteins in the mouth causes the sensation of astringency. Astringency is the mouth-puckering sensation experienced when one eats a green persimmon or a walnut, which, like most unripe fruits, contain tannins. The loss of astringency upon ripening is due to polymerization of the tannins, reducing their ability to react with proteins. Thus, tannins have pronounced effects on feed intake and palatability. Swain (1979) considers that tannins are the most significant plant components controlling the acceptability of plants to herbivores. Tannins are believed to have a major role in protection of plants from vertebrate herbivore, insect, bacterial, and viral attack by deterring herbivores with their astringent effects and by inactivating protein enzymes of insects and bacteria. The natural durability of many wood species may be due to the toxicity to microorganisms of phenolics deposited in the process of heartwood formation (Jung and Fahey 1983B).

Tannins and other phenolic compounds in plants are synthesized by two metabolic pathways, the shikimic acid pathway (Fig. 10.1) and the acetate–malonate pathway (Jung and Fahey 1983B). There are several categories of tannins, with the hydrolyzable and condensed tannins being the two main groups of nutritional interest. The condensed tannins, also known as proanthocyanidins, are the most widely distributed in plants. They are not readily hydrolyzed and are often of complex structure, formed from the condensation of flavanols such as catechin and epicatechin (Fig. 10.2). The tannins in sorghum grain and fava beans are of the condensed type.

Hydrolyzable tannins consist of esters of glucose with gallic acid or related compounds (Fig. 10.3). They can be readily hydrolyzed by hot mineral acids or esterases to yield the sugar core and the constituent acids. An example of a group of hydrolyzable tannins with detrimental effects on livestock are the tannins in oak.

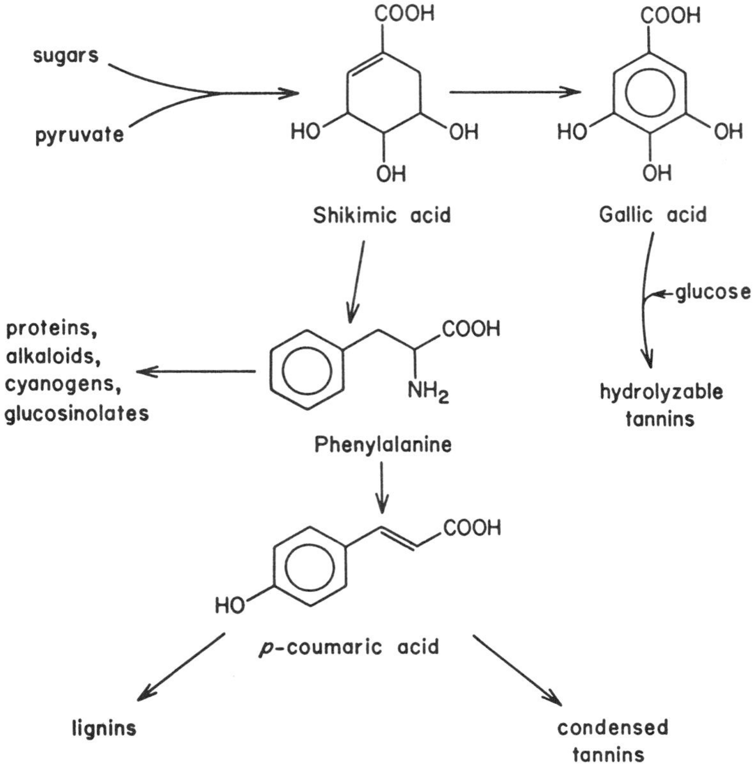

FIG. 10.1. Biosynthesis of tannins and lignins by plants.

There are other tannins which do not fit into either the condensed or hydrolyzable groups. An example is chlorogenic acid, found in sunflower meal and alfalfa.

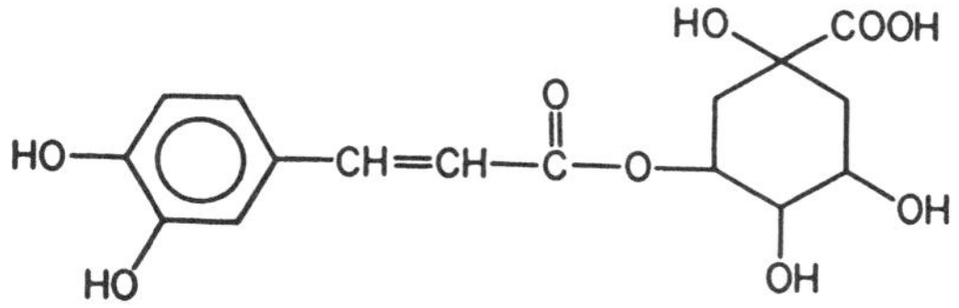

Chlorogenic acid

A number of phenolic acids other than gallic acid are widely distributed in plants. These include *p*-hydroxybenzoic acid, vanillic acid, caffeic acid, quinic acid, and sinapic acid. Some of these are shown in Fig. 10.4. These phenolic acids are widely distributed in oil seed meals, such as sunflower, soybean, cottonseed, peanut, rapeseed, crambe, and sesame meals. They have a variety of antinutritional properties. The

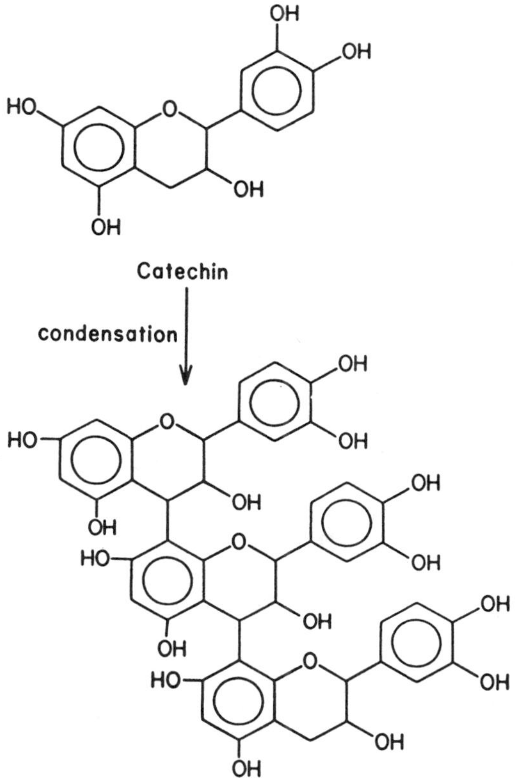

FIG. 10.2. Representative example of condensed tannin found in sorghum grain.
From Gupta and Haslam (1980).

phenolic groups provide an abundance of sites for hydrogen bonding with proteins. The nutritional properties of phenolic acids have been reviewed by Jung and Fahey (1983B).

Other phenolics important in livestock feeds include gossypol in cottonseed meal, hypericin in St.-John's-wort, and fagopyrin in buckwheat. The latter two are phenolic derivatives of naphthodianthrone.

Tannins are generally excreted by animals as conjugates of sulfuric and glucuronic acids. Methylation and dehydroxylation reactions also occur. Methylation increases the requirement for methyl donors; therefore, methionine and choline supplements may ameliorate some of the growth depression caused by tannins.

Ruminants excrete much larger quantities of aromatic acids, such as benzoic acid, in their urine than do nonruminants, particularly when they are fed a high-roughage diet (Martin 1969). The source of much of

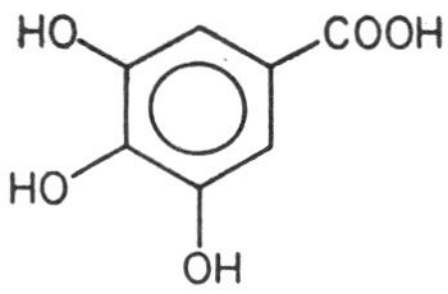

Gallic acid

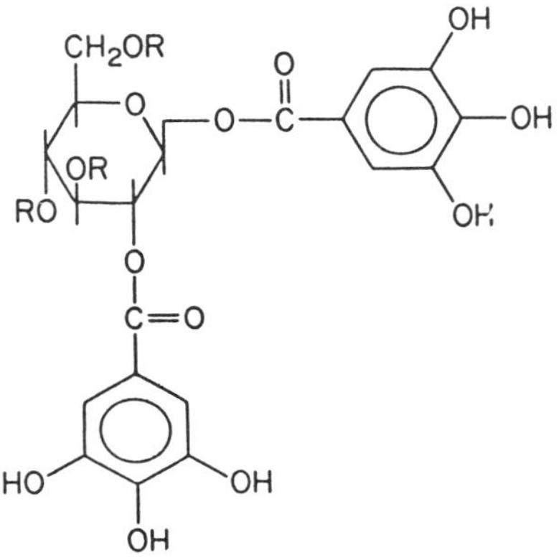

Tannic acid

FIG. 10.3. Structures of gallic acid and tannic acid. R = gallic acid, drawn in for C_1 and C_2 only.

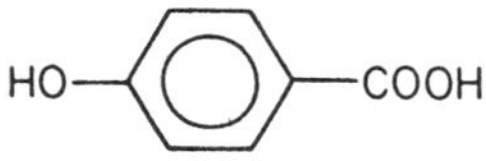

p-hydroxybenzoic acid

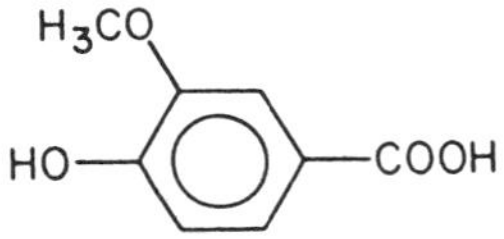

Vanillic acid

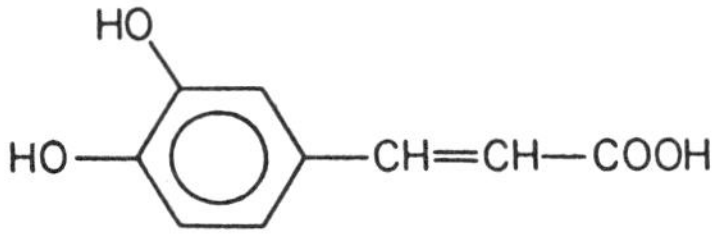

Caffeic acid

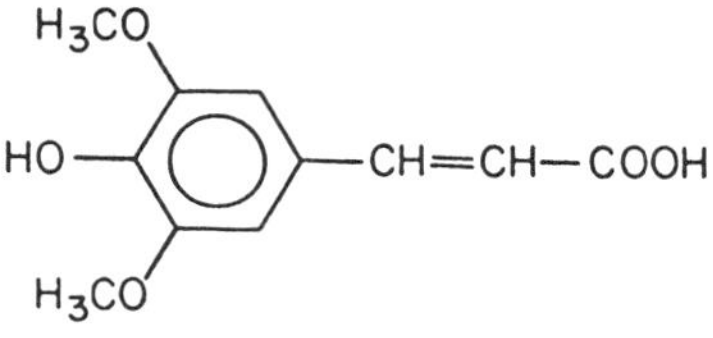

Sinapic acid

FIG. 10.4. Examples of some phenolic acids.

these aromatic acids is phenolic compounds in forages, such as the phenolic acids. They may be metabolized in the liver and conjugated. For example, benzoic acid is conjugated with glycine to produce hippuric acid, which is excreted in the urine:

$$\text{Benzoic acid} + \text{Glycine} \longrightarrow \text{Hippuric acid}$$

Hippuric acid (Greek *hippo* = horse) is found in large amounts in the urine of herbivores. Cinnamic acid in forages is a principal precursor of urinary aromatic compounds (Martin 1982A). Cinnamic acid, 3-phenylpropionic acid, and benzoic acid are the main conjugated aromatics in the urine of ruminants (Martin 1982B).

Morton (1980) has described a number of interesting correlations between the presence of tannins in foods and the incidence of esophageal cancer in humans. The consumption of high-tannin sorghum grain in several parts of the world, including the Netherlands Antilles in the West Indies and South Africa where Africans use "Kaffir corn," a bird-resistant brown sorghum, has been linked by Morton (1980) to a high incidence of esophageal cancer in these populations. Tannins in tea have also been implicated. The addition of cream or milk to tea eliminates this effect, presumably because of a protein–tannin complex formation.

SORGHUM TANNINS

Sorghum and millets are major food crops in the semiarid tropics, an ecological zone encircling the earth from China, India, most of Africa, and parts of the southern U.S. In the developed countries, about 96% of the total sorghum and millet grown is used for animal feed, whereas in the developing countries, only about 8% of these crops is used for livestock, with the rest used directly as human food. About 90% of the rural population of the Sahelian zone in Africa depends on these crops as their source of food energy. Sorghum is grown in these areas because it can tolerate relatively harsh ecological and growing conditions, including drought.

Sorghum seed grows in an open panicle that is very vulnerable to attack by wild birds (Fig. 10.5). In Africa, the red-billed weaver bird causes great destruction. It is considered the most numerous and destructive bird pest in the world, comparable to locusts in the amount of damage it causes. The weaver bird population base in Africa is estimated at 100 billion. In North America, blackbirds, starlings, and sparrows are a problem, while in Asia, sparrows, parakeets, crows, and mynas cause destruction of sorghum. In Latin America, doves, blackbirds, sparrows, parakeets, and parrots are the major problem species. Several bird-resistant lines of sorghums have been developed which have a high tannin content. The tannins are located in the pericarp and testa, near the surface of the seed (Fig. 10.6).

Four classes of sorghum grain—yellow, white, brown, and mixed—

FIG. 10.5. Seed heads of milo showing their susceptibility to bird damage.
Courtesy of Robert Dennis and W. H. Hale.

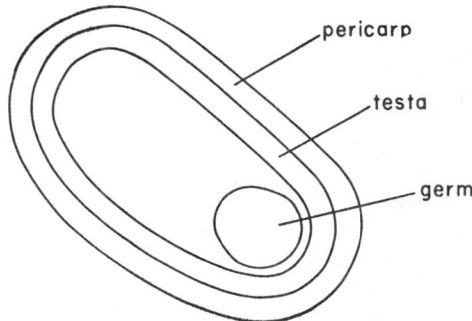

FIG. 10.6. Location of polyphenolic tannins in sorghum seeds.

are recognized in commerce. Yellow sorghum can contain grain of any color pericarp as long as it does not contain a pigmented testa, while white sorghum contains seeds without a testa. The brown sorghums have a pigmented testa, and these are bird-resistant. They also have a lower nutritional value than the others, so receive a lower price.

The yellow sorghum of U. S. origin has a very low level of condensed tannins and a nutritional value nearly equivalent to that of corn. Some countries produce high-tannin bird-resistant sorghum. The marketing of this grain in international channels has led to discrimination against sorghum in general. In areas where predation is not a major problem, low-tannin grain should be grown. In other areas, the crop cannot be successfully raised to maturity unless high-tannin bird-resistant varieties are used.

The sorghum tannins have an adverse effect on palatability, and protein and dry matter digestibility. They may interfere with the action of trypsin and amylase in the digestive tract or may bind to dietary protein to form an indigestible complex. Treatment of grain with dilute alkali improves its feeding value. Polyethylene glycol, which forms complexes with tannins, also appears to be a simple and cost-effective treatment for improving the nutritional value of high-tannin sorghum (Hewitt and Ford 1982).

Sorghum tannins seem to be of greatest nutritional significance in monogastric animals, particularly poultry. High-tannin sorghum may cause reduced growth and feed efficiency, leg abnormalities, and decreased egg production. Extra dietary protein partially overcomes these effects. In ruminants, there is no clear evidence that sorghum tannins are harmful. The association of protein with tannins reduces ruminal degradation of proteins, increasing the potential for high-quality protein to "bypass" the rumen. Tannin–protein complexes may be stable in the rumen (pH about 6.5) and dissociate in the small intestine, releasing the protein. This also raises the possibility of the liberated tannin causing damage to the intestinal tract.

The feeding of high-tannin sorghums to broiler chicks causes a leg abnormality characterized by an outward bowing of the hock joint (Elkin et al. 1978). It was suggested that absorbed tannins might increase the amount of cross-linking in collagen, thereby altering the organic matrix of the bone. Sell and Rogler (1983) demonstrated that high-tannin sorghums cause increased activities of liver UDPglucuronyltransferase, an enzyme involved in detoxification of phenolics. This enzyme induction indicates that phenolics are absorbed in the chick, and therefore would be available for reacting with protein (e.g., collagen) in vivo.

Barley contains anthocyanins in the green tissue and proanthocyanidins and catechins in the aleurone cells of the mature seed (Newman et al. 1984). These pigments are phenolic compounds which might reduce protein and amino acid availability in the gut. Proanthocyanidin-free barley cultivars have been developed and give superior chick growth than barley of conventional varieties (Newman et al. 1984). These workers suggest that phenolic compounds in barley are detrimental to nutritional quality and that reduction of these antinutritional constituents by plant breeding could improve the feeding value of this crop.

PHENOLICS IN PROTEIN SUPPLEMENTS

Chlorogenic acid is present in sunflower meal at levels of 1% or more of a total of 3–3.5% phenolic compounds. It inhibits the activity of digestive enzymes such as proteases (trypsin, chymotrypsin), amylase, and lipase. Supplementation with methyl donors, such as methionine and choline, tends to counteract the effects of chlorogenic acid. Chlorogenic acid and other phenolic acids are substrates for the ubiquitous copper-containing plant enzyme polyphenol oxidase, which converts these compounds to o-quinones. The o-quinones react nonenzymatically to polymerize or form covalent bonds with amino, thiol, and methylene groups. The epsilon amino of lysine and the thioether group of methionine can be attached in this manner to render these amino acids nutritionally unavailable, as shown in Fig. 10.7.

Rapeseed, mustard, and crambe meals contain sinapine, the choline ester of sinapic acid (Fig. 10.8). Rapeseed meal contains about 1% sinapine, about twice the level in crambe meal. Sinapine in rapeseed meal has an interesting effect in poultry. Hens that lay brown eggs,

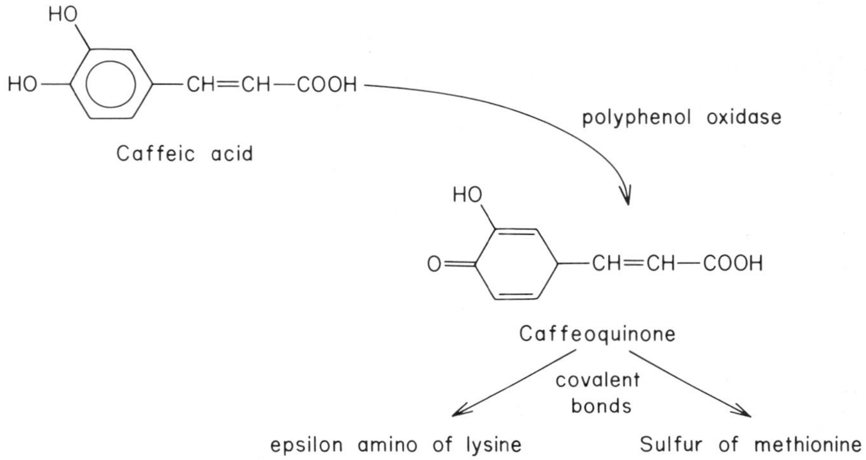

10. TANNINS AND POLYPHENOLIC COMPOUNDS

FIG. 10.7. Action of polyphenol oxidase in reactions of tannins with amino acids.

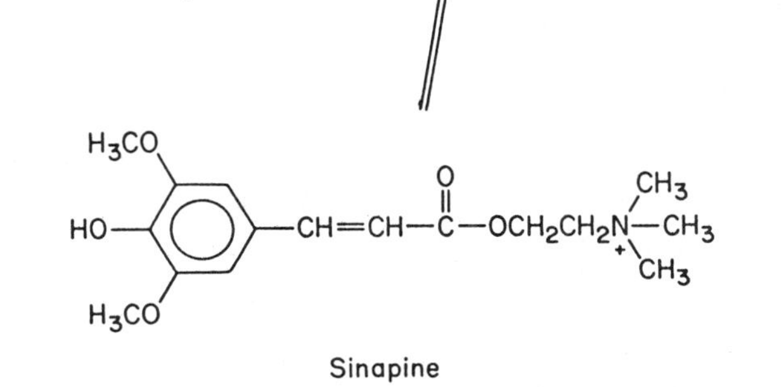

FIG. 10.8. Formation of sinapine from choline and sinapic acid.

particularly those of the Rhode Island Red breed, may produce eggs with a fishy odor when fed rapeseed meal. Bacteria in the ceca convert choline released from sinapine to trimethyamine which is then absorbed. Some birds are capable of oxidizing the absorbed trimethylamine to trimethyamine oxide, while others (that coincidentally lay brown eggs) cannot. They therefore deposit the amine in the eggs, giving them a fishy flavor (Goh et al. 1979).

Peanut skins have been evaluated as a protein source for swine (Hale and McCormick 1981). They contain about 17% protein, 26% ether extract, and 16–18% tannin. Although the proximate analysis indicates that they might be a satisfactory feed ingredient for swine, they gave unacceptable performance even when fed at low levels, causing reduced gains, feed conversion, and nitrogen digestibility. These effects were probably due to the high tannin content.

Fava beans (*Vicia faba*) contain condensed tannins (Marquardt et al. 1977). These have adverse effects on nutrient digestibility (Marquardt et al. 1978) and growth of chickens (Marquardt and Ward 1979). Heat treatment results in destruction of the condensed tannins and improved performance. Tannin-free varieties of fava beans have been developed (Marquardt and Ward 1979). Field beans (*Phaseolus* spp.) may contain tannins; these are particularly associated with beans that have colored flowers (Hewitt and Ford 1982).

Phenolic compounds such as chlorogenic acid and caffeic acid are potent antioxidants (Silvia Taga et al. 1984) and might have potential as food antioxidants. The synthetic antioxidants BHA and BHT are phenolics.

OAK POISONING

Poisoning of cattle from consumption of oak buds, leaves, twigs, and acorns occurs in many parts of the U.S. and Europe. Oak poisoning is generally seasonal, caused by ingestion of buds and leaves in the spring and acorns in the fall. Tannins, such as tannic acid and its phenolic acid constituent, gallic acid, are the causative agents. The tannin content of leaves and acorns tends to be highest in the immature stages. An unusually heavy crop of acorns, such as occurred in Ohio (Sandusky et al. 1977) and England (Dixon et al. 1979) in 1976, often results in a large number of cases of oak poisoning.

The initial signs of oak poisoning include anorexia, depression, clear watery nasal discharge, rumen stasis, excessive thirst, and frequent

FIG. 10.9. Typical signs of oak poisoning, with depression and gaunt appearance.
Courtesy of M. E. Fowler.

urination (Fig. 10.9). Initial constipation is followed by the excretion of dark, thin, mucoid, and often bloody feces.

The principal lesions are gastritis and nephritis. The abomasum and small intestine are often inflamed and hemorrhagic. The major lesion of oak poisoning is necrosis of the renal tubules. Affected kidneys are pale and swollen. There is impaired kidney function, with elevated blood urea nitrogen. Supplemental feeding with a mixture containing calcium hydroxide has been recommended as a preventative measure. A mixture of 30% alfalfa, 54% cottonseed meal, 6% vegetable oil, and 10% calcium hydroxide has been recommended by Dollahite *et al.* (1966).

Goats can utilize oak browse productively. Davis *et al.* (1975) observed that the total grazing capacity of a gambel oak (*Quercus gambelii*) range was almost doubled by including goats in a mixed grazing scheme with cattle. Goats are also used for brush control. Feeding high levels of immature gambel oak to goats did not produce any toxicological reactions in a Utah study (Nastis and Malechek 1981). These workers concluded that mature oak browse can contribute effectively to the nutrition of growing and lactating goats. Immature oak leaves, while not overtly toxic, had a low metabolizable energy content and were of low palatability.

TANNINS IN FORAGE LEGUMES

Legumes that do not cause bloat, such as sainfoin (*Onobrychis viciifolia*), crown vetch (*Coronilla coronata*), and bird's-foot trefoil (*Lotus corniculatus*) (Fig. 10.10), contain condensed tannins that in the rumen precipitate the soluble leaf proteins believed to cause bloat. Tannins do not occur in the leaves of legumes such as alfalfa that induce bloat. Tropical legumes do not cause bloat; this is attributed to their tannin content. Jones and Mangan (1977) found that tannin from forage legumes is firmly bound to soluble proteins in the rumen. They found a complete absence of soluble protein in rumen contents of cattle grazing pasture species that contain condensed tannins. Dissociation of the protein–tannin complex in the intestine would render the high-quality leaf protein available to the host, thus providing a means of rumen bypass of forage protein as well as avoiding bloat. Aqueous extracts of nonbloating legumes do not foam in vitro, but when polyvinylpyrrolidone (which complexes with tannins) is added, the ex-

FIG. 10.10. Bird's-foot trefoil (*Lotus corniculatus*), a nonbloating forage legume, contains condensed tannins that react with soluble proteins in the rumen.

tracts do produce large volumes of foam, further suggesting that forage tannins function in the rumen as antifoaming agents.

The tannins act by denaturing in the rumen the soluble leaf proteins that prevent the eructation of gas by forming a stable foam. Two approaches to the solution of the pasture bloat problem are evident. One is to replace bloat-inducing legumes with nonbloating species with leaf tannins. The second is to introduce tannins by plant breeding into the bloat-producing species. Marshall et al. (1979) concluded that the latter approach might be feasible by using artificial mutagens to induce tannin-containing variants.

Lespedeza spp. are important forages in the southern U.S. Sericea lespedeza (*Lespedeza cuneata*) contains condensed tannins that cause reduced digestibility of the forage because of inhibition of cellulolytic enzymes. The threshold tannin concentration at which a reduction in digestibility occurs is about 8–9% of the dry weight. Low palatability of sericea lespedeza is associated with its high tannin content.

In the preparation of leaf protein concentrates (LPC) from forages, tannins may reduce the extractability of protein from some plant species through formation of phenolic–protein complexes. Phenolics may also reduce the nutritive value of the extracted protein. Cheeke et al. (1980) reported very poor rat growth with LPC prepared from *Leucaena leucocephala* and *Manihot esculenta* (cassava), which was attributed to protein–phenolic reactions. Enzymatic activity in fresh leaves, such as polyphenol oxidase, may intensify the problem by producing highly reactive o-quinones which bind with thiol and amino groups of proteins (Horigome and Uchida 1983). The effects of phenolics on protein extractability, nutritive value of LPC, and interactions with lysine and methionine have been reviewed by Pirie (1978) and Pierpoint (1983). The presence of polyphenolic compounds can affect the quality of LPC by reducing digestibility, causing off-color, astringency, and bitterness, by altering storage life and stability, and by changing the behavior of LPC in food preparation systems (Barbeau and Kinsella 1983).

GOSSYPOL

Gossypol is a phenolic compound found in pigment glands of cottonseed (*Gossypium* spp.). It was isolated and named in 1899; the name is derived from *Gossypium* phenol. While gossypol is the major phenolic in cottonseed, there are at least 15 closely related compounds. The phenolic groups on the molecule are chemically reactive, so it reacts readily with various substances, including minerals and amino acids. The structure of gossypol is as follows:

$$\text{gossypol structure}$$

The pigment glands have a cell wall composed of cellulose, pectin, hemicellulose, and uronic acid, and are resistant to rupture. During the extraction of cottonseed oil, processors attempt to leave most of the gossypol in the meal, as it adversely affects the color and quality of the oil. Heat treatment of the seed causes gossypol to bind to protein and remain in the meal.

In animal feeding, the main concern from a toxicological point of view is with free gossypol. The bound gossypol is physiologically inactive, but because it is bound to protein and particularly lysine, it reduces the biological value of the protein. Some of the physiological effects of free gossypol are the following:

1. It causes olive green yolks in hen's eggs (Fig. 10.11). This is the most sensitive indicator of physiological activity. The cyclopropenoid

FIG. 10.11. Eggs from a control bird (left) and a hen given gossypol (right), showing the pigmentation of the yolk.

fatty acids in cottonseed meal, as previously discussed, cause pink albumins. They enhance the yolk discoloration by accelerating the increase in yolk pH during cold storage of eggs. The green pigmentation is due to a reaction between gossypol and yolk iron.

2. High dietary levels of gossypol in layer diets decrease the hatchability of eggs.

3. Gossypol causes a depressed appetite and loss of weight.

4. Liver and lung lesions, including edema and congestion of the lungs, and congestion and hemorrhages of the liver are caused by gossypol.

5. Ascites and tissue edema are observed.

6. The most common effect is cardiac irregularity; death is generally ascribed to cardiac failure.

7. "Thumps," a labored breathing, occurs in swine, with pulmonary edema a common feature in most animals.

8. Anemia may occur, with reduced erythrocytes, hemoglobin, and packed cell volume, due to a complexing of iron.

9. Male infertility is induced. This was discovered in China, where a low birth rate and a high incidence of heart problems were observed in a particular area. This was traced to the presence of gossypol in the cooking oil used in that region. In China, gossypol is being evaluated as a male birth control pill. At low levels, it blocks spermatogenesis and has minimal side effects. Gossypol also reduces sperm motility.

The biological effects are cumulative. For example, pigs fed on a cottonseed meal diet for a year may appear normal, but then abruptly begin to gasp for breath and die with severe anemia. For nonruminants, the free gossypol level in the diet should not exceed 0.01%, or about 9% cottonseed meal. The addition of iron salts, such as ferrous sulfate, inactivates gossypol, and can be used in a 1:1 ratio of iron to free gossypol. High dietary protein levels have a protective effect by reacting with free gossypol.

Glandless cotton, lacking the pigment glands that contain gossypol, has been developed, but it has not had a major impact. Gossypol appears to have protective effects against insects and diseases, so the glandless cotton is more susceptible to pests and requires use of more pesticides than regular cotton. Cottonseed meal is a by-product of the third order; cotton is grown primarily for its fiber, with the oil as a valuable by-product. Since the meal is the least valuable component, glandless cotton is not likely to be produced if it increases cultural problems in growing the crop.

Gossypol does not normally affect ruminant animals, presumably because of inactivation in the rumen where binding of gossypol to protein occurs. However, if high levels of whole cottonseed are fed for a prolonged period, cardiac damage can occur.

The presence of gossypol in cottonseed meal reduces the protein quality of the product. The bound gossypol is that which has reacted with amino acids, mainly lysine. The gossypol–protein interaction occurs through reaction of the formyl groups of gossypol with the epsilon amino groups of lysine and arginine. Gossypol may also react with the thiol group of cysteine. The products undergo various reactions and ultimately form insoluble, indigestible polymerization products. Both the phenolic and carbonyl groups are chemically very reactive. The phenolic groups can form hydrogen bonds or be oxidized to quinones that react with proteins.

HYPERICIN

The ingestion of St.-John's-wort, *Hypericum perforatum,* by grazing animals may cause the development of photosensitization reactions. There are reports of this condition from Europe, Australia, New Zealand, South America, and the U.S. St.-John's-wort is an erect perennial herb, 1–3 ft tall, with yellow flowers. It bears a superficial resemblance (mainly the yellow flowers) to tansy ragwort (*Senecio jacobaea*), with which it is sometimes confused by casual observers (Fig. 10.12). The leaves of St.-John's-wort contain numerous small dots, just visible to the naked eye, that are translucent pigment granules. The photodynamic agent is a derivative of naphthodianthrone called hypericin. Because of its numerous phenolic groups, it can be classed as a polyphenolic compound. The structure of hypericin is as follows:

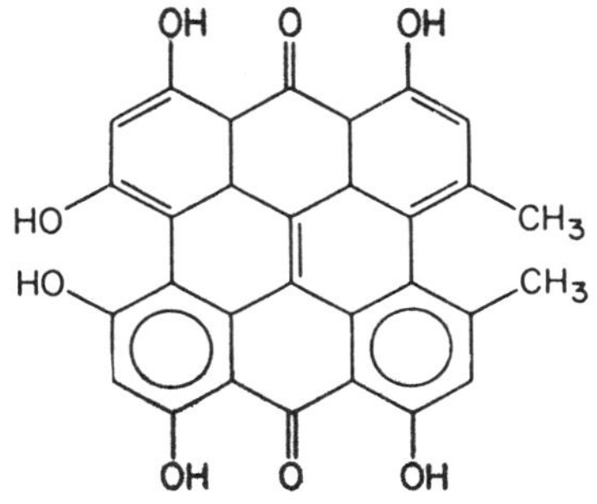

Hypericin is a primary photosensitizing agent. This means that it is the photodynamic pigment that reacts with light at the surface of the skin (Fig. 10.13). Secondary photosensitizers are compounds that induce mild liver damage and prevent the normal excretion of phylloery-

10. TANNINS AND POLYPHENOLIC COMPOUNDS 349

FIG. 10.12. A plant of St.-John's-wort (*Hypericum perforatum*) in bloom.

thrin, a degradation product of chlorophyll. Phylloerythrin is a photodynamic agent. Photodynamic agents react with ultraviolet light at the skin's surface and fluoresce when hit by a photon of light, thereby creating an unstable, energized molecule. Transfer of energy occurs when the unstable high-energy molecules collide with other cell constituents and create free radicals. These cause damage to cell membranes and lead to a breakdown of cellular structure.

Hypericum poisoning has occurred in all types of grazing livestock. An intake of about 3 g of dried hypericum per kilogram body weight will produce signs of photosensitization. Animals display photophobia with reluctance to be in sunlight. Erythema (reddening of the skin) occurs in white or lightly pigmented skin as does serous oozing and skin necrosis. Animals may die of starvation or of infection of the affected areas.

Black-skinned animals are rarely if ever affected. Kingsbury (1964) cites a case of *Hypericum* poisoning in Holstein cattle in which the entire herd was found "with the white skin hanging in rags and the

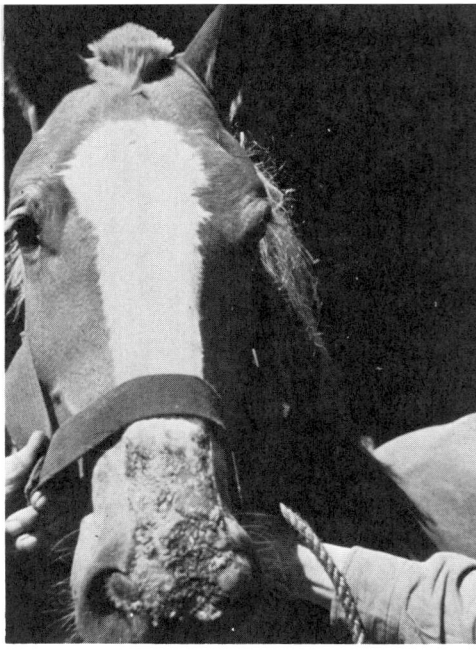

FIG. 10.13. Photosensitization exhibited by a horse due to ingestion of St.-John's-wort (*Hypericum perforatum*).

dark skin soft and supple as a glove." Intense itching may occur, and affected individuals may charge about as if demented. Newly shorn sheep are especially vulnerable.

Hypericum perforatum was once a major weed in California, Oregon, and Washington, but the beetle *Chrysolina quadrigemina* was introduced to control it. This was touted as a major success for biological control. The large stands of St.-John's-wort were virtually eliminated, as was the beetle with the loss of its food source. The plant has proliferated again and is now widespread, but does not seem to be causing significant livestock losses. The beetle uses hypericin as a feeding stimulant, and accumulates it, which apparently makes the insect distasteful to predators (Towers 1980).

"PROTECTED" OR "BYPASS" PROTEIN

Tannic acid has been used to treat high-quality proteins (e.g., casein) to render them indigestible in the rumen. The tannin–protein complex passes through (bypasses) the rumen, and in the intestine the complex dissociates, releasing the protein. This provides a means of supplementing the relatively low-quality bacterial protein with a higher quality dietary protein that does not get digested in the rumen. Formaldehyde has also been used to produce protected protein.

PHENOLIC COMPOUNDS AS REPRODUCTION INHIBITORS

Substances in plants may function in regulating reproduction in small herbivorous mammals, such as the meadow vole and the lemming, by providing dietary cues as to when food supplies are adequate for reproduction. Berger et al. (1977) found that cinnamic acids and their related vinyl phenols inhibited reproductive function in the meadow vole (*Microtus montanus*). When fed these compounds, the animals exhibited decreased uterine weight, inhibition of follicular development, and a cessation of breeding activity. The compounds were at highest concentration in native plants at the end of the vegetative growing season, possibly providing a dietary cue to turn off reproduction. Later studies by these workers (Sanders et al. 1981; Berger et al. 1981) demonstrated that a plant-derived cyclic carbamate, 6-methoxybenzoxazolinone (6-MBOA), stimulates reproductive activity in meadow voles. They suggest that 6-MBOA may trigger reproductive activity in the spring when food supplies are abundant. The significance, if any, of these compounds in livestock production is not known. However, if late-season forage contains phenolics that inhibit reproductive processes and spring forage has reproduction-stimulating compounds, potential implications for livestock production are obvious. 6-MBOA has a structural resemblance to melatonin, a hormone whose activity is related to light exposure.

Melatonin   6-MBOA

Melatonin is produced in the pineal gland and may have a role in the regulation of seasonal breeding. Thus, 6-MBOA may exert its effects through an interaction with melatonin.

BLACK WALNUT (JUGLONE) TOXICITY

Black walnuts (*Juglans nigra*) and other members of the walnut family (Persian or English walnuts, butternuts, hickories, and pecans) contain a phenolic derivative of naphthoquinone, juglone (5-hydroxy-1,4-naphthoquinone). It is an allelopathic substance which inhibits the growth of other plants. Tomatoes, potatoes, and other vegetables, as well as many other trees and shrubs are inhibited in their growth by juglone released into the soil from black walnut roots. As a result, competition to walnuts from other plants is reduced.

Toxicity of black walnuts to livestock, particularly horses, has been reported. True *et al.* (1978) reported that racehorses bedded down with black walnut chips or sawdust developed laminitis. MacDaniels (1983) discussed the toxic effect on herds of racehorses in Wisconsin and Michigan where severe allergenic reactions to the shedding of walnut pollen occurred as well as to the shedding of leaves in the autumn. In these instances, the problems were sufficiently serious that all the walnut trees were cut down, the stumps bulldozed, and the soil removed from the paddocks and stables. Ralston and Rich (1983) reported an incident in Colorado involving horse stalls bedded with pine shavings containing 20% black walnut shavings as a contaminant. Within 12 hr of exposure, the horses exhibited signs of toxicosis, including reluctance to move, early acute laminitis, slight edema of the limbs, and depression. There is some question as to whether juglone is the sole or major toxic factor, as administration of juglone to ponies and horses caused only minor pathology (True and Lowe 1980). Precursors of juglone may be implicated (MacDaniels 1983).

It is recommended that walnuts, particularly black walnuts, not be planted in horse pastures, and that black walnut chips and sawdust not be used for bedding.

The structure of juglone is as follows:

Juglone

RESORCINOLS

Resorcinol is *m*-dihydroxybenzene. Derivatives of resorcinol, the 5-alkyl resorcinols, occur in triticale and may account for some of the growth-depressing effects of this grain (Radcliffe *et al.* 1981).

$$\text{CH}_3(\text{CH}_2)_n\text{—}\underset{\text{OH}}{\underset{|}{\text{C}_6\text{H}_3}}\text{—OH}$$

5-alkyl resorcinol

According to Farrell *et al.* (1983), the resorcinols are of little if any significance in this regard, and adverse effects of triticale on performance of poultry and swine are likely due to trypsin inhibitors. However, Radcliff *et al.* (1983) found that differences in swine growth performance and efficiency of nutrient utilization were not associated with the trypsin inhibitor of triticale. These authors suggest that neither 5-alkyl resorcinols nor trypsin inhibitors are responsible for growth-inhibitory effects of triticale.

REFERENCES

Tannins (General)

FAHEY, G. C., JR., AL-HAYDARI, S. Y., HINDS, F. C., and SHORT, D. E. 1980. Phenolic compounds in roughages and their fate in the digestive system of sheep. J. Anim. Sci. *50,* 1165–1172.

GLICK, Z. 1981. Modes of action of gallic acid in suppressing food intake of rats. J. Nutr. *111,* 1910–1916.

GUPTA, R. K., and HASLAM, E. 1980. Vegetable tannins—structure and biosynthesis. *In* Polyphenols in Cereals and Legumes. J. H. Hulse (Editor), pp. 15–24. International Development Research Centre, Ottawa, Canada.

JUNG, H. G., and FAHEY, G. C., JR. 1983A. Effects of phenolic monomers on rat performance and metabolism. J. Nutr. *113,* 546–556.

JUNG, H. G., and FAHEY, G. C., JR. 1983B. Nutritional implications of phenolic monomers and lignin: A review. J. Anim. Sci. *57,* 206–219.

JUNG, H. G., FAHEY, G. C., JR., and GARST, J. E. 1983. Simple phenolic monomers of roughages and effects of in vitro fermentation on cell wall phenolics. J. Anim. Sci. *57,* 1294–1305.

MARTIN, A. K. 1969. Urinary excretion of aromatic acids by sheep given diets containing different amounts of protein and roughage. Br. J. Nutr. *23,* 389–399.

MARTIN, A. K. 1982A. The origin of urinary aromatic compounds excreted by ruminants. I. The metabolism of quinic, cyclohexanecarboxylic and non-phenolic aromatic acids to benzoic acid. Br. J. Nutr. *47,* 139–154.

MARTIN, A. K. 1982B. The origin of aromatic compounds excreted by ruminants. II. The metabolism of phenolic cinnamic acids to benzoic acid. Br. J. Nutr. *47,* 155–164.

MAYER, A. M., and HAREL, E. 1979. Polyphenol oxidases in plants. Phytochemistry *18,* 193–215.

MORTON, J. F. 1980. Search for carcinogenic principles. Recent Adv. Phytochem. *14,* 53–73.

SINGLETON, V. L., and KRATZER, F. H. 1969. Toxicity and related physiological activity of phenolic substances of plant origin. J. Agric. Food Chem. *17,* 497–512.

SWAIN, T. 1979. Tannins and lignins. *In* Herbivores: Their Interaction with Secondary Plant Metabolites. G. A. Rosenthal and D. H. Janzen (Editors), pp. 657–682. Academic Press, NY.

Sorghum Tannins

BULLARD, R. W., and ELIAS, D. J. 1980. Sorghum polyphenols and bird resistance. *In* Polyphenols in Cereals and Legumes. J. H. Hulse (Editor), pp. 43–49. International Development Research Centre, Ottawa, Canada.

BUTLER, L. G., HAGGERMAN, A. E., and PRICE, M. L. 1980. Biochemical effects of sorghum polyphenols. *In* Polyphenols in Cereals and Legumes. J. H. Hulse (Editor), pp. 36–38. International Development Research Centre, Ottawa, Canada.

ELKIN, R. G., FEATHERSTON, W. R., and ROGLER, J. C. 1978. Investigations of leg abnormalities in chicks consuming high tannin sorghum grains. Poult. Sci. *57*, 757–762.

HEWITT, D., and FORD, J. E. 1982. Influence of tannins on the protein nutritional quality of food grains. Proc. Nutr. Soc. *41*, 7–17.

HULSE, J. H. (Editor) 1980. Polyphenols in Cereals and Legumes. International Development Research Centre, Ottawa, Canada.

NEWMAN, R. K., NEWMAN, C. W., EL-NEGOUMY, A. M., and AASTRUP, S. 1984. Nutritional quality of proanthocyanidin-free barley. Nutr. Rep. Int. *30*, 809–816.

PRICE, M. L., BUTLER, L. G., ROGLER, J. C., and FEATHERSTON, W. R. 1979. Overcoming the nutritionally harmful effects of tannin in sorghum grain by treatment with inexpensive chemicals. Agric. Food Chem. *27*, 441–445.

ROONEY, L. W., BLACKLEY, M. E., MILLER, F. R., and ROSENOW, D. T. 1980. Factors affecting the polyphenols of sorghum and their development and location in the sorghum kernel. *In* Polyphenols in Cereals and Legumes. J. H. Hulse (Editor), pp. 25–35. International Development Research Centre, Ottawa, Canada.

SELL, D. R., and ROGLER, J. C. 1983. Effects of sorghum grain tannins and dietary protein on the activity of liver UDP-glucuronyltransferase. Proc. Soc. Exp. Biol. Med. *174*, 93–101.

Phenolics in Protein Supplements

AFZALPURKAR, A. B., and LAKSHMINARAYANA, G. 1981. Changes in chlorogenic, caffeic, and quinic acid contents during sunflower seed maturation. J. Agric. Food Chem. *29*, 203–204.

GOH, Y. K., CLANDININ, D. R., ROBBLEE, A. R., and DARLINGTON, K. 1979. The effect of level of sinapine in a laying ration on the incidence of fishy odor in eggs from brown-shelled layers. Can. J. Anim. Sci. *59*, 313–316.

HALE, O. M., and McCORMICK, W. C. 1981. Value of peanut skins (testa) as a feed ingredient for growing-finishing swine. J. Anim. Sci. *53*, 1006–1010.

HEWITT, D., and FORD, J. E. 1982. Influence of tannins on the protein nutritional quality of food grains. Proc. Nutr. Soc. *41*, 7–17.

MARQUARDT, R. R., and WARD, A. T. 1979. Chick performance as affected by autoclave treatment of tannin-containing and tannin-free cultivars of fababeans. Can. J. Anim. Sci. *59*, 781–789.

MARQUARDT, R. R., WARD, A. T., CAMPBELL, L. D., and CANSFIELD, P. E. 1977. Purification, identification and characterization of a growth inhibitor in faba beans (*Vicia faba* L. var. *minor*). J. Nutr. *107*, 1313–1324.

MARQUARDT, R. R., McKIRDY, J. A., and WARD, A. T. 1978. Comparative cell wall constituent levels of tannin-free and tannin-containing cultivars of fababeans (*Vicia faba* L.). Can. J. Anim. Sci. *58,* 775–781.
PEARSON, A. W., BUTLER, E. J., and FENWICK, G. R. 1980. Rapeseed meal and egg taint: The role of sinapine. J. Sci. Food Agric. *31,* 898–904.
SILVA TAGA, M., MILLER, E. E., and PRATT, D. E. 1984. Chia seeds as a source of natural antioxidants. J. Am. Oil Chem. Soc. *61,* 928–931.
SOSULSKI, F. 1979. Organoleptic and nutritional effects of phenolic compounds of oilseed protein products: A review. J. Am. Oil Chem. Soc. *56,* 711–716.

Oak Poisoning

DAVIS, G. G., BARTEL, L. E., and COOK, C. W. 1975. Control of gambel oak sprouts by goats. J. Range Manage. *28,* 216–218.
DIXON, P. M., McPHERSON, E. A., ROWLAND, A. C., and MACLENNAN, W. 1979. Acorn poisoning in cattle. Vet. Rec. *104,* 284–285.
DOLLAHITE, J. W., HOUSHOLDER, G. T., and CAMP, B. J. 1966. Effect of calcium hydroxide on the toxicity of post oak (*Quercus stellata*) in calves. J. Am. Vet. Med. Assoc. *148,* 908–912.
NASTIS, A. S., and MALECHEK, J. C. 1981. Digestion and utilization of nutrients in oak browse by goats. J. Anim. Sci. *53,* 283–290.
NESER, J. A., COETZER, J. A. W., BOOMKER, J., and CABLE, H. 1982. Oak (*Quercus ruber*) poisoning in cattle. J. S. Afr. Vet. Assoc. *53,* 151–155.
SANDUSKY, G. E., FOSNAUGH, C. J., SMITH, J. B., and MOHAN, R. 1977. Oak poisoning of cattle in Ohio. J. Am. Vet. Med. Assoc. *171,* 627–629.

Forage Tannins

BARBEAU, W. E., and KINSELLA, J. E. 1983. Factors affecting the binding of chlorogenic acid to fraction 1 leaf protein. J. Agric. Food Chem. *31,* 993–998.
BARRY, T. N., and DUNCAN, S. J. 1984. The role of condensed tannins in the nutritional value of *Lotus pedunculatus* for sheep. I. Voluntary intake. Br. J. Nutr. *51,* 485–491.
BARRY, T. N., and FORSS, D. A. 1983. The condensed tannin content of vegetative *Lotus pedunculatus,* its regulation by fertiliser application, and effect upon protein solubility. J. Sci. Food Agric. *34,* 1047–1056.
BARRY, T. N., and MANLEY, T. R. 1984. The role of condensed tannins in the nutritional value of *Lotus pedunculatus* for sheep. 2. Quantitative digestion of carbohydrates and proteins. Br. J. Nutr. *51,* 493–504.
CHEEKE, P. R., TELEK, L., CARLSSON, R., and EVANS, J. J. 1980. Nutritional evaluation of leaf protein concentrates prepared from selected tropical plants. Nutr. Rep. Int. *22,* 717–721.
GOPLEN, B. T., HOWARTH, R. E., SARKAR, S. K., and LESINS, K. 1980. A search for condensed tannins in annual and perennial species of *Medicago, Trigonella,* and *Onobrychis.* Crop Sci. *20,* 801–804.
HORIGONE, T., and UCHIDA, S. 1983. Nutritional quality of leaf proteins prepared from crops containing phenolic compounds and polyphenolase. *In* Proceedings of the Fourteenth International Grassland Congress, Lexington, Kentucky, 1981. J. A. Smith and V. W. Hays (Editors), pp. 362–364. Westview Press, Boulder, CO.

JONES, W. T., and MANGAN, J. L. 1977. Complexes of the condensed tannins of sainfoin (*Onobrychia viciifolia* Scop) with fraction 1 leaf protein and with submaxillary mucoprotein, and their reversal by polyethylene glycols and pH. J. Sci. Food Agric. *28*, 126–136.
JONES, W. T., BROADHURST, R. B., and LYTTLETON, J. W. 1976. The condensed tannins of pasture legume species. Phytochemistry *15*, 1407–1409.
KUMAR, R., and SINGH, M. 1984. Tannins: Their adverse role in ruminant nutrition. J. Agric. Food Chem. *32*, 447–453.
MARSHALL, D. R., BROVE, P., and MUNDAY, J. 1979. Tannins in pasture legumes. Aust. J. Exp. Agric. Anim. Husb. *19*, 192–197.
MARSHALL, D. R., BORVE, P., GRACE, J., and MUNDAY, J. 1981. Tannins in pasture legumes. 2. The annual and perennial Medicago species. Aust. J. Exp. Agric. Anim. Husb. *21*, 55–58.
McLEOD, M. N. 1974. Plant tannins—their role in forage quality. Nutr. Abst. Rev. *44*, 803–815.
PIERPOINT, W. S. 1983. Reactions of phenolic compounds with proteins, and their relevance to the production of leaf protein. *In* Leaf Protein Concentrates. L. Telek and H. D. Graham (Editors), pp. 235–267. AVI Publishing Co., Westport, CT.
PIRIE, N. W. 1978. Leaf Protein and Other Aspects of Fodder Fractionation. Cambridge Univ. Press, London and New York.

Gossypol

ABOU-DONIA, M. B. 1976. Physiological effects and metabolism of gossypol. Residue Rev. *61*, 125–160.
ANON. 1979. Gossypol—a new antifertility agent for males. Gynecol. Obstet. Invest. *10*, 163–176.
BERARDI, L. C., and GOLDBLATT, L. A. 1980. Gossypol. *In* Toxic Constituents of Plant Foodstuffs. I. E. Liener (Editor), 2nd Edition, pp. 184–237. Academic Press, NY.
CHANG, M. C., GU, Z., and SAKSENA, S. K. 1980. Effect of gossypol on the fertility of male rats, hamsters, and rabbits. Contraception *21*, 461–469.
JONES, L. A. 1979. Gossypol and some other terpenoids, flavonoids and phenols that affect quality of cottonseed protein. J. Am. Oil Chem. Soc. *56*, 727–730.
LINDSEY, T. O., HAWKINS, G. E., and GUTHRIE, L. D. 1980. Physiological responses of lactating cows to gossypol from cottonseed meal rations. J. Dairy Sci. *63*, 562–573.
NADAKAVUKAREN, M. J., SORENSEN, R. H., and TONE, J. N. 1979. Effect of gossypol on the ultrastructure of rat spermatozoa. Cell Tissue Res. *204*, 293–296.
QIAN, S.-Z., and WANG, Z.-G. 1984. Gossypol: A potential antifertility agent for males. Annu. Rev. Pharmacol. Toxicol. *24*, 329–360.

Hypericin

ARAYA, O. S., and FORD, E. J. H. 1981. An investigation of the type of photosensitization caused by the ingestion of St. John's wort (*Hypericum perforatum*) by calves. J. Comp. Pathol. *91*, 135–141.
GARRETT, B. J., CHEEKE, P. R., MIRANDA, C. L., GOEGER, D. E., and BUHLER, D. R. 1982. Consumption of poisonous plants (*Senecio jacobaea, Symphytum officinale, Pteridium aquilinum, Hypericum perforatum*) by rats: Chronic toxicity,

mineral metabolism and hepatic drug-metabolizing enzymes. Toxicol. Lett. *10,* 183–188.
KINGSBURY, J. M. 1964. Poisonous Plants of the United States and Canada. Prentice-Hall, Englewood Cliffs, NJ.
SMITH, B. L., and O'HARA, P. J. 1978. Bovine photosensitization in New Zealand. N. Z. Vet. J. *26,* 2–5.
TOWERS, G. H. N. 1980. Photosensitizers from plants and their photodynamic action. Prog. Phytochem. *6,* 183–202.
VICKERY, A. R. 1981. Traditional uses and folklore of *Hypericum* in the British Isles. Econ. Bot. *35,* 289–295.

Phenolic Compounds as Reproduction Inhibitors

BERGER, P. J., SANDERS, E. H., GARDNER, P. D., and NEGUS, N. C. 1977. Phenolic plant compounds functioning as reproductive inhibitors in *Microtus montanus.* Science *195,* 575–577.
BERGER, P. J., NEGUS, N. C., SANDERS, E. H., and GARDNER, P. D. 1981. Chemical triggering of reproduction in *Microtus montanus.* Science *214,* 69–70.
SANDERS, E. H., GARDNER, P. D., BERGER, P. J., and NEGUS, N. C. 1981. 6-Methoxybenzoxazolinone: A plant derivative that stimulates reproduction in *Microtus montanus.* Science *214,* 67–69.

Black Walnut (Juglone) Toxicity

CODER, K. M. 1983. Seasonal changes of juglone potential in leaves of black walnut (*Juglans nigra* L.) J. Chem. Ecol. *9,* 1203–1212.
MacDANIELS, L. H. 1983. Perspective on the black walnut toxicity problem—Apparent allergies to man and horse. Cornell Vet. *73,* 204–207.
RALSTON, S. L., and RICH, V. A. 1983. Black walnut toxicosis in horses. J. Am. Vet. Med. Assoc. *183,* 1095.
TRUE, R. G., and LOWE, J. E. 1980. Induced jugalone toxicosis in ponies and horses. Am. J. Vet. Res. *41,* 944–945.
TRUE, R. G., LOWE, J. E., HEISSEN, J. E., and BRADLEY, W. 1978. Black walnut shavings as a cause of acute laminitis. Proc. Am. Assoc. Equine Pract. *24,* 511–515.

Resorcinols

FARRELL, D. J., CHAN, C., and McCRAE, F. 1983. A nutritional evaluation of triticale with pigs. Anim. Feed Sci. Technol. *9,* 49–62.
RADCLIFFE, B. C., DRISCOLL, C. F., and EGAN, A. R. 1981. Content of 5-alkyl resorcinols in selection lines of triticale grown in South Australia. Aust. J. Exp. Agric. Anim. Husb. *21,* 71–74.
RADCLIFFE, B. C., EGAN, A. R., and DRISCOLL, C. J. 1983. Nutritional evaluation of triticale grain as an animal feedstuff. Aust. J. Exp. Agric. Anim. Husb. *23,* 419–425.

11

Other Plant Toxins and Poisonous Plants

SLEEPY GRASS (*Stipa robusta*)

Sleepy grass (*Stipa robusta*) is a tall perennial needlegrass, forming erect clumps, that grows in range areas of Colorado, Arizona, New Mexico, and Texas (Fig. 11.1). Reports of livestock poisoning have come mainly from New Mexico. Sleepy grass has a very interesting effect on animals. Ingestion of a moderate but nonlethal amount of the grass induces a profound stuporous condition which may last several days. When the horse was the principal means of travel, considerable inconvenience and sometimes danger occurred when animals consumed the plant. Reportedly the U.S. Cavalry was sometimes pinned down in Indian attacks when their horses were "spaced out" (USDA 1976). Horse traders sometimes fed sleepy grass to wild ponies and then sold them as halterbroken (Smalley and Crookshank 1976). Kingsbury (1964) summarized early research on the toxic signs. Mildly poisoned animals are dejected, inactive, and withdrawn. With a large dose they become somnolent, with a drooping head, closed eyes, and irregularity of gait if forced to move. They slobber copiously and may urinate frequently, even while lying down. Severely poisoned animals lie flat on their side with their head on the ground. They are in profound slumber and can be woken only momentarily. A type of catatonia as well as sleepiness is induced; animals may freeze in one position, with a hoof raised, for instance, and remain absolutely motionless, oblivious to pestering flies, for as long as 45 min (USDA 1976). In the past, horses were the main animals affected; sleepy grass is now mainly a problem with cattle (Anon. 1976). Sheep are not affected as severely as cattle and horses. The narcotic-like substance is diacetone

FIG. 11.1. Sleepy grass (*Stipa robusta*).

alcohol (Epstein *et al.* 1964). The soporific effects may be due to acetone, a known depressant, released in the digestive tract. Cows fed sleepy grass have been observed to lie in a position similar to that seen in ketosis, or acetonemia (Smalley and Crockshank 1976). In general, livestock do not consume the plant again after having experienced a bout of sleepy grass poisoning.

SESQUITERPENE LACTONES

Various Compositae contain toxicants called sesquiterpene lactones. These include a variety of plants common on U.S. rangelands, such as bitterweed or bitter rubberweed (*Hymenoxys odorata*), pingue or Colorado rubberweed (*Hymenoxys richardsonii*) (Fig. 11.2), and the sneezeweeds (*Helenium* spp.), including orange sneezeweed (*H. hoopesii*) (Fig. 11.3), southeastern bitterweed (*H. amarum*), and bitterweed (*H. autumnale*). The sesquiterpene lactones are highly irritating to the nose, eyes, and gastrointestinal tract. Sheep and goats are the main livestock species affected, primarily because the plants are unpalatable and rarely consumed in toxic quantities by cattle and horses.

FIG. 11.2. Pingue or Colorado rubberweed (*Hymenoyxs richardsonii*).

FIG. 11.3. Orange sneezeweed (*Helenium hoopesii*).
Courtesy of Oregon State University, Department of Rangeland Resources.

Sneezeweed poisoning is often referred to as "spewing sickness" because of the characteristic vomiting seen. Affected sheep may have a green stain around the mouth and stand with an upturned head attempting to retain regurgitated material. Vomited material is often inhaled into the lungs, causing either death from inhalation pneumonia or permanent lung damage accompanied by chronic coughing. Primary lesions are gastrointestinal tract irritation, congestion of the liver and kidney, and pulmonary damage.

A variety of sesquiterpene lactones have been identified in these species. A typical example is hymenoxon, found in *Hymenoxys odorata*. Helenalin is the major toxin in *Helenium autumnale*.

Hymenoxon (Hymenovin)   Helenalin

Note that these compounds contain a seven-membered ring, a lactone structure, and an exocyclic methylene group.

The annual loss of sheep and goats in Texas from *H. odorata* (bitterweed) has been estimated at several million dollars annually (Jones and Kim 1981A). It is an annual weed that grows on disturbed soils and overgrazed pastures in the semiarid regions of the U.S. Southwest. Poisoning generally occurs from late December to early May when bitterweed may be the only green plant available, particularly during droughts. The oral LD_{50} of dried bitterweed in sheep is 2.9–8.5 g/kg body weight (Kim et al. 1981).

The toxicity of sesquiterpene lactones is due to binding of the exocyclic methylene group with tissue constituents, such as sulfhydryl groups and other nucleophilic components (Fig. 11.4).

Kim and colleagues at Texas A&M University have shown protective effects of certain dietary additives against hymenoxon and bitterweed toxicity. Kim et al. (1974) reported that simultaneous administration of cysteine with a lactone preparation from *H. odorata* gave up to 80% protection against an LD_{90} dose of the poisonous lactone in dogs. In later studies Rowe et al. (1980) obtained similar results with sheep, with cysteine giving an increased survival rate of sheep injected with hymenoxon. Jones and Kim (1981A) pretreated mice with microsomal enzyme inducers (phenobarbital, polychlorinated biphenyl) or an enzyme inhibitor (chloramphenicol) and found no effect on the toxicity of

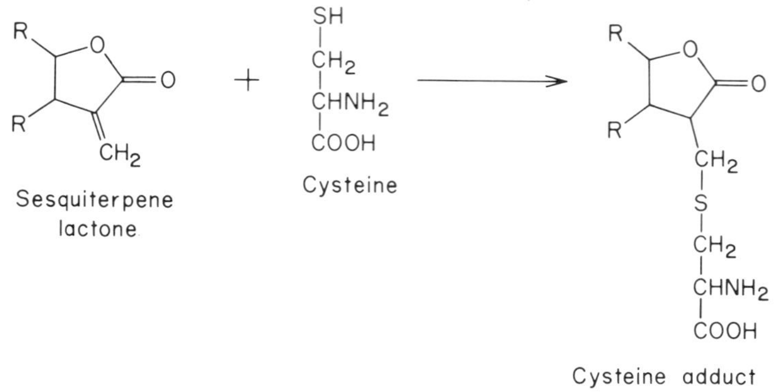

FIG. 11.4. Reaction of sesquiterpene lactones with thiols.

injected hymenoxon. This suggests that mixed function oxidases play little or no role in the detoxification of hymenoxon in mice. Prior administration of carbon tetrachloride did provide protection against hymenoxon; the LD_{50} was 241 mg/kg in controls and 630 mg/kg in carbon tetrachloride-treated mice. This suggests that metabolism of hymenoxon to a metabolite occurs, but not by the mixed function oxidase system. Carbon tetrachloride pretreatment would reduce the number of viable hepatocytes that could function in activation of the hymenoxon. Kim et al. (1981) also investigated the effects of dietary butylated hydroxyanisole (BHA), ethoxyquin, and methionine on the acute toxicity of hymenoxon in mice. Their results (Table 11.1) show that the synthetic antioxidants BHA and ethoxyquin offer protection against the toxicity of hymenoxon. Methionine also had protective activity and appeared to have synergistic effects with BHA. The role of methionine is probably to provide a precursor of cysteine or reduced glutathione. BHA increases the level of reduced glutathione and the

TABLE 11.1. Effects of Dietary Pretreatment on the Mortality (Mortality/Number Treated) of Mice Given Hymenoxon[a]

Pretreatment	Hymenoxon (mg/kg)						LD_{50}
	150	200	300	400	600	776	
None	1/9	1/9	4/9	8/9	—	—	286.5
BHA	0/4	0/9	1/9	1/9	1/5	4/5	662.8
Methionine	—	0/5	0/5	2/5	3/5	3/5	529.1
BHA-methionine	—	0/5	0/5	0/5	0/5	4/5	[b]
Ethoxyquin	—	1/5	0/5	0/5	1/5	4/5	701.5

[a] From Kim et al. (1981).
[b] Not able to be determined because of mortality in only one group.

activity of glutathione transferase (Miranda et al. 1981). In a preliminary study with sheep (Kim et al. 1981), the apparent LD_{50} of bitterweed was 4 and 8 g/kg for control and BHA-fed sheep, respectively, indicating some protective activity.

Hymenoxon and its metabolites appear to be excreted as glucuronides, (Hill et al. 1980) and mercaptic acids (Terry et al. 1983).

Cicuta (WATER HEMLOCK)

Various *Cicuta* spp. grow throughout North America. They are considered the most violently poisonous plants of the North Temperate Zone (Kingsbury 1964). Water hemlock is the common name usually applied to *Cicuta* (Fig. 11.5). It is similar in appearance to numerous other Umbelliferae, including poison hemlock (*Conium maculatum*) and wild carrot (*Daucus carota*). It is found *only* in swampy or wet habitats, such as along streams and in swampy areas and marshes. The plant may grow from 5 to 10 ft tall with a jointed, hollow stem. A

FIG. 11.5. Leaves and flowers of water hemlock (*Cicuta maculata*).

very characteristic feature of water hemlock is that it has a thickened storage organ at the base of the stem which is divided into chambers (Fig. 11.6). This chambered rootstock contains a yellowish oily liquid with the characteristic pungent odor of raw parsnip. The liquid is the poisonous principle, which is also found to a lesser extent in the lower portions of the stem. The toxin is called cicutoxin, a highly unsaturated higher alcohol:

$$HO-CH_2CH_2CH_2C\equiv C-C\equiv C-CH=CH-CH=CH-CH=CH\overset{OH}{\underset{|}{C}}HC_3H_7$$

Many people have died from the consumption of water hemlock. Cases are reported sporadically when fishermen, hikers, and others in wilderness areas mistake it for an edible plant such as wild parsnip. Children may mistake the plant for an edible one. Cicutoxin acts directly on the central nervous system and is a violent convulsant. The description by Kingsbury (1964) conveys its violent nature:

Symptoms appear within 15 minutes to more than an hour, but usually within about a half-hour after ingestion of a lethal dose. Excessive salivation is first noted. This is

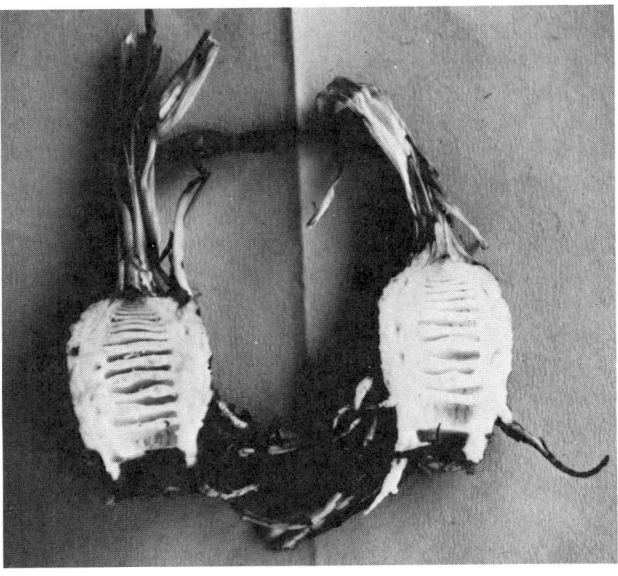

FIG. 11.6. Roots of water hemlock (*Cicuta maculata*), showing the characteristic chambers where the toxin is located.

quickly followed by tremors and then by spasmodic convulsions interspersed intermittently with periods of relaxation. The convulsions are extremely violent; head and neck are thrown rigidly back, legs may flex as though running, and clamping or chewing motions of the jaw and grinding of the teeth occur. Abdominal pain is evident. In some cases the tongue is chewed to shreds; in others, teeth have been broken in an unsuccessful attempt to pry the mouth open to administer treatment.

Most losses of livestock occur in early spring when the new growth appears before other forage is available. When the ground is soft, the rootstock may be pulled up and consumed along with the tops. The plants usually grow in small patches, which can be eliminated manually or with chemical sprays.

Garden carrots and celery contain a similar acetylinic alcohol, which is much less toxic than cicutoxin, called carotatoxin:

$$(C_9H_{17})-C\equiv C-C\equiv C-CH_2-\overset{\overset{\displaystyle OH}{|}}{CH}=CH_2$$

TETRADYMIA–ARTEMISIA POISONING

Tetradymia canescens (spineless horsebrush) and *T. glabrata* (little leaf horsebrush, spring rabbit brush, coal oil brush) are densely branched woody shrubs of the dry desert and sagebrush ranges of the western U.S. *Tetradymia glabrata* is one of the first plants to become green in the spring and one of the major feeds available when sheep are moved from winter to summer range (Fig. 11.7). Thousands of sheep have died as a result of consumption of horsebrush. Cattle do not seem to be affected. Sheep may die from acute liver dysfunction or as a result of photosensitization. *Tetradymia* contains toxins that cause liver damage, with centrolobular necrosis and fatty degeneration. If acute toxicity occurs, symptoms observed are anorexia, depression, twitching, incoordination, a rapid weak pulse, dyspnea, prostration, coma, and death. If the liver damage is less severe, secondary photosensitization may occur. Secondary photosensitization is caused by the reaction of phylloerythrin with light at the surface of unpigmented skin, causing tissue damage. Phylloerythrin is the normal breakdown product of chlorophyll in animals. It normally is absorbed and excreted via the bile. If liver damage has occurred, reducing biliary excretion, some of the phylloerythrin may enter the general circulation, and cause photosensitization. This is called secondary photosensitization because it occurs secondary to hepatic damage, whereas in primary photosensitization (e.g., hypericum) the plant toxin itself is the photodynamic agent.

FIG. 11.7. Little leaf horsebrush (*Tetradymia glabrata*).
Courtesy of Earl Johnson.

The photosensitization condition in sheep consuming *Tetradymia* spp. is commonly called "bighead" (Fig. 11.8). A day or so after consumption of the plant, redness of the skin about the head accompanied by uneasiness and itching develops. The tissues swell as a result of edema. When the edema is severe, the head may become greatly enlarged. Secondary infections and blindness may occur.

Johnson (1978) has conducted a number of feeding experiments with *T. canescens* and *T. glabrata* and was unable to routinely induce toxicity. He concluded that a number of predisposing factors are involved. One of these is prior consumption of *Artemisia* (sagebrush) spp., such as *A. nova* (black sagebrush) (Fig. 11.9). In field experiments, exposing sheep first to sagebrush and then to *Tetradymia* effectively induced bighead in range sheep, but not in farm-reared sheep (Johnson 1978), implicating additional predisposing conditions. Another factor that could be involved is the amount of green plant matter consumed, as this would influence the amount of chlorophyll available to produce phylloerythrin.

Jennings *et al.* (1978) conducted studies to identify the toxic factor(s) in *Tetradymia* spp. They isolated several furanosesquiterpenes (furan-

FIG. 11.8. A sheep with "bighead" (right) caused by *Tetradymia*-induced photosensitization.
Courtesy of M. E. Fowler.

oeremophilanes) that appear to be the toxic agents. The major one was given the name tetradymol:

Administration of tetradymol to laboratory animals revealed that the liver was the only affected organ, with centrolobular necrosis observed. Studies with microsomal mixed function oxidase inducers and inhibitors indicated that tetradymol is bioactivated in the liver to at least two different metabolites. These metabolites appear to disrupt energy metabolism by uncoupling oxidative phosphorylation from electron transport, producing a cellular ATP deficiency (Jennings *et al.* 1978).

FIG. 11.9. Black sage (*Artemisia nova*).
Courtesy of Earl Johnson.

The mode of action of black sagebrush in potentiating *Tetradymia* toxicity is not understood. *Artemisia* spp. are recognized as being potentially toxic (Kingsbury 1964); sesquiterpene lactones are almost universal constituents of *Artemisia* (Herz 1978) and are a possibility to account for the *Artemisia–Tetradymia* interaction. Sagebrush also contains monoterpenes or essential oils. Welch and Pederson (1981) demonstrated large differences in in vitro digestibility of a number of sagebrush species, but there was no apparent correlation with monoterpene content. Sagebrush species vary widely in palatability to grazing animals (Sheehy and Winward 1981); the constituents influencing preference by animals such as sheep and deer which browse on sagebrush have not been identified. Welch *et al.* (1983) reported that the monoterpenoid concentration in big sagebrush was not correlated with mule deer feeding preferences.

AMARANTHUS POISONING

Amaranthus retroflexus (redroot pigweed) is a ubiquitous weed in North America. It is commonly found in vegetable gardens, in cultivated crops, such as corn, and in barnyards, fence rows, and the unsprayed edges of wheat fields. Consumption of pigweed by livestock, particularly cattle and swine, has resulted in numerous poisonings. Buck et al. (1976) have provided a good description of amaranthus poisoning. The principal sign is perirenal edema, occurring a few days after stock is provided access to pigweed-infested areas. Typical signs are weakness, trembling, and incoordination, followed by knuckling of the pastern joints and paralysis of the hind limbs. In pigs, sternal recumbancy is a characteristic posture of affected animals. Death usually occurs within 2 days of the appearance of clinical signs, but in cases where survival is a week or more, the signs of acute nephrosis progress to a chronic fibrosing nephritis. As a result of kidney failure, blood urea nitrogen, serum creatinine, and potassium are elevated; death is probably due to hyperalkemic heart failure. The perirenal edema syndrome in cattle is very similar to that seen in oak poisoning, which is caused by phenolic compounds. *Amaranthus* spp. contain phenolics (Cheeke et al. 1981), but it is not known if these are involved in redroot pigweed poisoning. Other toxins known to be present in *Aramanthus* spp. include saponins, nitrates, and oxalates. The role of these, if any, in the perirenal edema syndrome is not known. The toxic principle seems to act specifically on the renal tubules.

There is interest in the development of various *Amaranthus* spp. as grain and forage crops. Grain amaranth was at one time, before the Spanish conquest, a major food crop in Latin America (Cole 1979). Its major attribute is that the seeds are of high protein quality and are particularly high in lysine (Fig. 11.10). In spite of its excellent amino acid composition (see Table 11.2) feeding trials with rats (Cheeke and Bronson 1980), poultry (Connor et al. 1980), rabbits (Harris et al. 1981), and swine have given poor results. Connor et al. (1980) have demonstrated the presence of a heat-labile hepatotoxin in *A. edulis*

TABLE 11.2. Comparative Composition (Dry Matter Basis) of *Amaranthus edulis* Seed with Other Grains[a]

Analysis	A. edulis	Yellow corn	Milo	Wheat	Barley
Crude protein	16.5	10.0	12.4	14.3	13.0
Ether extract	6.9	4.4	3.1	1.9	2.1
Crude fiber	5.8	2.2	2.2	3.4	5.6
Ash	4.4	1.2	1.9	1.8	2.7
Lysine	0.94	0.20	0.30	0.51	0.60
Methionine	0.32	0.10	0.10	0.20	0.20
Cystine	0.34	0.10	0.20	0.20	0.20
Calcium	0.25	0.02	0.04	0.06	0.09
Phosphorus	0.71	0.35	0.33	0.41	0.47

[a] Adapted from Connor et al. (1980).

FIG. 11.10. A seed head of grain amaranth (*Amaranthus hypochondriacus*).
Courtesy of D. J. Harris.

grain; severe ataxia was seen in chicks fed the raw grain. Swine fed a high level of *A. edulis* grain developed severe myocardial degeneration, and ascites (A. Takken and J. K. Connor 1984, personal communication, Animal Research Institute, Yeerongpilly, Qld., Australia). Steam pelleting the feed did not overcome the effects. The foliage of *Amaranthus* spp. also has given poor results in feeding trials. Cheeke and Carlsson (1978) reported severe growth depression in rats fed a diet containing amaranth leaf meal, and growth of rabbits fed amaranth forage was poor (Harris *et al.* 1980, 1981). Identification of the toxic components is necessary before progress can be made in utilization of amaranth as a grain and forage crop. As with other crops, such as rapeseed, it is likely that the levels of the toxic component(s) could be reduced by plant breeding and selection.

TREMETONE TOXICITY

White Snakeroot

White snakeroot (*Eupatorium rugosum*) (Fig. 11.11) is a showy herbacious perennial that grows over much of eastern North America. It

FIG. 11.11. White snakeroot (*Eupatorium rugosum*).
Courtesy of S. S. Nicholson.

generally grows in low, moist areas, near streams, and in open woods. It may form dense stands after an area is logged. It grows in late summer and early autumn, reaching 3–4 ft in height, with a white composite-type flower. During dry weather when other forage becomes scarce, livestock are likely to move into wooded areas and graze the plant. It often stays green and succulent late in the fall because it is protected from frost in its woodland habitat. After consuming white snakeroot for several days, livestock become depressed and develop a condition referred to as "trembles," in which there is muscle tremoring around the neck, shoulders, and legs. Affected animals stand in a hunched position. Often there is the odor of acetone on the breath, labored breathing, urinary incontinence, and constipation. In horses, there may be partial throat paralysis. Affected animals may recover, though there may be a long period of inappetence and muscular weakness. In severe poisoning, prostration and death occur. Congestion and fatty degeneration of the liver and kidneys are noted. Myocardial problems, including ascites of the pericardial sac and massive degeneration of the myocardium, have been noted in horses (Olson *et al.* 1984). Centrilobular necrosis of the liver and elevated serum enzymes, such as SGOT, CPK, LDH (see Chapter 2), also occur. Oral administration of activated charcoal, at 1 g/kg body weight, shows promise as a treatment (W. B. Buck 1982, Personal communication, University of Illinois, Urbana). White snakeroot causes significant livestock losses in the Midwest. In states such as Illinois, it is one of the major poisonous plants.

White snakeroot provides a classic example of a milk-transferred toxin. The toxin is called tremetol (alcohol form) or tremetone (ketone form). It can be transferred in cow's milk to humans, producing a condition called milk sickness. Symptoms are weakness, nausea, prostration, ketosis, delirium, coma, and death. In the pioneer days, it sometimes reached epidemic proportions, and whole villages were abandoned because of milk sickness. It is said to be responsible for the death of Abraham Lincoln's mother. For many years, the cause was not known, until in the early 1900s it was shown conclusively to be due to a toxin in white snakeroot. The structure of tremetone is as follows:

The disease was associated with pioneer living conditions, when most families had their own cow that grazed on newly cleared land, along stream banks, and so on, where *Eupatorium* would often be found. People consumed milk and butter from their own cow, and if it happened to be grazing on white snakeroot, the toxin was transferred to the consumers. Milk sickness has gradually disappeared from the American scene as more intensive dairying, with milk pooled from many different herds, occurred. However, with the resurgence of small-scale farming, often associated with lack of spraying for weed control, outbreaks of white snakeroot toxicity may occur. Stotts (1984) reported such an incident in which a calf receiving milk from a family cow developed trembles.

Eupatorium adenophorum or Crofton weed is known to cause respiratory disease in horses, characterized by coughing, rapid heaving respiration, and lung lesions (O'Sullivan 1979). The condition has occurred in Australia and Hawaii. The toxic agent has not been identified.

Rayless Goldenrod

Rayless goldenrod or jimmyweed (*Haplopappus heterophyllus*) is a tall, erect perennial plant common on dry rangelands in southern Colorado, Texas, New Mexico, and Arizona. It is usually found along irrigation canals, ditches, and in river valleys. Significant losses of range livestock have occurred in various parts of the southwestern U.S. as a result of consumption of this plant. The toxic principle is tremetol, causing the same toxicity signs as in white snakeroot poisoning, with trembles a predominant sign. Ingestion of 1–1.5% of body weight of the green plant over 1–3 weeks may cause toxicity in horses, cattle, and sheep.

PINE NEEDLE ABORTION

Consumption of the needles of ponderosa pine (*Pinus ponderosa*) can cause abortion in cattle (Fig 11.12). This had been suspected by ranchers for many years. In 1952, McDonald at Kamloops, British Columbia, experimentally induced abortion in cows fed pine needles. Allen and Kitts (1961) at the University of British Columbia showed that extracts of ponderosa pine needles had antiestrogenic properties in mice, causing a diminution in uterine size. Numerous studies since that time have confirmed the abortifacient effects of pine needles, but

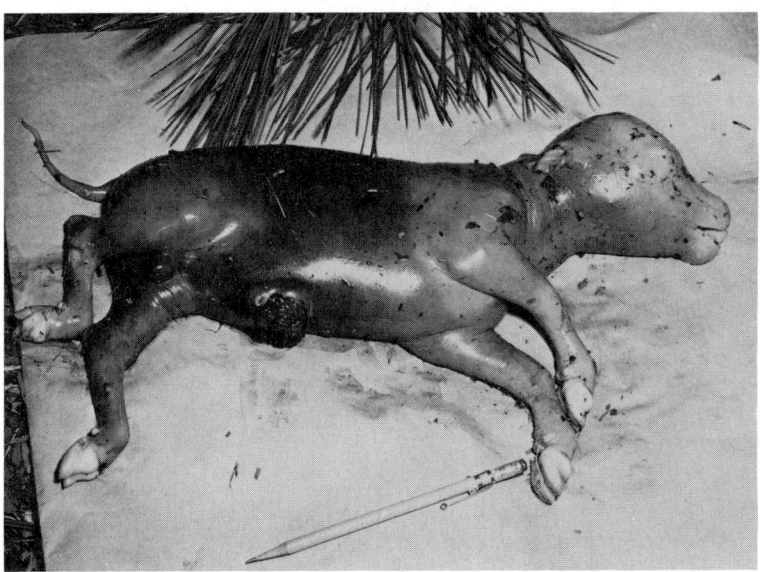

FIG. 11.12. A case of pine needle abortion caused by maternal consumption of ponderosa pine needles.
Courtesy of W. Majak.

the active principle remains unidentified, although diterpene resin acids have been implicated (Kubik and Jackson 1981).

Pine needle abortion is a common problem in the range areas of the western U.S. and Canada. In addition to the abortion loss, affected cows have an increased incidence of retained placenta and impaired breeding performance. Cows in the last trimester of pregnancy are susceptible, and the problem is generally observed only in the winter and spring when other forage is scarce. Green needles, needles from logging slash, or dried needles that have fallen from trees are all potentially toxic. Pine needles are not normally consumed; when the problem occurs, there are generally some environmental factors that cause cows to consume them. These include winter storms that force cattle to seek shelter under pine trees when grass and other forage is in short supply, sudden access to pine needles, feeding of poor-quality hay, or boredom. Abortions may occur within 48 hr of consumption and may be observed as much as 2 weeks after needles are eaten. Affected cows become depressed and dull in appearance, and edema of the genitalia and udder occurs prior to abortion. There may be a bloody discharge from the vulva. Cows may have a toxemia prior to and after the abortion. The aborted fetuses show autolysis, indicating fetal death in utero, and necrosis of the kidney tubules and pulmonary congestion.

Numerous attempts have been made to isolate the toxic substances. Allen and Kitts (1961) showed that administration of aqueous and/or acetone extracts of pine needles to pregnant mice caused a high degree of embryonic mortality; further studies (Allison and Kitts 1964; Cook and Kitts 1964) demonstrated that aqueous and chloroform extracts of pine needles had antiestrogenic properties as detected by a reduction in uterine weights of immature mice. Chow *et al.* (1972) and Anderson and Lozana (1979) demonstrated that the toxic factor(s) caused a disruption of gestation during early fetal development in mice, with subsequent resorptions. Kubik and Jackson (1981) found that a hexane extract of pine needles caused reproductive failure in mice during early pregnancy. Embryonic resorption was observed. The active components of the hexane extract were isolated and identified as a mixture of diterpene resin acids, including pimaric, isopimaric, sandaracopimaric, palustric/levopimaric, abietic, dehydroabietic, and neoabietic acids. Confirmation that these acids cause abortion in cattle has not yet been reported.

Adams *et al.* (1979) advanced a theory that pine needle abortion was linked to an infectious microorganism, *Listeria monocytogenes*. Pine needle-induced abortions have a number of similarities to *Listeria*-induced abortions, including involvement of a similar stage of gestation, depression, retained placenta, and genital exudates. Adams *et al.*

(1979) fed mice pine needles and demonstrated a high incidence (60–100%) of *Listeria monocytogenes* in the blood of treated mice, with no detectable microorganisms in the blood of control mice. They concluded that *Listeria* is likely involved in the pathology.

FLUOROACETATE (1080)

Sodium monofluoroacetate is Compound 1080, which has been widely used as a poison to kill coyotes in the U.S. and other vertebrate pests elsewhere. Certain plants, particularly in Australia and South Africa, contain 1080. It is a very potent toxin, inhibiting aconitate hydratase, one of the tricarboxylic acid (TCA) cycle enzymes. Cellular respiration is thus brought to a standstill, and death occurs from cellular lack of ATP.

Fluoroacetate occurs in a variety of Australian plants of the family Leguminosae. These include various species of *Acacia*, *Gastrolobium*, and *Oxylobium*. They are shrubby browse plants which may be consumed by livestock. In most cases they are quite palatable. Because of the high toxicity of 1080, only a few leaves (about 30 g wet weight) can kill a sheep. Brazilian and South African plants containing 1080 are known as well.

The oral lethal dose for livestock is about 0.3–1.75 mg 1080 per kilogram of body weight. An interesting finding in Australia is that certain native herbivores in Western Australia, where most of these plants are found, have developed a resistance to 1080 and can safely consume leaves of fluoroacetate-containing plants. For example, the LD_{50} for 1080 in the brush-tailed possum from western Australia is about 100 mg/kg, whereas for the same species from eastern Australia, the LD_{50} is 0.68 mg/kg, about a 150-fold difference (Mead *et al.* 1979).

Fluoroacetate can be utilized as acetate, but the fluorocitrate so formed cannot be converted to isocitrate (Fig. 11.13).

Kirsten *et al.* (1978) have advanced the suggestion that the principal mode of action of fluoroacetate may be to inhibit citrate transport through mitochondrial membranes, rather than the inhibition of aconitate hydratase. Fluorocitrate forms a thiol-ester bond with the sulfhydryl groups of two enzymes in the mitochondrial membrane. These enzymes function in the transfer of citrate across the membrane.

Mead *et al.* (1979), in their study of western Australian mammals resistant to 1080, found that detoxification of fluoroacetate resulted in an elevation of fluoride in the blood. They implicated a glutathione-dependent reaction, shown in Fig. 11.14.

It is interesting that while fluoroacetate is very toxic, fluoropropio-

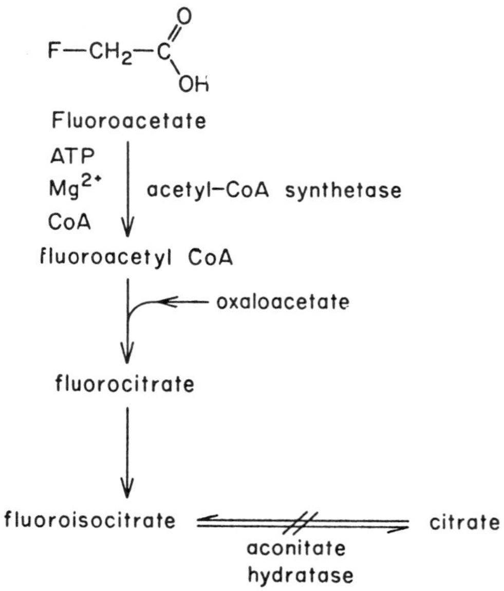

$$F-CH_2-C\overset{O}{\underset{OH}{\nwarrow}}$$

Fluoroacetate

$\downarrow$ ATP, Mg^{2+}, CoA — acetyl-CoA synthetase

fluoroacetyl CoA

$\downarrow$ ← oxaloacetate

fluorocitrate

$\downarrow$

fluoroisocitrate ⇌ citrate
aconitate hydratase

FIG. 11.13. Metabolism of fluoroacetate.

nate is nontoxic. Higher fluoro fatty acids are toxic if they have an odd number of carbons, and nontoxic if they have an even number. This can be explained on the basis of the products of β-oxidation (Fig. 11.15).

Because fluoroacetate blocks the TCA cycle, glucose metabolism is impaired and hyperglycemia occurs. In ruminants, cardiac failure occurs, with evidence of damage to the myocardium apparent on necropsy. Other findings include general cyanosis of the mucous membranes and other tissues, and the liver and kidney are dark and congested. There is no antidote to 1080 poisoning.

King et al. (1981) have suggested an interesting application in wildlife management of species differences in fluoroacetate tolerance. Certain native western Australian mammals, including rat kangaroos and rock wallabies, are threatened with extinction by introduced predators such as cats and foxes. The predators can be controlled by using 1080, to which the native fauna are resistant.

BRACKEN POISONING

Bracken fern (*Pteridium aquilinum*) contains a thiaminase enzyme and causes an induced thiamine deficiency in nonruminants, as previ-

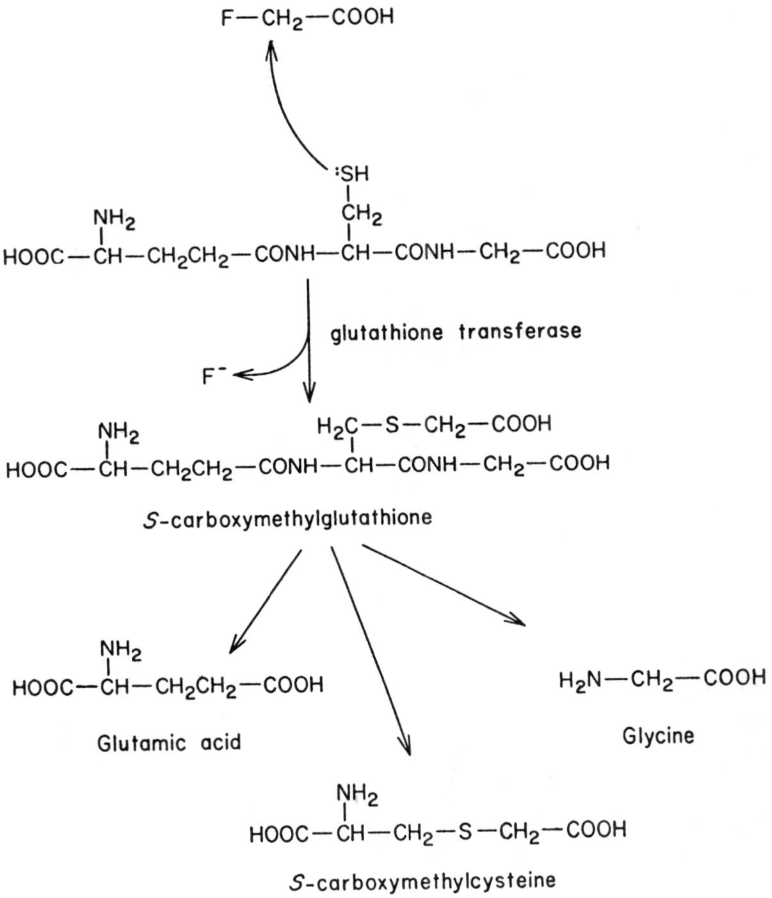

FIG. 11.14. Proposed mechanism of fluoroacetate detoxification by the Western Australian brush-tailed possum.

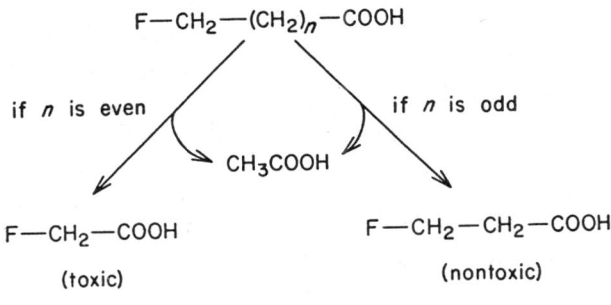

FIG. 11.15. β-Oxidation of higher fluoro fatty acids.

ously described (Chapter 7). Bracken also causes a fatal hemorrhagic syndrome in cattle, cancer in cattle, and possibly cancer in humans.

Bracken poisoning in cattle occurs after animals have consumed significant quantities of the fern. This may occur when other feed is scarce, as in early spring when the bracken "fiddleheads" are emerging. Conversely, when there is abundant lush, succulent forage, cattle may consume bracken as a source of roughage. The palatable young green fronds are about five times as toxic as the mature fronds, and the rhizomes are highly toxic. Bracken poisoning involves severe damage to the bone marrow, resembling radiation damage, with a consequent loss of cellular blood components, leading to severe leukopenia and thrombocytopenia. Hemorrhaging occurs, with blood in the feces, bleeding from the nose and vagina, and bleeding from the membranes of the eyes and mouth. A high fever develops (107°–109°F) in the terminal stages. On postmortem, much hemorrhaging in the stomach, intestines, lungs, and heart can be seen. Cattle are the most susceptible; sheep have been poisoned experimentally, while nonruminants such as horses are highly resistant.

Bracken fern also causes cancer in cattle. The disease is known as chronic bovine enzootic hematuria or redwater disease. It is believed to be due to the consumption of low levels of bracken over a prolonged period of time. It is seen most frequently in cattle 7 years or older. Cancerous tumors in the form of small polyps or nodules develop in the mucosa of the urinary bladder. Hematuria is due to bleeding from these polyps. Death occurs either from anemia and loss of blood, or from the spread of the cancer to other tissues. This condition has been observed in Great Britain, Western Europe, Brazil, Australia, and many parts of the U.S. It is suspected, but not proved, that both bracken poisoning and enzootic hematuria are caused by the same as yet unidentified factor in bracken.

Bracken fed to a variety of laboratory animals produces cancer. Rats and Japanese quail develop malignant intestinal adenocarcinomas, while guinea pigs develop hematuria nearly identical to the bovine condition. Mice are very susceptible and develop malignant tumors in a variety of tissues when given even a single dose of bracken extract. Efforts to identify the causative factor(s) have not been successful. Shikimic acid has been identified by Evans (1976) in Wales as one of the carcinogens in bracken, but this has not been confirmed by other investigators. Pamukcu *et al.* (1971) reported that dietary phenothiazine, an inducer of mixed function oxidase activity, reduced the incidence of intestinal and bladder neoplasms in rats. This may have been a result of enhanced hepatic detoxification of the bracken carcinogen(s). However, treatment of sheep with phenothiazine did not prevent bracken-induced changes in blood leukocytes (Idrus et al. 1977). Pamukcu et al. (1977) observed that dietary BHA reduced the incidence of intestinal tumors in rats. As reported in other sections, BHA protects against a variety of toxins, including pyrrolizidine alkaloids and hymenoxon, presumably by modifications of liver metabolism.

W. C. Evans et al. (1982) employed a bioassay involving calves to test bracken fractions for the cattle bracken-poisoning factor. Calves were administered bracken extracts, and the presence of the toxin was assessed by the depression in circulating leukocytes and thrombocytes after daily administration of bracken fractions for 20–30 days. With the use of this technique, a number of potential candidates for the bracken-poisoning factor have been tested. Quercetin and other tannins, which have been suggested by Pamucku et al. (1980A,B,C) as carcinogens in bracken, were negative. W. C. Evans et al. (1982) speculated that sesquiterpenes in bracken, such as the pterosides (glycosides) and their aglycones, pterosins, may be the active principles. These compounds, of which 30 or more are found in bracken, are somewhat unique to this plant. An example of a pterosin is pterosin B:

However, bracken extracts containing pterosins and pterosides were incapable of producing bracken poisoning in calves (Evans et al. 1983), so that these compounds do not appear to be the poisonous principles. Hirono et al. (1984) isolated a norsesquiterpene glucoside named ptaquiloside from a carcinogenic extract of bracken. This compound, when administered to a calf daily for 42 days, produced typical signs of cattle bracken poisoning, including greatly elevated leukocyte counts and a severe reduction in neutrophilic granulocytes. The bone marrow showed marked degeneration typical of acute bracken poisoning. Ptaquiloside thus appears to be a causative factor in bracken poisoning.

Bracken fern is a potential human health hazard. The young fiddleheads or croziers are consumed as a delicacy in Japan. (Fig. 11.16). The young fronds are high in carcinogenic activity. It has been suggested that the high incidence of stomach cancer in Japan could be partially

FIG. 11.16. Fresh young bracken fronds prepared for market in Tokyo, Japan, where bracken is consumed in large amounts by humans.
Courtesy of Iwao Hirono.

due to consumption of bracken. Transfer of the carcinogen through cow's milk has been demonstrated. Studies in England and Wales have shown a correlation of 0.7% of area of bracken coverage of the land with the stomach cancer index rate, leading to the suggestion that pollution of water supplies in bracken-infested areas could be significant.

In the U.S., there has been interest in "back-to-nature" lifestyles, with foraging for wild edible plants, including bracken. The comment of Evans (1979) is probably pertinent: "It should be stressed that the advice of anyone, however well intentioned, to eat bracken should be strongly resisted, whether it is applied to man or his dometic animals."

Finally, bracken has been linked to a condition in sheep grazing the North Yorkshire moors in England, called the "bright blindness factor." The sheep develop a degeneration of the neuroepithelium of the retina, rendering them permanently blind. The condition is accompanied by leukopenia and thrombocytopenia and has been definitely linked with bracken consumption in experimental studies (Barnett and Watson 1970).

BUCKWHEAT TOXICITY

Buckwheat (*Fagopyrum esculentum*) is a summer annual, widely grown as a green manure crop and a "catch" crop to be seeded after failure of another crop (Fig. 11.17). It reaches maturity 10–12 weeks from planting and often yields better on poor soil than cereal grains. The grain is used for animal feed and human consumption, while the green plant may be used for forage. The seed is a small, pyramidal, hard, brownish-black structure which has been used quite extensively in pancake mixes and flour.

Buckwheat seed and forage may cause photosensitization (fagopyrism), when light-skinned animals which have consumed buckwheat are exposed to sunlight. In mild cases, erythema of unpigmented areas of skin is observed, while in acute cases, nervous symptoms such as excitement, running, squealing, or bellowing, convulsions, and prostration may occur. The active principle in buckwheat is a derivative of nephthodianthrone called fagopyrin:

Fagopyrin

Mulholland and Coombe (1979) in Australia reported an outbreak of fagopyrism in sheep grazing buckwheat stubble. About a third of the flock was affected. Body weight gains and wool growth were less than for sheep grazing sorghum stubble. It was concluded that buckwheat stubble is of low palatability, has a low nutritive value, and provides a high risk of photosensitization reactions. Farrell (1978), also in Australia, has made a nutritive evaluation of buckwheat grain as a livestock feed. Buckwheat has higher concentrations of essential amino acids than other grains, with lysine and methionine contents of 0.58% and 0.32%, respectively. Buckwheat gave better animal performance than wheat or oats when no protein supplement was provided. However, attempts to improve performance with added energy, protein, or

FIG. 11.17. A field of buckwheat grown for seed. Buckwheat may cause photosensitization.

amino acids were unsuccessful. Thus, buckwheat gave fairly good performance when used as the major dietary item, with little response to supplementary nutrients. This is an unusual response. Farrell (1978) observed that buckwheat was less palatable than cereal grains and suggested that it may contain a toxin which may affect acceptability. In a Canadian study, Anderson and Bowland (1981) reported satisfactory results with growing pigs when buckwheat was used as a replacement for wheat or barley.

ALSIKE CLOVER POISONING

Alsike clover (*Trifolium hybridum*) is a short-lived perennial that has been widely grown in the eastern and northern midwestern states of the U.S. and is widely grown in the northern farming areas of Canada. It is especially well adapted to cool climates and heavy, poorly drained clay soils. In the early part of this century, alsike clover was responsible for widespread toxicity problems in livestock, particularly in horses. In most cases, photosensitization was the major problem. The condition was called clover sickness or trifoliosis. The decline in acreage of alsike clover and in the use of draft horses explains why the condition is not now a common problem in the U.S. However, it is still seen in Canada and is a significant problem in northern British Columbia. Some unusual contradictions have been noted. In some cases, the liver of affected animals becomes extremely enlarged, reaching a weight of 50–60 lb in horses. In other cases, the liver is small and fibrotic. Signs of liver failure are evident, including icterus, depression, stupor, and "head pushing" in horses, indicative of neurological involvement of elevated blood ammonia due to liver failure. Photosensitization without evidence of liver damage is seen in some instances, while in other cases there is severe liver damage but no photosensitization. The causative agent in alsike clover has not been identified.

YELLOW STAR THISTLE
(*Centaurea solstitialis*)

The prolonged consumption by horses of yellow star thistle (*Centaurea solstitialis*) or Russian knapweed (*C. repens*) causes a nervous disorder called equine nigropallidal encephalomalacia (ENE). Yellow star thistle causes significant ENE problems in northern California, with a lesser incidence in southern California and southern Oregon. It has also been reported in Argentina and Australia. Russian knapweed has been linked to ENE incidence in several western states, including Colorado, Utah, and Washington (Cordy 1978). In northern California, two annual peaks of ENE occur: June–July, and October–November. Generally, yellow star thistle grows in weedy horse paddocks or fallow fields.

Toxicity of *Centaurea* species seems restricted to horses. A variety of other domestic and laboratory animals have been fed the plants with no ill effects (Cordy 1978). Clinical signs of toxicity occur abruptly in horses, characterized by drowsiness, difficulty in eating and drinking, and aimless walking with the muzzle to the ground, or else total inactivity. The animals have particular difficulty in eating and drinking, apparently due to impairment of neural activity of the fifth, seventh, and twelfth cranial nerves (Cordy 1978). Lesions are typically found in any of four sites in the brain: the globus pallidus and substantia nigra of the left and right sides. The lesions are basically foci of necrotic tissue. Cordy (1978) has suggested that ENE is caused by disruption of the dopaminergic nigrostriatal pathway. The initial toxicity signs are due to release of dopamine from the nigrostriatal nerve endings, and the later signs may reflect dopamine deficiency. Affected horses generally die of starvation or dehydration.

The toxic principle of *Centaurea* species has not been identified. The lethal dose of green plant material is about 1.8–2.5 kg/100 kg body weight/day for *C. repens* and 2.3–2.6 kg for *C. solstitialis*.

Stypandra imbricata (BLIND GRASS)

Stypandra imbricata, a perennial grasslike plant which grows in Western Australia, is commonly referred to as blind grass. It has been responsible for numerous losses of sheep; animals either die after an acute illness with signs of neurological disturbance, or survive, but are permanently blind. Studies with sheep and goats (Main *et al.* 1981) and with rats (Huxtable *et al.* 1980) have shown that the lesions associated with blind grass toxicity are confined to the nervous system, with vacuolation of myelin in the brain and spinal cord. The optic nerves suffer a severe loss of myelinated axons, with a progressive disintegration until they are reduced to thin strands of scar tissue (Fig. 11.18). The toxic agent has not yet been identified.

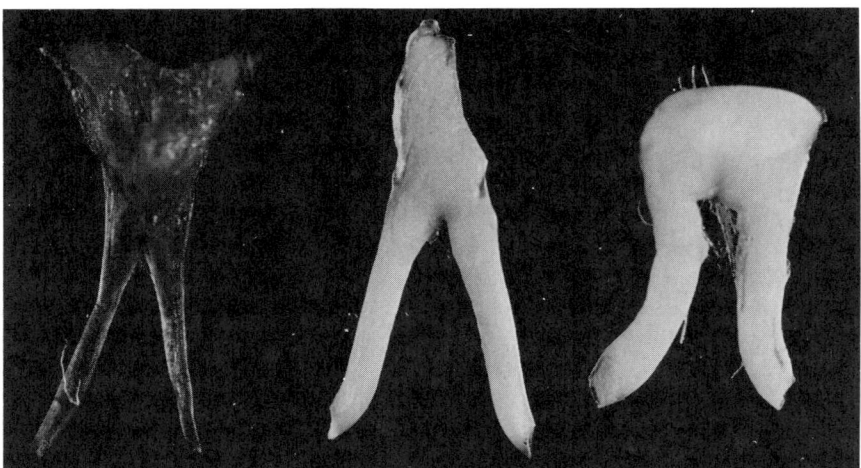

FIG. 11.18. Optic nerves from a control rat (center), a rat acutely intoxicated with *Stypandra imbricata* (right), and a rat 12 weeks after acute intoxication (left). Note the swelling with acute intoxication (right) followed by severe atrophy (left).
From Huxtable et al. (1980). Courtesy of P. R. Dorling and Neuropathology and Applied Neurobiology.

WEED SEEDS IN GRAINS AND SCREENINGS

Weeds seeds may be significant contaminants of grain and grain screenings. Such material is often used for livestock feed, leading to a potential for toxicoses. The problems, if any, are a function of what particular weed seeds are present, but a few general comments are appropriate. As indicated earlier (Chapter 4), this problem has been markedly reduced in industrialized countries due to the control of broad-leaved weeds in grain through the use of herbicides.

Harrold and Nalewaja (1977) in North Dakota investigated the nutritive value of 15 common weed seeds in grains. The bulk of the weeds present were wild buckwheat (*Polygonum convolvulus*), green and yellow foxtail (*Setaria viridis* and *S. lutescens*), and wild oats (*Avena fatua*). These did not present toxicological hazards. In subsequent work (Harrold et al. 1980), various weed seeds were fed to rats, including green and yellow foxtail, wild oats, wild buckwheat, and redroot pigweed (*Amaranthus retroflexus*). Toxicity problems were not observed except for pigweed seeds, which are unpalatable and resulted in lowered gains. *Amaranthus* spp. contain heat-labile toxins (see the section Amaranthus Poisoning, this chapter). Nephrotoxic effects might occur with a high level of pigweed contamination.

Shires et al. (1982) examined the effect of weed seed contamination of rapeseed on its feeding value. Wild mustard (*Brassica kaber*), stinkweed (*Thlaspi arvense*), and lamb's-quarters (*Chenopodium album*) were the major contaminants. Wild mustard and stinkweed contain glucosinolates, and mustard also contains erucic acid. Feeding trials (Shires et al. 1982) indicated that contamination with these seeds did not appreciably affect the feeding value of rapeseed meal.

Smartweeds (*Polygonum* spp.) are common weeds of North America, closely related to buckwheat. There is no indication of smartweed seeds causing livestock problems.

BLUE-GREEN ALGAE

Blue-green algae of a number of species have caused livestock mortalities in many countries when animals consumed algae-infected water. Algae blooms occur in summer and autumn when stock watering ponds are low. Anaerobic bacteria in the bottom mud raise the soluble phosphorus and nitrogen levels and increase the availability of carbon dioxide. These factors, combined with long days, bright sunlight, and warm water, favor the proliferation of algae. The algal cells develop gas bubbles, causing the algae colonies to rise to the surface, after which they are blown by wind into dense algal blooms. Among the toxic blue-green algae are *Microcystis aeruginosa, Anabaena circinalis,* and *Nodularia spumingena.*

The toxic principles of blue-green algae are cyclopeptides (Elleman *et al.* 1978). The primary site of toxicity is the liver, which shows centrilobular necrosis. Pale, swollen hemorrhagic livers, ascites, and numerous small hemorrhages are observed at necropsy (Jackson *et al.* 1984). Serum enzymes indicative of liver damage are elevated. The LD_{50} for purified toxin in mice is 0.056 mg/kg (Elleman *et al.* 1978). In a trial with sheep (Jackson *et al.* 1984), a dosage of 730–950 mg dried *M. aeruginosa* per kilogram body weight caused no lesions, while 990–1040 mg/kg caused mild, sublethal changes, and dosages in excess of 1040 mg/kg were lethal.

Algae blooms tend to be most toxic when the cells are in the rapidly proliferating stage or when the bloom is disintegrating (Seawright 1982). Pollution of water sources or fertilizer run-off increase the likelihood of algae blooms. Algae in ponds can be controlled by use of copper sulfate at about 1 kg/4,000,000 liters (Everist 1981). Prevention is best achieved by restricting pollution of water sources.

REFERENCES

Sleepy Grass (*Stipa robusta*)

EPSTEIN, W., GERBER, K., and KARLER, R. 1964. The hypnotic constituent of *Stipa vasey,* sleepy grass. Experientia *20,* 390.

KINGSBURY, J. M. 1964. Poisonous Plants of the United States and Canada. Prentice-Hall, Englewood Cliffs, NJ.

SEARS, P. B. 1977. Sleepygrass. Rangeman's J. *4,* 67–68.

SMALLEY, H. E., and CROOKSHANK, H. R. 1976. Toxicity studies on sleepy grass, *Stipa robusta* (Vasey) Scribn. Southwest. Vet. *29* (1), 35–39.

USDA. 1976. Spaced-out cows swear off grass. Agric. Res. January, pp. 1–16.

WOLFE, M. H. 1979. More notes on sleepygrass. Rangelands *1,* 144.

Sesquiterpene Lactones

CALHOUN, M. C., UECKERT, D. N., LIVINGSTON, C. W., and BALDWIN, B. C. 1981. Effects of bitterweed (*Hymenoxys odorata*) on voluntary feed intake and serum constituents of sheep. Am. J. Vet. Res. *42,* 1713–1717.

HERZ, W. 1978. Sesquiterpene lactones from livestock poisons. *In* Effects of Poisonous Plants on Livestock. R. F. Keeler, K. R. Van Kampen, and L. F. James (Editors), pp. 487–497. Academic Press, NY.

HILL, D. W., BAILEY, E. M., and CAMP, B. J. 1980. Tissue distribution and disposition of hymenoxon. J. Agric. Food Chem. *28,* 1269–1273.

JONES, D. H., and KIM, H. L. 1981A. Toxicity of hymenoxon in Swiss white mice following pretreatment with microsomal enzyme inducers, inhibitors and carbon tetrachloride. Res. Commun. Chem. Pathol. Pharmacol. *33*, 361–364.

JONES, D. H., and KIM, H. L. 1981B. Toxicity and mutagenicity of hymenoxon, a sesquiterpene lactone. Toxicol. Lett. *9*, 395–401.

KIM, H. L., SZABUNIEWICZ, M., ROWE, L. D., CAMP, B. J., DOLLAHITE, J. W., and BRIDGES, C. H. 1974. L-Cysteine, an antagonist to the toxic effects of an α-methylene-γ-lactone isolated from *Hymenoxys odorata* D. C. (Bitterweed). Res. Commun. Chem. Pathol. Pharmacol. *8*, 381–384.

KIM, H. L., ANDERSON, A. C., TERRY, M. K., and BAILEY, E. M. 1981. Protective effect of butylated hydroxyanisole on acute hymenoxon and bitterweed poisoning. Res. Commun. Chem. Pathol. Pharmacol. *33*, 365–368.

MIRANDA, C. L., REED, P. L., CHEEKE, P. R., and BUHLER, D. R. 1981. Protective effects of butylated hydroxyanisole against the acute toxicity of monocrotaline in mice. Toxicol. Appl. Pharmacol. *59*, 424–430.

ROWE, L. D., KIM, H. L., and CAMP, B. J. 1980. The antagonistic effect of L-cysteine in experimental hymenoxon intoxication in sheep. Am. J. Vet. Res. *41*, 484–486.

TERRY, M. K., WILLIAMS, H. A., KIM, H. L., POST, L. O., and BAILEY, E. M. 1983. Ovine urinary metabolites of hymenoxon, a toxic sesquiterpene lactone isolated from *Hymenoxys odorata* DC. J. Agric. Food Chem. *31*, 1208–1210.

Cicuta (Water Hemlock)

KINGSBURY, J. M. 1964. Poisonous Plants of the United States and Canada. Prentice-Hall, Englewood Cliffs, NJ.

Tetradymia–Artemisia Poisoning

CEDARLEAF, J. D., WELCH, B. L., and BROTHERSON, J. D. 1983. Seasonal variation in monoterpenoids in big sagebrush (*Artemisia tridentata*). J. Range Manage. *36*, 492–494.

HERZ, W. 1978. Sesquiterpene lactones from livestock poisons. *In* Effects of Poisonous Plants on Livestock. R. F. Keller, K. R. Van Kampen, and L. F. James (Editors), pp. 487–497. Academic Press, NY.

JENNINGS, P. W., REEDER, S. K., HURLEY, J. C., ROBBINS, J. E., HOLIAN, S. K., HOLIAN, A., LEE, P., PRIBANIC, J. A. S., and HULL, M. W. 1978. Toxic constituents and hepatotoxicity of the plant *Tetradymia glabrata* (Asteroceae). *In* Effects of Poisonous Plants on Livestock. R. F. Keeler, K. R. Van Kampen, and L. F. James (Editors), pp. 217–228. Academic Press, NY.

JOHNSON, A. E. 1978. Tetradymia toxicity—a new look at an old problem. *In* Effects of Poisonous Plants on Livestock. R. F. Keeler, K. R. Van Kampen, and L. F. James (Editors), pp. 209–216. Academic Press, NY.

JOHNSON, A. E., JAMES, L. F., and SPILLETT, J. 1976. The abortifacient and toxic effects of big sagebrush (*Artemisia tridentata*) and juniper (*Juniperus osteosperma*) on domestic sheep. J. Range Manage. *29*, 278–280.

KINGSBURY, J. M. 1964. Poisonous Plants of the United States and Canada. Prentice-Hall, Englewood Cliffs, NJ.

SHEEHY, D. P., and WINWARD, A. H. 1981. Relative palatability of seven *Artemisia* taxa to mule deer and sheep. J. Range Manage. *34*, 397–399.

WELCH, B. L., and PEDERSON, J. C. 1981. *In vitro* digestibility among accessions of big sagebrush by wild mule deer and its relationship to monoterpenoid content. J. Range Manage. *34,* 497–500.

WELCH, B. L., McARTHUR, E. D., and DAVIS, J. N. 1983. Mule deer preference and monoterpenoids (essential oils). J. Range Manage. *36,* 495.

Amaranthus Poisoning

BUCK, W. B., OSWEILER, G. D., and VAN GELDER, G. A. 1976. Clinical and Diagnostic Veterinary Toxicology. Kendall-Hunt Publishing Co., Dubuque, IA.

CHEEKE, P. R., and BRONSON, J. 1980. Feeding trials with amaranthus grain, forage and leaf protein concentrates. *In* Proceedings of the Second Amaranth Conference, pp. 5–11. Rodale Press, Emmaus, PA.

CHEEKE, P. R., and CARLSSON, R. 1978. Evaluation of several crops as sources of leaf meal: Composition, effect of drying procedure, and rat growth response. Nutr. Rep. Int. *18,* 465–473.

CHEEKE, P. R., CARLSSON, R., and KOHLER, G. O. 1981. Nutritive value of leaf protein concentrates prepared from *Amaranthus* species. Can. J. Anim. Sci. *61,* 199–204.

COLE, J. N. 1979. Amaranth. From the Past, for the Future. Rodale Press, Emmaus, PA.

CONNOR, J. K., GARTNER, R. J. W., RUNGE, B. M., and AMOS, R. N. 1980. *Amaranthus eduli:* An ancient food source re-examined. Aust. J. Exp. Agric. Anim. Husb. *20,* 156–161.

HARRIS, D. J., CHEEKE, P. R., and PATTON, N. M. 1980. A note on the feeding value of amaranthus (pigweed) and chenopodium (lamb's quarters) to rabbits. J. Appl. Rabbit Res. *3* (3), 11–13.

HARRIS, D. J., CHEEKE, P. R., and PATTON, N. M. 1981. Effect of feeding amaranthus, sunflower leaves, Kentucky bluegrass and alfalfa to rabbits. J. Appl. Rabbit Res. *4,* 48–50.

SAUNDERS, R. M., and BECKER, R. 1983. Amaranthus: A potential food and feed resource. Adv. Cereal Sci. Technol. *6,* 357–396.

STUART, B. P., NICHOLSON, S. S., and SMITH, J. B. 1975. Perirenal edema and toxic nephrosis in cattle, associated with ingestion of pigweed. J. Am. Vet. Med. Assoc. *167,* 949–950.

White Snakeroot

OLSON, C. T., KELLER, W. C., GERKEN, D. F., and REED, S. M. 1984. Suspected tremetol poisoning in horses. J. Am. Vet. Med. Assoc. *185,* 1001–1003.

O'SULLIVAN, B. M. 1979. Crofton weed (*Eupatorium adenophorum*) toxicity in horses. Aust. Vet. J. *55,* 19–21.

STOTTS, R. 1984. White snakeroot toxicity in dairy cattle. VM/SAC, Vet. Med. Small An. Clin. *79,* 118–120.

Pine Needle Abortion

ADAMS, C. J., NEFF, T. N., and JACKSON, L. L. 1979. Induction of *Listeria monocytogenes* infection by the consumption of ponderosa pine needles. Infect. Immun. *25,* 117–120.

ALLEN, M. R., and KITTS, W. D. 1961. The effects of yellow pine (*Pinus ponderosa* Laws.) needles on the reproductivity of the laboratory mouse. Can. J. Anim. Sci. *41*, 1–9.

ALLISON, C. A., and KITTS, W. D. 1964. Further studies on the antiestrogenic activity of yellow pine needles. J. Anim. Sci. *23*, 1155–1159.

ANDERSON, C. K., and LOZANO, E. A. 1979. Embryotoxic effects of pine needles and pine needle extracts. Cornell Vet. *69*, 169–175.

CALL, J. W., and JAMES, L. F. 1978. Pine needle abortion in cattle. In Effects of Poisonous Plants on Livestock. R. F. Keeler, K. R. Van Kampen, and L. F. James (Editors), pp. 587–590. Academic Press, NY.

CHOW, F. C., HANSON, K. J., HAMAR, D. W., and UDALL, R. H. 1972. Reproductive failure of mice caused by pine needle ingestion. J. Reprod. Fertil. *30*, 169–172.

COGSWELL, C., and KAMSTRA, L. D. 1980. Toxic extracts in ponderosa pine needles that produce abortion in mice. J. Range Manage. *33*, 46–48.

COOK, H. and KITTS, W. D. 1964. Anti-oestrogenic activity in yellow pine needles (*Pinus ponderosa*). Acta Endocrinol. (Copenhagen) *45*, 33–39.

KUBIK, Y. M., and JACKSON, L. L. 1981. Embryo resorptions in mice induced by diterpene resin acids of *Pinus ponderosa* needles. Cornell Vet. *71*, 34–42.

MANNERS, G. D., PENN, D. D., LURD, L., and JAMES, L. E. 1982. Chemistry of toxic range plants. Water-soluble lignols of ponderosa pine needles. J. Agric. Food Chem. *30*, 401–404.

McDONALD, M. A. 1952. Pine needle abortion in range beef cattle. J. Range Manage. *5*, 150–155.

NEFF, T. E., ADAMS, C. J., and JACKSON, L. L. 1982. Pathological effects of pine needle ingestion in pregnant mice. Cornell Vet. *72*, 128–136.

Fluoroacetate (1080)

KING, D. R., OLIVER, A. J., and MEAD, R. J. 1978. The adaptation of some Western Australian mammals to food plants containing fluoroacetate. Aust. J. Zool. *26*, 699–712.

KING, D. R., OLIVER, A. J., and MEAD, R. J. 1981. *Bettongia* and fluoroacetate: a role for 1080 in fauna management. Aust. Wildl. Res. *8*, 529–536.

KIRSTEN, E., SHARMA, M. L., and KUN, K. 1978. Molecular toxicology of (−)erythro-fluorocitrate: Selective inhibition of citrate transport in mitochondria and the binding of fluorocitrate in mitochondrial proteins. Mol. Pharmacol. *14*, 172–184.

McEWEN, T. 1978. Organo-fluorine compounds in plants. In Effects of Poisonous Plants on Livestock. R. F. Keeler, K. R. Van Kampen, and L. F. James (Editors), pp. 147–158. Academic Press, NY.

McILROY, J. C. 1982A. The sensitivity of Australian animals to 1080 poison. III. Marsupial and eutherian herbivores. Aust. Wildl. J. *9*, 487–504.

McILROY, J. C. 1982B. The sensitivity of Australian animals to 1080 poison. IV. Native and introduced rodents. Aust. Wildl. J. *9*, 505–517.

MEAD, R. J., OLIVER, A. J., and KING, D. R. 1979. Metabolism and defluorination of fluoroacetate in the brush-tailed possum (*Trichosurus vulpecula*). Aust. J. Biol. Sci. *32*, 15–26.

OLIVER, A. J., KING, D. R., and MEAD, R. J. 1979. Fluoroacetate tolerance, a genetic marker in some Australian mammals. Aust. J. Zool. *27*, 363–372.

Bracken Poisoning

BARNETT, K. C., and WATSON, W. A. 1970. Bright blindness in sheep. A primary retinopathy due to feeding bracken (*Pteris aquilina*). Res. Vet. Sci. *11*, 289–290.

EVANS, I. A. 1976. Relationship between bracken and cancer. Bot. J. Linn. Soc. *73*, 105–112.

EVANS, I. A. 1979. Bracken carcinogenicity. Res. Vet. Sci. *26*, 339–348.

EVANS, I. A., PROROK, J. H., COLE, R. C., AL-SALMANI, M. H., AL-SAMARRAI, A. M. H., PATEL, M. C., and SMITH, R. M. M. 1982. The carcinogenic, mutagenic and teratogenic toxicity of bracken. Proc.—R. Soc. Edinburgh, Sect. B: Biol. Sci. *81*, 65–77.

EVANS, W. C., PATEL, M. C., and KOOHY, Y. 1982. Acute bracken poisoning in homogastric and ruminant animals. Proc.—R. Soc. Edinburgh, Sect. B: Biol. Sci. *81*, 29–64.

EVANS, W. C., KORN, T., NATORI, S., YOSHIHIRA, K., and FUKUSKA, M. 1983. Chemical and toxicological studies on bracken fern, *Pteridium aquilinum* var. *latiusculum*. VIII. The inability of bracken extracts containing pterosins to cause cattle bracken poisoning. J. Pharm.-Dyn. *6*, 938–940.

GARRETT, B. J., CHEEKE, P. R., MIRANDA, C. L., GOEGER, D. E., and BUHLER, D. R. 1982. Consumption of poisonous plants (*Senecio jacobaea, Symphytum officinale, Pteridium aquilinum, Hypericum perforatum*) by rats: Chronic toxicity, mineral metabolism and hepatic drug-metabolizing enzymes. Toxicol. Lett. *10*, 183–188.

HIRONO, I. 1981. Natural carcinogenic products of plant origin. CRC Crit. Rev. Toxicol. *8*, 235–276.

HIRONO, I., FUSHIMI, K., MORI, H., MIWA, T., and HAGA, M. 1971. Comparative study of carcinogenic activity in each part of bracken. J. Natl. Cancer. Inst. (U.S.) *50*, 1367–1371.

HIRONO, I., KONO, Y., TAKAHASHI, K., YAMADA, K., NIWA, H., OJIKA, M., KIGOSHI, K., NIIYAMA, K., and UOSAKI, Y. 1984. Reproduction of acute bracken poisoning in a calf with ptaquiloside, a bracken constituent. Vet. Rec. 115: 375–378.

IDRUS, A. Z., WALKER, H. F., MacDONALD, D. C., and TOPPS, J. H. 1977. The effect of phenothiazine on the toxicity of bracken to sheep. Proc. Nutr. Soc. *36*, 83A.

PAMUKCU, A. M., WATTENBERG, L. W., PRICE, J. M., and BRYAN, G. T. 1971. Phenothiazine inhibition of intestinal and urinary bladder tumors induced in rats by bracken fern. J. Natl. Cancer Inst. (U.S.) *47*, 155–159.

PAMUKCU, A. M., YALCINER, S., and BRYAN, G. T. 1977. Inhibition of carcinogenic effect of bracken fern (*Pteridium aquilinum*) by various chemicals. Cancer *40*, 2450–2454.

PAMUCKU, A. M., YALCINER, S., HATCHER, J. F., and BRYAN, G. T. 1980A. Quercetin: A rat intestinal and bladder carcinogen present in bracken fern. Cancer Res. *40*, 3466–3472.

PAMUCKU, A. M., HATCHER, J., TAGUCHI, H., and BRYAN, G. T. 1980B. Quercetin. An intestinal and bladder carcinogen present in bracken fern. Proc. Am. Assoc. Cancer Res. *21*, 74.

PAMUCKU, A. M., WANG, C. Y., HATCHER, J., and BRYAN, G. T. 1980C. Oncogenicity of tannin from bracken fern. JNCI, J. Natl. Cancer Inst. *65*, 131–136.

PRICE, J. M., and PAMUCKU, A. M. 1968. The induction of neoplasms of the urinary bladder of the cow and the small intestine of the rat by feeding bracken fern. Cancer Res. *28*, 2247–2251.

Buckwheat Toxicity

ANDERSON, D. M., and BOWLAND, J. P. 1981. Evaluation of buckwheat (*Fagopyrum esculentum*) in diets of growing pigs. Proc.—Annu. Meet.—Am. Soc. Anim. Sci., West. Sect. *32*, 422–425.

EGGUM, B. O., KREFT, I., and JAVORNIK, B. 1981. Chemical composition and protein quality of buckwheat. Qual. Plant.—Plant Foods Hum. Nutr. *30*, 175–179.

FARRELL, D. J. 1978. A nutritional evaluation of buckwheat (*Fagopyrum esculentum*). Anim. Feed Sci. Technol. *3*, 95–108.

MULHOLLAND, J. G., and COOMBE, J. B. 1979. A comparison of the forage value for sheep of buckwheat and sorghum stubbles grown on the Southern Tablelands of New South Wales. Aust. J. Exp. Agric. Anim. Husb. *19*, 297–302.

NICHOLSON, J. W.. McQUEEN, R., GRANT, E. A., and BURGESS, P. L. 1976. The feeding value of tartary buckwheat for ruminants. Can. J. Anim. Sci. *56*, 803–808.

THACKER, P. A., ANDERSON, D. M. and BOWLAND, J. P. 1983A. Nutritive value of common buckwheat as a supplement to cereal grains when fed to laboratory rats. Can. J. Anim. Sci. *63*, 213–219.

THACKER, P. A., ANDERSON, D. M., and BOWLAND, J. P. 1983B. Chemical composition and nutritive value of buckwheat cultivars for laboratory rats. Can. J. Anim. Sci. *63*, 949–956.

Alsike Clover Poisoning

TAUB, J. L., POTTER, K. A., BAYLEY, W. M., and REED, S. M. 1982. Alsike clover poisoning. Mod. Vet. Pract. *63*, 307–309.

Yellow Star Thistle (*Centaurea* spp.)

CORDY, D. R. 1978. *Centaurea* species and equine nigropallidal encephalomalacia. In Effects of Poisonous Plants on Livestock. R. F. Keeler, K. R. Van Kamen, and L. F. James (Editors), pp. 327–336. Academic Press, NY.

Stypandra imbricata (Blind Grass)

HUXTABLE, C. R., DORLING, P. R., and SLATTER, D. H. 1980. Myelin oedema, optic neuropathy and retinopathy in experimental *Stypandra imbricata* toxicosis. Neuropathol. Appl. Neurobiol. *6*, 221–232.

MAIN, D. C., SLATTER, D. H., HUXTABLE, C. R., CONSTABLE, I. C., and DORLING, P. R. 1981. *Stypandra imbricata* (blindgrass) toxicosis in goats and sheep—clinical and pathologic findings in 4 field cases. Aust. Vet. J. *57*, 132–135.

Weed Seeds in Grains and Screenings

HARROLD, R. L., and NALEWAJA, J. D. 1977. Proximate, mineral and amino acid composition of 15 weed seeds. J. Anim. Sci. *44*, 389–394.

HARROLD, R. L., CRAIG, D. L., NALEWAJA, J. D., and NORTH, B. B. 1980. Nutritive value of green or yellow foxtail, wild oats, wild buckwheat or redroot pigweed seed as determined with the rat. J. Anim. Sci. *51*, 127–131.

SCHROEDER, M., DELI, J., SCHALL, E. D., and WARREN, G. F. 1974. Seed composition of 66 weed and crop species. Weed Sci. *22,* 345–348.
SHIRES, A., BELL, J. M., KEITH, M. O., and McGREGOR, D. I. 1982. Rapeseed dockage. Effects of feeding raw and processed wild mustard and stinkweed seed on growth and feed utilization of mice. Can. J. Anim. Sci. *62,* 275–285.
TKACHUK, R., and MELLISH, V. J. 1977. Amino acid and proximate analyses of weed seeds. Can. J. Plant Sci. *57,* 243–249.

Blue-Green Algae

ELLEMAN, T. C., FALCONER, I. R., JACKSON, A. R. B., and RUNNEGAR, M. T. 1978. Isolation, characterization and pathology of the toxin from a *Microcystis aeruginosa* (=*Anacystis cyanea*) bloom. Aust. J. Biol. Sci. *31,* 209–218.
EVERIST, S. L. 1981. Poisonous Plants of Australia. Angus & Robertson Publishers, Melbourne, Australia.
JACKSON, A. R. B., McINNES, A., FALCONER, I. R., and RUNNEGAR, M. T. C. 1984. Clinical and pathological changes in sheep experimentally poisoned by the blue-green algae *Microcystis aeruginosa*. Vet. Pathol. *21,* 102–113.
SEAWRIGHT, A. A. 1982. Animal Health in Australia. Vol. 2. Chemical and Plant Problems. Australian Government Publishing Service, Canberra.

12

Mycotoxins

The term "mycotoxin" is derived from *myco* meaning fungi and *toxin* meaning toxicant of biological origin. Mycotoxins are secondary substances or metabolites produced by a wide range of fungi, principally molds. Fungi that produce mycotoxins are generally called toxigenic species. There are over 100 known species of molds that produce mycotoxins. Most mycotoxins of importance in American agriculture are produced by species in the genera *Aspergillus, Penicillium,* and *Fusarium*. Some species produce a single mycotoxin, while others produce multiple toxins that are chemically related. When a sufficient quantity of food or feed containing mycotoxins is ingested by animals, poisoning can occur. Mycotoxin-induced disease states in animals are called "mycotoxicoses." This term should not be confused with "mycoses" which are diseases caused by fungi, not mycotoxins. Several mycotoxicoses in livestock and poultry have been clearly associated with the presence of feed-borne mycotoxins (Table 12.1).

Individual mycotoxins are not named in a standardized manner. Some derive their names from the producing mold or the type of toxic effect manifested. Others are known by their actual chemical name.

Mycotoxicoses are important on a worldwide basis. Their impact on animal agriculture has been recognized for many centuries. Until about 1960, concern centered mainly on poisoning of animals and associated economic loss. More recently, the issue of food safety due to food-chain transfer of residues has demanded greater attention. The change of focus stemmed primarily from the discovery of aflatoxin as the cause of widespread losses among turkey poults, swine, and cattle in the early 1960s. This event stimulated public concern and scientific inquiry into such toxic effects as carcinogenicity, teratagenicity, and immunosuppression. During the same era, advancements in analytical

TABLE 12.1. Mycotoxicoses in Livestock and Poultry

Mycotoxicosis/Causative mycotoxin(s)	Animal species	Main toxigenic fungi	Main susceptible crops
I. Toxicoses associated with concentrate grains			
Aflatoxicosis B_1, B_2, G_1, G_2	Poultry, swine, cattle	Aspergillus flavus, A. parasiticus	Corn, small grains, peanuts, cottonseeds, cassava, copra, most nut crops, mixed feeds
Ergotism Ergotamine, ergotoxin, and ergometrine alkaloids	Poultry, cattle, sheep, swine	Claviceps purpurea	Rye, wheat, barley, rice, millet, sorghum, triticale
Zearalenone (F-2) toxicosis	Swine	Fusarium roseum and several other Fusarium spp.	Corn, wheat, barley, sorghum, oats
Trichothecene toxicoses T-2 toxin, diacetoxyscirpenol, vomitoxin, and several others (>40 known)	Swine, cattle, poultry	Fusarium spp. (8 known), e.g., F. trichinctium and F. roseum, and 4 other genera	Corn, wheat, barley, oats, millet, rye, mixed feeds
Leucoencephalomalacia, toxin unknown	Horse	Fusarium moniliforme	Corn
Ochratoxicosis Ochratoxin A, B (7 others not found naturally)	Swine, poultry	Penicillium viridicatum, Aspergillus ochraceus	Corn, wheat, oats, barley, beans, peanuts, mixed feeds
Citrinin toxicosis	Swine, poultry	Penicillium spp. (13) and	Wheat, rye, oats, barley

II. Mycotoxicoses associated with roughages

Disease/Toxin	Animals	Fungus	Substrate
Lupinosis Phomopsin A, other unknown toxins	Sheep, cattle	*Phomopsin leptostromiformis*	Lupines
Facial eczema Sporidesmin	Sheep, cattle	*Pithomyces chartarum*	Pasture grasses chiefly ryegrass
Paspalum staggers Paspalitrems, e.g., paxilline, paspalanine	Cattle, sheep, horses	*Claviceps pasali*	*Paspalum* grasses, e.g., Dallis grass
Tremors (staggers) Penitrem A, lolitrem A, janthitrem; >20 tremorgens known	Cattle, sheep, horses	*Penicillium* spp., e.g. *P. crustocium*, *P. cyclopium*	Ryegrass, silage, corn
Slobbers Slaframine	Cattle, sheep, horses	*Rhizoctonia leguminecala*	Legumes, principally red clover
Sweet clover poisoning Dicoumarol	Cattle, sheep, horses	Unknown	Sweet clover
Stachybotryotoxicosis Stachybotryotoxins, e.g., satratoxin	Horse	*Stachybotrys* spp., e.g., *S. atra* and *Mycothecium* spp.	Hay, straw

methodologies provided a means for identifying and quantitating mycotoxins and their metabolites in meat, dairy, and poultry products.

GENERAL ELEMENTS OF MYCOTOXICOLOGY

The subject of mycotoxins can be subdivided into three general areas: mycology, mycotoxin chemistry, and toxicology. Mycology encompasses mold taxonomy as well as all the requirements, processes, and factors necessary for colonization (spore germination and mold growth) and biosynthesis of mycotoxins in foods and feeds.

Taxonomic classification of fungi has historically been based almost exclusively on morphological rather than physiological considerations. Consequently, there is little correlation between the pattern of toxin production by particular fungi and their phylogenetic classification (Samuels 1984). Simpson and Batra (1984) divided toxigenic fungi into two broad nonphylogenetic groups: (1) endomycotoxic fungi and (2) exomycotoxic fungi. Species in the first group produce intracellular toxins that have typically acute and sometimes fatal toxic effects on animals. Certain species of the genus *Amanita* (toxic mushrooms) are endomycotoxic fungi. Members of this group have had a negligible impact on animal agriculture. The second group, exomycotoxic fungi, produces toxins that are excreted into the substratum. There are over forty known genera of fungi that fall into this group. The most toxicologically significant of these genera can be assigned to two phylogenetic orders: Eurotiales which includes *Aspergillus* and *Penicillium,* and Hypocreales which includes *Fusarium.* Although different genera within each order may produce the same or similar toxins, specific toxins are not shared between these two orders (Samuels 1984). Toxins produced by genera of Eurotiales are quite diverse and representative of several different classes of secondary metabolites including the polyketides (aflatoxins, ochratoxins, citrinin, patulin, and sterigmatocystins) and the bisanhydrides (rubratoxin). More toxigenic genera are classified under Hypocreales than any other order. Toxigenic genera include *Fusarium, Myrothecium,* and *Trichoderma.* Compared to Eurotiales a larger number of hypocrealean genera have toxigenic species, but the number of toxins produced is fewer and includes the polyketide zearalenone and the sesquiterpene trichothecenes (Samuels 1984). Most of the hypocrealean toxins are produced by species of *Fusarium,* a genus that has undergone much taxonomic revision in recent years. Other notable orders, but of less toxicological significance, include: (1) Clavicipitales, e.g., *Claviceps* spp. that produce ergot alkaloids; (2) Pleosporales, e.g., *Bipolaris* spp. that produce sterigmatocystins; (3)

Sordariales, e.g., *Stachybotrys* spp. that produce macrocyclic trichothecenes; and Diaporthales, e.g., *Phomopsis* spp. that produce toxins associated with lupinosis.

Toxigenic species within the three dominant genera, *Aspergillus, Penicillium,* and *Fusarium,* occupy a wide range of habitats, sometimes alone, sometimes in competition. Pitt and Udagawa (1984) distinguished these three genera in a general way as follows: *Fusarium* species attack plant crops as an active pathogen, but are seldom found in a saprophytic role. *Penicillium* species are active pathogens on a few kinds of fruit, but their dominant role is in saprophytic habitats, especially in cooler climates. *Aspergillus* species are occasional pathogens in plant or animal tissues, and are dominant saprophytes in warmer regions, and under conditions of reduced water activity.

Spores of most toxigenic fungi are ubiquitously distributed throughout the world. Colonization and mycotoxin elaboration can occur in growing crops or harvested and stored feeds when certain nutritional and environmental requirements are satisfied. Although substantial variation exists among fungal species, the basic requirements for colonization include (1) a substrate (e.g., feedstuff) which will supply energy and nutrients, (2) a moisture content of about 14–24% and relative humidity greater than 70%, (3) appropriate temperature which varies widely among different species, and (4) oxygen. Whether the requirements for toxin elaboration are the same as those for mold growth is species specific.

The susceptibility of different feedstuffs to mold infection is directly related to their adequacy as substrates. The seed or kernel, because of its richness in carbohydrates, is a frequently targeted plant part. However, other species utilize cellulose and therefore prefer the more fibrous parts of the plant. Many factors that enhance mycotoxin production do so by increasing the availability of substrate to the fungus. Damage to the seed coat (pericarp) of corn induced by insects, drought, hail, frost, or mechanical harvesting favors fungal invasion. Moreover, insects can also serve as carriers of fungal spores.

The occurrence of preharvest mycotoxins in certain geographic regions is primarily a function of climate. Toxigenic fungi that produce aflatoxins thrive in warm, humid environments which exist in the southern U.S. Others require both warm and cool temperatures such as the *Fusarium* spp. that produce trichothecenes and zearalenone. These climatic conditions are found in the north-central region of the U.S. In farm-stored feeds, a combination of climate and specific storage conditions is influential in both colonization and toxin production; therefore, less regionality is expected. Also, transport or incorporation of grains into commercial feeds removes regionality.

Another important concern in mycology is establishing whether moldy feed is inhabited by toxigenic species. Visual or cultural evidence of mold growth in feed is not a reliable measure, neither qualitatively nor quantitatively, of the presence of mycotoxins. Although many molds can cause deleterious compositional, textural, organoleptic, or palatability changes, they may not produce mycotoxins. On the other hand, mycotoxins may be present in feeds without visual evidence of moldiness. This is because mycotoxins are generally much more stable and persistent in feeds than the fungi that produced them. Feed processing, particularly pelleting which kills fungi due to increased temperature, usually does not destroy the toxin.

Progress in mycology should help provide solutions to mycotoxin problems in the future. Many of the processes and factors that cause fungi to synthesize mycotoxins have not been elucidated. A better understanding of these factors may prompt discovery of preventative or control measures in the future. Also, there are suspected mycotoxicoses for which neither a fungal species nor a mycotoxin has yet been identified. Several examples are given in Table 12.2.

Through research, causative organisms and the mycotoxins involved will eventually be discovered. Mycological research has demonstrated the existence of toxigenic fungi and associated mycotoxins prior to recognition of a field problem. For example, T-2 toxin was isolated and identified in 1968, but its natural existence was not recognized until 1971 when moldy corn containing T-2 toxin was diagnosed as the cause of dairy cattle deaths in Wisconsin (Hsu et al. 1972). Several toxigenic fungi and their mycotoxins have been demonstrated in the laboratory, but their involvement in field disease remains to be established (Table 12.3).

The second subject area, mycotoxin chemistry, encompasses composition and analysis of mycotoxins. A great deal of progress has been

TABLE 12.2. Diseases Suspected as Mycotoxicoses[a]

Disease	Affected animal	Major effect
Ryegrass staggers	Sheep, cattle	Ataxia
Bermuda grass staggers	Cattle	Ataxia
Myrotheciotoxicosis	Sheep, cattle	Depression, enteritis, salivation
Fescue foot	Cattle	Necrosis of fetlock
Deg Nala	Water buffalo	Necrosis of extremities
Endemic Balkan nephropathy	Man	Nephritis
Onyalai	Man	Thrombocytopenic purpura
Kashin-Beck	Man	Osteoarthritis

[a] From Pier (1981).

TABLE 12.3. Mycotoxins Not Yet Definitely Associated with Field Disease[a]

Mycotoxin	Experimental effects
Alternaria toxin	Depression, fetal death, hemorrhage
Citreovirdin	Weakness, paralysis
Cyclopiazonic acid	Hepatic, renal, and pancreatic necrosis
Islanditoxin (cyclochlorotine)	Periportal hepatic necrosis
Luteoskyrin	Centrilobular hepatic necrosis; carcinogenic
Patulin	Convulsions, edema, hemorrhage; teratogenic
Rubratoxin	Hepatitis, coagulopathy; teratogenic
Sterigmatocystin	Hepatic and renal necrosis; carcinogenic

[a] From Pier (1981).

made in recent years toward characterizing the chemistry of mycotoxins. A knowledge of the physicochemical properties (e.g., solubility, ionization, and molecular size and shape), established with purified mycotoxins, provides a basis for understanding their fate (absorption, distribution, metabolism, and elimination) in the body. Both qualitative and quantitative aspects of metabolism are dependent on these physicochemical properties. The availability of molecular sites for metabolism is an important determinant of whether metabolites are more or less toxic or persistent than parent mycotoxins. Similarly, the reactivity of a few mycotoxins in biological systems can be explained from their molecular structure. This subject is broadly referred to in toxicology as the "structure–activity" relationships of toxicants. For example, the reactivity of all trichothecenes is attributed to the 12,13-epoxide and 9,10 double bond in the nucleus of the molecule (Chu 1977). The type of R-group substitution accounts for the variation in toxicity of structurally related mycotoxins such as the trichothecenes. The toxicology of many mycotoxins can be explained at the chemical level. For example, the chemical basis of the carcinogenicity of aflatoxin B_1 has been elucidated (Wogen and Busby 1980).

A related area in mycotoxin chemistry is analysis. Methods and instrumentation are now available for identification and quantification of most of the common mycotoxins in animal feedstuffs. Chemical testing of feeds is the only definitive way of determining the presence of mycotoxins. It provides the basis for establishing exposure and determining whether mycotoxin residues are present in meat, dairy, and poultry products. Although this chapter does not address the subject of analyzing mycotoxins, its importance should not be underestimated. Also, the method of sampling feeds is of utmost importance. For more information on these subjects (sampling and analysis), readers are referred to the scientific literature, in particular, the *Journal of the Association of Official Analytical Chemists* and the *Journal of Agricul-*

tural and Food Chemistry. Specific recommended references include Stahr (1977), Rodricks *et al.* (1977), Lillehoj (1979), *Official Methods of Analysis* (1980), Davis *et al.* (1980), Gregory and Manley (1981), Chu (1983), and Shotwell (1983).

The third subject area, toxicology, is the primary emphasis of this chapter. Mycotoxins can cause a broad spectrum of biological effects, including both acute and chronic toxicoses. Evidence of mycotoxicoses was apparent long before their identities were known. For example, in 1908 a disease characterized by photosensitization in ruminants in New Zealand was described. This was probably the first description of facial eczema caused by sporidesmin, which was not identified until 1963 (Hesseltine 1979). Similarly, evidence of kidney degeneration in Danish swine first described in 1928 was probably caused by ochratoxin, which was not identified until 1965 (Hesseltine 1979).

Since 1960, an extensive amount of mycotoxin research has been conducted in domestic and experimental animals. The range of biological effects is as varied as the chemical structures of the different mycotoxins and, as would be expected, their individual toxicology is greatly dependent on dose and frequency and duration of exposure. In general, toxic effects are initiated at the subcellular level and may or may not be accompanied by discernible pathological changes. Aflatoxins induce numerous morphological changes in the liver which coincide with observed clinical effects whereas tremorgens produce neurological effects and perhaps death without discernible lesions. Mycotoxins are frequently classified on the basis of the organs or tissues they affect (Table 12.4). Manifestation of these effects is subject to a wide array of modifying influences which are broadly classified as genetic (e.g., species), physiological (e.g., age, sex, reproductive status), and environmental (e.g., diet, disease, other toxicants). While Table 12.4 is not an exhaustive listing, it does indicate the wide diversity of toxic effects caused by mycotoxins to which livestock may be exposed.

The most common chronic effects in livestock are decreased growth rate and feed intake. Subclinical effects occurring in the liver, kidneys, or gastrointestinal tract may be responsible. Diagnosis of mycotoxicoses is generally impossible from clinical signs or postmortem examination without confirmation by chemical analysis of feed contamination or residues in tissues or excrement. Some toxic effects are secondary, such as for aflatoxin B_1 and T-2 toxin which interfere with the immune system, increasing the susceptibility to infectious diseases. Similarly, indirect influences of nutritional status, environmental factors, and synergistic interactions among mycotoxins have been observed experimentally, but the extent of their natural existence is generally unknown.

TABLE 12.4. Classification of Mycotoxins Based on Biological Effects

Hepatotoxins
 Aflatoxin, sporidesmin, rubratoxin B, sterigmatocystin, trichothecenes, ochratoxin A, phomopsin A
Nephrotoxins
 Ochratoxin A, citrinin, aflatoxin, oxalate
Neurotoxins
 Trichothecenes (e.g., vomitoxin, satratoxin), slaframine, penitrem A (and other tremorgens), ochratoxin A (poultry only), ergot alkaloids, lupinosis (ammonia accumulation)
Genitoxins
 Zearalenone, ergot alkaloids, aflatoxin (possibly in ruminants)
Pulmonary toxins
 4-Ipomeanol
Dermitoxins
 12,13-Epoxytrichothecenes (e.g., T-2 toxin, nivalenol)
Photosensitizing agents
 Sporedesmin (facial eczema)
Carcinogens
 Aflatoxin, sterigmatocystin, luteoskyrin, patulin, penicillic acid, T-2 toxin (possibly), ochratoxin A (possibly), citrinin (possibly)
Teratogens
 Aflatoxin (possibly), ochratoxin A (possibly)
Immunosuppressants
 Aflatoxin, trichothecenes (T-2 toxin), ochratoxin (possibly)
Hematologic agents
 Dicoumarol, aflatoxin, trichothecenes

Although to date mycotoxins have provided very few benefits compared to the losses, some commercial applications have been found. Zearalenol, which is a natural metabolite of zearalenone, is marketed as a growth promotant for cattle and has possible application as a chemotherapeutant for humans. Other mycotoxins are known to have antibiotic and herbicidal activity which may be developed in the future.

This chapter discusses only those mycotoxins confirmed or suspected of causing mycotoxicoses in livestock and poultry. Some mycotoxins belong to chemical categories previously discussed and were included in those chapters. These are slaframine (Chapter 5), an indolizidine alkaloid produced by fungi growing on red clover, ergot alkaloids (Chapter 5), which are produced by *Claviceps* spp. of fungi, and dicoumarol associated with sweet clover poisoning (Chapter 6). Fescue foot and summer fescue toxicosis are examples of suspected mycotoxicoses for which causative mycotoxins have not been definitively identified; because fescue contains alkaloids, a discussion of these toxicoses was presented in Chapter 5.

The mycotoxins discussed in the following sections are broadly classified as to whether they are produced by fungi growing on concentrate-type feeds or on roughages. Those in the former category include the aflatoxins, zearolenone, trichothecenes, ochratoxin, citrinin, rubratoxin, and patulin. The order of their presentation reflects the current information base and apparent real or potential significance to animal agriculture. In the latter group are the tremorgens, lupine mycotoxins, sporidesmin, and stachybotryotoxins. Two mycotoxicoses associated with roughages for which causative fungi have not been delineated, namely, moldy straw-induced photosensitization and kikuyu poisoning, are also discussed.

MYCOTOXINS ASSOCIATED WITH CONCENTRATE FEEDS

Aflatoxin

Aflatoxins are a family of related bisfuranocoumarin compounds produced primarily by toxigenic strains of *Aspergillus flavus* and *Aspergillus parasiticus*. Only about one-half of the known strains of *A. flavus* and *A. parasiticus* produce aflatoxins. Although other fungi such as *Penicillium* spp., *Rhizopus* spp., *Mucor* spp., and *Streptomyces* spp. are capable of producing aflatoxins, their relevance to livestock production has not been established. The name "aflatoxin" derives from *Aspergillus* (a-), *flavus* (-fla-), and toxin. *Aspergillus flavus* and *A. parasiticus* produce four major toxins: B_1, B_2, G_1, and G_2 (Fig. 12.1). These were named according to their fluorescence properties under shortwave ultraviolet light on thin-layer chromatographic (TLC) plates; B_1 and B_2 fluoresce blue, whereas G_1 and G_2 fluoresce green. Fourteen other aflatoxins are known but most of these are metabolites formed endogenously in animals administered one or more of the four major aflatoxins. Metabolites of toxicological significance include: aflatoxin B_1 2,3-oxide (AFB_1 2,3-oxide); aflatoxin M_1 (AFM_1); aflatoxicol; and aflatoxin B_{2a} (AFB_{2a}) (Fig. 12.1).

Aflatoxins were discovered as a result of massive losses of turkeys in Great Britain in 1960. Over 100,000 turkeys died in the outbreak, which was called "turkey X disease." Intensive investigation revealed the cause to be mycotoxins in moldy groundnut (peanut) meal that was imported from Brazil for use as a protein supplement in animal diets.

Occurrence. Aflatoxin-producing strains of *Aspergillus* are distributed worldwide in soil and air. When environmental conditions are favorable and a substrate (feed or seed) is accessible as a nutrient source, colonization and mold growth can easily occur. The resulting profile of aflatoxins and their individual concentrations will vary greatly according to the existing environmental conditions (temperature, moisture, aeration), the substrate, and the type of mold involved. For example, *A. flavus* growing on corn produces primarily B_1 and B_2 whereas *A. parasiticus* on corn produces all four major alfatoxins: B_1, B_2, G_1, and G_2. On soybeans, only negligible concentrations of B_1 are produced by both species. *Aspergillus flavus* is primarily a seed-colonizing mold and is usually referred to as a storage mold. It is capable of colonizing most of the important grain crops including corn, small grains, peanuts, cottonseed, cassava, copra, and most nut crops if moisture and temperature conditions are favorable. The moisture content of the seed is probably the most important factor. In general, mold growth and aflatoxin formation require a moisture content of greater than 14%, a temperature of at least 25°C, and some degree of aeration (O_2). When these requirements are met, mold infestation followed by aflatoxin formation in target crops are likely to occur.

Three major feedstuffs with high potential for invasion by *Aspergillus* spp. during growth, harvest, transportation, or storage are corn, cottonseed, and peanuts. Colonization of soybeans and small grains generally occurs in storage. Storage conditions for soybeans that promote aflatoxin formation, aside from optimal moisture and temperature conditions, are lack of aeration systems or their improper use (temperature differentials can cause moisture migration), kernel damage and spore dissemination caused by storage insects, presence of fines (dust, weed seeds, and broken kernels), and poor sanitary practices in feed areas.

In corn, elevated aflatoxin content in any particular year is usually the result of increased invasion by molds prior to harvest. There are several possible contributing factors. Drought-stressed corn is susceptible to damage from insects such as corn earworms or borers that feed on the husks or kernels of the ear. Kernels with a disrupted seed coat (pericarp) are more accessible to fungal spores that may be present on the silks or that are carried on the bodies of insects. Once in the kernel, the spores germinate and grow utilizing the nutrients afforded by the kernel. Other factors or stresses known to be associated with increased aflatoxin contamination of corn include corn left in the field beyond maturity, close planting, competition from weeds, and inadequate fertilization. Stored corn, particularly ground, high-moisture corn, has an explosive potential for aflatoxin production. Storage in airtight silos or incorporation of certain preservatives can effectively retard mold

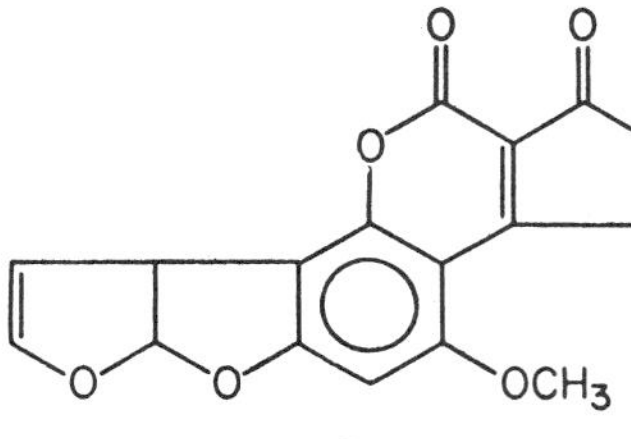

FIG. 12.1. Chemical structures of naturally occurring aflatoxins (B_1, B_2, G_1, G_2) and their metabolites. (I) Aflatoxin B_1; (II) aflatoxin B_2; (III) aflatoxin G_1; (IV) aflatoxin G_2; (V) aflatoxin M_1; (VI) aflatoxin M_2; (VII) aflatoxin B_{2a}; (VIII) aflatoxin B_1 2,3-oxide; (IX) aflatoxicol; (X) aflatoxin H_1; (XI) aflatoxin P_1; (XII) aflatoxin Q_1.

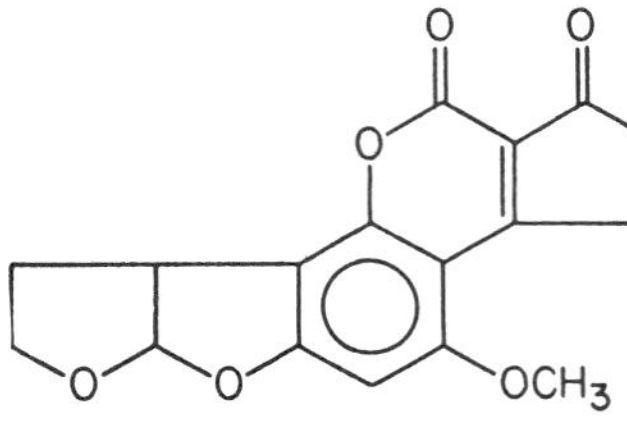

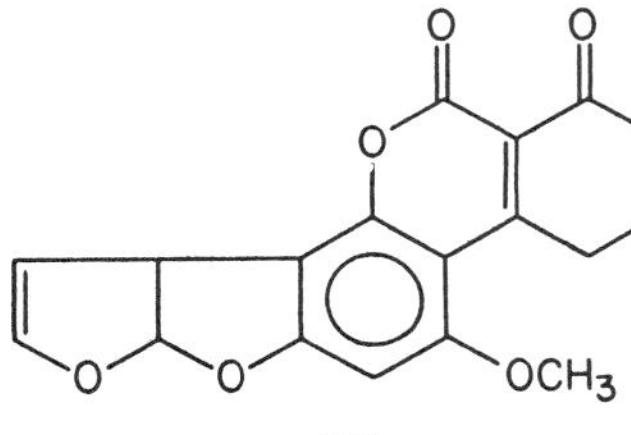

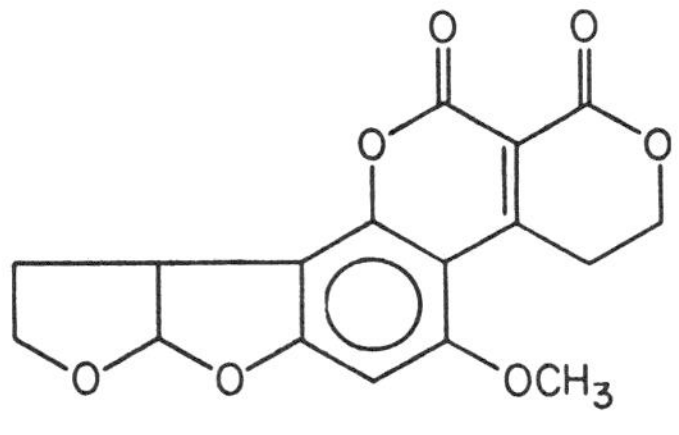

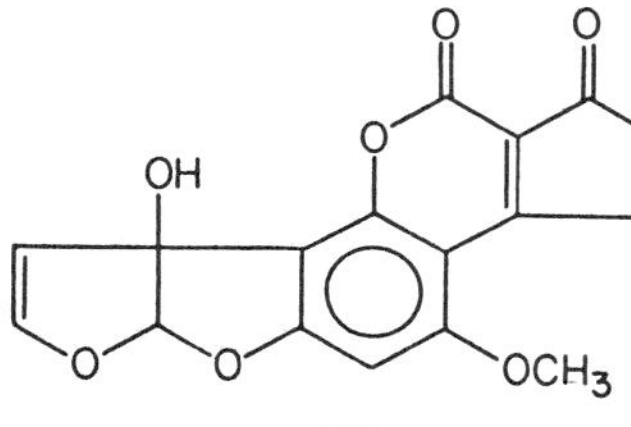

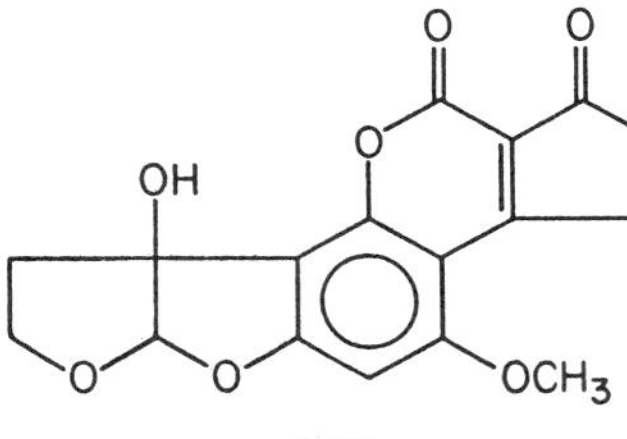

growth. Inadequately dried corn not anaerobically stored will probably be invaded by various molds. Prepared feeds left for more than a day or two in feeding bins and troughs are also susceptible.

In cotton, aflatoxin production is primarily a field problem, with insect damage an important factor. *Aspergillus flavus* penetrates the carpel wall of bolls through damaged areas such as exit holes of the pink bollworm. The lygus bug and stinkbug contribute to mold colonization by serving as spore carriers. Chronic field contamination is usu-

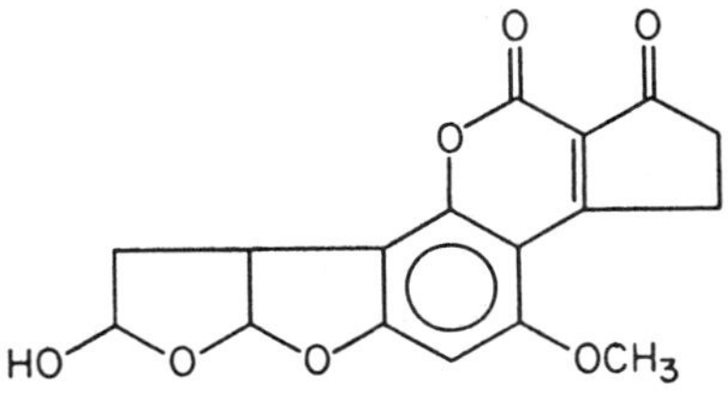

VII

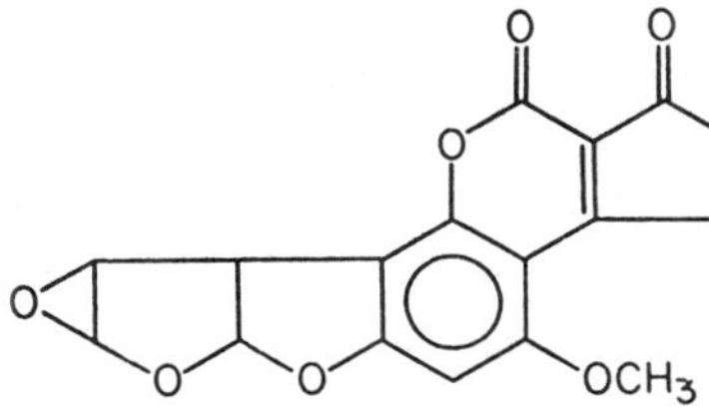

VIII

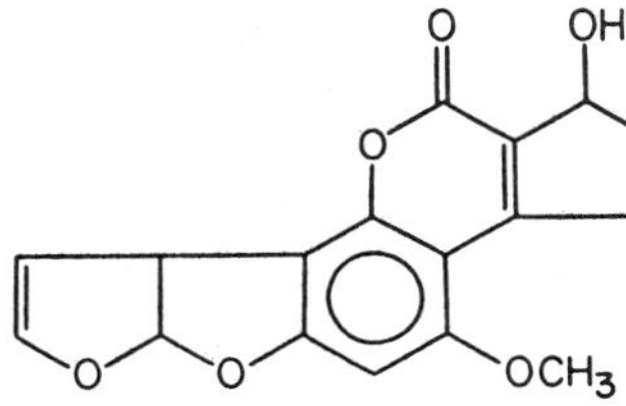

IX

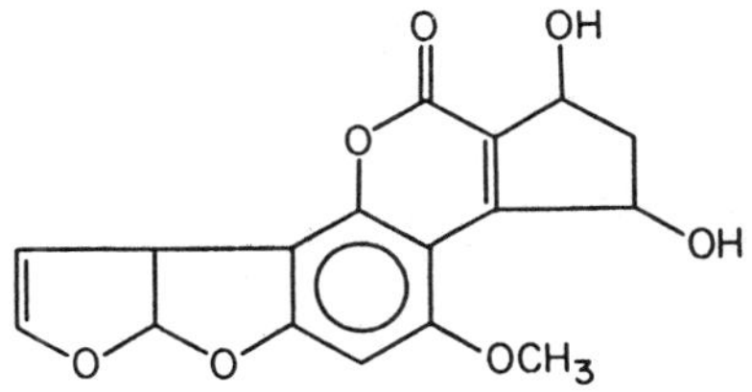

X

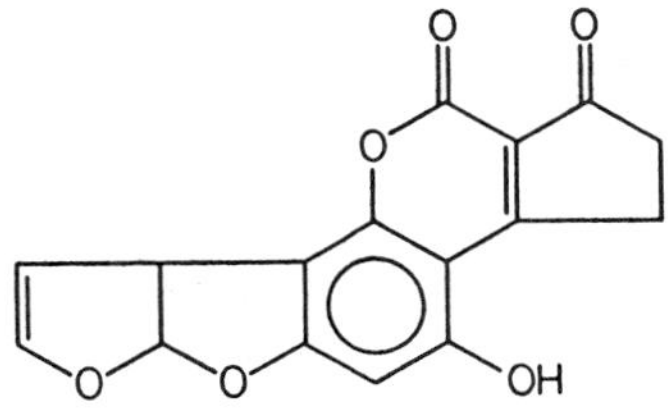

XI

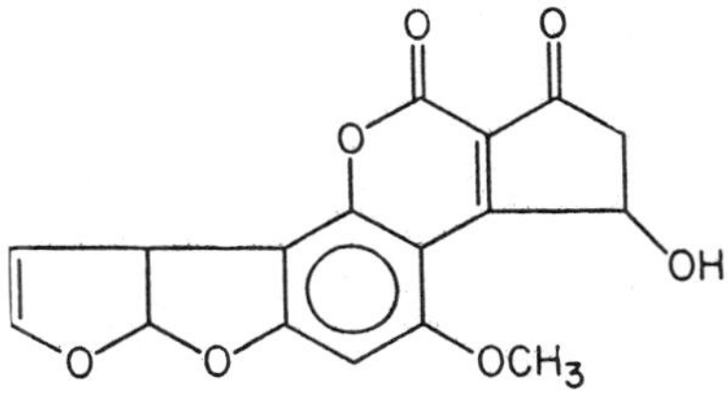

XII

FIG. 12.1. *(Continued)*

ally associated with daily mean temperatures of 34°C or greater late in the growing season (July and August in the U.S.) coupled with unusually heavy rainfall. If the cotton is harvested before the moisture has evaporated, aflatoxin production in storage is likely to follow. When aflatoxin-containing cottonseed is processed for oil, most of the toxin is concentrated in the meal. Cottonseed meal (CSM) is a common protein supplement for livestock and poultry. A serious problem of liver cancer in hatchery trout fed aflatoxin-containing CSM occurred in the 1960s

(Halver 1969). Feeding dairy cattle contaminated CSM is a problem because of possible translocation of the metabolite aflatoxin M_1 into milk.

Peanuts can be colonized by *Aspergillus* spp. in the ground before digging, during curing and drying in windrows or stacks, and in storage. Before digging, invasion has been attributed to drought-induced stress, damaged pods, or overmaturity. After digging, a moisture content of 14–30% is conducive to mold invasion and subsequent aflatoxin formation, but is prevented when the moisture content is very high. The optimum temperature range for aflatoxin production in peanut kernels of undamaged pods is 25°–35°C. When peanuts are being cured, retardation of drying by rainfall or humid weather usually results in some degree of aflatoxin production. A moisture content greater than 30% or less than 10% or a temperature greater than 41°C or less than 12°C restricts growth of mold. Prompt, steady drying to a moisture content of 7–8% within 3–5 days prevents formation of aflatoxins in peanuts. In storage, growth of *A. flavus* and production of aflatoxins are optimal at >90% relative humidity at about 30°C for either damaged or sound kernels or kernels in the shell (Diener and Davis 1969). Peanut meal (or groundnut meal) used for feed commonly carries large numbers of *Aspergillus* spores. Consequently, under favorable moisture and temperature conditions, spore germination followed by aflatoxin production can easily occur.

Toxicity in Different Species. *Acute Intoxication.* Acute aflatoxin poisoning of farm animals is less likely to occur than chronic aflatoxicosis. However, it is noteworthy that the massive poisoning of turkeys in Great Britain in 1960 which led to the initial discovery of aflatoxins was a case of acute poisoning. The principal target organ in all species is the liver. Numerous liver functions are affected, and the cumulative impact can be fatal to animals. After administration of a large single dose of AFB_1, hepatocytes undergo progressive changes, which include infiltration with lipids, eventually ending in necrosis (cell death). These toxic effects are believed to be the result of widespread and nonspecific interactions between AFB_1 or its activated metabolites and various cell proteins. Interaction with key enzymes can disrupt basic metabolic processes in the cell, such as carbohydrate or lipid metabolism, and protein synthesis. Modification of permeability characteristics of hepatocytes or subcellular organelles, primarily the mitochondria, contributes to the necrosis. As the liver loses its functionality, other effects appear such as derangement of blood-clotting mechanisms (coagulopathy), icterus (jaundice), and reduction of essential serum proteins which are synthesized in the liver. Impairment of blood clotting and increased capillary fragility result in widespread hemorrhaging, including accumulation of blood in the gastrointestinal tract. In addition to liver damage, higher doses may cause necrosis of kidney tubules in some species. Although the thymus is a target organ, the effects on immunosuppression are more associated with chronic aflatoxicosis than with acute intoxication.

In young swine, which are highly susceptible to acute poisoning, early hepatocytic changes occur within about 6 hr after exposure. Decreased liver function as indicated by sulfobromophthalein (BSP) clearance at about 3 hr may be the earliest toxic effect measurable. Hemorrhaging and cell necrosis occur by 9–12 hr, elevated serum glutamic oxaloacetic transaminase (SGOT) at 12–24 hr, and death by 24–36 hr (Pier 1981). In general, these same biological changes occur in all acutely intoxicated species. However, the susceptibility (toxicity) among different species is highly variable (Table 12.5).

These differences are believed to be directly related to the ability of animals to metabolize AFB_1 to the reactive metabolite AFB-dihydrodiol via AFB_{2a} or AFB_1 2,3-oxide intermediates (Fig. 12.2). Rabbits and ducks have a high rate of AFB_1 metabolism and are highly sensitive to aflatoxins, whereas sheep and rats have a slower rate of AFB_1 metabo-

TABLE 12.5. Comparative Toxicity of Aflatoxin B_1[a]

Species	Age or weight	Sex	Route[b]	LD_{50} (mg/kg)
Rabbit (Dutch Belted)	3 months	M/F	Ip or po	0.30
Duck (White Pekin Khaki Campbell)	1 day	M/F	Po	0.33–0.36
Cat (mixed breed)	Adult	M/F	Po	0.55
Pig (Poland–China)	Weanling	M	Po	0.62
Dog (mixed breed)	Adult	M/F	Po	0.5–1.0
Trout (rainbow)	9 months	M/F	Ip	0.81
Guinea pig	250 g	M	Ip	1.4
Sheep (cross breed)	2 years	M	Po	2.0
Baboon	5 kg	—	Po	2.02
Monkey (Cynomolgus)	2 years	M	Po	2.2
Monkey (Macaque)	38–44 months	F	Po	7.8
Rat (Fisher)	0–4 days	M/F	Ip	1.1–1.36
	12 days	M/F	Ip	12–15
	21 days	M/F	Ip	8.0
	42 days	M/F	Ip	4.0–5.0
	70 days	M/F	Ip	0.75–1.3
Hamster	30 days	M	Po	10.2
Mouse (CFW Swiss)	30 days	M	Ip	>150.0
	58 days	M	Ip	40.0
	100 days	M	Ip	12.0

[a] Adapted from Busby and Wogan (1979).
[b] Ip, intraperitoneally; po, per os (by mouth).

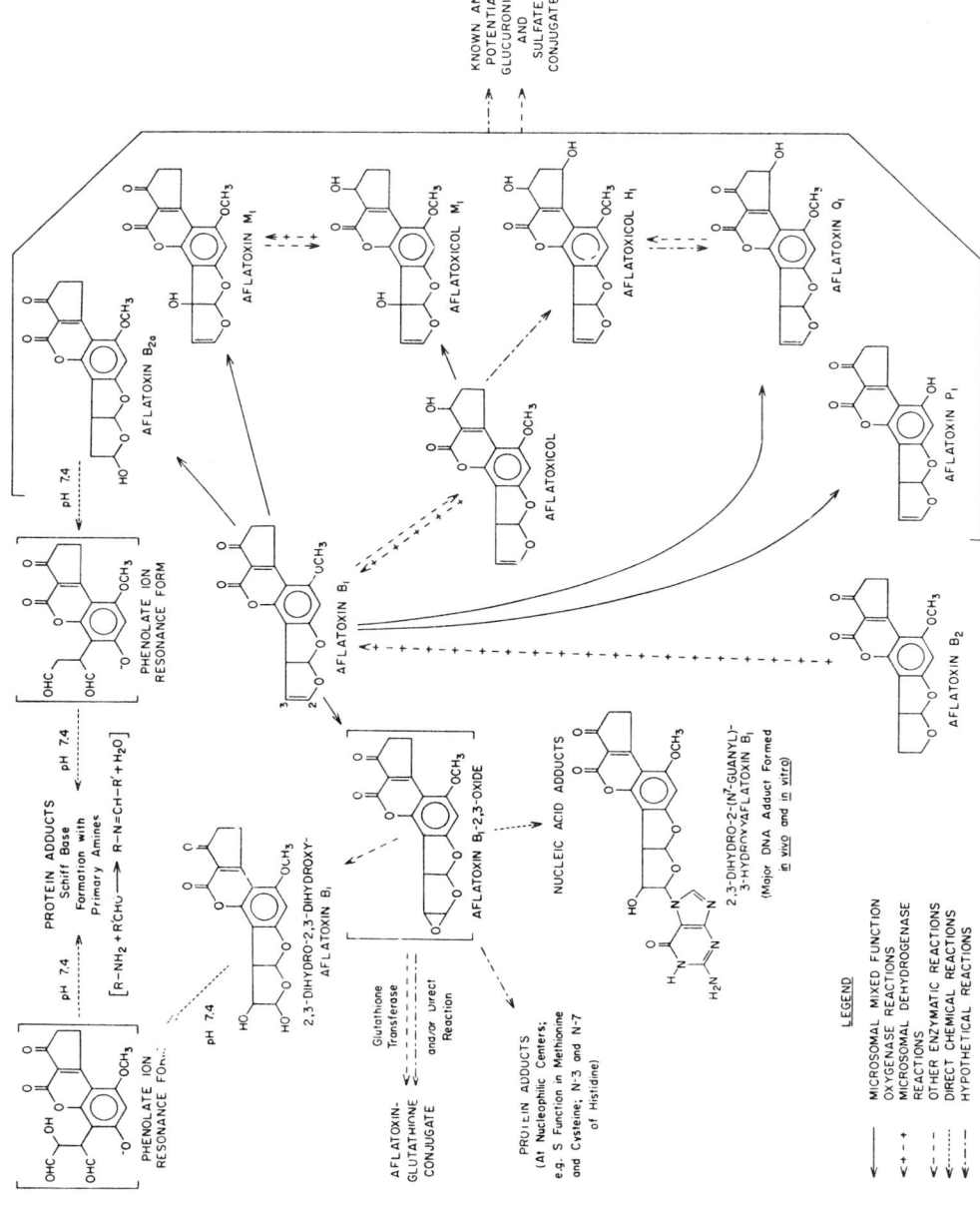

lism and are less sensitive (Hsieh et al. 1977). Species with the ability to form greater quantities of aflatoxicol also tend to be more sensitive to acute poisoning. Because the determinants of acute intoxication are specific metabolic pathways and interactions with cellular receptors that are different from chronic aflatoxicosis, the acute LD_{50} is not very useful in predicting actual field contamination problems.

Chronic Intoxication. Chronic poisoning, or aflatoxicosis, can result when low levels of toxin are ingested over a prolonged period. The toxic effects are not nearly as specific or clinically evident as in acute intoxication. In general, affected livestock exhibit decreased growth rate, lowered productivity (milk or eggs), and immunosuppression. Also, carcinogenicity has been observed and studied extensively in several nonfarm species. Although both immunosuppression and carcinogenicity are chronic toxicoses, they are discussed separately in following sections.

Reduced growth rate is considered the most common effect associated with chronic aflatoxicosis in farm animals. In young animals fed low levels of toxin this may be the only detectable abnormality. The lack of other clinical signs frequently causes aflatoxicosis to remain undiagnosed, resulting in serious economic loss. Although the actual mechanism(s) by which aflatoxins decrease growth rate is not known, subtle disturbances of one or more basic metabolic processes (carbohydrate, lipid, or protein metabolism) in the liver are probably involved. Supporting evidence is that aflatoxins increase the dietary requirement for nutrients such as protein. Without the necessary dietary adjustments, suboptimal growth would occur. In addition, a dose-related loss of appetite usually accompanies the decreased growth rate, accounting for at least part of the growth depression.

Liver damage is also prevalent in chronic aflatoxicosis in all species. At necropsy, the liver is usually pale to yellow, and the gall bladder may be enlarged. The occurrence of icterus and hemorrhaging is unpredictable and varies greatly with species and dose. Histological changes include subcellular accumulation of lipid, fibrosis, and extensive bile duct proliferation.

Swine. Since corn is a major component of swine diets, particularly in growing and finishing pigs, chronic exposure is a common occurrence in the southern and midwestern U.S. Numerous cases of porcine aflatoxicosis have been reported by veterinary diagnostic laboratories located in these regions (Hoerr and D'Andrea 1983). Feed containing 0.4 ppm or greater of AFB_1 fed from weaning to market weight can

FIG. 12.2. Metabolism of aflatoxin B_1.
From Busby and Wogan (1981): reprinted with permission, CRC Press, Boca Raton, FL.

adversely affect the health of pigs. Among the mildest effects are decreased feed efficiency and poor rate of gain. More severe effects, that may result in death, include acute hepatitis, systemic hemorrhage, and nephrosis.

Although piglets are more sensitive than older pigs, exposure during gestation or during sucking is generally not great enough to cause adverse effects. However, stunted growth has been observed in piglets that nursed on sows fed contaminated feed; AFM_1, which is a toxic metabolite of AFB_1 produced in the liver, is transferred into milk. There is no evidence that aflatoxin is detrimental to the reproductive performance of either sows or boars. However, there is indication that aflatoxins suppress the immune system of pigs causing decreased resistance to infectious diseases in the field.

The main toxic effects in growing and finishing pigs are decreased growth rate, liver damage, icterus, and hemorrhaging. Stress from handling can suddenly trigger widespread subcutaneous hemorrhaging; increased pressure in the gluteal muscles of the ham can result in ataxia exemplified by the pig assuming a doglike sitting position. Administration of vitamin K is an effective treatment. Prerequisite to hemorrhage is prolongation of blood-clotting time (coagulopathy). Deficiency of vitamin K exacerbates the condition. Both growth-rate and blood-coagulation processes return to normal when aflatoxin exposure is removed, demonstrating the reversibility of these toxic effects. Moreover, supplementation of the diet with additional vitamin K or menadione has been shown to improve weight gain during this recovery period. Also, supplementation with additional protein affords some protection during exposure.

The lesions of porcine aflatoxicosis generally reflect changes occurring directly in the liver and vasculature or indirectly from emaciation. In mild chronic aflatoxicosis, the liver becomes yellow and firm which coincides with icterus. Hepatic lesions can be subtle enough to allow the liver to pass inspection at slaughter. Histopathological hepatic changes may include variation in hepatocytic size, atypical (enlarged and multiple) nuclei, cytoplasmic lipid vacuolization, and mild fibrosis. Larger doses can cause severe fibrosis, disorganization of the lobular pattern, hyperplasia of biliary ductules, hypertrophy of biliary ducts, dilation of lymphatic vessels, and eventually hepatocytic necrosis beginning with centrilobular hepatocytes. In the latter stages of lethal chronic aflatoxicosis, the liver effects generally include severe fibrosis, necrosis, nodules of hyperplastic hepatocytes, extensive hyperplasia of biliary ductules, lipid accumulation, inflammation, and markedly atypical hepatocytes. Extrahepatic effects include edema in the lungs, at the base of the heart, and in the spinal folds of the colon.

Also, hemorrhages may be present in the gastrointestinal tract, ventricular endocardium, lymph nodes, kidneys, and the fascia and subcutis of axillary and groin tissue. Compositional changes in blood accompany organ effects. Some of these are useful diagnostic criteria.

Poultry. Avian species are quite variable in sensitivity to chronic aflatoxicosis. Turkey poults and ducklings are the most sensitive; a dietary level of 0.25 ppm impairs their growth, whereas levels of 1.5 ppm in broilers and 4 ppm in Japanese quail are required to reduce growth rate. Age is another factor causing variable toxicity. Poults and ducklings are more sensitive than their adult counterparts, whereas no age effect is apparent in chickens. In general, the toxic effects manifested in avian species are the same as those observed in mammals, including reduced growth rate, impaired blood coagulation, hemorrhage, hepatic necrosis, decreased resistance to infection, and death. While growth retardation is usually the first dose-dependent adverse effect to appear in mammals, this is not the case in birds. Impaired blood coagulation, increased susceptibility of carcasses to bruising, and decreased resistance to infection result from dosages below those affecting growth. Dietary concentrations of aflatoxin greater than 2 ppm can significantly diminish egg production in layers, with production decreased to 50% with 10 ppm and to 0% at 20 ppm.

Several interesting nutritional interactions involving protein, lipid, and vitamin metabolism have been demonstrated in birds administered aflatoxins chronically. Increasing the dietary protein content counteracts growth impairment in broilers. Increasing the lipid content of the diet reduces mortality in broilers but does not overcome other toxic effects associated with aflatoxicosis. Feeding diets deficient in certain nutrients, such as methionine, lysine, vitamin D_3, and riboflavin, exacerbates chronic aflatoxicosis. In contrast, a deficiency of thiamine or vitamin A affords some protection against aflatoxicosis. Some of the mechanisms underlying these interactions have been elucidated. In the case of vitamin D_3, it is believed that aflatoxins impair the availability of bile salts in the gut, resulting in decreased absorption of fat-soluble vitamins. Decreased liver vitamin A in intoxicated chickens is a direct result of this effect. Because of impaired vitamin D assimilation, secondary effects such as lowered blood calcium have been noted, probably due to lowered calcium absorption from the digestive tract. Rickets were observed in broiler chicks fed aflatoxin-containing diets which were also marginally deficient in vitamin D (Hamilton 1977). Decreased bone strength in broilers fed aflatoxins was attributed to inadequate mineralization of bone (Huff 1980).

Ruminants. The metabolic and physiological responses during chronic aflatoxicosis are generally the same as those that occur in nonruminants. Early signs of poisoning include reduction in feed intake followed by weight loss or decreased rate of gain. Other effects associated with chronic intake of aflatoxin-containing feedstuffs in dairy and beef cattle are decreased feed efficiency, immunosuppression, increased susceptibility to stress, and decreased reproductive performance. Calves are more sensitive than adult cattle. A dosage level of about 0.2 mg/kg body weight/day causes reduced rate of gain and impaired blood coagulation in calves. Early metabolic indicators of aflatoxicosis in calves are poor feed utilization and a rapid rise in serum alkaline phosphatase (APT) activity. In young beef cattle (400–500 lb) and in yearling steers, diets containing 7–10 ppm of aflatoxins cause decreased growth and feed efficiency. Recovery of normal rates of feed intake and growth after exposure ceases is very slow. Hepatic changes similar to those in other species are observed.

Although adult cattle are not as sensitive and the signs are not as evident as in growing animals, chronic aflatoxicosis is characterized by unthriftiness, anorexia, a drying and peeling of skin on the muzzle, prolapse of the rectum, liver damage, elevated levels of blood constituents such as cholesterol, bilirubin, and serum indicator enzymes, and edema in the abdominal cavity. Milk production may be dramatically decreased in dairy cows fed aflatoxin-contaminated feed. However, it is not clear to what degree this effect is the direct result of decreased feed intake. No effect on milk fat has been observed.

There is growing evidence of antireproductive effects of aflatoxins in ruminants. Observed effects include decreased fertility in sheep and abortion and birth of underweight calves in cattle. The mechanisms of these effects have not been delineated, but two unrelated toxic actions have been suggested: (1) an indirect effect on the dam mediated by aflatoxin-induced hypovitaminosis A, and (2) due to structural similarity of aflatoxins and steroid hormones, a direct antagonistic interaction with steroid hormone receptors (Bodine and Mertens 1983).

There is some evidence that ruminal microorganisms are relatively sensitive to aflatoxins. Results from in vitro systems indicate decreased cellulolysis, volatile fatty acid (VFA) production, and ammonia formation. Both in vitro and in vivo findings are suggestive of a shift in the proportions of VFA produced, resulting in decreased acetic acid and increased butyric acid, and a decreased total VFA production. There are several antimicrobial agents in crude extracts of *A. flavus*, but it is assumed AFB_1 is the principal toxin. Mertens (1979) speculated that the growth and function of ruminal microorganisms could be disrupted by levels of aflatoxins commonly encountered in the field. Moreover, the relative resistance of the ruminant in contrast to monogastrics points to the premise that the rumen microflora may possess some detoxification processes for aflatoxin (Bodine and Mertens 1983).

Biological Effects. *Effects on Protein Synthesis.* Aflatoxins interact with a number of different cell components involved in protein synthesis. These interactions are considered to be the major biochemical lesion in aflatoxin intoxication accounting for the majority of the acute, chronic, or carcinogenic responses in animals. The interactions can be classified into two groups: those inhibiting nucleic acid synthesis (i.e., interference with transcription processes), and those interfering with protein synthesis (i.e., translational processes).

Interference with nucleic acid synthesis is indicated by a substantial decrease in RNA content in liver nuclei after administration of AFB_1. Direct modification of the DNA template by the binding of AFB_1 metabolites to nucleophilic sites on DNA interferes with the transcription process, thereby decreasing RNA synthesis. Also, there is some evidence indicating that RNA polymerase (the enzyme that catalyzes the synthesis of RNA on the DNA template) is inhibited by AFB_1 metabolites. Another explanation for the reduced RNA content involves a posttranscriptional site of interaction. AFB_1 is apparently capable of increasing the rate of RNA breakdown by altering the lysosomal membrane, resulting in the release of ribonuclease (RNase). Of these three effects, the experimental evidence strongly favors the first, interference with template function, as the primary one.

Interference with protein synthesis at the translational level is evidenced by decreased incorporation of amino acids into proteins after administration of AFB_1. It appears that part of the effect involves disaggregation of polyribosomes in the endoplasmic reticulum (ER). Also, there is evidence suggesting that AFB_1 metabolites inhibit the elongation and/or termination steps in protein synthesis. It is generally believed that decreased protein synthesis at a low dose and at an early stage after toxin administration is due primarily to inhibitory effects at the translational level rather than RNA synthesis. At higher doses and longer exposure times all the steps in protein synthesis are affected, but interference in transcriptional processes is considered more important to the overall affect. AFB_1 metabolites may also interfere with mitochondrial protein synthesis.

Effects on Metabolic Processes. Many of the toxic responses in animals intoxicated with aflatoxins can be attributed to alterations of basic metabolic processes. These include interference with carbohydrate and lipid metabolism and mitochondrial respiration. However, the contribution of alterations in these essential processes to the overall toxicity has been difficult to assess.

The primary effect on carbohydrate metabolism is decreased hepatic glycogen content simultaneous with increased serum glucose. The decrease in glycogen apparently results from inhibition of key glycogenic enzymes and increased activity of enzymes that deplete glycogen precursors. It is believed that inhibition of enzyme synthesis in the liver accounts for the decreased glycogenesis rather than a direct inhibitory reaction between the enzymes and aflatoxins. In vivo studies on glycogenesis have been complicated somewhat by the normal depletion of glycogen associated with aflatoxin-induced anorexia.

Interference with lipid metabolism occurs in several ways. The most prominent effect is the accumulation of lipid in the liver, resulting in a condition known as "fatty liver." The effect is believed to be due to impaired transport of lipids (triglycerides, phospholipids, and cholesterol) out of the liver after their synthesis rather than an increase in their synthesis. It is not clear if the effect on lipid transport is secondary (e.g., decreased protein synthesis) or is a primary lesion during aflatoxicosis. In chickens, impaired lipid transport occurs at a dosage considerably lower than that required for reduction of growth rate and RNA synthesis. A second effect of aflatoxins on lipid metabolism is inhibition of fatty acid and cholesterol biosynthesis. Evidence suggests two types of interference can occur: decreased synthesis of enzymes such as fatty acid synthetase, and direct interaction between aflatoxins and enzymes such as thiokinase. A third effect on lipid metabolism is interference with absorption of lipids. In chickens, decreased absorption from the gut is believed to result from decreased bile salts and pancreatic lipase. Increased excretion of lipids in feces supports this conclusion.

Several mitochondrial changes are induced by aflatoxins. Inhibition of Krebs cycle enzymes, particularly succinic dehydrogenase, results in a reduction of substrate phosphorylation. A second effect which is exemplified by a reduction in oxygen consumption in isolated mitochondrial preparations after aflatoxin is introduced either in vitro or in vivo is inhibition of electron transport apparently at a site between cytochrome b and cytochrome c_1 or c. AFB_1 is the most potent inhibitor, followed by AFM_1, and then AFG_1. Also, AFB_1 has been shown to inhibit cytochrome oxidase. A third effect is the uncoupling of oxidative phosphorylation and reduction of the ADP:O ratio, also measured in vitro. Inhibition of various hepatic ATPase activities has also been observed.

Interaction with Cellular Macromolecules and Organelles. The ability of aflatoxins to alter vital biochemical processes lies in their interaction with macromolecules, such as nucleic acids or proteins, and with subcellular organelles, such as mitochondria and ribosomes.

These interactions may be either covalent or noncovalent; metabolic activation of aflatoxin is prerequisite to covalent binding to macromolecules.

Aflatoxins have a high affinity for nucleic acids and polynucleotides (RNA, DNA), particularly at nucleophilic sites on the guanine base. Formation of these aflatoxin–nucleic acid adducts (covalent bonds) can lead to a number of different biological effects, including cancer. In general, the toxicologic and carcinogenic potency of different aflatoxins appears to correlate well with their affinity for nucleic acids, after metabolic activation.

The dialdehyde groups of AFB_{2a} or AFG_{2a} have the potential of reacting covalently with amino groups in proteins (enzymes) forming Schiff bases. The significance of this reaction on the activity of most enzymes is poorly understood. Aflatoxin B_1 binds noncovalently with serum albumin at nucleophilic centers, including cysteine, histidine, and lysine. Similarly, binding to lysine-rich histone has been demonstrated. Although interactions between aflatoxins and regulatory proteins such as histone, steroid hormone receptors, or glucocorticoid receptors have been observed, the toxicologic significance is generally unknown. Aflatoxins are also capable of binding to subcellular organelles such as ribosomes and other sites in the endoplasmic reticulum. Noncovalent binding to polyribosomes is believed to be an important part of the inhibitory effect on protein synthesis.

Carcinogenesis, Teratogenesis, and Mutagenesis. Aflatoxin is carcinogenic in several species including rats, ducks, mice, trout, and subhuman primates. AFB_1 is considered the most potent carcinogen known, with AFG_1, AFB_2, and AFG_2 in their order of decreasing potency. Dietary levels of AFB_1 as low as 15 ppb fed chronically to rats cause a high rate of hepatic carcinomas (or tumors). Carcinogenesis in livestock and poultry appears to be a rarity, but has been reported in swine fed contaminated feed for long periods (>2 yr). Pigs that survived the acute toxicity phase when fed highly contaminated cottonseed and peanut meal all developed hepatic carcinomas much later (Carnaghan and Crawford 1964). The lack of observed carcinogenesis in livestock could also be because they are usually marketed long before tumors would become clinically apparent. Bodine and Mertens (1983) concluded that although the carcinogenic event, per se, as related to farm animals appears at present to be of little scientific interest, the mutagenic threat of aflatoxin or its active metabolite(s), however, could be of extreme importance in considering the productive and reproductive performance of animals.

Activation of AFB_1 by the formation of the 2,3-oxide and subsequent covalent binding to DNA is considered the first stage in carcinogenesis.

Almost all of the AFB_1 metabolite bound to DNA is with the seventh nitrogen of guanine.

2,3-dehydro-2-(N^7-guanyl)-3-hydroxyaflatoxin B_1

This interaction provides the basis for altering gene expression which results in the development of carcinomas. Exactly how the structural alteration of DNA changes its physiological function is not known.

As in other aspects of aflatoxicosis the susceptibility to aflatoxin-induced carcinogenesis is subject to various physiological or environmental factors. Most of these variables are related to alterations in metabolism. Because metabolic activation is essential to produce the carcinogenic response, any change that shifts the balance between activating and detoxifying pathways will affect carcinogenicity. Most of the investigations on this subject have been conducted in rats, and the results are generally inconsistent. For example, feeding diets low in protein has both increased and decreased carcinogenicity in separate studies. Factors found to enhance carcinogenicity include feeding a lipotrope-deficient diet or a diet containing 0.2% methionine. Inhibiting factors include low dietary vitamin A, pretreating with phenobarbital or diethylstilbestrol, and hypophysectomy. In rainbow trout, the most susceptible species to carcinogenesis, feeding diets containing cyclopropene fatty acids increases the carcinogenicity of AFB_1, AFM_1, aflatoxicol, and AFQ because of an effect on liver metabolism (Fig. 12.3) (see Chapter 8).

Aflatoxins are known to have mutagenic activity. Testing for mutagenesis in vitro using *Salmonella typhimurium* (the Ames assay) has shown that AFB_1 followed by aflatoxicol are the most potent mutagens (Table 12.6). A liver enzyme system (microsomal fraction) is required for AFB_1 mutagenic activity, indicating that activated metabolites are the mutagens. Other indications of AFB_1 mutagenesis have also been shown, such as chromosomal aberrations. Whether aflatoxins are teratogenic is inconclusive. Birth defects and fetal resorption were observed in studies with hamsters administered the toxin intraperitone-

FIG. 12.3. Liver cancer (hepatoma) in a rainbow trout as a result of dietary exposure to 1 ppb aflatoxin.
Courtesy of R. O. Sinnhuber.

ally. Studies in rats indicate large doses (>1 mg/kg) are required before fetal anomalies are observed. Among the effects seen in the offspring at higher doses are tumors and substantially decreased longevity.

Immunosuppression. Impairment of the immune system has been observed in chronic aflatoxicosis in several species. Two primary ef-

TABLE 12.6. Relative Mutagenic Potency of Aflatoxins by the Ames *Salmonella typhimurium* Test with Activation by Rat Liver Enzymes[a]

Aflatoxin	Relative mutagencity
AFB_1	100.0
Aflatoxicol	23.0
AFG_1	3.0
AFM_1	3.0
Aflatoxicol H_1	2.0
AFQ_1	1.0
AFB_2	0.2
AFP_1	0.1
AFG_2	0.1
AFB_{2a}	0
AFG_{2a}	0

[a] Adapted from Wong and Hsieh (1976).

fects can occur. First, an interference in the development of acquired immunity (immunogenesis) can develop, and this involves effects on the cell-mediated immune system. Although the mechanism of this toxic response has not yet been fully characterized, the thymus and thymus-derived lymphocytes (T cells) are apparently sensitive to AFB_1 and AFM_1. Antigen–cell interactions and phagocytosis appear to be modified as well. Antibody formation is normal unless very high dosages are administered. Experiments with guinea pigs show that both growth rate and cell-mediated immunity as measured by delayed hypersensitivity are decreased by approximately the same dosage of aflatoxins. A serious consequence of impaired immunogenesis is diminished effectiveness of vaccination procedures commonly carried out in swine and poultry.

The second main immunosuppressive effect is impairment of native resistance to infection. Decreased phagocytic activity of macrophages and concentrations of nonspecific humoral substances such as complement and interferon have been implicated. Such deficiencies seriously hinder the animal's ability to defend against invading pathogens. For example, in cows given aflatoxin and then challenged intramammarily with mastitis-causing microorganisms, there was more pronounced teat inflammation and higher bacteria counts in milk than in mastitic cows receiving no aflatoxin (Brown et al. 1981). Changes in susceptibility to several infectious organisms in various species fed aflatoxin are given in Table 12.7. The data from chickens clearly show that enhanced susceptibility is dependent on the type of invading organism.

Metabolic Fate. Metabolic fate encompasses all the biological processes related to absorption, distribution throughout the body, metabolism, and excretion (see Chapter 3). The metabolic fate of aflatoxins is

TABLE 12.7. Native Resistance to Infection in Aflatoxin-Intoxicated Animals[a]

Host	Infectious agent	Dose of aflatoxin	Enhanced susceptibility
Chicken	Eimeria tenella	0.2–2.5 ppm	+
	Salmonella typhimurium	0.625–10 ppm	+
	Candida albicans	0.625–10 ppm	+
	S. gallinarum	5 ppm	−
	S. worthington, S. derby, and S. thompson	10 ppm	+
Turkey	Aspergillus fumigatus	5 ppm B_1	−
Hamster	Mycobacterium paratuberculosis	1.88 mg B_1/kg (3 times/week)	−
Calf	Fasciola hepatica	0.5–1.0 mg B_1/kg (single dose)	+

[a] Adapted from Pier et al. (1979).

not yet completely understood. In general, it can be stated that aflatoxins are absorbed easily from the gastrotintestinal tract, are extensively metabolized, and the majority of a dose is eliminated relatively rapidly from the body. Substantial variation in these processes (e.g., rate of disappearance of AFB_1) occurs among species. The metabolic fate of aflatoxins in livestock and poultry is highly pertinent for two reasons: (1) metabolism (activation) accounts for much of the extreme toxicity and carcinogenicity of aflatoxins, and (2) distribution of aflatoxins or their metabolites to various body locations of food animals can impart hazardous residues to products used as food for man.

Metabolism. The metabolism of aflatoxins has been the focus of much research in recent years. This effort has been prompted largely by findings that have linked metabolism and liver cancer in exposed animals. Several metabolites and associated enzymatic reactions have been identified using a combination of in vitro and in vivo experimental methods. These are summarized in Fig. 12.2. It is now quite evident that the liver is the primary organ affected by aflatoxins because of its direct involvement in their metabolism. Most of the phase 1 reactions are catalyzed by mixed function oxidase (MFO) enzymes located in the ER of the liver cell. Conjugation or phase 2 reactions occur either in the ER or in the cytosol, depending on the reaction.

There are four metabolic reactions of toxicologic significance in animals. First, AFB_1 2,3-oxide is an unstable, highly reactive intermediate formed by the MFO enzymes. This oxide (or epoxide) is strongly electrophilic and consequently binds covalently to various nucleophilic sites in the cell such as nucleic acids (RNA and DNA) or proteins (methionine, cysteine, and histidine) forming adducts. Interference with the normal function of these cellular components may result. For example, present evidence is quite conclusive that AFB_1 2,3-oxide is the reactive precursor leading ultimately to cancer. The second significant metabolic reaction leads to the formation of protein–aflatoxin adducts. One pathway involves the further metabolism of AFB_1 2,3-oxide to the dihydrodiol followed by spontaneous molecular rearrangement at physiological pH to an ionic form which interacts covalently with proteins to form a Schiff base. An alternative pathway relies directly on MFO-catalyzed ring hydroxylation at the 2 position, forming AFB_{2a}, which also rearranges molecularly to form Schiff bases with the primary amino groups of proteins. This rather nonspecific interaction with proteins including key enzymes can be fatal to liver cells.

The third toxicologically significant metabolic reaction does not involve MFO, but rather occurs in the cytosol and is catalyzed by an NADPH-dependent reductase enzyme. The reduced product formed is aflatoxicol (AFL). It appears that the susceptibility of different species to AFB_1 is directly correlated with the rate of AFL production. However, whether the toxic effects observed in acute or chronic aflatoxicosis are directly caused by AFL has not been definitively determined. Since AFB_1 can be resynthesized from AFL via a microsomal NADP-utilizing dehydrogenase, there is some speculation that AFL may serve as an intracellular reservoir for AFB_1, and the ultimate conversion to AFB_{2a} is the active metabolite in acute aflatoxicosis. Similarly, the high mutagenic activity of AFL may be due to its conversion to AFB_1 and then to AFB_1 2,3-oxide.

The fourth important metabolic reaction is the hydroxylation of AFB_1, forming AFM_1. Although this hydroxylated metabolite is not as toxic or carcinogenic as AFB_1, it is significant because it is the major metabolite found in edible animal products such as milk. Because of its inherent carcinogenicity, it is considered an important food chain contaminant.

Studies in vitro have elucidated the production by MFO of several other metabolites, all considered detoxication products of AFB_1. AFQ_1 is formed by ring hydroxylation and AFP_1 by O-demethylation. AFL can be detoxified by oxidation to $AFLH_1$ or via dehydrogenation back to AFB_1 and then on to various detoxication products. Depending on species and other factors, all of the metabolites formed by oxidation are candidates for conjugation to glucuronides and/or sulfates prior to excretion. After biliary excretion of these water-soluble conjugates, microflora of the gastrointestinal tract such as in the cecum appear to be able to cleave (hydrolyze) the bond linking the metabolites and conjugates. After deconjugation, reabsorption of the metabolites would complete the process of enterohepatic circulation, thereby prolonging retention of the toxins in the body. The degree that this actually occurs is not known. Also, there is conflicting evidence regarding the metabolism of AFB_1 in the gastrointestinal tract. In the rat, cecal microflora have been shown to produce an array of metabolites different from those produced by the liver. There is some indication that transformation in the rumen may occur, but the extent and the identity of metabolites have not been elucidated.

Residues. Metabolic fate studies have been conducted in most livestock species utilizing radiolabeled (^{14}C tracer in the aromatic ring) AFB_1 (i.e., [^{14}C]AFB_1). These studies indicate whether there may be hazardous residues in tissues, milk, and eggs. Although there is considerable variation among species, the principal route of excretion is the feces ($\geq 75\%$). The remainder ($\leq 20\%$) is excreted in the urine. Only a small portion of AFB_1 administered orally is found unaltered in excretions (feces and urine) or secretions (milk and eggs), indicating extensive metabolism. The majority of the hydroxylated metabolites produced in the liver are conjugated and secreted into the gastrointestinal tract by way of the bile. In general, more than 90% of a dose of AFB_1 is eliminated from the body within 24 hr. However, a small amount of aflatoxin is not eliminated quickly and may be taken up by tissues and retained for longer periods of time. Highest residue concentrations are in liver, with lesser amounts in kidneys and possibly muscle. In pigs, the amount in muscle is only about 1% that in liver; in chickens, there is a more equal distribution between muscle and liver. Although these tissue residues have not been completely identified, there is evidence implicating AFM_1 and AFB_{2a}. In both blood and tissues, the residues appear to be bound to proteins; this is probably the primary factor delaying their elimination. The actual toxicity of these bound residues has been difficult to measure. As shown in Table 12.8, the feed-to-liver ratio of AFB_1 concentration varies considerably between beef cattle and swine. In poultry, AFM_1 in liver is a better indicator of feed exposure.

Transmission of AFM_1 into milk in lactating dairy cattle is a major food safety concern because of the carcinogenicity of AFM_1 and the potential for contamination of dairy cattle rations. This concern grew out of an observation that when milk from AFB_1 fed cows was fed to ducklings, hepatotoxic effects developed. Concentrations of AFM_1 in milk increase proportionally with amount of AFB_1 in the diet. When ingestion is continuous, milk concentrations will increase until an equilibrium with intake is established; this will occur within a week. At this point, the percentage of the daily dietary dose appearing in milk as AFM_1 is about 1% (range of from 0.4% to 3%), and the feed-to-milk ratio is about 300; i.e., an AFB_1 level of 300 µg/kg of feed results in an AFM_1 level of about 1 µg/liter of milk (Table 12.8). AFM_1 generally appears in milk within 24 hr after exposure of cows to AFB_1 and is no longer detectable 4–5 days after removal of AFB_1 exposure. In

TABLE 12.8. Expected Feed-to-Tissues Ratios of Aflatoxin B_1 in Food Animals[a]

Animal	Tissue	Aflatoxin in tissue	Feed-to-tissues ratio
Beef cattle	Liver	B_1	14,000
Dairy cattle	Milk	M_1	300
Swine	Liver	B_1	800
Layers	Eggs	B_1	2,200
Broilers	Liver	M_1	1,200

[a] Adapted from Stoloff (1980).

processed milk, AFM_1 is not affected by pasteurization, cheesemaking, or yogurtmaking. It is found equally distributed between the curd and whey fractions and is insoluble in milk fat.

Transmission of aflatoxins into eggs when layers are fed contaminated feed can occur. The residue which is distributed mostly into the yolk has not been completely characterized. Based on AFB_1, the feed-to-egg ratio as shown in Table 12.8 is greater than 2000.

Zearalenone

The zearalenones are a family of phenolic compounds produced by several species of *Fusarium* that can cause estrogenic effects and infertility in animals. These compounds are one of two major categories of *Fusarium* toxins; the other group is the trichothecenes which are discussed in the following section. The chemical name of zearalenone, sometimes called F-2 toxin, is 6-(10-hydroxy-6-oxo-*trans*-1-undecenyl)-β-resorcyclic acid lactone. The chemical structures of zearalenone and zearalenol, the primary metabolite, are as follows:

	R
Zearalenone	O
Zearalenol	OH

The name zea-ral-en-one is derived from the primary host, corn (*Zea mays*), or *zea-*, from resorcylic acid lactone, abbreviated *-ral-*, from the double bond at C-1' and C-2', or *-en-*, and from the ketone moiety at C-6', or *-one*.

Occurrence. *Fusarium* spp. are distributed worldwide and have been shown to infect many important crops and feeds, such as corn, wheat, sorghum, barley, oats, sesame seed, hay, corn silage, and commercial feeds. Corn is clearly the most frequently contaminated crop. Several species of *Fusarium* produce zearalenone, most notably *F. roseum* (the name of the sexual stage is *Gibberella zeae*). Others include *F. avenaceum, F. nivale, F. sambucinum,* and *F. moniliforme.* Production of zearalenone by *Fusarium* spp. usually occurs in storage when moisture and temperature conditions are optimal. However, in the field, infected corn ears may develop a rot of the crown or cob, aptly named "Gibberella rot." Toxins other than zearalenone, such as deoxynivalenol, which causes feed refusal and vomiting in swine, are likely to be produced simultaneously but in greater quantities. The ear is most susceptible to Gibberella rot during silking. It is believed that ideal conditions for development of Gibberella rot are chronic rainfall combined with mean temperatures of ≥70°F during silking. Since these climatic conditions are infrequent during the silking stage, Gibberella rot is not an annual problem. However, in the past 25 years there have been three major outbreaks in the midcentral U.S. at about 7-year intervals.

The equivalent of Gibberella rot in wheat, barley, and oats in the field is called "scab," which is characterized by a dark discoloration of the kernel. Scab is more prevalent on wheat grown in humid and semihumid areas or when sufficient moisture is available during the flowering and early postflowering stages. The amounts of zearalenone produced by either Gibberella rot or scab are usually less than in stored feed. In addition, hyperestrogenic effects are usually not seen in swine fed infected feed because of their sensitivity to the refusal toxins (trichothecenes) which will limit intake.

In stored feeds, copious amounts of zearalenone may be produced by toxigenic molds when conditions are optimal. Whether colonization by *Fusarium* occurs in the field or in storage, growth is optimum at 20°–25°C and a high moisture content (greater than 23%; 45% is optimum). Zearalenone production is stimulated when the temperature drops to about 15°C while the moisture content remains high. These conditions may be encountered in areas such as the midcentral U.S. where cob corn is frequently stored in open cribs. During a wet or humid autumn, if temperatures are warm during the day so that *Fusarium* can become established, the cool night may stimulate the production of zearalenone.

Biological Effects. The primary biological effect induced by zearalenone in animals is hyperestrogenism which is a chronic toxicosis. Swine, particularly prepubertal females (gilts), are the most commonly affected species, but there is also some evidence for effects in cattle. Estrogenic responses in gilts, which are indistinguishable from those caused by administering excess estradiol, can be caused by feeds containing 1 ppm zearalenone. The syndrome is characterized by swelling of the vulva (vulvovaginitis) and mammary glands, enlargement of the uterus, atrophy of the ovaries, anal prolapse, and vaginal prolapse. Studies in laboratory species indicate that the uterine and mammary effects are induced by an interaction of zearalenone with estrogenic cytosolic receptors in these organs. Moreover, the action of zearalenone on the hypothalamus and pituitary glands appears to be the same as estrogen. In young male swine, zearalenone can cause feminization which includes testicular atrophy, swelling of the prepuce, and enlargement of the mammary glands. In sows, reproductive problems may occur, but greater levels of zearalenone (50–100 ppm) in feed are required. Disorders include infertility characterized by constant estrus (nymphomania), pseudopregnancy, reduced litter size, smaller and occasionally malformed offspring, and juvenile hyperestrogenism. Other effects in the sow include ovarian abnormalities, death of ova, proliferation of uterine mucosal glands, and ductile development in the mammary gland. In the boar, there is no evidence of reproductive effects. However, reduced spermatogenesis has been noted in other species such as geese.

The sensitivity of gilts in comparison to rats is illustrated by the fact that a 25 times greater dose is required to induce uterine enlargement in young rats. It has been speculated that the lower capacity for metabolism of zearalenone to zearalenol may account for the greater sensitivity of young swine. Although the hydroxylated metabolite zearalenol possesses estrogenic activity, it is excreted more quickly than zearalenone.

Metabolic Fate and Residues. Zearalenone is absorbed quite easily from the gastrointestinal tract, as would be expected from its high lipid solubility. Dietary supplementation with binding substances such as anion exchange resins is sometimes used to reduce exposure by decreasing absorption and increasing fecal excretion of zearalenone. Metabolic reduction in the liver gives rise to two stereoisomers of a single metabolite called α- and β-zearalenol; the keto group in the sixth position is reduced to a hydroxyl group. This reduction is catalyzed by an enzyme called 3α-hydroxysteroid dehydrogenase, which exists in several different forms and subcellular locations. This enzyme not only metabolizes zearalenone but is also inhibited by it. There is evidence that feeding alfalfa counteracts this inhibitory effect (James and Smith 1982).

The extent of zearalenone metabolism to zearalenol and the proportions of the α- and β-isomers are quite variable among species. Compared to rats, pigs metabolize zearalenone more slowly because they possess a lesser amount of the 3α-hydroxysteroid dehydrogenase enzyme. Consequently, pigs excrete zearalenone primarily as conjugated parent compound in feces. In all species tested, more zearalenone is excreted unchanged (as a combination of free and conjugated forms) than as zearalenol. α-Zearalenol is usually the major metabolite except in dairy cattle where β-zearalenol predominates. The physiological importance of α- and β-zearalenol is still unclear, but in terms of their estrogenic potency compared to zearalenone, β-zearalenol is three times more potent and α-zearalenol is considered less potent. Both glucuronide and sulfate conjugates of zearalenone and zearalenol are detected in urine and feces.

Because zearalenone is metabolized and conjugated it is eliminated relatively quickly (within a few days) from the body. With high doses, residues of both zearalenone and zearalenol are measurable in liver but do not persist. In lactating cows, less than 1% of a dose is transmitted into milk as free or conjugated forms of zearalenone and zearalenol. In general, neither zearalenone nor its metabolites are considered significant contaminants of the food chain.

Trichothecenes

The trichothecenes are a family of tetracyclic sesquiterpenoid substances produced by several species of *Fusarium* and at least five other genera of fungi suggesting that the potential for tricothecene production is quite widespread in nature. Of the more than 40 naturally occurring trichothecenes that have been identified, the most notable with regard to animal agriculture are T-2 toxin, diacetoxyscirpenol (DAS), and vomitoxin (deoxynivalenol or DON). The trichothecene name was derived from a fungus called *Trichothecium roseum,* from which the first of these compounds was isolated. All members are derivatives of the trichothecane ring system containing an olefinic bond between C-9 and -10 and an epoxy group at C-12 and C-13. The latter is the basis for the name 12,13-epoxytrichothecenes frequently used for these toxins.

Classification of Trichothecenes. Because a wide range of fungi produce trichothecenes worldwide and each species usually produces more than one toxin, problems have been encountered with nomenclature and classification. For example, *Fusarium tricinctum* is synonymous with *Fusarium sporotrioides* in the U.S., but not in other parts of the world. The same fungi producing a different pattern of toxins in one geographical location versus another due to different environmental stimuli has added to the confusion. The situation was clarified somewhat by Ueno (1977) who derived a classification system for trichothecenes, placing them into four groups according to chemical structure. Group A toxins (Fig. 12.4), the largest group, are differentiated by various combinations of hydroxyl or acyloxyl (OAc) substitutions at R_1–R_5 of trichothecene. The simplest member of this group, trichodermol, has one hydroxyl group at R_2. Important members of this class include T-2 toxin, DAS, and monoacetoxyscirpenol.

Group B trichothecenes (Fig. 12.5) are distinguished by possessing a

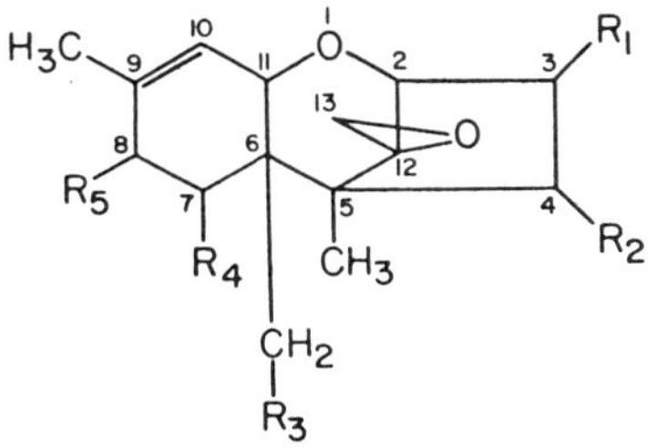

FIG. 12.4. Group A trichothecenes.

	R_1	R_2	R_3	R_4	R_5
T-2 toxin	OH	OAc	OAc	H	Isovaleroxy
HT-2 toxin	OH	OH	OAc	H	Isovaleroxy
Diacetoxyscirpenol	OH	OAc	OAc	H	H
Monoacetoxyscirpenol	OH	OH	OAc	H	H
Trichodermol (Roridin C)	H	OH	H	H	H

carbonyl group at R_5 (C-8). Like group A toxins, members of this group possess different combinations of hydroxyl and acyloxyl substitutions at R_1–R_4. The nivalenol analog deoxynivalenol (DON or vomitoxin) is an important member of this group.

Group C is composed of macrocyclic trichothecenes with a smaller number of members owing partially to the lesser number of substitution sites (Fig. 12.6). Genera that produce macrocyclic trichothecenes include *Myrothecium* spp. and *Stachybotrys* spp. but not *Fusarium* spp. A well-known member of this group is satratoxin which is produced by *Stachybotrys atra* and believed to be the cause of the disease in horses called stachybotryotoxicosis (discussed in the section on Stachybotryotoxins, this chapter).

Group D is comprised of a single member, crotocin, distinguished from other trichothecenes due to the presence of a second epoxide moiety at C-7 and C-8 (Fig. 12.7).

Fusarium spp. that produce group A toxins generally do not produce group B toxins, and vice versa. Although there are exceptions, as shown in Table 12.9, these fungi tend to be minor contributors to the total production of trichothecenes. Also, *F. roseum*, which is the primary producer of zearalenone is also known to synthesize both group A and B trichothecenes. However, most species can be subdivided into varieties that produce either group A or group B toxins.

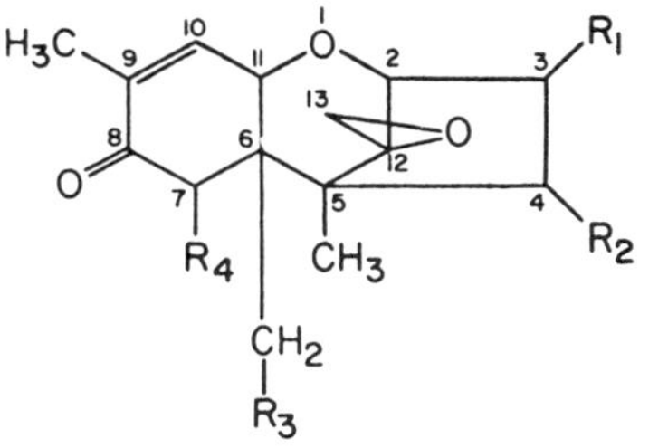

FIG 12.5. Group B trichothecenes.

	R_1	R_2	R_3	R_4
Nivalenol	OH	OH	OH	OH
Monoacetylnivalenol (fusarenon-X)	OH	OAc	OH	OH
Diacetylnivalenol	OH	OAc	OAc	OH
Deoxynivalenol (vomitoxin)	OH	H	OH	OH

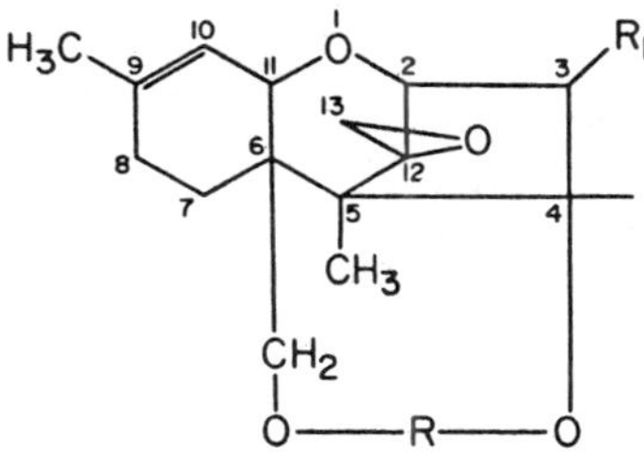

FIG. 12.6. Group C macrocyclic trichothecenes.

	R
Verrucarin A	—C(O)CH(OH)CH(CH$_3$)CH$_2$CH$_2$OC(O)CH=CHCH=CHC(O)—
Satratoxin	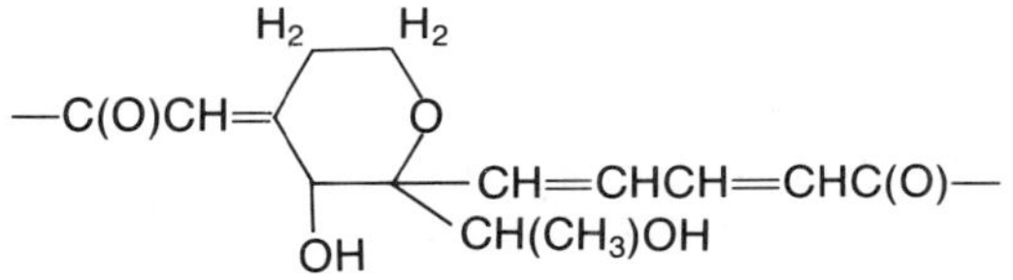 —C(O)CH=⟨ring with H$_2$, H$_2$, OH, CH(CH$_3$)OH⟩—CH=CHCH=CHC(O)—

Occurrence. Several diseases of farm animals in different parts of the world are attributed to trichothecenes. The more classical of these are moldy corn toxicosis (T-2 toxin) in North America, red mold disease or Akakabi (nivalenol, vomitoxin, and fusarenon-X) associated with moldy wheat and barley in Japan, bean hull poisoning (neosolaniol and T-2 toxin) in horses in Japan, and stachybotryotoxicosis (satratoxin) in horses, swine, and poultry fed moldy feed and hay in Russia and Hungary. A disease in humans, alimentary toxic aleukia (ATA), was recognized before 1900. Although the cause was not known until recently, several thousand people in the USSR developed ATA and many died after consuming cereals that overwintered in the fields. It is now believed T-2 toxin and/or its derivatives were the cause of ATA.

As indicated by the geographic distribution of these diseases, trichothecenes are generally found in the cooler climates of the world. As described in the previous section on zearalenone, *Fusarium* spp. generally colonize grains or cereals in the field, but most toxin production

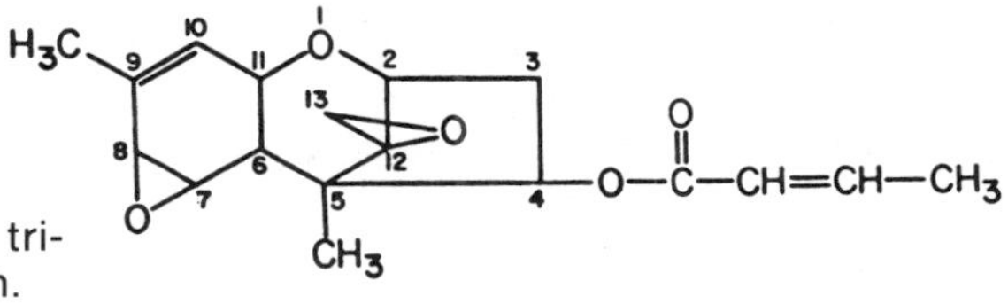

FIG. 12.7. Group D trichothecenes: crotocin.

TABLE 12.9. Species of *Fusarium* That Produce Trichothecenes[a]

Species	Group A		Group B	
	T-2	DAS[b]	NIV[b]	DON[b]
F. tricinctum	+	+		
F. sporotrichiodes	+	+		
F. poae	+	+		
F. graminearum			+	+
F. nivale			+	
F. lateritium		+	+	
F. equiseti		+	+	
F. semitectum		+	+	

[a] Adapted from Ueno (1977).
[b] DAS, diacetoxyscirpenol; NIV, nivalenol; DON, deoxynivalenol.

occurs in storage triggered by cool temperatures (0°–15°C). Head blight disease is the name given to wheat grains infected during the flowering stage. Important environmental factors include high moisture or humidity and both warm and cool temperatures. Greater trichothecene production is usually found in years when the autumn is cool and wet and harvest is delayed. Exposure of cob corn stored in open cribs to moisture coupled with alternating cooling and warming in autumn can result in both fungal colonization (warm temperature) and toxin production (cool temperature). Frost-damaged corn ears tend to be more susceptible to fungal invasion. Trichothecenes are relatively stable, persisting in feeds long after the fungi that produced them have disappeared.

The importance of environmental factors as determinants of the geographical distribution of trichothecenes is exemplified by findings in Japan showing that group A *Fusarium* spp. tend to predominate in the northern districts, whereas group B species tend to frequent the more southern districts. Similarly, laboratory studies indicate the pattern of trichothecenes varies greatly with temperature. For example, *F. tricinctum* produces T-2 toxin and DAS at 8°C and then begins production of HT-2 toxin at 25°C. Comparative growth rates of the various toxigenic species of fungi vary with temperature, so the profile of trichothecenes produced will vary accordingly.

In the U.S., the extent to which farm animals are exposed to trichothecenes, other than DON, is unclear. Consumption of either contaminated corn or wheat is considered the greatest route of potential exposure. In addition to direct feeding of contaminated grains, feed fractions derived from either wet-milling or dry-milling of grains can contain levels of trichothecenes several times greater than the starting grain.

Mode of Action. Trichothecenes cause a wide variety of biological effects owing to the diversity of chemical structures within the group. Whereas these different responses such as neurotoxic, cytotoxic, or immunotoxic effects are related to the type of side-chain substitution, the reactivity of all trichothecenes is due to the presence of both the 12,13-epoxide and the 9-10 double bond in the molecule.

Trichothecenes are considered the most potent inhibitors of protein synthesis in eucaryotic cells. The specific site of interference is the translational stage which takes place on the polysomes in the endoplasmic reticulum. Trichothecenes can interfere with all three translational processes, namely, initiation, elongation, and termination. In initiation, ribosomal subunits from a cellular pool join the mRNA at the initiation region to form the first peptide bond. Several initiation factors and an enzyme, peptidyl transferase, are involved. Some trichothecenes inhibit peptidyl transferase and others cause polyribosomes to break down, thereby imparing protein synthesis. After initiation and during elongation, a variable number of amino acids are added to the growing polypeptide chain. This process requires peptidyl transferase and other factors. Termination occurs in response to a termination codon on mRNA followed by release of the polypeptide chain (protein) and mRNA from the ribosome. Trichothecene inhibition of peptidyl transferase and possibly other effects can impair the elongation or termination steps. The site of action of some trichothecenes is preferentially the initiation steps; for others, it is elongation or termination steps (Table 12.10.). The inhibitory potency varies considerably among the trichothecenes and appears to be closely correlated with the type of R-group substitution on the trichothecene rings.

Although most of the relationships between impaired protein synthesis and clinical effects have not been characterized, one example is the degeneration seen in actively dividing cells such as the thymus, bone marrow, small intestine, testis, and ovary. These effects are in a

TABLE 12.10. Site of Action in Protein Synthesis of Different Trichothecenes[a]

Initiation	Elongation or termination
Scirpenol	Trichodermin
15-Acetoxyscirpenol	Trichodermol
Diacetoxyscripenol	Crotocol
Verrucarin A	Trichothecolone
T-2 toxin	Crotocin
	Trichothecin
	Verrucarol

[a] Adapted from McLaughlin *et al.* (1977). The test system was rabbit reticulocytes.

category of toxic responses sometimes called "radiomimetic" injury, so named because of the similarity to radiation poisoning. Interference with immunologic responses results from damage incurred on the thymus and other lymphoid tissues.

The epoxy group of trichothecenes also can react with sulfhydryl groups on enzymes. This has the potential for numerous biological disturbances. Also, there is some evidence that certain trichothecenes bind to membrane components.

Biological Effects in Different Species. Although trichothecene toxicoses involve a broad spectrum of clinical disorders, those that tend to be common among most species include nausea, vomiting, feed refusal, inflammation, epithelial necrosis, diarrhea, abortion, hemorrhage, hematological changes, and nervous disturbances. The degree to which these effects are manifested varies greatly among different species. Young or immature animals are generally more susceptible than adults, and there does not appear to be a sex difference. Both acute and chronic effects can occur and are generally quite similar in nature.

In swine, DAS and T-2 toxin administered intravenously have LD_{50} values of 0.3 and 1.3 mg/kg, respectively. Acute toxic effects in swine include vomiting, lethargy, frequent defecation, and posterior paresis. Refusal of feed containing greater than 16 ppm T-2 toxin and 10 ppm DAS has been observed. Feed levels from 1 to 10 ppm DAS may cause decreased growth rate and oral necrosis. Levels of T-2 toxin in the range of 8–12 ppm may produce decreased rate of gain and reduced pig and litter size.

In poultry, as in swine, DAS is more toxic than T-2 toxin; these two are the most toxic trichothecenes. Some evidence suggests chickens are less sensitive to oral administration than other animals. Feed containing about 4 ppm of T-2 toxin or DAS can cause oral necrosis, nervous disorders, hepatic hematoma, and reduced weight gain in broilers. A similar dose in young chicks may also cause abnormal feathering. In turkey poults, but not in broilers, reduced feed efficiency accompanies the adverse effect on growth. Poults are not as sensitive as broilers. Greater levels of exposure in broilers can result in severe hematopoietic damage and clinical blood changes such as decreased number of leukocytes and impaired blood coagulation. In laying hens, the toxicosis is similar to that in young chickens except a greater dosage is required. A level of about 20 ppm in feed can cause decreased egg production, egg shell thickness, and shell strength. Return to normal body weight and egg laying occurs gradually over about 2 to 3 weeks, whereas healing of necrotic lesions may take only half as long.

The first natural case of trichothecene toxicosis was discovered in dairy cattle fed moldy corn in 1971 (Hsu *et al.* 1972). These cattle exhibited a hemorrhagic syndrome after consuming corn containing 1 ppm T-2 toxin. Experimental administration of 0.2 mg T-2 toxin per kilogram body weight to adult cattle caused feed refusal and ruminal ulcers. Hemorrhaging is not produced experimentally by pure T-2, indicating that the 1971 problem may not have been entirely due to T-2 toxin. Other mycotoxins may have been present, but were not analyzed. Calves fed rations containing 10–50 ppm T-2 toxin developed necrotic lesions of the lips, denudation of ruminal papilla, abomasal ulcers, and severe diarrhea. Monoacetoxyscirpenol, a group A toxin (Fig. 12.5), is suspected of causing illness and death in dairy cattle in the north-central U.S. Rumen function does not appear to be adversely affected by trichothecenes. Moreover, there is no evidence that these toxins are metabolized in the rumen.

Feed Refusal and Vomiting. Vomiting (emesis) and feed refusal are associated with most trichothecenes, particularly DON (vomitoxin). Swine are especially sensitive to these toxic effects. Rejection of contaminated feed (feed refusal) causes decreased weight gain and, if enough is consumed, vomiting (emesis) will result. As little as 5% infected kernels or about 10 ppm DON in feed causes refusal in swine. Whether the refusal is due to a localized irritant action in the upper gastrointestinal tract or to palatability is not known. It is not advisable to reverse the refusal by masking moldy feed (e.g., with molasses) because of the possible development of toxic reactions from other trichothecens that are probably present in the feed.

Vomiting is an acute reaction occurring within a few minutes after either parenteral or oral administration of toxin and may reoccur intermittently over several hours. Swine, dogs, cats, and ducklings appear to be especially sensitive. Ducklings are even sometimes used as a bioassay tool to test for the presence of trichothecenes in feed. Even though DON, a group B toxin (Fig. 12.5), is more associated with vomiting, group A toxins (Fig. 12.4) have greater emetic potency. T-2 and HT-2 toxins are about 100 times more potent than DON in ducklings. Whether the vomiting results from gastric irritation or central nervous system stimulation is not known. However, because pretreatment with central depressants such as chlorpromazine suppresses vomiting after fusarenon-X administration, the toxin may act by stimulation of the chemoreceptor trigger zone in the medulla oblongata. After removal of contaminated feed, animals rapidly recover from both feed refusal and vomiting with no apparent permanent damage.

Cytoxic Effects. Both T-2 and DAS react upon contact with epithelial tissues such as skin and mucous membranes, causing local cytotoxic effects. Application to the skin causes inflammation, scaling, subepidermal hemorrhaging, and necrosis (epithelionecrosis). Guinea pigs are particularly sensitive. Direct contact with contaminated feed has caused local necrotic effects on the snout and legs of pigs. When administered orally, epithelionecrotic effects may occur on the lips or beak, mouth, esophagus, and stomach in all species. The epithelium of the gastrointestinal tract generally undergoes erosion and ulceration, sometimes accompanied by hemorrhaging, resulting in severe gastroenteritis and possibly death. The clinical lesions are exacerbated by the invasion of the necrotic areas by the normal microflora in the area causing infection. Necrotic effects in the intestines are variable among animal species, but diarrhea is common and rectal hemorrhaging sometimes occurs.

The mechanism of epithelionecrosis has not been elucidated. However, there is some evidence that the toxins may increase capillary permeability by direct action on the vessels or by disrupting mast cells, resulting in the release of chemical mediator substances which are biologically reactive and may affect vessel integrity. Group A toxins such as T-2 toxin and DAS have the greatest epitheliotoxicity, with DAS being about five times more toxic than T-2 toxin. Edematous damage of skin tissue is associated with group C macrocyclic trichothecenes, whereas skin reddening is more characteristic of group B toxins.

Blood Effects. Blood effects associated with trichothecene intoxication include hemorrhagic diathesis, decreased hematopoesis (anemia), and decreased numbers of white blood cells (leukopenia), thrombocytes (thrombocytopenia), and red blood cells (erthrocytopenia). All of these clinical changes are observed in humans suffering from chronic ATA and in cats and chickens administered trichothecenes. Cellular degeneration in bone marrow and spleen, as in other proliferative tissues, is believed to be responsible for the decreased cell numbers in blood. Hemorrhagic diathesis, characterized by blood in the gastrointestinal tract and widespread petechial (focal) hemorrhaging, is thought to be associated with defective blood coagulation. All these effects are probably manifestations of the inhibitory action of trichothecenes on protein synthesis.

In cattle and swine, hemorrhagic diathesis attributed to trichothecene exposure has been observed under field conditions but has been difficult to produce experimentally. However, hemorrhagic bowel lesions usually accompanied by the presence of blood in the feces occurs in cattle, swine, and poultry following acute parenteral administration of T-2 toxin or DAS.

Neurotoxic Effects. Neurological signs resulting from disturbances in the central and autonomic nervous systems include ataxia, impaired righting reflex, hysteroid seizures, and abnormal positioning of wings in poultry treated with T-2 toxin. A decline in body temperature, elevation of blood pressure, and decreased respiratory rate have been measured in laboratory rodents during acute trichothecene intoxication. Posterior paresis (paralysis) occurs in swine acutely intoxicated with DAS. Bean hull poisoning in horses is characterized by staggering and other neurological disturbances. In cats, visual disturbances and meningeal hemorrhage are caused by T-2 toxin. The vomiting reaction in several species may also be a result of a neurological disturbance. A number of nervous disorders associated with ATA in humans are believed to be due to destructive changes in the neurons of the third ventricle and the sympathetic ganglia. These include impaired reflexes and neuropsychiatric symptoms.

Leukoencephalomalacia is a neurological syndrome limited to horses and donkeys, which is sometimes called moldy corn poisoning. A causative mycotoxin has not been identified, but the syndrome is produced by feeding cultures of *F. moniliforme* isolated from moldy corn. Clinical signs in affected horses and donkeys include sleepiness, circling, blindness, altered gait, ataxia, paralysis of facial, oral, or pharyngeal muscles, lateral recumbency, paddling, coma, and death. Postmortem examination has revealed cerebral lesions with areas of liquefactive necrosis and cavitation in the white matter. Some hepatotoxic effects have been noted experimentally.

Other Biological Effects. Trichothecenes are capable of modifying the immune response. Thymic involution, impaired antibody production, and decreased capacity for clearance of inoculated bacteria are caused by T-2 toxin and DAS. Furthermore, T-2 toxin is able to cross the placental barrier in mice and cause thymic atrophy in the fetus. Stachybotryotoxins impair both cellular and humoral immunity.

There is some evidence in laboratory species that T-2 toxin is both mutagenic and teratogenic. Abortion and retarded growth rate of offspring have been described in cattle, swine, and laboratory species. Long-term feeding studies in rats have not provided consistent results on the question of carcinogenicity. However, benign and malignant tumors of the gastrointestinal tract and brain were found in rats given T-2 toxin in one study, indicating that the possibility does at least exist.

Metabolic Fate and Residues. Except for T-2 toxin, there is relatively little known on the metabolic fate of trichothecenes in animals. Most of the information on T-2 toxin has come from studies utilizing tritium-labeled (^{3}H) toxin, which can be prepared by chemical synthesis. In general, after an oral administration of [^{3}H]T-2 toxin, the radiotracer is rapidly absorbed, almost completely metabolized, distributed widely throughout the body, and eliminated relatively quickly in both urine and feces. Of course, there are differences among species in all of these processes.

The metabolism of T-2 toxin has not been completely worked out. Several metabolites are known to exist, but not all have been chemically identified. Those that have been identified are produced by the stepwise removal of acyloxyl groups (Fig. 12.8). The most thoroughly studied metabolite, HT-2 toxin, arises from a deacetylation at C-4, resulting in the addition of a hydroxyl group. This reaction is catalyzed by an enzyme called carboxyesterase, which is located primarily in the ER of cells. Several tissues have the capacity to deacetylate T-2 toxin, but liver and kidney tend to be the most active. There is some evidence that the intestinal mucosal cells produce HT-2 toxin during absorption. This has been advanced as the explanation for the rapid appearance of HT-2 toxin in blood shortly after oral administration of T-2 toxin. There appears to be no preabsorption metabolism of T-2 toxin in the gastrointestinal tract or rumen. The widely varying profile of metabolites detected in excretory products indicates there are metabolic differences among various species. It is notable that the fraction of unidentified metabolites is substantial; in dairy cattle and chickens, these are the major metabolites.

The metabolites of T-2 toxin are circulated in the blood and distributed to many tissues and organs. There does not, however, appear to be a significant accumulation of any metabolites. In swine, for example, the percentage of administered radioactivity found 18 hr after dosing was 0.7% in muscle and 0.29–0.43% in liver (Robinson *et al.* 1979). Because of the high toxicity (low LD_{50}) of trichothecenes there has been concern about human exposure through consumption of edible tissues contaminated with T-2 toxin or its metabolites. Table 12.11 shows that the ratios between tritium (^{3}H) levels in edible tissues and feed are quite low. In general, levels of exposure in animals great enough to result in hazardous residues would be clearly toxic to farm animals. Therefore, the hazard to humans is believed to be quite low. Table 12.11 also shows that the tissue-to-feed ratios are higher in the chicken than in swine, suggesting that the potential for a residue problem is greater in chickens. The majority of the retained residue is unidentified metabolites. The more polar metabolites such as neosolaniol or T-2 tetraol are apparently excreted very rapidly.

Small amounts of tritium radiolabel have been found in milk and eggs after experimental administration of [^{3}H]T-2. In dairy cattle, about 0.2% of a dose was secreted into milk over several days following

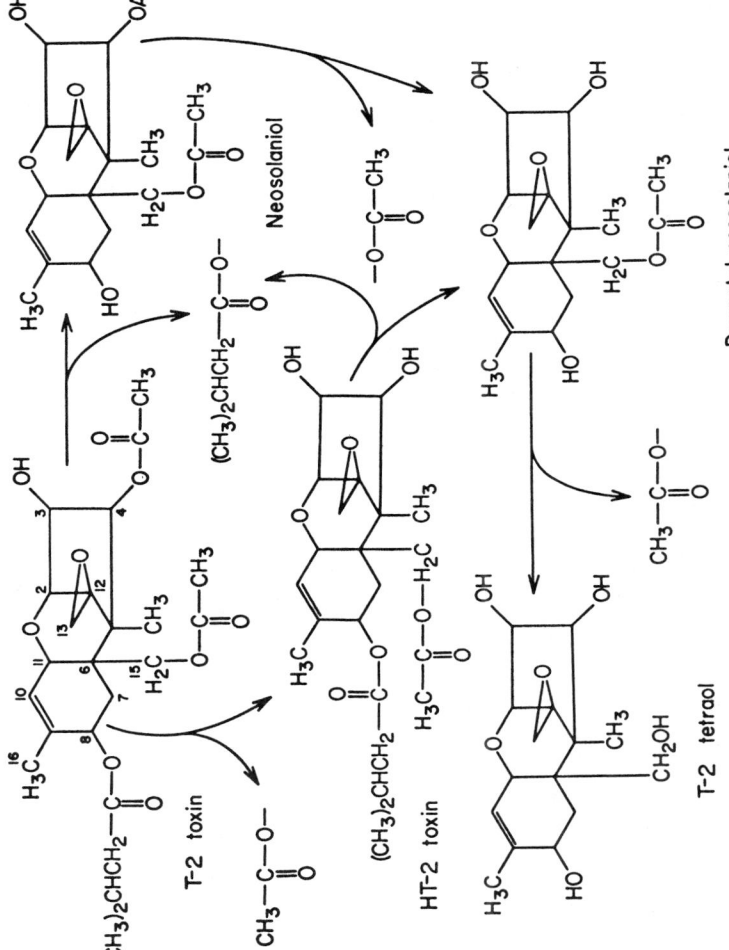

FIG. 12.8. Metabolism of T-2 toxin in rats.

TABLE 12.11. The Relationship between the Levels of [^{3}H]T-2 Toxin in Feed and Edible Tissues of Chicks and Swine[a]

Tissue	Animal	Time after dosing (hr)	Feed level (ppm)	Tissue level (ppb)	Tissue:feed ratio
Muscle	Chick	24	1.26	17.3	0.014
	Swine	18	1.25	3.1	0.002
Heart	Chick	24	1.26	13.7	0.011
	Swine	18	1.25	3.9	0.003
Liver	Chick	24	1.26	34.0	0.027
	Swine	18	1.25	13.8	0.011

[a] Adapted from Yoshizawa et al. (1981).

a single administration (Yoshizawa et al. 1981). Again, the majority of this residue is unknown metabolites.

Most (70–80%) of T-2 toxin and its metabolites are excreted via the biliary system into the gastrointestinal tract, and the remainder is excreted in urine. There is some evidence for reabsorption of toxins from the intestines, resulting in enterohepatic recirculation. Moreover, it has been suggested that metabolites of T-2 toxin excreted in bile may be the cause of histopathological lesions in the intestinal tract.

Ochratoxins

The ochratoxins are a family of isocoumarin derivatives of the amino acid phenylalanine. Nine ochratoxins have been identified, but only ochratoxin A (OA) and very rarely ochratoxin B (Fig. 12.9) are produced under natural conditions. The ochratoxins are produced by several species of *Aspergillus* and *Penicillium*. The toxins were first isolated from *A. ochraceus,* from which their name was derived. The most important producers of ochratoxins are *P. viridicatum* and *A. ochraceus*. Ochratoxin A is a potent nephrotoxin. It has been confirmed as the cause of kidney disease in livestock (e.g., porcine nephropathy) and suspected as the cause of Balkan nephropathy in humans.

Occurrence. The fungi that produce ochratoxins are distributed worldwide and are common in soil, decaying plants, insects, stored

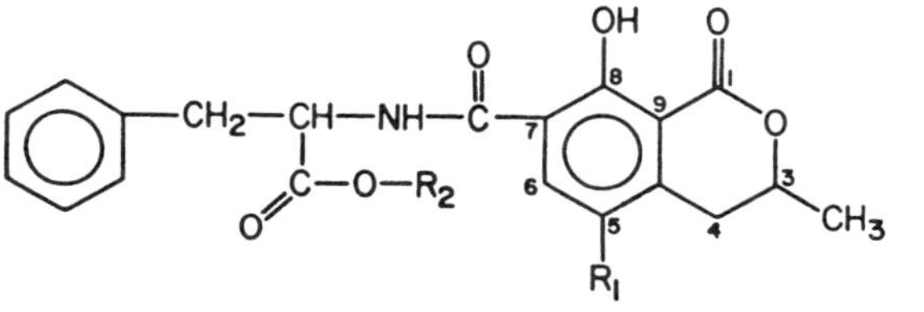

FIG. 12.9. Structures of ochratoxins A, B, and C.

	R_1	R_2
Ochratoxin A	Cl	H
Ochratoxin B	H	H
Ochratoxin C	Cl	C_2H_5

seeds, and cereals. Ochratoxicosis, particularly nephropathy (kidney disease) in swine, is quite prevalent in certain regions and is usually associated with the feeding of contaminated barley. Ochratoxin A has also been detected in wheat, oats, barley, corn, beans, peanuts, hay, green coffee beans, and mixed feeds. Up to 27 ppm has been measured in Canadian wheat and Danish barley.

Not all strains of *P. viridicatum* and *A. ochraceus* are toxigenic. Also, some strains produce citrinin as well as OA. Production of OA in the toxigenic strains is primarily governed by temperature and moisture. In general, optimal conditions for OA production are a moisture content of 19–22% and a temperature of about 24°C. Toxin production can still occur at temperatures as low as 4°C. Because the environmental conditions conducive to OA production are not particularly unusual and because of the ubiquity of the fungal strains, surveys of feedstuffs for OA generally reveal frequent contamination in the cooler climates of the world such as North America and northern Europe. However, usual concentrations of OA are below toxic levels.

Mode of Action. Ochratoxins interact with a number of macromolecules in the body, but the toxicological significance or relationship to ochratoxicosis of most of these interactions have not been established. The toxin has a high affinity for a number of enzymes, but does not in all cases result in inhibition. It is a competitive inhibitor of protein synthesis, particularly in the spleen, due to its ability to inhibit the enzyme phenylalonyl-tRNA synthetase. OA also binds strongly to serum albumins. The hydrolysis metabolites of OA, Oα and Oβ, also bind to proteins but with much less affinity than OA. Binding to serum proteins may serve to decrease the amount of OA available to other sites of toxic action or affect the binding of other biologically active substrates in serum, possibly resulting in secondary effects. Interestingly, the affinities of different ochratoxins with serum albumins correlate quite well with their acute toxicities.

Interference with carbohydrate metabolism is considered to play a key role in ochratoxicosis. Depending on the species, glycogen concentration in organs may either increase or decrease during ochratoxicosis. In rats, liver glycogen is depleted after a single dose of OA, whereas in chickens, glycogen accumulates in liver and muscle during chronic exposure, resulting in glycogen storage disease, or hyperglycogenation. The depletion effect in rats has been attributed to decreased glycogenesis (glycogen formation) due to inhibition by OA of a key enzyme system, glycogen synthetase. Impaired transport of glucose into liver and acceleration of glycogenolysis (glucose utilization) have also been suggested as possible causes. In chickens, the accumulated glycogen in liver is not metabolically available due to apparent OA inhibition of the glucagon-mediated mobilization process, whereas the muscle glycogen mobilized by epinephrine is available. Neither the cause of hyperglycogenation nor its role in ochratoxicosis in chickens has been determined. There is some evidence that OA inhibits a cyclic AMP-dependent protein kinase which initiates a series of enzyme reactions leading to glycogenolysis (catabolism to glucose). In the kidney, but not in liver, OA also inhibits gluconeogenesis by inhibiting an enzyme called carboxykinase.

Cell energetics may also be disturbed during ochratoxicosis, as evidenced by in vitro studies indicating that both OA and Oα are respiratory chain inhibitors. This effect is believed to result from competitive interference of the toxin on the uptake of metabolic anions such as ADP and P_i into mitochondria. Additionally, the ADP:O ratio and O_2 consumption are both decreased by OA, indicating impaired ATP production. Succinic dehydrogenase and other Krebs cycle enzymes are also inhibited by OA.

Biological Effects. The kidneys undergo several morphological and functional changes during both acute and chronic intoxication. These changes are quite comparable among all animals tested and also in humans with Balkan nephropathy. At necropsy, the kidneys appear gray in color, have a granular surface, and are usually enlarged. Widespread subcutaneous, mesenteric, and perirenal edema occurs in swine. Histological changes include atrophy of the proximal tubules accompanied by disintegration of the brush border, swollen mitochondria, and disorganization of the endoplasmic reticulum. Interstitial cortical fibrosis and glomerular fibrosis are characteristic of chronic ochratoxicosis. These morphological changes are accompanied by various impaired renal functions indicated by elevation of potassium, protein (proteinuria), and glucose (glycosuria) in urine. The earliest indicator is an increased appearance in urine of leucine aminopeptidase (LAP), an enzyme normally located in the brush border of the proximal tubule. Other measurements indicating nephropathy are decreased glomerular filtration rate (GFR), decreased clearance of test substances, increased blood urea nitrogen (BUN), and inability to concentrate urine (decreased osmolality).

Although the liver is not a major target organ, some toxic effects occur there. Fatty degeneration, accumulation of glycogen (glycogen storage disease), especially in chickens, and centrolobular necrosis including lesions in the intrahepatic biliary system have been observed. Decreases in serum protein content and coagulation of blood in chickens are indicative of liver damage.

The gastrointestinal tract of several species is affected by ochratoxins and is characterized by enteritis. Also, necrosis of various lymphoid tissues such as the spleen and lymph nodes occurs. The implications of this effect on the immune system are apparent but have not been thoroughly investigated. There is some indication of suppressed antibody production in mice and decreased phagocytic activity in poultry. A case of apparent immunosuppression in breeding sows in Germany was linked to consumption of barley and oats containing two *Penicillium* mycotoxin: viomellein and ochratoxin B (Leistner 1984).

Teratogenic and embryocidal (e.g., fetal resorption) effects of OA have been observed experimentally in several laboratory species but not in food-producing animals. The studies in mice revealed deformities of the skull, vertebrae, and ribs after an intraperitoneal injection but not after oral administration. Carcinogenicity of ochratoxins has not been demonstrated except in an especially sensitive strain of mouse known as the ddy strain. In these mice, both renal and hepatic tumors were detected with very high levels of OA. Interestingly, simultaneous administration of OA and citrinin enhanced the incidence of renal tumors but not that of liver tumors (Kanisawa 1984).

Effects in Different Species. Among farm animals, monogastric species (horses, swine, poultry) are much more sensitive to ochratoxins than ruminants. As with other mycotoxins, impaired growth in young animals including calves is the first observable sign of intoxication. Young animals are clearly more sensitive than adults. Ochratoxicosis is generally not diagnosable until postmortem examination of kidneys is conducted.

Adult ruminants are afforded protection by rumen microorganisms, particularly protozoa which hydrolyze the peptide bond of OA, thereby forming less toxic ochratoxin α and phenylalanine (Fig. 12.10). This reaction is catalyzed by the enzymes carboxypeptidase A and chymostrysinogen. There is some evidence that ochratoxin C (OC), which is equally toxic to OA, may be formed in minor quantities by rumen microorganisms. The beneficial role of the rumen in protecting against ochratoxicosis in cattle is evidenced by the wide difference in toxicity of OA when administered orally (13 mg/kg body weight) versus intravenously (1 mg/kg body weight). Even though there is substantial evidence that OA has bacteriostatic activity against gram-positive bacteria, rumen fermentation is not impaired by levels of OA normally found in feed. However, in one experiment weaned calves administered OA equivalent to 2–40 ppm in feed developed nephritis, enteritis, coagulopathy, and reduced growth, suggesting the rumen may not be as protective in the less-developed ruminant (Pier et al. 1976). Also,

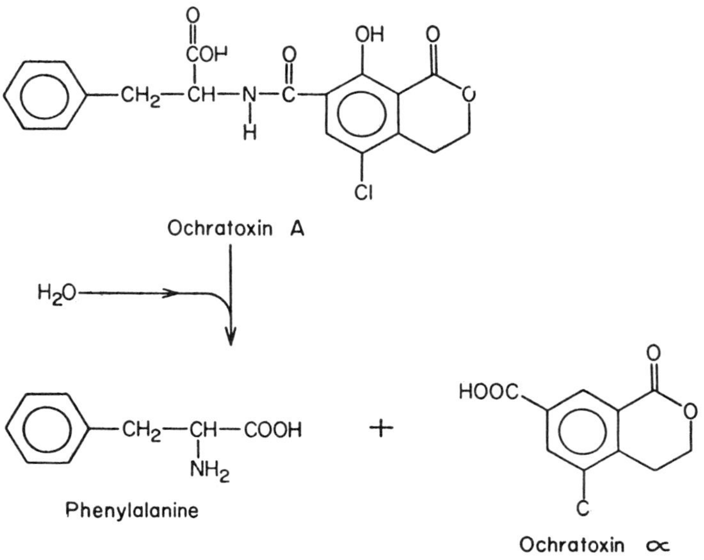

FIG. 12.10. Hydrolysis of ochratoxin A to phenylalanine and ochratoxin α.

there is some indication that OA causes abortion or contributes to other diseases in cattle. However, the general consensus is that natural ochratoxicosis in ruminants is rare.

In swine, the primary syndrome is called porcine nephropathy which generally occurs after chronic ingestion of diets containing from 0.2 to 4 ppm OA. The effects on the kidney are comparable to those seen in a variety of other species (e.g., poultry, rats, fish, and monkeys) and are also similar to the human syndrome called Balkan nephropathy. Both the swine and human diseases are endemic to certain parts of Europe. Other effects include weight loss or decreased growth and depression. A number of extrarenal effects have been observed experimentally, but these are only induced by high dosages. The kidneys are the sole target organ when naturally occurring levels of exposure are administered. The rat is a useful model for studying chronic but not acute porcine nephropathy. A single, acutely toxic dose administered to rats can be fatal without producing renal effects, but this does not occur in swine.

In poultry, ochratoxicosis has been documented in broilers, layers, and turkeys. As in other species the principal effect is nephropathy. The oral LD_{50} varies among different species (e.g., broilers, 3.4 mg/kg; poults, 5.0 mg/kg; quail, 16.5 mg/kg). Interestingly, the sensitivity of poultry (acute and chronic) is greater to ochratoxin A than to either T-2 toxin or aflatoxin B_1; the minimum dietary dose that impairs growth in young broiler chicks is 2 ppm for ochratoxins, 2.5 ppm for AFB_1, and 4 ppm for T-2 toxin. The only other clinical signs that may be observed are neurological effects which include tremors, flailing, and loss of righting reflex. At necropsy, birds may appear dehydrated or emaciated, with some indication of preventricular hemorrhage, which is a result of decreased coagulation. Coagulopathy can occur in broilers fed 1 ppm OA for a few weeks. Another unique effect in poultry is visceral gout characterized by white urate deposits throughout the body cavity and internal organs. Kidney and liver effects are typically the same as in other species. Hepatic accumulation of glycogen is particularly notable in poultry. Like AFB_1, OA decreases bone strength in young broiler chicks. The bones may become rubbery apparently from poor mineralization. In layers, levels of OA as low as 0.5 ppm can decrease egg production and feed consumption. Egg fertility is not particularly susceptible; however, the egg embryo appears to be rather sensitive to OA. As little as 0.01 μg/egg is embryocidal.

In turkeys, ochratoxicosis is characterized by mortality, decreased growth rate and feed efficiency, nephropathy, and decreased carcass pigmentation. The latter effect may result in an increase in the number of carcass condemnations.

Metabolic Fate and Residues. Ochratoxin A and aflatoxin B_1 are the two mycotoxins considered to be the most serious food-chain residue problems (Stoloff 1979). Poultry and swine fed OA-containing feed retain some of the toxin in their tissues. Survey data from slaughterhouses in Europe indicate that 25–35% of pigs suffering from porcine nephropathy contain measurable OA residues. Moreover, these carcasses generally pass the inspection process, thereby gaining access to the human food chain.

Ochratoxin A is absorbed from the stomach and also throughout the remainder of the gastrointestinal tract. In pigs given a single experimental dose of OA, about 66% was absorbed, and the peak concentration in plasma occurred at about 8 hr postadministration (Galtier et al. 1981). In the blood, much of the OA is found bound to serum albumin, especially in cattle, pigs, and man. This sequestration of OA in blood decreases its distribution in the body and also retards urinary excretion. The highest tissue levels of OA are found in blood and kidneys, with lower levels in liver and muscle. In pigs fed 1 ppm dietary OA, the ratio of residue concentrations in kidney versus feed is 1 to 38 at 24 hr after administration (Krogh 1978A). This is a relatively small ratio compared to other mycotoxins. Moreover, the fact OA is more toxic than most other mycotoxins intensifies the hazard from residues in edible tissues, particularly kidneys.

Compared to other species, the pig eliminates OA residues quite slowly; for example, the biological half-life of OA is about 90 hr in pigs compared to about 4 hr in chickens. The rate of elimination is even slower (e.g., 100–110 hr) from kidneys and liver (Galtier et al. 1981). Therefore, the residue problem in swine compared to other species can be attributed to such factors as greater protein binding and greater retention in tissues. Levels of OA as high as 100 ppb have been measured in kidneys of nephropathic swine. Swine known to have OA exposure should be fed OA-free feed for at least 4 weeks before slaughter to allow for OA elimination from tissues. The residue problem in chickens is not considered as great, but levels of up to 30 ppb have been found in chickens with avian nephropathy. In ruminants, OA hydrolysis in the rumen decreases the possibility of residues as well as toxicosis.

Citrinin

Citrinin is a nephrotoxin produced by several species of the genus *Penicillium* and three species of the genus *Aspergillus*. The principal toxigenic fungi in animal feeds is *P. viridicatum*. Citrinin is a quinone methide that was first identified as a secondary metabolite of *P. citrinum*, from which it derived its name. In crystalline form, citrinin is lemon yellow in appearance and is practically insoluble in water. The structure of citrinin is as follows:

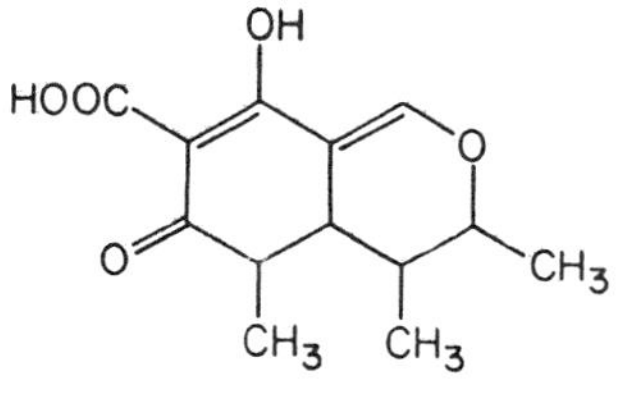

Occurrence. The fungi that produce citrinin are found in the temperate regions of the world. Feedstuffs known to be colonized by the toxigenic strains include most grains, such as wheat, oats, barley, rye, and corn. Concentrations as high as 80 ppm have been found in Canadian wheat, but the frequency of contamination at levels high enough to be toxic is rather rare. Contamination of Danish and Canadian grains fed to livestock could be a cause of disease in these regions. Toxicologic assessment of natural contamination of feeds with citrinin has been complicated due to its occurrence with other mycotoxins, including ochratoxin A, patulin, penicillic acid, and aflatoxins.

Fungal infestation and subsequent production of citrinin in feedstuffs are influenced by temperature, substrate, and moisture. In moist grains, citrinin production is maximal at about 25°C but may also occur at lower temperatures (5–12°C). In peanut pods in culture, kernel moisture content and pod damage are major determinants of fungal growth; citrinin production can range up to 1200 ppm. The toxin is not very stable and degrades under certain conditions such as increased heat (60°–70°C) or when the moisture content is increased very much above the optimum for growth. These factors and others undoubtedly impact on the environmental occurrence of citrinin.

Mode of Action. The mode of action of citrinin at the biochemical level has not been completely established, but experimental findings in laboratory rodents have revealed a number of possibilities. The site of these effects is primarily the kidneys and liver. Within a few hours after citrinin administration, DNA, protein, and glutathione (GSH) content in these tissues are decreased. In addition, the respiratory capacity (i.e., uptake of O_2) and the metabolic enzyme, succinic dehydrogenase, are inhibited. Other hepatic changes include decreased glycogen content, increased lipid content (fatty liver), and decreased cholesterol synthesis.

The reduction of GSH concentrations in renal and hepatic tissues was found to occur within 2–4 hr after administration of citrinin to rats (Berndt et al. 1980). Although GSH depletion is clearly related to the hepatoxicity of some chemicals (e.g., acetaminophen, bromobenzene), such a relationship has not been established between citrinin and nephrotoxicity. Interestingly, renal concentrations of GSH returned to normal within 2 to 3 days postadministration and then became elevated by 30–40%. The biological significance of this response is also unclear. Comparatively, ochratoxin A apparently has a less profound effect on tissue GSH content.

Biological Effects. Although citrinin toxicosis in domestic animals is not well characterized, experimental evidence clearly shows that the principal toxic effect in all species tested is in the kidneys (nephropathy). In swine, experimental citrinin-induced kidney damage is similar to that observed in field cases of porcine nephropathy, which is attributed to ochratoxin A (OA). This plus the fact that citrinin and OA frequently coexist in moldy feeds has created some serious difficulties in delineating whether citrinin represents a real threat to animal health. Because of its less common occurrence and lower concentrations in feedstuffs, the present consensus of opinion is that citrinin is a contributor rather than a primary cause of porcine and avian nephropathy. There is experimental evidence that citrinin and OA act synergistically, but this phenomenon has not been explored in domestic livestock.

The nephrotoxic effects of citrinin are similar in all species. In general, acute tubular necrosis possibly followed by renal failure occur. Grossly, the kidneys are enlarged and appear pale and tan colored. Within the tubular nephron the area most commonly damaged is the proximal tubule. It degenerates, becoming necrotic and even mineralized in some species. These morphological changes are accompanied by alterations in functional indicators, including elevated BUN, decreased GFR, proteinuria, glycosuria, and creatinuria. In addition, daily urinary volume may increase substantially (polyuria) and remain elevated for 2 to 3 days after a single dose of citrinin. Decreased osmolality of urine coincides with the polyuria. Watery excreta occurs in chickens. Depending on the dose, the lesions in the tubules are reversible. However, in rats, severe tubular damage indicated by massive glycosuria is followed by renal failure and death.

The liver is moderately affected by citrinin, particularly in guinea pigs and chickens. Grossly, the livers are enlarged, mottled, and friable. Changes in the gastrointestinal tract include gastric inflammation and cecal ulceration in guinea pigs and hemorrhagic jejunums in chickens. Death of citrinin-intoxicated dogs is due to intestinal intussusception (twisted intestines resulting in obstruction), indicating a neurologic effect.

Citrinin does not appear to be carcinogenic but does promote the renal carcinogenicity of potent carcinogens such as dimethylnitrosamine and perhaps OA. It is mutagenic in some microbial systems, but is negative in the *Salmonella typhimurium* Ames assay. Citrinin is not teratogenic in mice, but has potent embryocidal activity in the chicken embryo where the LD_{50} was 80.5 µg/egg.

Although citrinin and OA have similar nephrotoxic effects (both pathologically and functionally), there are specific differences distinguishing the two toxins. First, the effects from citrinin are not cumulative. Repeated low dosages of citrinin are not nephrotoxic, whereas OA administered in the same way would produce toxic effects in the kidneys. Second, citrinin does not usually cause enteritis as does OA. There are changes in the gastrointestinal tract such as in the guinea pig, but the tract is generally unaffected in most species.

Toxicity in Different Species. Citrinin is not nearly as toxic to farm animals as aflatoxin B_1, ochratoxin A, or T-2 toxin. In poultry, dietary levels in excess of 130 ppm are required to produce clinical changes which include growth depression, increased water consumption, and diarrhea as well as nephrotoxic and hepatotoxic effects. The increased water consumption is transitory, occurring after initial administration, and is postulated to be due to an action of citrinin on the sensory thirst mechanism. Layers tolerate dietary levels of citrinin in excess of 250 ppm without reductions in body weight, egg production, or egg quality. In swine, levels in excess of 20 mg/kg of body weight can cause reduced growth rate and nephropathy. The toxicity of citrinin has been studied in a variety of laboratory species. In the rat and mouse, the intraperitoneal LD_{50}, which is much greater than the oral toxicity, is 60–80 mg/kg body weight. Other clinical signs observed in nonfarm animals include salivation, lacrimation secretions, vomiting (dog), and hyperemia of the ears and mucous membranes. The major effects in rabbits given acute doses are diarrhea and kidney lesions (Hanika *et al.* 1983).

Because citrinin possesses antibiotic activity against Staphylococci and other gram-positive bacteria, it could possibly interfere with rumen function. This subject has not been experimentally explored primarily because there is no field evidence that citrinin toxicosis occurs in ruminants.

Metabolic Fate and Residues. Although its metabolic fate in animals has not been extensively studied, it is clearly evident that citrinin is readily absorbed from the gastrointestinal tract, metabolized, and rapidly eliminated. Metabolism to more water-soluble metabolites has been demonstrated, but their identification has not yet been achieved. In rats, the major part of a dose is excreted in urine within 24 hr as metabolites; these are partially conjugated to GSH which correlates with the depletion of GSH from hepatic and renal tissue (Berndt *et al.* 1980). Urine is normally the primary route of excretion; however, in citrinin-damaged kidneys the toxin is poorly excreted due to decreased GFR. To compensate, biliary excretion becomes the primary excretory route. The rapid elimination of citrinin and its metabolites prevents accumulation in tissues, thereby decreasing the toxicologic threat to animals. In addition, the possibility of retained residues in edible tissues is reduced.

Rubratoxins

Rubratoxins are bisanhydride metabolites produced by *Penicillium rubrum* and *P. purpurogenum*. The toxins derive their name from *P. rubrum*, from which they were first isolated. There are two known toxins, rubratoxin A (RA) and B (RB). Comparatively RB is about twice as toxic as RA. The structure of rubratoxin is as follows:

	R
Rubratoxin A	OH, H
Rubratoxin B	O

These fungal toxins were originally found in moldy corn and caused a disease characterized by hepatitis, nephrosis, and hemorrhage when fed to cattle and swine. Because *Aspergillus flavus* was isolated along with *P. rubrum,* the primary attention was directed toward the former and the production of aflatoxins. However, isolates of *P. rubrum* grown on corn and fed to animals are more toxic than *A. flavus* isolates. Even though no field cases of rubratoxicosis have been reported, Pier (1981) speculated that toxic effects may occasionally occur in cattle and swine.

Fungi that produce RA and RB are commonly found in food and feedstuffs, but the toxins themselves have never been found naturally. This could be due in part to the lack of an adequate analytical method (Mirocha 1980). *Penicillium rubrum* has been isolated from peanuts, corn, bran, and sunflower seeds. The fungi can be readily grown in the laboratory on a number of synthetic media, but the production of RA and RB is limited to specific types.

Although the mechanism of action has not been clearly established, biochemical changes at the subcellular level include inhibition of mitochrondrial respiration, ATPase activity, and protein synthesis. The latter involves binding to DNA, decreased RNA polymerase activity, decreased levels of RNA, and disaggregation of polysomes. A structure–activity relationship is apparent from studies with laboratory species. Alteration of functional groups such as hydrogenation of the α,β unsaturated lactone moiety greatly decreases toxicity. Therefore, toxicity is attributed primarily to the parent compound.

The liver is the primary target organ although congestion and hemorrhaging may occur in several other organs including kidneys, spleen, lungs, and gastrointestinal tract. Liver cells show degenerative changes and possibly necrosis. Alterations of hepatic function are indicated by increased prothrombin time (coagulopathy) and bilirubinemia (icterus). Mild degenerative changes in renal tubule epithelium are seen in some species. Clinical signs induced experimentally include depression, anorexia, decreased weight gain, coagulopathy, hemorrhaging, bloody feces, and death.

Experimental rubratoxicosis has been produced in a number of species, including several farm animals. In calves, daily doses of 8 mg/kg of RB decreased liver function, 12 mg/kg caused depression and anorexia, and 16 mg/kg caused acute hepatic failure and death (Pier et al. 1976). Acute toxicity is much less in monogastric species; the LD_{50} is about 400 mg/kg in rats and 83 mg/kg in chicks. The lesser toxicity in these species is thought to be related to less absorption from the gastrointestinal tract, possibly due to greater degradation in the tract. This is also apparent from the greater (10 times) toxicity of RB administered parenterally versus orally in rats. Similarly, chronic feeding studies show that rubratoxins are not very toxic to chickens; 500 ppm in the diet fed for 3 weeks was required to produce weight loss (Wogan et al. 1971). Rubratoxin B is not carcinogenic but is mutagenic, teratogenic, and embryotoxic in mice and egg embryos; the α,β unsaturated lactone ring is necessary for this activity (Mirocha 1980). Also, there is the potential for immunosuppression due to impairment of complement formation in the liver.

Perhaps the most significant environmental feature of rubratoxins is their ability to act synergistically with other mycotoxins. Experimental evidence in calves, guinea pigs, rats, and dogs indicates that AFB_1 and RB administered together produce toxic effects not seen with either toxin alone. Table 12.12 illustrates synergistic action between RB and AFB_1 in rats. Synergism between ochratoxin A and RB in rats has also been demonstrated.

Much of the low oral toxicity of rubratoxins appears to be due to their slow absorption from the gastrointestinal tract, which may include detoxification prior to absorption. Studies in rats show 80% elimination of ^{14}C labeled RB in 7 days in urine and feces equally (Unger and Hayes 1979). Partial metabolism including conjugation to glucuronides and sulfates precedes elimination. There is some evidence that conjugates excreted in bile are recirculated through the enterohepatic system; this is based on biphasic elimination from liver (half-lives are 14 and 100 hr). Evidence for involvement of hepatic MFO in their metabolism includes localization of ^{14}C labeled RB in the ER after administration and decreased toxicity of RB in rodents after stimulation of MFO with barbiturates. The latter finding also demonstrates that RB is more toxic than its metabolites. There is no

TABLE 12.12. Synergistic Lethal Action of Rubratoxin B and Aflatoxin B_1 Administered Simultaneously to Male Rats[a,b]

Treatment regimen	Total toxin (mg/rat)	Body weight at end of dosing (% of control)	Mortality	Preneo-plastic liver lesions
1. Controls (DMSO 3×/week for 5 weeks)	—	100	0/10	0/7
2. Rubratoxin B (25 mg/kg, 3×/week for 4 weeks)	39.7	95	0/10	0/7
3. Aflatoxin B_1 (0.2 ppm in diet for 6 weeks)	0.11	102	0/10	6/7
4. Rubratoxin B and aflatoxin B_1 simultaneously	—[c]	86	9/20	5/8

[a] From Wogan et al. (1971).
[b] Note the synergistic response in lethal activity (mortality), but not carcinogenicity.
[c] Combined dosing regimens of groups 2 and 3.

evidence that rubratoxins should be considered a residue problem in food animals.

Patulin

Patulin is a hemiacetal lactone produced by several species in the genera *Aspergillus, Penicillium,* and *Byssochlamys.* This toxin is toxic to both plants and animals and has potent antibiotic activity. Most of the known fungal producers of patulin were identified in the 1940s at a time when the search for antibiotics was quite intense. Its chemical structure is as follows:

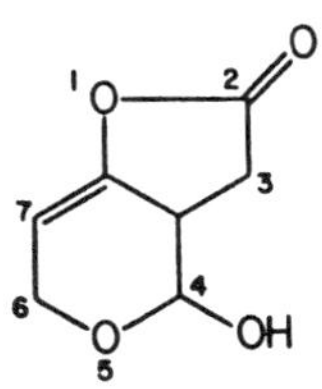

Patulin was originally called claviformin, named for *Penicillium claviforme,* from which it was first isolated. The name "patulin" was derived because structural characterization was made in *Penicillium patulum.*

The occurrence of patulin in foods and feeds is not well understood. To date, proven natural contamination in agricultural products has been limited to rotted apples and apple juice or cider. Concentrations as high as 1000 ppm have been found in these commodities. The toxin has been implicated in several cases of toxicoses involving cattle and sheep. From the standpoint of potential contamination of foods and feeds, Ciegler (1977) listed the following fungi: *P. urticae, P. expansum, P. melinii, P. cyclopium, A. clavatus, A. terreus,* and *B. nivea.* Even though these patulin-producing fungi are periodically found in foodstuffs such as cereals and legumes, the toxin itself has not been detected.

Patulin is relatively unstable under alkaline conditions but is quite stable in an acidic environment. This accounts for its long-term stability in some foods and feeds but not in others. During its breakdown, patulin reacts with sulfhydryl-containing amino acids or proteins, forming patulin–cysteine adducts. Although less reactive than patulin, these adducts appear to retain some of the toxic potency of the parent toxin.

No toxicologic studies with patulin have been conducted in domestic animals, but there is some indirect evidence for patulin toxicosis (Ciegler 1977). Feeding of mold-infested malt was believed to have caused the deaths of over 100 dairy cattle in Japan. The fungi *P. urticae* that produces patulin was identified, and experimental feeding of malt inoculated with this fungi to a bull resulted in neurological signs, brain hemorrhage, and death. Similar effects were observed in rodents injected with patulin. In France, cattle fed wheat contaminated with *A. clavatus* (a patulin producer) developed pulmonary edema and congestion, and several died. Sheep fed extracts of *B. nivea* (also a patulin producer) exhibited anemia, decreased serum protein concentrations, nasal discharge, loss of rumination, pain in the sternal area, anorexia, and weight loss (Lynch 1979). Postmortem findings included abomasal hemorrhaging and lesions in liver and kidneys.

Patulin is a relatively potent mycotoxin. Experimental findings in laboratory species indicate LD_{50} values ranging from 10 to 35 mg/kg of body weight, depending on animal species and route of administration. Toxicity is lower by the oral route than when injected. In the chick, the LD_{50} is 170 mg/kg. The primary toxic effects are ascites, hydrothorax, and pulmonary edema. It is a dermal irritant when applied topically. Also, lesions develop at the site of subcutaneous injection. At the molecular level, patulin inhibits aerobic respiration, membrane permeability, and ATPase activity. When it interacts with sulfhydryl-containing amino acids such as cysteine during breakdown, the resulting adducts may be toxic.

Patulin is also toxic to bacteria, protozoa, and fungi. In fact, it was tested extensively for possible antibiotic use in humans but proved to be too toxic. It inhibits the growth of bacteria and protozoa, but its toxicity on ruminal or intestinal microflora has not been determined. Mertens (1979) speculated that patulin ingestion would be detrimental to gastrointestinal tract microflora.

No metabolism studies in farm animals have been conducted. Studies in rats indicate rapid metabolism and elimination. Therefore, one could speculate patulin has a low potential for residues in food products of animal origin.

MYCOTOXINS ASSOCIATED WITH ROUGHAGES

Fungal Tremorgens

Ryegrass staggers is a neuromuscular disorder affecting sheep and cattle grazing permanent pastures in which perennial ryegrass (*Lolium perenne*) is a dominant species. It occurs mainly in Australia and New Zealand. It occurred in Oregon in the past, but is no longer commonly observed. This is probably due to the almost universal practice of field burning of grass seed fields.

The toxic agents appear to be a class of fungal toxins called tremorgens. These are produced by *Penicillium* spp. such as *P. paxilli*. A variety of tremorgens have been isolated; two of the major ones are verruculogen and paxilline.

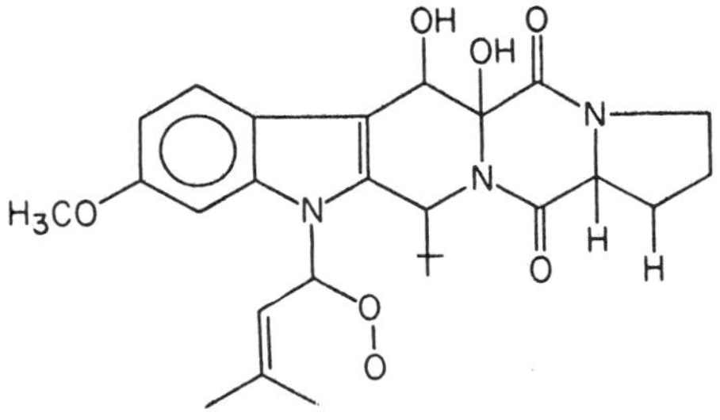

Verruculogen

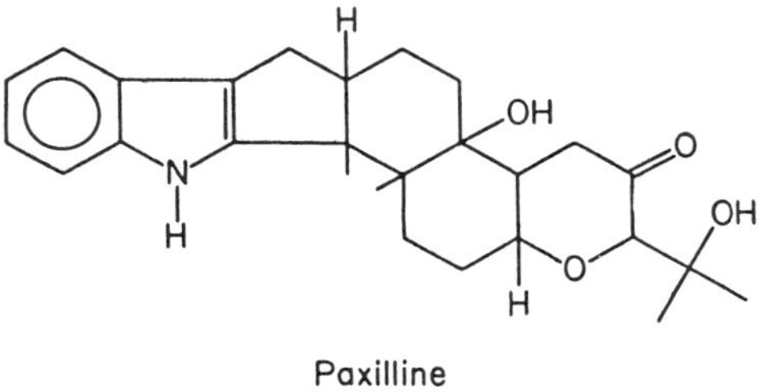

Paxilline

12. MYCOTOXINS

Ryegrass staggers generally occurs when the pasture is short. After exposure to the toxic pasture, animals may be observed to experience a tremoring syndrome. The disorder, also characterized by locomotor incoordination, may not be apparent until the affected animals are disturbed or forced to run, at which time they exhibit various degrees of incoordination, with an abnormal staggering gait, head shaking, stumbling and collapse, sometimes followed by severe muscular spasms (Fig. 12.11). After a time the animal will regain its feet and walk away. Morbidity may be very high, with up to 80% of a flock of sheep affected, but mortality is very low. Mortality is generally due to misadventure during an attack of staggers, such as drowning in streams or ponds or falling over embankments. With an affected herd or flock, normal management procedures may become difficult, as any attempt to move the animals results in occurrence of staggers. Even in severe cases, there are no pathological tissue changes. This combination of lack of tissue damage and dramatic neuromuscular dysfunction suggests the presence of a potent neurotoxic agent.

Gallagher et al. (1977) have demonstrated that the administration of fungal tremorgens such as verruculogen and paxilline to sheep results in symptoms of ryegrass staggers similar to those seen in field cases. The tremorgens are remarkably potent. A dose of 0.005 mg/kg body weight given intravenously was sufficient to cause moderate trembling and ataxia. It is believed that the predominant means by which livestock acquire the toxins is through ingestion of soil when grazing short pasture. The toxic pasture grass in general does not seem to contain the tremorgens. Apparently the presence of ryegrass is necessary for

FIG. 12.11. A lamb exhibiting neurological effects of ryegrass staggers.
Courtesy of R. E. G. Keogh and M. E. Soulsby.

the growth of the *Penicillium* spp. in the soil, where it produces toxin. Heavy grazing intensity increases the likelihood that the grass supply will be depleted, causing close grazing that increases soil, and hence tremorgen, ingestion. White *et al.* (1980) have shown that the tremorgens can be absorbed by ryegrass plants from the soil and translocated to the leaves.

Fletcher and Harvey (1981) advanced the suggestion that an endophytic seed-borne fungus was associated with the occurrence of ryegrass staggers. The endophyte was concentrated in the leaf sheath of the plants, which could explain why ryegrass staggers is generally associated with grazing the base of ryegrass pasture. Gallagher *et al.* (1981) have isolated two potent neurotoxins from forage that caused ryegrass staggers. They proposed the general name lolitrem A and lolitrem B. The neurotoxic effects of the lolitrems in mice are distinctive from those produced by the *Penicillium* tremorgens; the tremors and incoordination are slower in onset and are more protracted. The origin of the lolitrems is presently unknown.

Another group of tremorgens, called janthitrems, has been isolated from *Penicillium janthinellem,* a fungus associated with toxic ryegrass pastures (Gallagher *et al.* 1980). It is evident that a wide variety of tremorgens may be involved in development of ryegrass staggers.

Ryegrass staggers is normally associated with the grazing of ryegrass pastures, but can result from consumption of conserved forage as well. Hunt *et al.* (1983) described an outbreak of ryegrass staggers in ponies fed a pelleted diet containing 60% perennial ryegrass straw. The ponies exhibited classic signs of ryegrass staggers, including excitability, abnormalities in gait, exaggerated limb action, weaving when standing still, and tetany in extreme cases. *Penicillium cyclopium* was found in both the ryegrass straw and in fecal samples.

A condition called paspalum staggers occurs in cattle, and occasionally in sheep and horses, in several parts of the world where *Paspalum* spp. are grown as pasture grasses, including New Zealand, Australia, South Africa, the U.S., Portugal, and Italy (Mantle *et al.* 1977). The grass is frequently infected by ergot (*Claviceps paspali*), and for many years ergot was believed to be the causative agent of paspalum staggers. Mantle *et al.* (1977) dosed cattle and sheep with *Claviceps paspali* sclerotia and produced the condition, characterized by head tremors, incoordination, and collapse when disturbed. The sclerotia were subjected to various extraction procedures to isolate the tremorgenic fraction. The ergot component was without activity. A mixture of tremorgens was isolated that appeared to be the active fraction. Thus, it appears that paspalum staggers is caused by tremorgens produced by *Claviceps paspali* infection of *Paspalum* seed heads.

Lupinosis

Lupinosis is caused by the ingestion of toxins produced by the fungus *Phomopsis leptostromiformis,* which grows on lupines (Fig. 12.12). It infects the green lupine plant initially, but persists on the stubble. Lupinosis is characterized by severe liver damage. It has been known for over a century and has been observed in Germany, Poland, South Africa, New Zealand, and Australia. It is of most importance in Australia, where lupines are widely grown as a grain crop, and sheep are grazed on the stubble.

The first signs of development of lupinosis are a loss of appetite and a loss of weight and condition (Fig. 12.13). With acute toxicity, following a high intake of toxic lupine, the liver increases in weight due to a massive accumulation of fat in the liver cells. The liver is greatly enlarged, bright yellow or orange in color, and very greasy when cut. The gall bladder is increased in size severalfold over normal. The subcutaneous tissue and body fat are tinged with orange or yellow.

Chronic lupinosis is characterized by liver necrosis (Fig 12.14) and the appearance of signs of liver dysfunction such as jaundice. The membranes of the eye and mouth may be intensely yellow. Affected animals appear dejected and depressed and lag behind the rest of the flock. The detection of these stragglers, by regularly moving the flock, provides an early indication of a lupinosis problem and is a management technique recommended whenever lupine stubble is grazed. In cases of chronic lupinosis, the liver is coppery or tan colored and is

FIG. 12.12. Seeds of grain lupine (left) showing their comparative size and shape to soybeans (right).

FIG. 12.13. The effects of chronic lupinosis. These sheep were the same age (12 months) on the same property. The sheep on the left had grazed lupine stubble for 3 months, while the sheep on the right had been on clover pasture.
Courtesy of J. G. Allen.

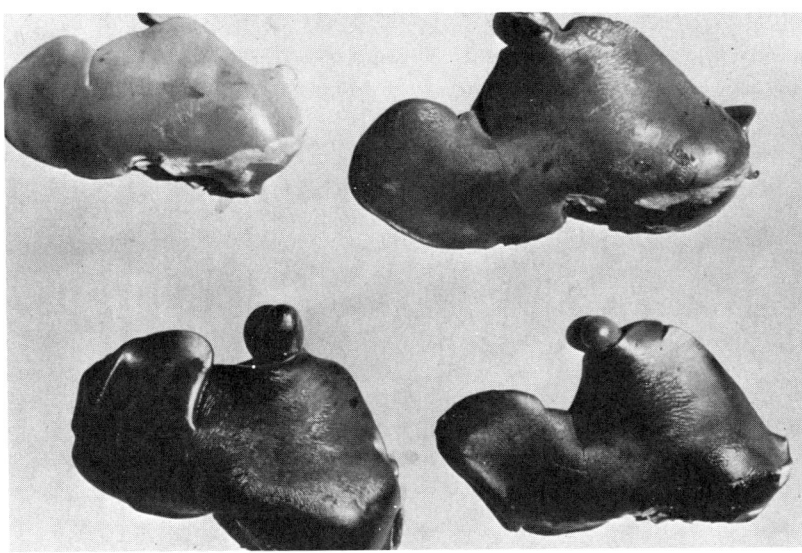

FIG. 12.14. Livers from sheep with chronic lupinosis. Note the extreme paleness of the top left liver. All showed significant microscopic damage.
Courtesy of J. G. Allen.

smaller than normal. It feels hard and fibrotic and has a granular appearance. In Australia, sheep exposed to lupinosis toxins may also consume pyrrolizidine alkaloids in such plants as *Heliotropium europaeum*. Because both have hepatoxic effects, it is likely that they could have additive or synergistic activity.

The growth of the *Phomopsis* fungus on lupine may be weather related, with warm, wet, humid conditions favoring its growth. In Australia it was formerly believed that lupinosis was associated with summer rains, but this was because the varieties then grown had coarse, unpalatable stems which were only consumed when they were moistened and softened by rain. All lupine stubble is potentially toxic. High sheep stocking rates seem to increase the lupinosis incidence.

Sheep that develop lupinosis develop high liver copper levels. This is a reflection of liver damage; similar elevations in liver copper are observed in pyrrolizidine alkaloid poisoning. Allen *et al.* (1979) found that liver copper and selenium levels are increased and zinc levels are decreased in lupinosis. Oral administration of zinc sulfate to sheep treated with toxic lupine extracts had protective activity against liver damage (Allen and Masters 1980) and reduced the liver copper levels, suggesting that zinc supplementation may have potential practical application. However, at the levels of zinc used, zinc toxicity may be encountered.

Allen (1978) has reported a lupine-associated myopathy in sheep that is identical in appearance to white muscle disease (a selenium–vitamin E-responsive myopathy), but which does not respond to and is not prevented by selenium or vitamin E supplementation. It is not known if the lupine-associated myopathy is caused by a direct action of the lupinosis toxins on the muscles or by these toxins interfering with selenium availability to the muscles. Van Rensburg *et al.* (1975) have observed cardiac lesions in lupinosis that they attributed to direct action of the toxin.

Lupinosis also affects cattle. Allen (1981) described two conditions in cattle affected with lupinosis. First, there is a fatty liver syndrome that affects only cows in late pregnancy or those that have recently calved and is the more commonly observed condition. The animals show fatty livers and signs of ketosis. Allen (1981) suggests that the lupinosis toxins induce a mild depression of appetite, which is sufficient to produce a glucose deficit. Thus, this syndrome appears to be a secondary nutritional ketosis.

The second syndrome in cattle is a cirrhotic liver condition similar to that seen in chronic lupinosis in sheep. In sheep, the lupinosis is generally obvious while the animals are still grazing the lupine stubble, and photosensitization is rarely seen. Conversely, in cattle the animals may not exhibit signs of lupinosis for some time after grazing on lupines ceases. Generally, the condition occurs when new grass growth begins at the beginning of the pasture season. Photosensitization is commonly seen. The udder is often affected, and the cow refuses to let the calf nurse. The photosensitization is probably a result of the increased chlorophyll intake when grass growth begins. The increased phylloerythrin load cannot be excreted by the damaged liver, and secondary photosensitization occurs.

The toxic agents have been isolated and named phomopsin A and phomopsin B. The phomopsins are cyclic hexapeptides involving didehydro- and hydroxyamino acids and a chlorine-containing phenylalanine-derived subunit (J. A. Edgar and C. C. J. Culvenor, 1984, Personal communication, CSIRO Division of Animal Health, Parkville, Australia). They have a colchicine-like action in arresting mitosis in liver cells, inhibiting microtubule polymerization. The aromatic ring of phomopsin is the active site of the toxicologic effects. Phomopsin is extremely toxic; Jago et al. (1982) found the minimum lethal dose in sheep to be 10 μg/kg body weight.

Prevention of Lupinosis. Some varieties of lupine, such as Ultra, are resistant to infection by *Phomopsis*. Unfortunately, *Phomopsis*-resistant cultivars do not thrive in all areas where lupines are grown. Animal management techniques are required to prevent stock losses. Sheep introduced onto lupine stubbles should be herded daily and closely observed for signs of incipient lupinosis. Late-pregnant or recently calved cows should not be grazed on lupine stubbles.

Sporidesmin

Facial eczema is a dermatitic condition of sheep and cattle that occurs extensively in the northern part of the North Island of New Zealand, and has also been observed in Australia and South Africa. Facial eczema is caused by spores of a fungus (*Pithomyces chartarum*) which grows in the dead litter of ryegrass pastures. The spores contain a toxin called sporidesmin, which causes liver damage leading to secondary photosensitization (Fig. 12.15). The condition causes extensive stock losses in sheep and dairy cattle, from mortality and reduced production.

The danger periods for facial eczema follow warm wet weather that favors fungal growth. The spore numbers in the pasture litter rise very rapidly under favorable fungal growth conditions and can be seen as a

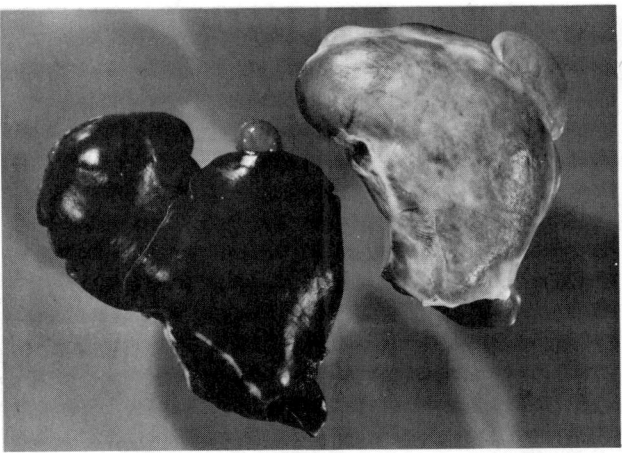

FIG. 12.15. Liver and gall bladder from an animal with facial eczema (right) contrasted with those from a nonaffected animal (left). The liver may be almost white in a severe case of facial eczema.
Courtesy of H. H. Meyer.

cloud of black dust when the pasture is disturbed (e.g., mowing). The major toxin produced in the fungal spores is sporidesmin (Fig. 12.16). There are several others (sporidesmin B, C, D, E, F, G, and H), but they are of low biological activity. The structure of sporidesmin is as follows:

In sheep, the first observable signs are photodynamic lesions on the ears and face. The animals become restless, shake their heads, and may rub their eyes and ears against solid objects and on the ground. The ears become swollen, red, and droopy, and the lips and eyelids swell. Scabs may form over these areas. The sheep stop grazing, particularly if the lips are badly irritated. All parts of the skin exposed to sunlight, including the line where the wool parts down the back, may

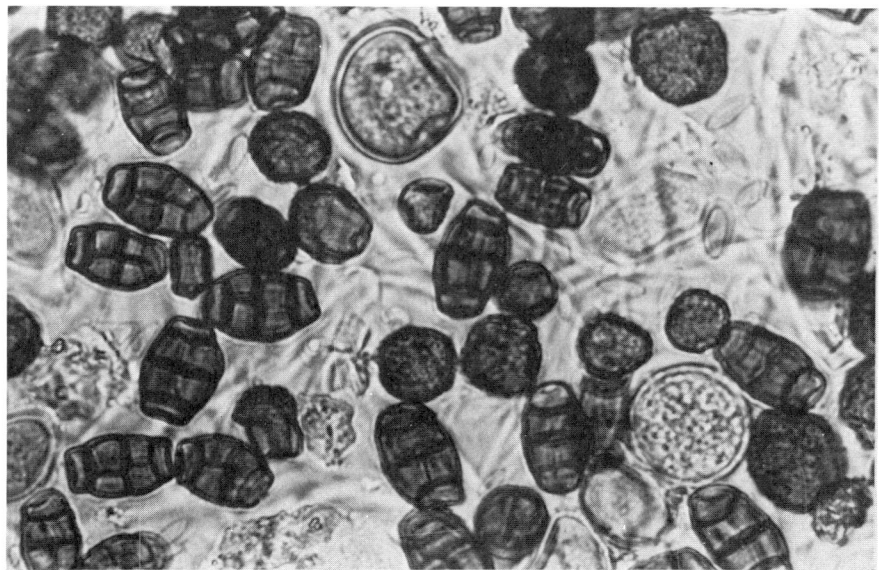

FIG. 12.16. Spores of *Pithomyces chartarum*, the fungus which produces sporidesmin. The spores are often referred to as being shaped like hand grenades.
Courtesy of H. H. Meyer.

be affected. The animals show photophobia and seek out any shade available. Newly shorn sheep are especially vulnerable.

In dairy cattle, the lesions are most frequent on the udder and teats and down the inside of the hind legs, especially in Jerseys, and on the white areas of Holsteins. Sometimes the skin of the face peels off over a large area (Fig. 12.17). Prior to observable signs of facial eczema, milk production falls off markedly.

Most affected animals recover, especially if treatment programs are initiated. Adequate shade should be provided. Even a few minutes of exposure to strong sunlight can cause intense skin damage. Affected dairy cows should be dried off (cessation of lactation) immediately. This reduces appetite, and hence the intake of toxin, and allows maximum diversion of nutrients to tissue repair. It also lessens the intake of chlorophyll, thus reducing the phylloerythrin load on the liver. Affected animals should be treated to prevent fly strike.

New Zealand researchers have shown that zinc supplementation may have protective effects against facial eczema. Dosing with zinc salts at the time animals are exposed to toxic pastures will reduce both

FIG. 12.17. A severe case of facial eczema showing the skin peeling off the face.
Courtesy of H. H. Meyer.

the number of animals that develop facial eczema and the amount of liver damage. To achieve protection, dose rates with zinc are high, equivalent to 20–25 mg/kg body weight/day. Zinc oxide or chelated (e.g., with EDTA) zinc are recommended; zinc sulfate has corrosive effects on the gastrointestinal tract and should not be used. There is only a three- or fourfold margin of safety between protective and toxic dose rates. In ruminants, fibrosis of the pancreas occurs with high zinc intakes. Therefore, zinc dosing is not recommended as a routine treatment, but can be used for short-term protection on farms with occasional facial eczema problems.

Munday (1982, 1984) has demonstrated that the metabolic defects induced by sporidesmin may be initiated by the intrahepatic generation of oxygen free radicals (superoxide anions). Further, he demonstrated that of a variety of metals, zinc was among the most effective inhibitors of active oxygen generation from sporidesmin, suggesting that the protective effect of zinc against sporidesmin toxicity is likely a result of inhibition of superoxide radical generation.

Poor productive performance of sheep (ovine ill-thrift) in Nova Scotia (Brewer *et al.* 1971) has been associated with mycotoxins similiar to sporidesmin. Brewer *et al.* (1972) reported that *Chaetomium* spp. of fungi on pasture grasses produce an antibiotic, chetomin, that appears to be involved in ovine ill-thrift. They proposed that chetomin is poorly absorbed from the rumen and thus accumulates, where it exerts an antibiotic effect, impairing rumen fermentation.

Stachybotryotoxins

Stachybotryotoxicosis is a trichothecene-induced disease occurring in animals, particularly horses, after consumption of moldy straw or hay. It has been observed in the USSR, Hungary, Czechoslovakia, Romania, Finland, France, and South Africa. The causative fungi are *Stachybotrys atra* (or *S. alternans*) and possibly *Myrothecium* spp. and *Dendrodochium* spp. Because *Stachybotrys atra* utilizes cellulose, it may be found on any organic matter rich in this nutrient including paper, cotton, plant debris, sugar cane roots, cereal grains, straw and hay. Several toxins collectively called stachybotryotoxins are produced, but not all have been identified. However, the disease is most likely caused by one or more group C macrocyclic trichothecenes (Fig. 12.6) including roridin E, verrucarin J, and satratoxin E, G, and H.

Although the disease is best characterized in horses, it is known to occur in cattle, sheep, swine, poultry, and man. In horses, stachybotryotoxicosis progresses in stages after consumption of mold-infested straw or hay over several weeks. Initially, the lips, tongue, and buccal mucosa are irritated due to the epithelionecrotic effects of the toxins. Those areas can become severely necrotic and edematous. The second stage is characterized by coagulopathy, leukopenia, and thrombocytopenia due to toxic effects on hematopoietic processes. Additionally, the necrosis in the mouth area becomes worsened, diarrhea usually develops due to gastrointestinal tract irritation, and animals become quite weak. Death from hemorrhaging and septicemia may occur. A different form of strachybotryotoxicosis develops in horses that ingest a large quantity of toxic fodder. In these cases, clinical signs progress rapidly and include nervous disorders, loss of reflex responses, loss of vision, evidence of circulatory collapse, and finally death.

Other farm animals affected with stachybotryotoxicosis generally exhibit the same toxic signs as horses. In calves, hemorrhaging is widespread throughout the body. Toxic effects in swine include vomiting, tremors, anemia, and abortion. Dermatitis develops around the teat area as well as the mouth area in nursing sows. Skin rash and necrosis have been observed in humans after handling contaminated straw. Farm workers in Hungary were also shown to suffer respiratory distress, nose bleed, and eye irritation after contact with infected litter and fodder.

Studies in laboratory species indicate stachybotryotoxins are also immunosuppressive. Both antibody production and delayed cutaneous hypersensitivity are impaired. These effects have not been studied in domestic animals.

Moldy Straw-Induced Photosensitization

Bagley *et al.* (1983) described a photosensitization condition in cattle in Utah which appears to be associated with the consumption of moldy straw. It is commonly referred to as "straw-pile disease." In the dryland farming and ranching areas of the western U.S., it is a common practice to use a straw catcher on wheat combines to periodically discharge the straw and chaff in large piles. Wintering beef cattle are turned into the fields and may consume the straw throughout the winter. A heavy infestation of mold is often evident in the top several inches of the piles. Straw-pile disease is associated with this type of wintering program. Affected cattle show the usual clinical signs of photosensitization, including ulcerated lesions on nonpigmented skin. The eyes and skin around the eyes are particularly affected, as are the muzzle, underside of the jaw, the teats, and the lower limbs. These areas have the shortest hair coat and receive both direct light and light reflected from snow. Evidence of liver damage, including elevated serum γ-glutamyl transpeptidase and glutamic-oxaloacetic transaminase, are observed. The affected cattle recover when removed from access to the straw piles, although a recurrence can occur when they are placed on green spring pasture, indicating sufficient liver damage to interfere with biliary excretion of pylloerythrin. The presumed etiology is mycotoxin-induced hepatic damage, leading to secondary photosensitization. The causative fungi have not been identified.

Kikuyu Poisoning

Kikuyu grass (*Pennisetum clandestinum*) is a tropical forage species used extensively in pastures. In New Zealand, South Africa, and Australia, a condition called kikuyu poisoning has been observed in cattle, sheep, and horses. It was first reported from New Zealand. Signs of toxicity generally occur 24–48 hr after animals are given access to the toxic pasture. Clinical signs include anorexia, depression, pilo-erection, drooling, colic, grinding of teeth, cessation of ruminal and intestinal movement, and lack of fecal excretion. Muscle twitching, a high-stepping gait, and occasionally convulsions are seen. A distinctive feature is "sham drinking"; cattle will congregate at water and even put their mouths into or onto the water, but fail to drink (Newsholme et al. 1983). The most striking lesion is an intensive necrosis of the rumen and omasum mucosa. Mortality of affected animals is about 80%. Selective involvement of the superficial epithelial layers suggests the direct action of a toxin acting from within the lumen of the alimentary tract (Newsholme et al. 1983). In many but not all cases, kikuyu poisoning is associated with army worm infestation of the pasture. Bryson (1982) suggests that there is an alteration of the composition of the grass by the army worm, perhaps exacerbated by fungi. Circumstantial evidence points to trichothecene mycotoxins produced by *Myrothecium verrucaria* as the causative agents. No satisfactory treatment for affected animals has been developed.

DECREASING EXPOSURE TO MYCOTOXINS

Because the problem of mycotoxin contamination in agricultural crops is so complex, no practical methods have been developed for completely eliminating any of the mycotoxins. Consequently, the realistic goal in commercial agriculture is to minimize exposure of livestock to mycotoxins. This goal can be accomplished by using a series of methods that fall into two general categories: (1) prevention of mycotoxin contamination and (2) exclusion of feeds that are inadvertently contaminated. Prevention of mold growth and toxin elaboration in feeds is clearly the best approach. Farmers must follow proper preharvest and postharvest methods, particularly during times when environmental conditions favor mold infestation. Moreover, the prevention program must be specifically designed for each crop, taking into account the growing, harvesting, drying, processing, and storage procedures. When prevention fails, exposure can be minimized by incorporating exclusion methods such as diverting the feed from being fed, by blending with uncontaminated feed, or by removing the mycotoxin. However, before an appropriate exclusion strategy is formulated, the concentration of the mycotoxin in the feed should be established if possible. This requires proper sampling and analytical determination.

Prevention and control methods have been more extensively developed and tested for aflatoxin than the other mycotoxins, but many of these methods are applicable to crops targeted by toxigenic fungi. Several excellent reviews of the various methods are available (Lillehoj and Hesseltine 1977; Ciegler 1978; Goldblatt and Dollear 1979; Dickens 1983; Jemmali 1983; Draughon 1983; Anderson 1983).

Aflatoxin

Knowing that certain crops in particular geographic locations are susceptible to invasion by aflatoxigenic fungi should prompt farmers to take certain precautions. First, it must be recognized that fungal invasion and toxin production can occur in the field and during harvest as well as in storage. At each stage, multiple factors are involved. Some that are inherent to the production system or that are a result of bad management can be alleviated. There are others that are not controllable, such as climatic conditions. Regardless, it is essential that the entire production system is carefully managed. A second important element to be recognized is that the quantity of aflatoxin produced in seeds or kernels is cumulative depending on the frequency of favorable conditions. Such conditions may occur in several short periods or one prolonged period. Less than 0.1% of a population of seeds or kernels may contain an amount of aflatoxin to give an unacceptable concentration in the entire population (Dickens 1983).

Prevention. Preharvest contamination of susceptible crops appears to be promulgated by plant stress during critical periods of seed formation in combination with hot weather. The two most troublesome forms of stress favoring fungal colonization are drought and insect damage. Reasonably effective prevention has been accomplished by proper irrigation and insect control measures during the period of seed formation. Other recommended practices include planting fungus-free seed, controlling plant diseases, optimizing plant density, and not leaving crops in the field beyond maturity. An additional preharvest problem in U.S. corn production is that hydrids adopted for the corn belt are grown in the South where they are less adopted and believed to be subject to fungal attack (Widstrom and Zuber 1983).

The plant breeding approach to decreasing preharvest aflatoxin contamination is still in the experimental stages. The goal is to develop cultivars that would resist infection by aflatoxigenic fungi or inhibit aflatoxin production. To date, resistant cultivars have not been identified for most crops. There is substantial evidence that aflatoxin contamination in corn is under genetic control, i.e., the level of aflatoxin produced in some corn varieties is quite different from that in other varieties, and the difference is heritable. However, genetic studies have been hampered, particularly by environmental interactions. Although resistant corn hybrids are not presently available, plant breeders remain optimistic that genetic control of the aflatoxin problem in corn might be possible (Widstrom and Zuber 1983). Some success has been achieved with pistachio nuts; cultivars with closed husks may be more resistant to *A. flavus* infection than those cultivars that produce nuts which split the husk and expose the kernel before the nuts are harvested (Dickens 1983).

Pesticide application to prevent preharvest aflatoxin production is another method that may be feasible in the future. Research into finding a chemical control method was prompted by the discovery that dichlorvos, an organophosphate insecticide, inhibited aflatoxin production in cultures of rice, corn, wheat, and peanuts (Rao and Harein 1972). Since then, several pesticides have been field tested with marginal success. Application to corn of the insecticides Bux, carbaryl, and Dyfonate decreased AFB_1 levels from 28.9 ppb to 4 ppb, 1 ppb, and 2 ppb, respectively (Draughon 1983). Although application of pesticides would pose an additional hazard, those that serve the multiple functions of killing both insects and fungi and preventing mycotoxin production would afford enough benefit to offset the risk associated with their use.

During harvesting, equipment should be properly adjusted and operated so that crop damage (e.g., disrupting the seed coat) or picking up excessive amounts of dirt and debris are avoided. Processing of cottonseeds and peanuts or drying of corn should not be delayed. Because aflatoxin contamination can occur in less than 48 hours, high-moisture crops should be dried immediately after harvest. When preharvest fungal infection occurs, early harvest and drying will help curtail postharvest aflatoxin contamination. If an agricultural commodity is suspected of being contaminated with aflatoxins, it should be sampled and analyzed for AFB_1 if possible. Confirmation of the presence of AFB_1 would allow diversion of the material to avoid contamination of larger lots of feed.

During handling, processing, or storage, maintaining dryness is essential. Wet product should not be blended with dry product in an attempt to achieve dryness. The wet product will mold even though the average moisture content of the mixture may be less than the recommended level. Dried commodities that become too moist during storage can also become contaminated. Dickens (1983) cited several causes including leaking roofs, wetting of the product with insecticide sprays, conveyance of water from flooded elevator dump pits into warehouses, seepage of water onto warehouse floors, and storage of the product on concrete floors without vapor barriers. Dickens (1983) also described a problem inherent to modern bulk storage facilities, that being moisture migration and condensation. For example, in a large pile of warm grain, cool weather will cause convection currents of warm moist air through the pile. When the warm air reaches the surface of the pile, moisture will condense, wetting a portion of the grain, thereby creating favorable conditions for mold growth. Also, the cool surface of a storage structure can cause condensate drip. To help prevent these problems, fans should be employed to exhaust warm moist air and to replace it with cool air. Moreover, aeration will cool stored grains and help prevent moisture migration.

For crops such as ground high-moisture corn, which is highly susceptible to fungal invasion, oxygen exclusion is absolutely necessary. Storing high-moisture corn in bunkers is particularly problematic. Topping the corn with a silage cap to minimize aeration is one practice that is used. However, this may be a problem since it can take several days to accomplish this packing and capping, thereby allowing toxin formation in oxygenated pockets.

Chemical prevention of aflatoxin production in corn has proved somewhat successful. Various fumigants, fungicides, and insecticides have been tested. Whereas insecticides are neither economically feasible nor ecologically wise, fumigants such as methyl bromide and fungicides such as the propionates are both effective and relatively safe for treating animal feeds (Draughon 1983). Addition of some chemicals such as propionic acid both preserve the corn and enhance (by about 4%) its value as an animal feed. Both mold growth and toxin formation are reduced by addition of either 2% ammonia or 1% propionic acid. Other additives that have proven somewhat effective are sodium diacetate, sorbic acid, and Gentian Violet. Draughon (1983) stated that chemicals for postharvest control of aflatoxin production should be discouraged, since they are a poor substitute for proper drying and good management practices and are not economically feasible. Furthermore, chemicals may offer only an effective short-term solution such as protection during fermentation or until drying. Propionates can afford protection for at least 90 days; it has been postulated that the acid is eventually removed by microbial growth.

Exclusion. Although prevention is the best method of decreasing the exposure of animals and humans to aflatoxins, there will be times when contamination cannot be avoided. Unfavorable climatic conditions (wet or humid) coupled with excessive insect or disease invasion may completely overwhelm any benefit derived from preventative measures. During such years it is especially important that feeds are sampled and analyzed so that aflatoxin concentrations can be established. Damaged commodities such as peanuts can be culled from the market by using a quality control and batch testing program.

Aflatoxin-contaminated products must be excluded from the food chain, but there are alternative measures that can be taken to salvage the damaged product. Many approaches have been suggested. These can be broadly classified as either physical separation methods or detoxification by biological or chemical means. Physical separation may consist of diverting an entire lot of contaminated product or removing the contaminated portion of a lot. To date, various types of physical separation methods have been employed with marginal success, particularly in peanuts and other types of nuts. Segregation of contaminated and uncontaminated kernels has been based on different criteria, including discoloration, the appearance of bright-greenish yellow (BGY) fluorescence under ultraviolet light, or density. BGY fluoresence has also been used to separate aflatoxin-contaminated cottonseeds. In corn, physical separation methods have generally proved to be ineffective and too expensive. Dilution of contaminated corn with clean corn is practiced on farms but is not advisable unless the concentration of AFB_1 has been conclusively established. Moreover, this practice is prohibited by law for corn moving in interstate commerce or corn used for human consumption. During either dry- or wet-milling of corn, aflatoxin is physically partitioned out of the starch fraction and becomes concentrated in the fractions destined for animal feeding.

Of the various methods intended to remove aflatoxins from contaminated products by molecular degradation, chemical treatment is considered the most effective and practical. Two parts of the aflatoxin molecule seem to be susceptible to chemical attack: the internal ester of the coumarin moiety and the double bond of the terminal furan when it is present (Anderson 1983). Chemicals shown to destroy alfatoxins without leaving deleterious residues or damaging nutrients include ammonia, methylamine, sodium hydroxide, calcium hydroxide, and formaldehyde. Of these, ammonia appears to be the most promising in oil seed meals and corn. The combination of ammonia (0.6–4%), heat, and moisture (10–20%) is particularly effective. For example, the concentration of aflatoxin in contaminated cottonseed was decreased from 1915 ppb to 78 ppb by treatment with 1.5% ammonium hydroxide and 17% moisture in giant polyethylene bags placed in direct sunlight (Lough et al. 1979). Feeding studies indicate the aflatoxin-contaminated corn treated by the ammonia process has had no adverse effects on cattle, swine, or chickens (Anderson 1983). Ammonia concentrations in finished feed must be <0.1% to avoid feed refusal in swine (Jensen et al. 1977). Although chemicals other than ammonia are effective, various drawbacks have limited their application. For example, Chakrabarti (1981) found that 3% hydrogen peroxide, 7% methanol,

3% perchloric acid, or 5% dimethylamine hydrochloride reduced the AFB_1 content of corn from 397 ppb to less than 20 ppb, but it was necessary to grind the corn; and the chemicals are too expensive.

Other deactivating methods have proven effective but are not commercially practical or are too costly for feed crops. These include extraction of feeds with solvents, irradiation, or heat treatment. Oil or dry roasting of contaminated peanuts, which destroys most of the aflatoxin present, has some applicability in this industry. Heat treatment (160°–180°C) of corn is an effective deactivating method but causes lowered feed efficiency in swine due partly to loss of lysine and methionine (Hale and Wilson 1979).

Other Mycotoxins

As in the case of aflatoxin, prevention of fungal infestation, particularly during storage, is the best control measure. In general, mycotoxin removal or detoxification methods have not been developed or implemented as extensively as for aflatoxins.

Zearalenone concentrates in the nonstarch fractions during wet-milling. Also, it is destroyed at a $pH \geq 11$ or at temperatures above 120°C. However, like aflatoxins, treatment with chemicals (ammonia and hydrogen peroxide) followed by heating and drying appears to have the greatest potential. No modification processes are presently being used commercially.

Trichothecene-contaminated feed can be blended with clean feed and fed to animals. Because contaminated corn is refused by swine, blending with clean corn until the feed is accepted appears to be the most practical solution to the problem. Feeding the contaminated feed to less sensitive species such as cattle is an alternative approach. Even though DON can be extracted by solvents and T-2 toxin is degraded by mineral acids, chemical detoxification methods such as these have not been utilized to upgrade contaminated feed.

Ochratoxin A is rapidly destroyed in green coffee beans by roasting, but the method is impractical for animal feeds. A strict quality control program has been successful in minimizing the utilization of OA-contaminated barley malt in the manufacture of beer.

Slaframine-producing fungi cannot presently be prevented from growing on legumes, particularly red clover, during wet weather. As a consequence, utilization of red clover in livestock production in susceptible areas has greatly diminished as has the occurrence of the salivary syndrome. No chemical detoxification methods have been developed.

REFERENCES

Mycotoxicology

CHU, F. S. 1977. Mode of action of mycotoxins and related compounds. Adv. Appl. Microbiol. 22, 83–143.

CHU, F. S. 1983. Immunochemical methods for mycotoxin analysis. In Proceedings of the International Symposium on Mycotoxins. K. Naguib, M. M. Naguib, D. L. Park, and A. E. Pohland (Editors), pp. 177–194. The General Organization for Govt. Printing Offices, Cairo, Egypt.

DAVIS, N. D., DICKENS, J. W., FREIE, R. L., HAMILTON, P. B., SHOTWELL, V. L., and WYLLIE, T. D. 1980. Protocols for surveys, sampling, post-collection handling, and analysis of grain samples involved in mycotoxin problems. J. Assoc. Off. Anal. Chem. 63, 95–102.

GREGORY, J. F., and MANLEY, D. 1981. High performance liquid chromatographic determination of aflatoxins in animal tissues and products. J. Assoc. Off. Anal. Chem. 64, 144–151.

HESSELTINE, C. W. 1979. Introduction, definition, and history of mycotoxins of importance to animal production. In Interactions of Mycotoxins in Animal Production, pp. 3–18. Natl. Acad. Sci., Washington, DC.

HSU, I. C., SMALLEY, E. B., STRONG, F. M., and RIBELLIN, W. E. 1972. Identification of T-2 toxin in moldy corn associated with a lethal toxicosis in dairy cattle. Appl. Microbiol. 24, 682–690.

LILLEHOJ, E. B. 1979. Natural Occurrence of Mycotoxins in Feeds: Pitfalls in Determination, pp. 139–153. Natl. Acad. Sci., Washington, DC.

Official Methods of Analysis 1980. Natural Poisons: Mycotoxins, 13th Edition, pp. 414–434. AOAC, Arlington, VA.

PIER, A. C. 1981. Mycotoxins and animal health. Adv. Vet. Sci. Comp. Med. 54, 216A–224A.

PITT, J. I., and UDAGAWA, S. 1984. Introduction—taxonomy of mycotoxin-producing fungi. In Toxigenic Fungi—Their Toxins and Health Hazard. H. Kurata and Y. Ueno (Editors), pp. 75–77. Elsevier, NY.

RODRICKS, J. V., HESSELTINE, C. W., and MEHLMAN, M. A. (Editors) 1977. Mycotoxins in Human and Animal Health. Pathotox Publishers, Park Forest South, IL.

SAMUELS, G. J. 1984. Toxigenic fungi as Ascomycetes. In Toxigenic Fungi—Their Toxins and Health Hazard. H. Kurata and Y. Ueno (Editors), pp. 119–128. Elsevier, NY.

SHOTWELL, O. L. 1983. Aflatoxin detection and determination in corn. In Aflatoxin and *Aspergillus flavus* in Corn, U. L. Diener, R. L. Asquith, and J. W. Dickens (Editors), So. Cooperative Ser. Bull. 279, 38–45. Craftmaster Printers, Inc., Opelika, AL.

SIMPSON, M. E., and BATRA, L. R. 1984. Ecological relations in respect to a boll rot of cotton caused by *Aspergillus flavus*. In Toxigenic Fungi—Their Toxins and Health Hazard. H. Kurata and Y. Ueno (Editors), pp. 24–32. Elsevier, NY.

STAHR, H. M. 1977. Analytical Toxicology Methods Manual, pp. 163–175. Iowa State Univ. Press, Ames.

WOGAN, G. N., and BUSBY, W. F. 1980. Naturally occurring carcinogens. In Toxic Constituents of Plant Food Stuffs. I. E. Liener (Editor), 2nd Edition, pp. 329–369. Academic Press, NY.

Aflatoxin

ANGSUBHAHORN, S., POOMVISES, P. E., ROMRUEN, K., and NEWBERNE, P. M. 1981. Aflatoxicosis in horses. J. Am. Vet. Med. Assoc. *178,* 274–278.

BENNETT, G. A., and ANDERSON, R. A. 1978. Distribution of aflatoxin and/or zearalenone in wet-milled products—A review. J. Agric. Food Chem. *26,* 1055–1060.

BODINE, A. B., and MERTENS, D. R. 1983. Toxicology, metabolism, and physiological effects of aflatoxin in the bovine. *In* Aflatoxin and *Aspergillus flavus* in Corn. U. L. Diener, R. L. Asquith, and J. W. Dickens (Editors) So. Cooperative Ser. Bull. 279, 46–50. Craftmaster Printers, Inc., Opelika, AL.

BROWN, R. W., PIER, A. C., RICHARD, J. L., and KRUGSTAD. 1981. Effects of dietary aflatoxin on existing bacterial intramammary infections in dairy cows. J. Am. Vet. Res. *42,* 927–933.

BURGUERA, J. A., EDDS, G. T., and OSUNA, O. 1983. Influence of selenium on aflatoxin B_1 or crotalaria toxicity in turkey poults. Am. J. Vet. Res. *44,* 1718–1721.

BUSBY, W. F., and WOGAN, G. N. 1979. Food-borne mycotoxins and alimentary mycotoxicoses. *In* Food-borne Infections and Intoxication. H. Riemann, and F. L. Bryan, (Editors), 2nd Edition, pp. 519–610. Academic Press, NY.

BUSBY, W. F., and WOGAN, G. N. 1981. Aflatoxins. *In* Mycotoxins and *N*-Nitroso Compounds: Environmental Risks, R. C. Shank (Editor), Vol. II, pp. 3–28. CRC Press, Boca Raton, FL.

CAMPBELL, T. C., and HAYES, J. R. 1976. The role of aflatoxin metabolism in its toxic lesion. Toxicol. Appl. Pharmacol. *35,* 199–222.

CARNAGHAN, R. B. A., and CRAWFORD, M. 1964. Relationship between ingestion of aflatoxin and primary liver cancer. Br. Vet. J. *120,* 201–204.

DAVILA, J. C., EDDS, G. T., OSUNA, O., and SIMPSON, C. F. 1983. Modification of the effects of aflatoxin B_1 and warfarin in young pigs given selenium. Am. J. Vet. Res. *44,* 1877–1883.

DIENER, U. L., and DAVIS, N. D. 1969. Aflatoxin formation by *Aspergillus flavus*. *In* Aflatoxin. L. A. Goldblatt (Editor), pp. 13–54. Academic Press, NY.

GREGORY, J. F., GOLSTEIN, S. L., and EDDS, G. T. 1983. Metabolic distribution and rate of residue clearance in turkeys fed a diet containing aflatoxin B_1. Food Chem. Toxicol. *21,* 463–468.

HALVER, J. E. 1969. Aflatoxicosis and trout hepatoma. *In* Aflatoxin. L. A. Goldblatt (Editor), pp. 265–303. Academic Press, NY.

HAMILTON, P. B. 1977. Interrelationship of mycotoxins with nutrition. Fed. Proc., Fed. Am. Soc. Exp. Biol. *36,* 1899.

HOERR, F. J., and D'ANDREA, G. H. 1983. Biological effects of aflatoxin in swine. *In* Aflatoxin and *Aspergillus flavus* in Corn. U. L. Diener, R. L. Asquith, and J. W. Dickens (Editors), So. Cooperative Ser. Bull. 279, 51–55. Craftmaster Printers, Inc., Opelika, AB.

HSIEH, D. P. H. 1978. Comparative metabolism and toxicokinetics of aflatoxin B_1: An overview. Toxicol. Appl. Pharmacol. *45,* 272 (Abstr.).

HSIEH, D. P. H., WONG, Z. A., WONG, J. J., MICHAS, C., and RUEBNER, B. 1977. Comparative metabolism of aflatoxin. *In* Mycotoxins in Human and Animal Health. J. V. Rodricks, C. W. Hesseltine, and M. A. Mehlman (Editors), pp. 37–50. Pathotox Publishers, Park Forest South, IL.

HUFF, W. E. 1980. Discrepancies between bone ash and toe ash during aflatoxicosis. Poult. Sci. *59,* 2213–2215.

JONES, F. T., HAGLER, W. H., and HAMILTON, P. B. 1982. Association of low levels of aflatoxin in feed with productivity losses in commerical broiler operations. Poult. Sci. *61,* 861–868.

LOUGH, O. G., GINGG, C., and BILLOTTI, M. 1979. Detoxifying aflatoxin contaminated cottonseed. Am. Dairy Sci. Assoc. Meet., Abstract No. 117.

MERTENS, D. R. 1979. Biological effects of mycotoxins upon rumen fermentation and lactating dairy cows. In Interactions of Mycotoxins in Animal Production, pp. 118–136. Natl. Acad. Sci., Washington, DC.

PATTERSON, D. S. P. 1980. Metabolism and mode of action of aflatoxin in relation to the etiology of liver diseases in man and farm animals. In Proceedings of the Sixth International Symposium on Animal, Plant and Microbiol Toxins. D. Eaker and T. Wadstrom (Editors), pp. 681–689. Pergamon Press, Oxford, UK.

PIER, A. C., RICHARD, J. L., and THURSTON, J. R. 1979A. The influence of mycotoxins on resistance and immunity. In Interactions of mycotoxins on resistance and immunity. In Interactions of Mycotoxins in Animal Production, pp. 56–66. Natl. Acad. Sci., Washington, DC.

PIER, A. C. 1981. Mycotoxins and animal health. Adv. Vet. Sci. Comp. Med. 25, 185–243.

PIER, A. C., CYSEWSKI, S. J., RICHARD, J. L., BAETZ, A. L., and MITCHELL, L. 1976. Experimental mycotoxicosis in calves with aflatoxin, ochratoxin, rubratoxin, and T-2 toxin. Proc. U.S. Anim. Health Assoc. pp. 130–148.

SHOTWELL, O. L. 1977. Aflatoxin in corn. J. Am. Oil Chem. Soc. 54, 216A–224A.

STOLOFF, L. 1980. Aflatoxin M in milk. J. Food Prot. 43, 226–230.

WONG, J. J., and HSIEH, D. P. H. 1976. Mutagenicity of aflatoxins related to their metabolism and carcinogenic potential. Proc. Natl. Acad. Sci. U.S.A. 73, 2241–2244.

Fusarium Mycotoxins: Zearalenone and Trichothecenes

ALLEN, N. K., PEGURI, A., MIROCHA, C. J., and NEWMAN, J. A. 1983. Effects of *Fusarium* cultures, T-2 toxin and zearalenone on reproduction of turkey females. Poult. Sci. 62, 282–289.

BENNETT, G. A., and ANDERSON, R. A. 1978. Distribution of aflatoxin and/or zearalenone in wet-milled products—A review. J. Agric. Food Chem. 26, 1055–1060.

CARSON, M. S., and SMITH, T. K. 1983. Effect of feeding alfalfa and refined plant fibers on the toxicity and metabolism of T-2 toxin in rats. J. Nutr. 113, 304–313.

CHI, M. S., MIROCHA, C. J., KURTZ, H. J., WEAVER, G. A., BATES, F., ROBISON, T., and SHIMODA, W. 1980. Effect of dietary zearalenone on growing broiler chicks. Poult. Sci. 59, 531–536.

CIEGLER, A. 1978. Trichothecenes: Occurrence and toxicoses. J. Food Prot. 41, 399–403.

FRIEND, D. W., TRENHOLM, H. L., ELLIOT, J. I., THOMPSON, B. K., and HARTIN, K. E. 1982. Effect of feeding vomitoxin-contaminated wheat to pigs. Can. J. Anim. Sci. 62, 1211–1222.

HSU, I. C., SMALLEY, E. B., STRONG, F. M., and RIBELLIN, W. E. 1972. Identification of T-2 toxin in moldy corn associated with a lethal toxicosis in dairy cattle. Appl. Microbiol. 24, 682–690.

JAMES, L. J., and SMITH, T. K. 1982. Effect of dietary alfalfa on zearalenone toxicity and metabolism in rats and swine. J. Anim. Sci. 55, 110–118.

McLAUGHLIN, C. S., VAUGHAN, M. H., CAMPBELL, I. M., MER WEI, C., STAFFORD, M. E., and HANSEN, B. S. 1977. Inhibition of protein synthesis by trichothecenes. *In* Mycotoxins in Human and Animal Health. J. V. Rodricks, C. W. Hesseltine, and M. A. Mehlman (Editors), pp. 263–273. Pathotox Publishers, Park Forest South, IL.

MIROCHA, C. J., PATHRE, S. V., and ROBINSON, T. S. 1981. Comparative metabolism of zearalenone and transmission into bovine milk. Food Cosmet. Toxicol. *19,* 25–30.

MORAN, E. T., JR., HUNTER, B., FERKET, P., YOUNG, L. G., and McGIRR, L. G. 1982. High tolerance of broilers to vomitoxin from corn infected with *Fusarium graminearum*. Poult. Sci. *61,* 1828–1831.

PIER, A. C., CYSEWSKI, S. J., RICHARD, J. L., BAETZ, A. L., and MITCHELL, L. 1976. Experimental mycotoxicosis in calves with aflatoxin, ochratoxin, rubratoxin, and T-2 toxin. Proc. U.S. Anim. Health Assoc. pp. 130–148.

ROBISON, T. S., MIROCHA, C. J., KURTZ, H. J., BEHRENS, J., WEAVER, G. A., and CHI, M. S. 1979. Distribution of tritium-labelled T-2 toxin in swine. J. Agric. Food Chem. *27,* 1141–1413.

RUHR, L. P., OSWEILER, G. D., and FOLEY, C. W. 1983. Effect of the estrogenic mycotoxin zearalenone on reproductive potential in the boar. Am. J. Vet. Res. *44,* 483–485.

SMITH, T. K. 1980. Influence of dietary fiber, protein and zeolite on zearalenone toxicosis in rats. J. Anim. Sci. *50,* 278–285.

SMITH, T. K. 1981. Effect of dietary calcium, phosphorus and vitamin D on zearalenone toxicosis in rats. Can. J. Anim. Sci. *61,* 191–197.

UENO, Y. 1977. Trichothecenes: Overview address. *In* Mycotoxins in Human and Animal Health. J. V. Rodricks, C. W. Hesseltine, and M. A. Mehlman (Editors), pp. 189–207. Pathotox Publishers, Park Forest South, IL.

UENO, Y. 1979. Toxicological evaluation of trichothecene mycotoxins. *In* Proceedings of the Sixth International Symposium on Animal, Plant, and Microbiol Toxins. D. Eaker and T. Wadstrom (Editors), pp. 663–671. Pergamon Press, Oxford, UK.

UENO, Y., TASHIRO, F., and KOBAYASHI, T. 1983. Species differences in zearalenone-reductase activity. Food Chem. Toxicol. *21,* 167–173.

YOSHIZAWA, T., SWANSON, S. P., and MIROCHA, C. J. 1980. *In vitro* metabolism of T-2 toxin in rats. Appl. Environ. Microbiol. *40,* 901–906.

YOSHIZAWA, T., MIROCHA, C. J., BEHRENS, J. C., and SWANSON, S. P. 1981. Metabolic fate of T-2 toxin in a lactating cow. Food Cosmet. Toxicol. *19,* 31–39.

YOUNG, L. G., McGIRR, L., VALLI, V. E., LUMSDEN, J. H., and LUN, A. 1983. Vomitoxin in corn fed to young pigs. J. Anim. Sci. *57,* 655–664.

Ochratoxins

GALTIER, P., ALVINERIC, M., and CHARPENTEAU, J. L. 1981. The pharmacokinetic profiles of ochratoxin A in pigs, rabbits and chickens. Food Cosmet. Toxicol. *19,* 735–738.

HAMILTON, P. B., HUFF, W. E., HARRIS, J. R., and WYATT, R. D. 1982. Natural occurrences of ochratoxicosis in poultry. Poult. Sci. *61,* 1832–1841.

KANISAWA, M. 1984. Synergistic effect of citrinin on hepatorenal carcinogenesis of ochratoxin A in mice. *in* Toxigenic Fungi—Their Toxins and Health Hazard. H. Kurata and Y. Ueno (Editors) pp. 245–254. Elsevier, NY.

KROGH, P. 1978A. Mycotoxicoses of animals. Mycopathologia *65,* 43–45.

KROGH, P. 1978B. Casual associations of mycotoxic nephropathy. Acta Pathol. Microbiol., Suppl. *269,* 1–25.

KROGH, P. 1979. Ochratoxins: Occurrence, biological effects and casual role in diseases. *In* Proceedings of the Sixth International Symposium on Animal, Plant and Microbial Toxins. D. Eaker and T. Wadstrom (Editors), pp. 673–680. Pergamon Press, Oxford, UK.
KUBENA, L. F., PHILLIPS, T. D., CREGER, C. R., WITZEL, D. A., and HEIDELBAUGH, N. D. 1983. Toxicity of ochratoxin A and tannic acid to growing chicks. Poult. Sci. *62*, 1786–1792.
LEISTNER, L. 1984. Toxigenic penicillia occurring in feeds and foods. *In* Toxigenic Fungi—Their Toxins and Health Hazard. H. Kurata and Y. Ueno (Editors), pp. 162–171. Elsevier, NY.
PIER, A. C., CYSEWSKI, S. J., RICHARD, J. L., BAETZ, A. L., and MITCHELL, L. 1976. Experimental mycotoxicosis in calves with aflatoxin, ochratoxin, rubratoxin, and T-2 toxin. Proc. U.S. Anim. Health Assoc. pp. 130–148.
STOLOFF, L. 1979. Mycotoxin residues in edible animal tissues. *In* Interactions of Mycotoxins in Animal Production, pp. 157–166. Natl. Acad. Sci., Washington, DC.

Citrinin

BERNDT, W. O., HAYES, A. W., and PHILLIPS, R. D. 1980. Effects of mycotoxins on renal function: Mycotoxic nephropathy. Kidney Int. *18*, 656–664.
HANIKA, C., CARLTON, W. W., and TUITE, J. 1983. Citrinin mycotoxicosis in the rabbit. Food Chem. Toxicol. *21*, 487–494.
NELSON, T. S., BEASLEY, J. N., KIRBY, L. K., JOHNSON, Z. B., BALLAM, G. C., and CAMPBELL, M. M. 1981. Citrinin toxicity in growing chicks. Poult. Sci. *60*, 2165–2166.

Rubratoxins

MIROCHA, C. J. 1980. Rubratoxin, sterigmatocystin and stachybotrys mycotoxins. *In* Conference on Mycotoxins in Animal Feeds and Grains Related to Animal Health. W. Shimoda (Editor), pp. 152–176. National Technical Information Service, U.S. Dept. of Commerce, Springfield, VA.
PIER, A. C. 1981. Mycotoxins and animal health. Adv. Vet. Sci. Comp. Med. *25*, 185–243.
PIER, A. C., CYSEWSKI, S. J., RICHARD, J. L., BAETZ, A. L., and MITCHELL, L. 1976. Experimental mycotoxicosis in calves with aflatoxin, ochratoxin, rubratoxin, and T-2 toxin. Proc. U.S. Anim. Health Assoc. pp. 130–148.
RICHMOND, M. L., GRAY, J. I., and STINE, G. M. The rubratoxins: Causative agents in food/feed-borne disease. J. Food Prot. *43*, 579–586.
UNGER, P. D., and HAYES, A. W. 1979. Disposition of rubratoxin B in the rat. Toxicol. Appl. Pharmacol. *47*, 585–591.
WOGAN, G. N., EDWARDS, G. S., and NEWBERNE, P. M. 1971. Acute and chronic toxicity of rubratoxin B. Toxicol. Appl. Pharmacol. *19*, 712–720.

Patulin

CIEGLER, A. 1977. Patulin. *In* Mycotoxins in Human and Animal Health. J. V. Rodricks, C. W. Hesseltine and M. A. Mehlman (Editors), pp. 609–624. Pathotox Publishers, Park Forest South, IL.

LYNCH, G. P. 1979. Effects of mycotoxins on ruminants. *In* Interactions of Mycotoxins in Animal Production, pp. 96–117. Natl. Acad. Sci., Washington, DC.

MERTENS, D. R. 1979. Biological effects of mycotoxins upon rumen fermentation and lactating dairy cows. *In* Interactions of Mycotoxins in Animal Production, pp. 118–136. Natl. Acad. Sci., Washington, DC.

Fungal Tremorgens

COCKRUM, P. A., CULVENOR, C. C. J., EDGAR, J. A., and PAYNE, A. L. 1979. Chemically different tremorgenic mycotoxins in isolates of *Penicillium paxilli* from Australia and North America. J. Nat. Prod. *42*, 534–536.

DIMENNA, M. E., MANTLE, P. G., and MORTIMER, P. H. 1976. Experimental production of a staggers syndrome in ruminants by a tremorgenic *Penicillium* from soil. N. Z. Vet. J. *24*, 45–46.

FLETCHER, L. R. 1982. Observations of ryegrass staggers in weaned lambs grazing different ryegrass pastures. N. Z. Exp. Agric. *10*, 203–207.

FLETCHER, L. R., and HARVEY, I. C. 1981. An association of a *Lolium* endophyte with ryegrass staggers. N. Z. Vet. J. *29*, 185–186.

GALLAGHER, R. T., KEOGH, R. G., LATCH, G. C. M., and REID, C. S. W. 1977. The role of fungal tremorgens in ryegrass staggers. N. Z. J. Agric. Res. *20*, 431–440.

GALLAGHER, R. T., LATCH, G. C. M., and KEOGH, R. G. 1980. The Janthitrems: Fluorescent tremorgenic toxins produced by *Penicillium janthinellum* isolates from ryegrass pastures. Appl. Environ. Microbiol. *39*, 272–273.

GALLAGHER, R. T., WHITE, E. P., and MORTIMER, P. H. 1981. Ryegrass staggers: Isolation of potent neurotoxins lolitrem A and lolitrem B from staggers-producing pastures. N. Z. Vet. J. *29*, 189–190.

GIESECKE, P. R., LANIGAN, G. W., and PAYNE, A. L. 1979. Fungal tremorgens associated with ryegrass staggers in South Australia. Aust. Vet. J. *55*, 444.

HUNT, L. D., BLYTHE, L., and HOLTAN, D. W. 1983. Ryegrass staggers in ponies fed ryegrass straw. J. Am. Vet. Med. Assoc. *182*, 285–286.

LANIGAN, G. W., PAYNE, A. L., and COCKRUM, P. A. 1979. Production of tremorgenic toxins by *Penicillium janthinellum* Biourge: A possible aetiological factor in ryegrass staggers. Aust. J. Exp. Biol. Med. Sci. *57*, 31–37.

MANTLE, P. G., MORTIMER, P. H., and WHITE, E. P. 1977. Mycotoxic tremorgens of *Claviceps paspali* and *Penicillium cyclopium:* A comparative study of effects on sheep and cattle in relation to natural staggers syndromes. Res. Vet. Sci. *24*, 49–56.

SHAW, J. N., and MUTH, O. H. 1949. Some types of forage poisoning in Oregon cattle and sheep. J. Am. Vet. Med. Assoc. *114*, 315–317.

WHITE, E. P., SMITH, G. S., DIMENNA, M. E., and MORTIMER, P. H. 1980. Absorption and translocation of *Penicillium* by ryegrass plants. N. Z. Vet. J. *28*, 123–124.

Lupinosis

ALLEN, J. G. 1978. The emergence of a lupinosis-associated myopathy in sheep in Western Australia. Aust. Vet. J. *54*, 548–549.

ALLEN, J. G. 1981. An evaluation of lupinosis in cattle in Western Australia. Aust. Vet. J. *57*, 212–215.

ALLEN, J. G., and NOTTLE, F. K. 1979. Spongy transformation of the brain in sheep with lupinosis. Vet. Rec. *104*, 31–33.

ALLEN, J. G., and MASTERS, H. G. 1980. Prevention of ovine lupinosis by the oral administration of zinc sulfate and the effect of such therapy on liver and pancreas zinc and liver copper. Aust. Vet. J. *56*, 168–171.

ALLEN, J. G., MASTERS, H. G., and WALLACE, S. R. 1979. The effect of lupinosis on liver cooper, selenium and zinc concentrations in merino sheep. Vet. Rec. *105*, 434–436.

CROKER, K. P., ALLEN, J. G., PETTERSON, D. S., MASTERS, H. G., and FRAYNE, R. F. 1979. Utilization of lupin stubbles by merino sheep: Studies of animal performance, rates and time of stocking, lupinosis, liver copper and zinc, and circulating plasma enzymes. Aust. J. Agric. Res. *30*, 551–564.

JAGO, M. V., PETERSON, J. E., PAYNE, A. L., and CAMPBELL, D. G. 1982. Response of sheep to different doses of phomopsin. Aust. J. Exp. Biol. Sci. *60*, 239–251.

PETERSON, J. E. 1983. Embryotoxicity of phomopsin in rats. Aust. J. Exp. Biol. Med. Sci. *61*, 105–116.

VAN RENSBURG, I. B. J., MARASAS, W. F. O., and KELLERMAN, T. S. 1975. Experimental *Phomopsis leptostromiformis* mycotoxicosis of pigs. J. S. Afr. Vet. Med. Assoc. *46*, 197–204.

WOOD, P. M., and ALLEN, J. G. 1980. Control of ovine lupinosis: Use of resistant cultivar of *Lupinus albus*–cv. Ultra. Aust. J. Exp. Agric. Anim. Husb. *20*, 316–318.

Sporidesmin

BREWER, D., CALDER, F. W., MacINTYRE, T. M., and TAYLOR, A. 1971. Ovine ill-thrift in Nova Scotia. 1. The possible regulation of the rumen flora in sheep by the fungal flora of permanent pasture. J. Agric. Sci. *76*, 465–477.

BREWER, D., DUNCAN, J. M., JERRAM, W. A., LEACH, C. K., SAFE, S., TAYLOR, A., VINING, Z. C., ARCHIBALD, R. McG., STEVENSON, R. G., MIROCHA, C. J., and CHRISTENSEN, C. M. 1972. Ovine ill-thrift in Nova Scotia. 5. The production and toxicology of chetomin, a metabolite of *Chaetomium* spp. Can. J. Microbiol. *18*, 1129–1137.

CAMPBELL, A. G., and WESSELINK, C. 1973. Facial eczema liver damage and liver weight change in lambs. N. Z. J. Exp. Agric. *1*, 291–292.

MARASAS, W. F. O., ADELAAR, T. F., KELLERMAN, T. S., MINNE, J. A., VAN RENSBURG, I. B. J., and BURROUGHS, G. W. 1972. First report of facial eczema in sheep in South Africa. Onderstepoort J. Vet. Res *39*, 107–112.

MUNDAY, R. 1982. Studies on the mechanism of toxicity of the mycotoxin, sporidesmin. I. Generation of superoxide radical by sporidesmin. Chem.-Biol. Interact. *41*, 361–374.

MUNDAY. R. 1984 Studies on the mechanism of toxicity of the mycotoxin sporidesmin. 3. Inhibition by metals of the generation of superoxide radical by sporidesmin. J. Appl. Toxicol. *4*, 182–186.

SMITH, B. L., EMBLING, P. P., TOWERS, N. R., WRIGHT, D. E., and PAYNE, E. 1977. The protective effect of zinc sulphate in experimental sporidesmin poisoning of sheep. N. Z. Vet. J. *25*, 124–127.

SMITH, B. L., COE, B. D., and EMBLING, P. P. 1978. Protective effect of zinc sulphate in a natural facial eczema outbreak in dairy cows. N. Z. Vet. J. *26*, 314–315.

TOWERS, N. R., and SMITH, B. L. 1978. The protective effect of zinc sulphate in

experimental sporidesmin intoxication of lactating dairy cows. N. Z. Vet. J. *26,* 199–202.
TOWERS, N. R., and STRATTON, G. C. 1978. Serum gamma-glutamyltransferase as a measure of sporidesmin-induced liver damage in sheep. N. Z. Vet. J. *26,* 109–112.

Stachybotryotoxins

MIROCHA, C. J. 1980. Rubratoxin, sterigmatocystin and stachybotrys mycotoxins. *In* Conference on Mycotoxins in Animal Feeds and Grains Related to Animal Health. W. Shimoda (Editor), pp. 152–176. National Technical Information Service, U.S. Dept. of Commerce, Springfield, VA.
SCHNEIDER, D. J., MARASAS, W. F. O., DALE KUYS, J. C., KRIEK, N. P. J., and VAN SCHALKWYK, G. C. 1979. A field outbreak of suspected stachybotryotoxicosis in sheep. J. S. Afr. Vet. Assoc. *50,* 73–81.

Moldy Straw-Induced Photosensitization

BAGLEY, C. V., McKINNON, J. B., and ASAY, C. S. 1983. Photosensitization associated with exposure of cattle to moldy straw. J. Am. Vet. Med. Assoc. *183,* 802.

Kikuyu Poisoning

BRYSON, R. W. 1982. Kikuyu poisoning and the army worm. J. S. Afr. Vet. Assoc. *53,* 161–165.
BRYSON, R. W., and NEWSHOLME, S. J. 1978. Kikuyu grass poisoning of cattle in Natal. J. S. Afr. Vet. Assoc. *49,* 19–21.
MARTINOVICH, D., and SMITH, B. 1972. Kikuyu poisoning in sheep. N. Z. Vet. J. *20,* 169.
MARTINOVICH, D., and SMITH, B. 1973. Kikuyu poisoning of cattle. N. Z. Vet. J. *21,* 55–63.
NEWSHOLME, S. J., KELLEMAN, T. S., VAN DER WESTHIUZEN, G. C. A., and SOLEY, J. T. 1983. Intoxication of cattle on kikuyu grass following army worm (*Spondoptera exempta*) invasion. Onderstepoort J. Vet. Res. *50,* 157–167.

Decreasing Exposure to Mycotoxins

ANDERSON, R. A. 1983. Detoxification of aflatoxin-contaminated corn. *In* Aflatoxin and *Aspergillus flavus* in Corn. U. L. Diener, R. L. Asquith, and J. W. Dickens (Editors), So. Cooperative Ser. Bull. 279, 87–90. Craftmaster Printers, Inc., Opelika, AL.
CHAKRABARTI, A. G. 1981. Detoxification of corn. J. Food Prot. *44,* 591–592.
CIEGLER, A. 1978. Detoxification of aflatoxin-contaminated agricultural commodities. *In* Toxins: Animal, Plant and Microbial, pp. 729–738. Pergamon Press, NY.
DICKENS, J. W. 1983. Prevention and control of mycotoxins. *In* Proceedings of the International Symposium on Mycotoxins. K. Naguib, M. M. Naguib, D. L. Park, and A. E. Pohland (Editors), pp. 131–142. The General Organization for Govt. Printing Offices, Cairo, Egypt.

DRAUGHON, F. A. 1983. Control or suppression of aflatoxin production with pesticides. *In* Aflatoxin and *Aspergillus flavus* in Corn. U. L. Diener, R. L. Asquith, and J. W. Dickens (Editors), So. Cooperative Ser. Bull. 279, 81–86. Craftmaster Printers, Inc., Opelika, AB.

GOLDBLATT, L. A., and DOLLEAR, F. G. 1979. Modifying mycotoxin contamination in feeds—use of mold inhibitors, ammoniation, roasting. *In* Interactions of Mycotoxins in Animal Production, pp. 167–184. Natl. Acad. Sci., Washington, DC.

HALE, O. M., and WILSON, D. M. 1979. Performance of pigs on diets containing heated or unheated corn with or without aflatoxin J. Anim. Sci. 48, 1394–1400.

JEMMALI, M. 1983. Decontamination of mycotoxins. *In* Proceedings of the International Symposium on Mycotoxins. K. Naguib, M. M. Naguib, D. L. Park, and A. E. Pohland (Editors), pp. 143–150. The General Organization for Govt. Printing Offices, Cairo, Egypt.

JENSEN, A. H., BREKKE, O.L., FRANK, G. R., and PEPLINSKI. 1977. Acceptance and utilization by swine of aflatoxin-contaminated corn treated with aqueous or gaseous ammonia. J. Anim. Sci. 45, 8–12.

LILLEHOJ, E. B., and HESSELTINE, C. W. 1977. Aflatoxin control during plant growth and harvest of corn. *In* Mycotoxins in Human and Animal Health. J. V. Rodricks, C. W. Hesseltine, and M. A. Mehlman (Editors), pp. 107–119. Pathotox Publishers, Park Forest South, IL.

LOUGH, O. G., GINGG, C., and BILLOTTI, M. 1979. Detoxifying aflatoxin contaminated cottonseed. Am. Dairy Sci. Assoc. Mtg. Abstr. *117*.

RAO, H. R. G., and HAREIN, P. K. 1972. Dichlorvos as an inhibitor of aflatoxin production on wheat, corn, rice, and peanuts. J. Econ. Eutomol. 65, 988–989.

WIDSTROM, N. W., and ZUBER, M. S. 1983. Sources and mechanisms of genetic control in the plant. *In* Aflatoxin and *Aspergillus flavus* in Corn, U. L. Diener, R. L. Asquith, and J. W. Dickens (Editors), So. Cooperative Ser. Bull. 279, 72–76. Craftmaster Printers, Inc., Opelika, AL.

GENERAL REFERENCES

CHU, F. S. 1977. Mode of action of mycotoxins and related compounds. Adv. Appl. Microbiol. 22, 83–143.

COLE, R. J., and COX, R. H. 1981. Handbook of Toxic Fungal Metabolites. Academic Press, NY.

DIENER, U. L. 1976. Environmental factors influencing mycotoxin formation in the contamination of foods. Proc. Am. Phytopathol. Soc. 3, 126–139.

DIENER, U. L., ASQUITH, R. L., and DICKENS, J. W. 1983. Aflatoxin and *Aspergillus flavus* in Corn. So. Cooperative Ser. Bull. 279. Craftmaster Printers, Inc., Opelika, AL.

HAYES, A. W. 1980. Mycotoxins: A review of biological effects and their role in human diseases. Clin. Toxicol. 17, 45–83.

HAYES, A. W. 1981. Mycotoxin Teratogenicity and Mutagenicity. CRC Press, Boca Raton, FL.

KEELER, R. F., and TU, A. T. 1983. Handbook of Natural Toxins. Vol. 1. Plant and Fungal Toxins. Marcel Dekker, NY.

KURATA, H., and UENO, Y. 1984. Toxigenic Fungi—Their Toxins and Health Hazard. Elsevier, NY.

MILOSLAV, E. 1983. CRC Handbook of Naturally Occurring Food Toxicants. CRC Press, Boca Raton, FL.

NAGUIB, K., NAGUIB, M. M., PARK, D. L., and POHLAND, A. E. 1983. Proceedings of the International Symposium on Mycotoxins. The General Organization for Govt. Printing Offices, Cairo, Egypt.

NATIONAL ACADEMY OF SCIENCE 1979. Interactions of Mycotoxins in Animal Production. Natl. Acad. Sci., Washington, DC.

PIER, A. C. 1981. Mycotoxins and animal health. Adv. Vet. Sci. Comp. Med. *54*, 216A–224A.

PIER, A. C., RICHARD, J. L., and CYSEWSKI, S. J. 1980. Implications of mycotoxins in animal disease. J. Am. Vet. Med. Assoc. *176*, 719–724.

PIER, A. C., RICHARD, J. L., and THURSTON, J. R. 1979. Effects of mycotoxins on immunity and resistance of animals. *In* Proceedings of the Sixth International Symposium on Animal, Plant and Microbial Toxins. D. Eaker and T. Wadstrom (Editors), pp. 691–699. Pergamon Press, Oxford, UK.

PURCHASE, I. F. H. 1974. Mycotoxins. Elsevier Scientific Publishing Co., Amsterdam.

RODRICKS, J. V. 1976. Mycotoxins and Other Fungal Related Food Problems. Adv. Chem. Ser. *149*. Am. Chem. Soc., Washington, DC.

RODRICKS, J. V., HESSELTINE, C. W., and MEHLMAN, M. A. 1977. Mycotoxins in Human and Animal Health. Pathotox Publishers, Park Forest South, IL.

SALUNKHE, D. K., WU, M. T., DO, J. Y., and MAAS, M. R. 1980. Mycotoxins in foods and feeds. *In* Safety of Foods. H. D. Graham (Editor), pp. 198–264. AVI Publishing Co., Westport, CT.

SHANK, R. C. 1981. Mycotoxins and N-Nitroso Compounds: Environmental Risks. Vols. I and II. CRC Press, Boca Raton, FL.

SWICK, R. A. 1984. Hepatic metabolism and bioactivation of mycotoxins and plant toxins. J. Anim. Sci. *58*, 1017–1028.

URAGUCHI, K., and YAMAZAKI, M. 1978. Toxicology, Biochemistry, and Pathology of Mycotoxins. Halsted Press, New York.

WILSON, B. J. 1978. Hazards of mycotoxins to public health. J. Food Prot. *41*, 375–384.

WYLLIE, T. D., and MOREHOUSE, L. G. 1977. Mycotoxic Fungi, Mycotoxins, Mycotoxicoses. Vol. 1. Mycotoxic Fungi, and Chemistry of Mycotoxins. Marcel Dekker, NY.

WYLLIE, T.D., and MOREHOUSE, L. G. 1978A. Mycotoxic Fungi, Mycotoxins, Mycotoxicoses. Vol. 2. Mycotoxicoses of Domestic and Laboratory Animals, Poultry, and Aquatic Invertebrates and Vertebrates. Marcel Dekker, NY.

WYLLIE, T.D., and MOREHOUSE, L. G. 1978B. Mycotoxic Fungi, Mycotoxins, Mycotoxicoses. Vol. 3. Mycotoxicoses of Man and Plants: Mycotoxin Control and Regulatory Aspects. Marcel Dekker, NY.

Index

A

Abortifacients, in pine needles, 373–375
Abortion, 161
Abrin, 243
Abrus precatorius, 243
Absorption, of toxicants, 48
Acacia, 375
Acacia berlandieri, 161–162
Acer rubrum, 279–280
Acetone, 359
N-Acetyl loline, 157–158
Ackee, 280–282
trans-Aconitic acid, 296
Aconitium, 140
Acremonium coenophialum, see Epichloe typhina
Acute bovine pulmonary emphysema, 5, 152, 261–265
Adaptation, to toxins, 75
Aesculin, 86
Aesculus glabra, 86
Aesculus hippocastanum, 86
Aesculus octandra, 86
Aflatoxins, 69, 402–422
    carcinogenicity, 186, 415–417
    chemistry, 404–405
    effects on metabolic processes, 413–414
    effects on protein synthesis, 413
    immunosuppression, 417–418
    interactions with cellular components, 414–418
    metabolic fate, 418–419
    metabolism, 407–408, 419–420
    mutagenicity, 415–417
    occurrence, 402–406
    residues, 420–422
    teratogenicity, 415–417
Agalactia, 161
Agropyron repens, 160–161
Agrostemma githago, 198
Alfalfa, *see also Medicago sativa*
    meal, detoxification with, 49, 424
    protein concentrate, 197
    sprouts, 276
    tablets, 223
Alfombrilla, 198
Algae, blue green, 385–386
Alimentary toxic aleukia, 427
Alkali disease, 266–267
Alkaloids
    definition, 5–8
    synthesis, 92–93
Allelochemicals, 77–78
Allelopathic effects, 78, 352
Allergenic proteins, 239–240
S-Allylcysteine sulfoxide, 279
Alsike clover, *see Trifolium hybridum*
Amanita mushrooms, 15, 396
Amaranthus, 328, 368–370
Amaranthus edulis, 369–370
Amaranthus hypochondriacus, 370
Amaranthus retroflexus, 314, 318, 368–369, 385
Amelanchier alnifolia, 176
Ames test for mutagenicity, 37–39, 416
Amino acids, as toxicants, 14, 257–282
    1-amino-D-proline, 274

479

480 INDEX

β-amino propionitrile, 270
Ammi majus, 189
Amoracia lapathifolia, 181
Amsinckia intermedia, 66, 100, 104
Amygdalin, 173–174
Amylase inhibitors, 13, 239
Anabaena circinalis, 386
Anabasine, 120
Anagyrine, 129
Anemia
   from brassica anemia factor, 276–280
   from formaldehyde, 323–324
   from pyrrolizidine alkaloids, 109
   from trimethylamine, 323–324
Anemonin, 225
Anguina agrostis, 306
Annual ryegrass toxicity, 4, 16, 65, 304–308
Anthocyanins, 340
Anthoxanthum odoratum, 188
Antiestrogens
   in bird's-foot trefoil, 218–219
   in pine needles, 373–374
   of phytoestrogens, 223
Antioxidants, *see also* Butylated hydroxyanisole; Ethoxyquin
   synthetic, 62
Arrow grass, *see Triglochin*
Artemisia, 366, 368
Artemisia nova, 366
Arthrogryposis
   induced by
      Conium, 117
      Datura, 153–154
      Lupinus, 129
      Nicotiana, 119–121
Ascites, 107–109
Asclepias eriocarpa, 192–193
Asclepias labriformis, 192–193
Asclepias syriaca, 192
Aspergillus, habitats, 397
Astragalus, 3, 8, 15, 71, 72, 85, 142–144, 199–204, 266
   effects of herbicides on toxicity, 72
   and teratogenic effects, 71
   toxicity in wild animals, 80
   variation in toxin content, 75
Astragalus argillophilus, 200
Astragalus atropubescens, 200
Astragalus bisulcatus, 200, 266
Astragalus canadensis, 200
Astragalus cibaria, 200

Astragalus cicer, 199, 253
Astragalus diversifolius, 200
Astragalus earlei, 200
Astragalus falcatus, 85, 200, 203
Astragalus lentiginosus, 200, 203
Astragalus miser, 11, 75, 200
Astragalus mollissimus, 200
Astragalus nothoxys, 200
Astragalus pattersonii, 200
Astragalus pectinatus, 200
Astragalus pterocarpus, 200
Astragalus pubentissumus, 200
Astragalus racemosus, 200
Astragalus thurberi, 200
Astragalus wootonii, 200
Astringency, 333
Atropine, 86, 153
Australian mammals, resistance to 1080, 375
Autoxidation products of lipids, 303–304
Avena fatua, 385
Avidin, 16, 308–309
Azoxyglycosides, 211

B

Bacillus, 251
Balansia epichloe, 126
Barley, 296–298
Base strength, relation to toxicity, 48–49
Bean hull poisoning, 433
Bees, poisoning of, 203–204
Benzoic acid, 337
Bergapten, 190
BHA, *see* Butylated hydroxyanisole
Bighead
   caused by oxalates, 318–319
   caused by *Tetradymia*, 366–367
Bioassay
   of fescue foot toxins, 33
   of phytoestrogens, 33–34
   of saponins, 33
   of toxicants, 31–33
Biochanin A, 216–217, 220
Biogenic amines, 282–283
Biological control
   of *Echium plantagineum*, 76
   of *Hypericum perforatum*, 350
   of poisonous plants, 72–73
   of *Senecio jacobaea*, 114–115

INDEX

Biotin deficiency, induced by avidin, 308–309
Biotransformation, 38, 54–62
Bird's-foot trefoil, see *Lotus corniculatus*
Bishop's weed, see *Ammi majus*
Bitter rubberweed, see *Hymenoxys odorata*
Bitterweed, 3, 18, 26, see also *Hymenoxys odorata*; *Helenium autumnale*
Black locust, see *Robinia pseudoacacia*
Black walnut, see *Juglans nigra*
Blighia sapida, 23, 280–282
Blind grass, see *Stypandra imbricata*
Blind staggers, 266
Bloat, 252–257, 344
Bloat-producing proteins, 252–257
Blood clotting, 188
Blue-green algae, 385–386
Borago officinalis, 103
Bovine enzootic hematuria, 378
Bowman–Birk trypsin inhibitor, 235
Bracken fern, see *Pteridium aquilinum*
Brassica, 4, 5, 9, 180–186, 276–279
Brassica anemia factor, 276–279
Brassica campestris, 181
Brassica chinensis, 181
Brassica kaber, 385
Brassica napus, 181
Brassica oleracea, 181
Brassylic acid, 300
Breed differences, in susceptibility to toxins, 82
Bright blindness factor, in bracken, 380
Bromegrass, see *Bromus inermus*
Bromus inermus, 158
Bromus tectorum, 225
Broomweed, see *Gutierrezia sarothrae*
Browning reaction, 294
Buckwheat, see *Fagopyrum esculentum*
Bufadienolides, 194–195
Bur buttercup, see *Ceratocephalus testiculatus*
Burke and Wills expedition, 249–250
Burning
  field, 65, 307–308, 450
  for nematode control, 65
Buttercups, see *Ranunculus*
Butylated hydroxyanisole
  and bracken carcinogenicity, 379
  and hymenoxon toxicity, 362–363
  as inducing agent, 62
  and methylazoxymethanol, 212
  as phenolic compound, 342
  and pyrrolizidine alkaloids, 115
Bypass protein, 339, 350–351

C

Caffeic acid, 334, 336, 341
Calcinogenic glycosides, 12, 209–211
Calcinosis, 209
Canavalia ensiformis, 241–242, 275–276
Canavanine, 275–276
Canbra oil, 301
Cancer
  and aflatoxins, 69, 415–417
  bracken-induced, 378
  cycad-induced, 212
  effect of tannins on, 337
  nitrosamine-induced, 328
  protective effects of indole-3-carbinol, 186
Canola meal, 83, 184
Carbohydrates, as toxicants, 15
β-Carboline alkaloids, 150–152
Carbon tetrachloride, 38, 79–80
Carboxypeptidase inhibitors, 239
Carboxytractyloside, 13, 212–214
Carcinogenicity
  of aflatoxins, 415–417
  of trichothecenes, 433
Carcinogens
  definition, 38
  effect of cyclopropenoid fatty acids on, 301
  in plants, 18
Cardenolides, 192–194
Cardiac glycosides, 11, 190–192
Carotatoxin, 365
Carp, thiaminases in, 247
Carrots, 365
Cassava, 87, 178, 345
Castanospermine, 148
Castanospermum australe, 148
Castor bean, see *Ricinus communis*
Catechin, 335, 340
Celery
  carotatoxin in, 365
  dermatitis from, 190
Cenchrus ciliaris, 319
Centaurea, 383–384
Centaurea diffusa, 78
Centaurea nigra, 78

Centaurea repens, 383
Centaurea solstitialis, 383
Ceratocephalus testiculatus, 225–226
Cestrum diurnum, 12, 209–211
Chaconine, 131–132
Chaetomium, and ovine ill-thrift, 460
Chastek's paralysis, 247
Cheilanthes sieberi, 251
Chenopodium album, 318, 385
Chetomin, 460
Chlorogenic acid, 334, 340
Chlorophyll, 349
Chloroplasts, 254–255
Chokecherry, see *Prunus virginiana*
Cholesterol absorption, effect of saponins, 197
Cholinesterase inhibition, 133
Chrysanthemum morifolium, 85
Chrysolina gemelata, 73
Chrysolina hyperici, 73
Chrysolina quadrigemina, 350
Cicuta, 3, 17, 44, 76, 86, 363–365
Cicuta maculata, 363–365
Cicutoxin, 364
Cinnamic acid, 337, 351
Cinnibar moth, see *Tyria jacobaea*
Citrinin, 442–445
  biological effects, 443
  metabolic fate, 445
  mode of action, 443
  occurrence, 442–443
  residues, 445
  toxicity in different species, 444–445
Claviceps cinerea, 122
Claviceps paspali, 122, 125, 452
Claviceps purpurea, 122, 125
Clostridium sporogenes, 251
Clover disease, 4, 83, 217
Cocarcinogens, 38
Cocklebur, see *Xanthium*
Coevolution, of plants and herbivores, 77–79
Colorado rubberweed, see *Hymenoxys richardsonii*
Comfrey, see *Symphytum officinale*
Compositae, 359
Condensed tannins, 333, 335
Conhydrine, 118
γ-Coniceine, 118
Coniine, 117–119
Conium maculatum, 6, 44, 71, 86, 115–119, 363

and crooked calf disease, 71, 117–119
Convallaria majalis, 86
Convicine, 207
Cooper's Creek, 249
Copper
  effect of pyrrolizidine alkaloids, 100, 109, 110
  effect of tall fescue, 160–161
  in lupinosis, 455
Coronarian, 204
Coronilla, 12
Coronilla coronata, 344
Coronilla varia, 99, 204–205, 253
Corynebacterium rathayi, 306
Corynetoxins, 304–308
Cottonseed meal, 345–348
p-Coumaric acid, 334
Coumarin, 10, 186–189
Coumestans, 214–223
Coumestrol, 216–217
Cracker heels, 200
Crambe abyssinica 181, 183, 185
Crambe meal, sinapine in, 340
Crepenynic acid, 303
Crooked calf disease, 117–118, 129
Crotalaria
  contamination of sorghum, 66
  poisoning, 98–99
Crotalaria dura, 98
Crotalaria retussa, 66, 98
Crotalaria sagittalis, 98
Crotalaria spectabilis, 98
Crotocin, 426–427
Crown vetch, see *Coronilla varia*
Cyamopsis tetragonolobus, 299
Cyanide, see Hydrogen cyanide
β-Cyano-L-alanine, 271–272
Cyanogenic glycosides, 9, 173–180
Cycas, 211–212
Cycasin, 212
Cycas media, 87, 211
Cyclopamine, 137–138
Cycloposine, 137–138
Cyclopropenoid fatty acids, 16, 301–302
Cyclops lamb, 135
Cymopterus watsonii, 189–190
  effects of herbicides on toxicity, 72
Cynoglossum officinale, 102
Cysteine
  and hymenoxon toxicity, 362
  and pyrrolizidine alkaloid toxicity, 115
Cytisus scoparius, 127, 130

INDEX 483

Cytochrome P_{450}, 54–55
Cytoplasmic proteins, 14, 252–257

D

Daidzein, 50, 216, 217, 220
Dallis grass, see *Paspalum*
Daphne, 86
Datura, 8, 86
Datura stramonium, 86, 152–154
Datura tea, toxicity, 68
Daucus carota, 116, 363
Death camas, see *Zigadensus*
Deer
  and *Artemisia*, 368
  and bracken palatability, 177–178
Delphinium, 3, 8, 140–142
Delphinium andersonii, 140
Delphinium barbeyi, 72, 140
Delphinium brownii, 142
Delphinium glaucum, 140
Delphinium menziesii, 140
Delphinium occidentale, 140
Delphinium trolliifolium, 140
Deoxynivalenol, see Vomitoxin
Dermatitis, see Photosensitization
Dermatitis, celery, 190
Dhurrin, 173–174
Diacetone alcohol, 358–359
Diacetoxyscirpenol, 425
Dicoumarol, 186–188
Dieffenbachia sequine, 21, 85–86, 320–321
Diet supplements, to prevent toxicoses, 74–75
Digitalis, 11, 86
Digitalis purpurea, 86, 190–191
Digitalis tea, 68
Digitaria decumbens, 319
Digitonin, 190
3,4-Dihydroxypyridine, 259
Dilichos lablab, 253
Dimethyl disulfide, 277–278
Dimethyl tryptamine, 149
Diterpene resin acids, in Ponderosa pine needles, 374
Divicine, 207
Drymaria arenaroides, 198
Dutchman's breeches, see *Thamnosma*

E

Echimidine, 101, 104
Echinatine, 49, 102, 104
Echiumine, 104
Echium plantagineum, 4, 6, 76, 82, 99, 114
Economic losses, from poisonous plants, 70
Elaidic acid, 303
Emphysema, see Acute bovine pulmonary emphysema
Endophytes
  infection of grasses, 126
  of ryegrass, 77, 452
  of tall fescue, 157
Epichloe typhina, endophyte of tall fescue, 157
Epoxide hydrolases, 59
  and resistance of sheep to pyrrolizidine alkaloids, 59, 114
Equine nigropallidial encephalomalacia, 383–384
Equisetum arvense, 245, 247–248
Equol, 217, 220
Ergonovine, 124
Ergotamine, 124
Ergotism, 64, 120–127
Ergotoxamine, 124
Ergotoxine, 124
Erucic acid, 300–301
Essential oils, in *Artemisia*, 368
Estradiol, 216
Ethnobotany, 87
Ethoxyquin
  and hymenoxon toxicity, 62, 362–363
  and pyrrolizidine alkaloid toxicity, 62, 115
p-Ethyl phenol, 217–221
Eucalyptus caleyi, 80
Eucalyptus globulus, 80
Eucalyptus melanophloia, 15
Eupatorium adenophorum, 372
Eupatorium rugosum, 3, 19, 64, 370–372
Euphorbia pulcherrima, 85

F

Facial eczema, 456–460
Fagopyrin, 381
Fagopyrum esculentum, 381–382

False hellebore, *see Veratrum californicum*
Fat necrosis, 159–160
Fatty acids, trans, 302–303
Fatty liver, in lupinosis, 453–454
Fava beans, tannins in, 342
Favism, 204–208
Fescue alkaloids, 154–161
Fescue foot, 158–159
Festuca arundinacea, 154–161
Fibronectin, 305
Fiddleneck, *see Amsinckia intermedia*
Field burning, effects on ryegrass-associated toxicities, 65
Fish-flavored eggs, 340–342
Fluoroacetate, 4, 5, 19, 357–377
Fluorensia cernua, 81
Fog fever, 261
Formaldehyde, 19, 322–324
Formononetin, 216, 217, 220, 221
 metabolism in rumen, 50–51
N-Formyl loline, 157–158
Founder, 223
Foxglove, *see Digitalis purpurea*
Fractionation, of tissue preparations, 36
Furanosesquiterpenes, 366–367
Furocoumarins, 189–190
Fusarium, 422–423
 habitats of, 397

G

Gallic acid, 334, 342
Garden huckleberry, 135
Garlic, 279
Gastrolobium, 80, 375
Geigeria, 5
Genistein, 216, 217, 220
 metabolism in rumen, 52
Gibberella rot, 422
β-Glucanase, 298
β-Glucans, 296–298
Glucose-6-phosphate dehydrogenase, 277
 deficiency of, 206–207
Glucosinolates, 9, 10, 180–186
γ-Glutamyl transpeptidase, 35–36
Glutathione peroxidase, 207, 277
Glutathione-S-transferase, 60, 62
Glycoproteins, 16
Glycosides, 9–13
Goiter, 183, 259

Goitrin, 181–182
Goitrogens, 9, 81, 179, 180–186, 259
Gossypol, 17, 301, 345–348
Gramine, 150–152
Grass tetany, 296
Gums, 299
Gutierrezia sarothrae, 198

H

Halogeton, *see Halogeton glomeratus*
Halogeton glomeratus, 16, 74, 314–317
Haplopappus heterophyllus, 372–373
Heinz–Ehrlich bodies, 208, 276–279
Helenalin, 361
Helenium amarum, 26, 359
Helenium autumnale, 359, 361
Helenium hoopesii, 74, 359–360
 grazing management, 74
Heliotridine, 49, 103, 104
Heliotrine, 49, 93, 104–105, 113
Heliotropium europaeum, 6, 51, 99–100, 455
Hemagglutinins, 240–244
Hemlock, 44
Hemosiderin
 in *Brassica* poisoning, 276
 in pyrrolizidine alkaloid toxicosis, 109
Herbal teas, *see* Tea, herbal
Herbicides, 66, 71–72
High mountain disease, 146
High-performance liquid chromatography (HPLC), 31
Hippuric acid, 56, 337
Holly, *see* Ilex
Hordenine, 150, 152
Horsebrush, *see Tetradymia*
Horse chestnut, *see Aesculus hippocastanum*
Horsetail, *see Equisetum*
Hound's tongue, *see Cynoglossum officinale*
Humans, poisonous plants hazardous to, 85–87
Hydrogen cyanide, 173–177
Hydrolyzable tannins, 333
Hydroperoxides, 252, 304
Hydroxyanaidal, 107
p-Hydroxybenzoic acid, 334, 336
5-Hydroxydimethyl tryptamine, 149
trans-4-Hydroxy-2-hexenal, 107

Hydroxylated fatty acids, 303
Hymenoxon, 361–363
Hymenoxys odorata, 62, 359–363
Hymenoxys richardsonii, 359–360
Hyperglycemia, in fluoroacetate poisoning, 376
Hypericin, 17, 348–350
Hypericum perforatum, 17, 73, 348–350, 365
Hypocalcemia, 316
Hypoglycemia, 213–214, 281–282
Hypoglycin, 280–282
Hypomagnesemia, 296

I

Identification of toxins, 31
Ilex, 85
Immunosuppression
  in aflatoxicosis, 417–418
  in locoweed poisoning, 144
Indigofera spicata, 15, 84, 199, 260, 274
Indole alkaloids, 6, 120–127
Indole-3-carbinol, 186
Indolizidine alkaloids, 8, 142–148
Indospecine, 274
Inducers, of drug-metabolizing enzymes, 59
Intermedine, 49, 101, 104
Ipomeanol, 265
Iron
  absorption, 323–324
  effect of pyrrolizidine alkaloids on tissue levels, 109–110
Isoflavones, 13, 214–223
Isolation of plant toxins, 30
Isothiocyanates, 10, 182
Isouramil, 207
Ixiolaena brevicompta, 303

J

Jacobine, 93, 104
  effect on drug-metabolizing enzymes, 62
Jacoline, 104
Jacozine, 104
Janthitrems, 452
Jervine, 137–138

Jimmyweed, see *Haplopappus heterophyllus*
Jimsonweed, see *Datura stramonium*
Jojoba, see *Simmondsia californica*
Jojoba glycosides, 223–224
Juglans nigra, 78, 352
Juglone, 78, 352

K

Kalanchoe daigremontiana, 195
Kalanchoe lanceolata, 195
Kidney beans, 240–241
Kikuyu grass, see *Pennisetum clandestinum*
Kikuyu poisoning, 461–462
Koala, specialist feeding, 47
Kochia scoparia, 251, 314
Krimpsiekte, 195
Kunitz trypsin inhibitor, 235

L

Lablab purpureus, 253
Laburnum anagyroides, 127, 130
Lactobacillus, 224, 264
Lactose, 293
Laetrile, 173–174
Laminitis, 223, 352
Lantana camara, 22
Larkspur, see *Delphium*
Lasiocarpine, 49, 104–105
Lathyrogens, 268–273
Lathyrus, 15, 267
Lathyrus hirsutus, 269
Lathyrus latifolia, 269
Lathyrus odoratus, 269
Lathyrus sativa, 269–271
Lathyrus sylvestrus, 269
Leaf protein concentrates, phenolics in, 345
Lectins, 14, 240–244
Lespedeza, 345
Lespedeza cuneata, 345
Lespedeza stipulacea, 253
Lethal dose, 42
Leucaena leucocephala, 4, 14, 51, 81, 257–261, 299, 322, 345
Leukoencephalomalacia, 433

Leukopenia, in bracken poisoning, 379–380
Limnanthes alba, 181, 185
Linamarin, 173, 175
Linatine, 274
Linolenic acid, 303
Linseed meal, 179, 274
Linum usitatissimum, 274, 303
Linustatin, 179
Lipids as toxicants, 16, 300–304
Lipoxidases, 252
Lipoxygenases, 252
Listeria monocytogens, role in pine needle abortion, 374–375
Locoism, 199
Locoweed, *see Astragalus* and *Oxytropis*
Loline, 158
Longitarsus jacobaea, 114
Lophyrotomin, 15
Lotaustralin, 179
Lotus corniculatus, 253, 344
Lotus pedunculatus, tannins in, 75–76
Lucernic acid, 196
Lupine, *see Lupinus*
Lupinosis, 4, 453–456
Lupinus, 3, 85, 127–130
Lupinus albus, 128
Lupinus argenteus, 128
Lupinus caudatus, 129
Lupinus latifolius, 129
Lupinus laxiflorus, 129
Lupinus leucophyllus, 128
Lupinus leucopsis, 128
Lupinus luteus, 128
Lupinus sericeus, 128–129
Lycopsamine, 49, 101, 104
Lysyl oxidase, 270

M

Macrozamia, 211–212
Maillard reaction, 294
Male infertility, 347
Malvalic acid, 301
Management
 of animals, 73–76
 of range and pasture, 71–73
α-Mannosidase, 145
Mannosidosis, 145
Maple poisoning, 279–280

Marsilea drummondii, 87, 245, 249–250
Medicagenic acid, 196
Medicago sativa, 215, 253, 257, 344
 oxalate in, 317, 320
Megalocytosis, 107, 183
Melatonin, 351
Melilotoside, 186–187
Melilotus, 10, 186
Melitotus alba, 186
Melilotus officinalis, 186
Merluccius productus, 323
Mesquite, *see Prosopis glandulosa*
Metabolic fate, of aflatoxins, 418–419
Metabolism of toxicants, 47–62
Methemoglobin, 12, 202, 327
Methionine, and hymenoxon toxicity, 362
6-Methoxybenzoxazolinone, 351
5-Methoxydimethyltryptamine, 149
Methylazoxymethanol (MAM), 212
N-Methylconiine, 118
S-Methylcysteine sulfoxide, 76, 276–279
3-Methylindole, 36, 152, 262–263
Methyllycaconitine, 142
Microcystis aeruginosa, 386
Microsomal enzyme inducers, 58–59, 361
Microtus montanus, 351
Milk sickness, 64, 372
Milk-transferred toxins, 25
 aflatoxin, 420–422
 bitterweed, 26
 bracken, 380
 ergot alkaloids, 125
 glucosinolates, 185
 indolizidine alkaloids, 146
 pyrrolizidine alkaloids, 114
 quinolizidine (lupine) alkaloids, 129
 tremetol, 64, 370–372
Milkweed, 3, 11, *see also Asclepias*
Mimosine, 14, 17, 50, 257–261
Miserotoxin, 200–204
Mixed function oxidases, 54–59, 186, 264
 and aflatoxin metabolism, 419–420
 and bracken carcinogenicity, 378
 and hymenoxon toxicity, 362
 inducing agents, 58–59
 and tetradymol toxicity, 367
Molasses toxicity, 299
Moldy corn toxicosis, 427, 433
Moldy straw-induced photosensitization, 461
Monamine oxidase, 282–283

Monarch butterfly, 77, 193–194
Monocrotaline, 49, 93, 104
Monoterpenes, in *Artemisia*, 368
Morea polystachya, 75, 194–195
Moreton Bay chestnut, see *Castanospermum australe*
Mountain thermopsis, see *Thermopsis montana*
Muldamine, 138
Mustard meal, sinapine in, 340
Mutagenesis, 37–38
Mutagenicity
  of aflatoxins, 415–417
  of pyrrolizidine alkaloids, 101
  of trichothecenes, 433
Mycoses, definition of, 393
Mycotoxicosis, definition of, 393
Mycotoxins, 18, 69, 393–477
  analysis, 398
  biological effects, 400–401
  chemistry, 399
  decreasing the exposure, 462–467
Myopathy, in lupinosis, 455

N

Nardoo, see *Marsilea drummondii*
Nasturtium officinalis, 181
Nematodes, 306
Neolinustatin, 179
Nerium oleander, 86, 191
Neurolathyrism, 270–271
Nicotiana, 6, 86, 119–120
Nicotiana attenuata, 119
Nicotiana glauca, 120
Nicotiana tabacum, 119
Nicotiana trigonophylla, 119
Nicotine, 119
Nightshades, see *Solanum*
Nitrates
  in *Amaranthus*, 369
  in *Kochia scoparia*, 251
  toxicity, 327–328
Nitriles, 10, 183
Nitrite, 202, 327–328
Nitro-containing glycosides, 11, 199–204
Nitrogen dioxide, 328
Nitrogen tetraoxide, 328
3-Nitropropanol, 200–204
3-Nitropropionic acid, 200–204

Nitrosamides, 328–329
Nitrosamines, 328–329
Nodularia spumigena, 386
Nutritional secondary hyperparathyroidism, 318–319

O

Oak, see *Quercus*
Oak poisoning, 3, 342–343
Ochratoxins, 436–442
  biological effects, 438–439
  effects in different species, 439–441
  metabolic fate, 441–442
  mode of action, 437–438
  occurrence, 436–437
  residues, 441–442
Oleander, see *Nerium oleander*
Onion poisoning, 19, 27, 279
Onobrychis viciifolia, 253, 344
Opisthotonus, 246, 251
Orange sneezeweed, see *Helenium hoopesii*
Organofluorine compounds, see Fluoroacetate
Osteolathyrism, 269–270
Oubain, 193
Ovine ill-thrift, 459–460
Oxalates, 4, 16, 251, 314–321, 369
Oxalis pes-caprae, 314, 318
β-N-Oxalyl-D-α-diaminopropionic acid, 270–271
Oxylobium, 80, 375
Oxytropis, 3, 8, 142–143
Oxytropis lambertii, 200
Oxytropis sericea, 143, 146, 200

P

Palatability, of toxic plants, 79
Pancreatic hypertrophy, 236–237
Panicum, 314
Paspalum, 125, 452
Paspalum staggers, 452
Pasture management, 71–76
Paterson's curse, see *Echium plantagineum*
Patulin, 448–450
Paxilline, 450

Pea aphids, effects of alfalfa saponins on, 84
Pectins, 295–296
Pelargonic acid, 300
Pelargonium clonesticum, 85
Penicillium, habitats of, 397
Penicillium jensi, 186
Penicillium nigricans, 186
Pennisetum clandestinum, 319, 461–462
Perennial ryegrass staggers, 65, 450–452
Perilla frutescens, 265
Perlolidine, 155–158
Perloline, 155–158
Phalaris, 148–152
Phalaris arundinacea, 148, 150–152
Phalaris poisoning, 148–150
Phalaris staggers, 148–150
Phalaris tuberosa, 8, 148–150
Phaseolus, 240–241, 342
Phenethylamine alkaloids, 162
Phenol, 332
Phenolic acids, 334, 336
Phenolics, 17–18, 332–353, 369
Phenothiazine, and bracken carcinogenicity, 40, 378–379
3-Phenylpropionic acid, 337
Philodendron cordatum, 85–86
Phomopsin A, 456
Phomopsin B, 456
Phomopsis leptostromiformis, 453
Photosensitization, 4, 5, 26, 348–350
  induced by
    alsike clover, 383
    buckwheat, 381
    furocoumarins, 189–190
    hypericin, 348–350
    Kochia scoparia, 251
    lupinosis, 456
    moldy straw, 461
    phytoalexins in celery, 190
    psoralens, 190
    pyrrolizidine alkaloids, 107
    ranunculin, 225
    sporidesmin, 457–458
    Tetradymia, 365
  primary, 26, 365
  secondary, 26, 365
Phototoxicity, 78
Phylloerythrin, 26, 348–349, 365
Phytates, 16, 321–322
Phytoalexins, 189–190

Phytoestrogens, 13, 214–223
  bioassay, 33
  metabolism in rumen, 52–53
Phytolacca americana, 86
Pigweed, see *Amaranthus retroflexus*
Pine needle abortion, 373–375
Pinque, see *Hymenoxys richardsonii*
Pinus ponderosa, 373–375
Piperidine alkaloids, 6, 115–119
Piperonyl butoxide, 60–61
Pithomyces chartarum, 456, 458
Plant breeding, to reduce toxicants, 69, 82–85
Poinsetta, see *Euphorbia pulcherrima*
Poison Control Center, 43
Poison hemlock, see *Conium maculatum*
Poisonous plants, hazardous to humans, 85–87
Pokeweed, see *Phytolacca americana*
Polioencephalomalacia, 53, 250–251
Poloxalene, 255–256
Polycyclic diterpene alkaloids, 8, 140–142
Polyethylene glycol, 339
Polygonium, 385
Polygonium convolvulus, 385
Polypeptide toxicants, 15
Polyphenol oxidase, 340–341, 345
Ponderosa pine, see *Pinus ponderosa*
Potato poisoning, 133
Pressor amines, 282–283
Proanthocyanidins, 333, 340
Progoitrin, 181
N-Propyl disulfide, 19, 279
Prosopis glandulosa, 298
Protease inhibitors, 13, 235–238
Protein toxicants, 13
Protoanemonin, 225
Prunasin, 177
Prunus, 176–177
Prunus virginiana, 176–177
Pseudoconhydrine, 118
Psophocarpus tetragonolobus, 238, 241
Psoralens, 189
Ptaquiloside, as bracken carcinogen, 379
Pteridium aquilinum, 5, 14, 18, 68, 177, 245–247, 376–380
Pterosides, in bracken, 379
Pterosins, in bracken, 379
Pyridine alkaloids, 6, 119–120
Pyridoxine antagonist, 273

Pyrroles, 106
Pyrrolizidine alkaloids, 6, 58, 75, 93–115
  effects on drug-metabolizing enzymes, 62
  metabolites, 106–107
  poisonings of humans, 67
  rumen metabolism, 51–52
Pythagoras, death of, 205

Q

Quackgrass, see *Agropyron repens*
Quercetin, carcinogen in bracken, 379
Quercus, 342–343
Quercus gambelii, 343
Quinic acid, 334
Quinolizidine alkaloids, 7, 127–131
o-Quinones, 340

R

Raffinose, 293
Rancidity, 303
Range management, 71–76
Ranunculin, 224–226
Ranunculus acris, 86, 224
Ranunculus repens, 86, 224
Rapeseed meal, 183–185, 299, 340, see also Canola meal
Raphanus sativus, 181
Rayless goldenrod, see *Haplopappus heterophyllus*
Red mold disease, 427
Redwater disease, 378
Reed canarygrass, see *Phalaris arundinacea*
Residues
  of aflatoxins, 420–422
  of citrinin, 445
  of glucosinolates, 185
  of ochratoxins, 441–442
  of trichothecenes, 433–436
  of zearalenone, 424–425
Resins, 17
Resorcinols, 352–353
Retronecine, 103–104
Retrorsine, 103–105
Retusine, 104
Rheum rhaponeticum, 320
Rhizoctonia leguminicola, 147

Rhodanese, 175
Rhubarb, see *Rheum rhaponeticum*
Ricin, 243
Ricinoleic acid, 303
Ricinus communis, 243–244
Riddelliine, 104
Robinia pseudoacacia, 244
Rubratoxins, 445–448
Rumen, metabolism of toxicants, 51
Rumex, 318
Russian knapweed, see *Centaurea repens*
Ryegrass, see Annual ryegrass toxicity; Perennial ryegrass staggers
Ryegrass staggers, 4, 65, 450–452

S

Safrole, 18, 56
Sagebrush, see *Artemisia*
Sainfoin, see *Onobrychis viciifolia*
Saliva, effects on plants, 78
Salmonella, use in mutagenesis testing, 37–38
Salvation Jane, see *Echium plantagineum*
Saponaria officinalis, 198
Saponins, 11, 33, 84, 195–199, 369
Sarcobatus vermiculatus, 314, 317–318
Sarsaponin, 198
Saskatoon serviceberry, see *Amelanchier alnifolia*
Satratoxin, 427
Sawfly larvae, 15
Scab, of wheat, 423
Sclerotia, 123
Screening plants for toxicity, 84
Secondary compounds, 76
Selenium
  in glutathione peroxidase, 207
  in tall fescue, 161
  toxicity of, 179, 199
Selenium methylselenocysteine, 267, 268
Selenium methylselenomethionine, 268
Selenoamino acids, 15, 265–268
Selenocystathione, 268
Selenocystine, 268
Senecio alpinus, 115
Senecio jacobaea, 4, 44, 72, 74, 75, 94–98, 111–115
  effects of herbicides on toxicity, 72

epoxide hydrolase and toxicity, 59
resistance of sheep, 59
toxicity, 43
use of botanical name, 44
Senecio latifolius, 94, 95
Senecio longilobus, 80, 94, 95, 97
Senecionine, 49, 93, 104
Senecio poisoning, 74, 94–98
Senecio riddelli, 94, 95
Senecio vulgaris, 94, 111, 115
Seneciphylline, 49, 93, 104
Serotonin, 149, 283
Sesbania, 162, 229
Sesbanine, 162
Sesquiterpene lactones, 18, 359–363
Setaria lutescens, 385
Setaria sphacelata, 314, 318–319
Setaria viridis, 385
Shikimic acid
  in biosynthesis of tannins, 334
  in bracken carcinogenicity, 378
Silica, 325–327
Silica urolithiasis, 325–327
Simmondsia californica, 223
Simmondsin, 223, 224
Sinapic acid, 334, 336, 341
Sinapine, 340–341
Skeletal deformity, 24, *see also* Crooked calf disease; Teratogenesis; Teratogenic Effects
Slaframine, 146–148
Sleepy grass, *see Stipa robusta*
Slobbers, 146–147
Smartweeds, *see Polygonium*
Sneezeweeds, 3, 18, *see also Helenium*
Solanidine, 131–132
Solanine, 131–132
Solanum, 131–135, 146
Solanum alkaloids, 7, 131–135
Solanum dimidiatum, 135, 146
Solanum dulcamara, 135
Solanum eleagnifolium, 135
Solanum fastigiatum, 135, 146
Solanum kwebense, 135, 146
Solanum malacoxylon, 12, 209–211
Solanum nigrum, 134
Solanum pseudocapsicum, 85
Sorghum, 175
  tannins of, 337–340
Southeastern bitterweed, *see Helenium amarum*

Soyasapogenols, 196
Specialist feeders, 47
Species differences in response to toxins, 81–82
Spectabiline, 104
Spewing sickness, 361
Sporidesmin, 4, 456–460
Spring parsley, *see Cymopterus watsonii*
Stachybotryotoxins, 460–461
Stachyose, 293
Starch blockers, 13, 239
Sterculia foetida, 301
Sterculic acid, 301
Steroid alkaloids, 7, 131–139
Steroid glycosides, 190–198
Stipa robusta, 3, 358–359
St.-John's-wort, *see Hypericum perforatum*
Straw-pile disease, 461
Stypandra imbricata, 24, 384
Subterranean clover, *see Trifolium subterran*
Succinic dehydrogenase, inhibition of, 203
Sucrose, 298–299
Sudan grass, 175, 176, 180
Sulfobromophthalein (BSP) clearance rate, 33
Summer fescue toxicosis, 155–158
Sunflowers, tannins in, 335, 340
Supinidine, 104
Swainsona, 4, 8, 143–146
  grazing management of, 74
Swainsonine, 145
Sweet clover, *see Melilotus*
Sweet clover poisoning, 186–189
Sweet pea, *see Lathyrus*
Sweet potato poisoning, 265
Symphytine, 101, 104
Symphytum officinale, 86, 101, 102, 191
  in herbal teas, 68, 86
Synthetic antioxidants, and pyrrolizidine alkaloid toxicity, 75, *see also* Butylated hydroxyanisole; Ethyoxyquin

T

Tagetes, phototoxins in, 78
Tall fescue, *see Festuca arundinacea*
Tanacetum vulgare, 44, 98
Tannic acid, 336, 342

INDEX

Tannins, 18, 255, 332–345
Tarweed, see *Amsinckia intermedia*
Tea, bush, 68
Tea, herbal, 68
Teratogenesis, 40–42
Teratogenic effects
  of aflatoxins, 415–417
  of *Astragalus*, 71, 144
  of cyanogenic glycosides, 179–180
  of *Leucaena leucocephala*, 261
  of *Lupinus*, 70, 129–130
  of *Nicotiana* 119–121
  of *Solanum*, 133
  of Sudan grass, 180
  of trichothecenes, 433
  of *Veratrum californicum*, 71, 135–137
  of wild cherries, 179
Tetradymia, 82, 365
Tetradymia canescens, 365–368
Tetradymia glabrata, 365–368
Tetradymol, 61, 367
Thamnosma montana, 190
Thamnosma texana, 190
Thermopsis montana, 131
Thiaminase, 14, 245–251
Thiocyanate, 10, 175, 179, 182
Thiosulfate sulfurtransferase, 175
Thlaspi arvense, 181, 385
Thyroxine, 182
Tobacco, see *Nicotiana*
Tomatidine, 135
Tomatine, 135
Toxicants
  definition, 1
  distribution in feeds, 20
  historical aspects, 2–3
  metabolic effects, 20–27
  problems around the world, 4–5
  site of action, 20–27
Toxicoses, management, 43–44
Transferases, 60
Trembles, 371
Tremetol, 371
Tremetone, 370–373
Tremorgens, 450–452
Tricarballylic acid, 296
Trichoderma viride, bioassay of saponins, 33
Tricothecenes, 425–436
  biological effects, 430–431
  blood effects, 432

  classification, 425–426
  cytotoxic effects, 431–432
  feed refusal and vomiting, 431
  metabolic fate, 433–436
  mode of action, 429–430
  neurotoxic effects, 432–433
  occurrence, 427–428
  residues, 433–436
Trifolium hybridum, 382–383
Trifolium pratense, 215, 253
Trifolium repens, 78, 177, 253
Trifolium subterran, 13, 214, 253
Triglochin, 176
Trimethylamine oxide, 19, 322–324, 342
Trisetum flavescens, 12, 209
Trisoralen, 190
Triterpenoid glycosides, 190–198
Triticale, 352–353
Tropane alkaloids, 8, 152–154
Tropical ataxic neuropathy, 178
Tropical grasses, oxalate content, 319
Trypsin inhibitors, 13, 235–238
Tryptamine alkaloids, 8, 148–152
Tryptophan, 14, 261–265
T-2 toxin, 425, 434–436
Tunicamycin antibiotics, 305
Turneforcidine, 104
Tyria jacobaea, 47, 72, 114

U

Uncoupling of oxidative phosphorylation
  and carboxytractyloside, 214
  and tetradymol, 367
Urinary calculi, 325–327

V

Vanillic acid, 334, 336
Veratramine, 138
Veratrum alkaloids, 7, 135–139
Veratrum californicum, 7, 71, 135–138
  and cyclops lambs, 71
Verrucarin A, 427
Verruculogen, 450
Vicia faba, 12, 204–208
Vicia sativa, 271
Vicia villosa, 273
Vicine, 12, 204–208

Vitamin D, metabolism, 209
Vitamin E, inhibitor, 241
Vitamin K, deficiency, 186
Vomiting sickness, 281
Vomitoxin, 425, 426, 431
Vultures, immunity to botulinism of, 48

W

Warfarin, 187
Water hemlock, see *Cicuta*
Water solubility, relation to toxicity, 48–49
Weed seeds, in grains and screenings, 385
White clover, see *Trifolium repens*
White snakeroot, see *Eupatorium rugosum*
Wild animals, effects of toxicants on, 80–81
Wild parsnip, 364
Winged bean, see *Psophocarpus tetragonolobus*
Wisteria, 87

X

Xanthium, 212–214
Xanthium strumarium, 13, 212–214

Xanthotoxin, 190
Xylose, 292

Y

Yellow star thistle, see *Centaurea solstitialis*
Yucca shidegria, 198–199

Z

Zearalenone, 422–425
  biological effects, 423–424
  metabolic fate, 424–425
  occurrence, 422–423
  residues, 424–425
Zigadensus, 3, 7, 138–139
Zigadensus paniculatus, 139
Zinc
  protection against
    lupinosis, 455
    pyrrolizidine alkaloids, 75
    facial eczema, 458–459
  tissue levels of
    and lupinosis, 455
    and pyrrolizidine alkaloids, 27
Zygacine, 139